STUDENT'S SOLUTIONS MANUAL

BEVERLY FUSFIELD

CALCULUS & ITS APPLICATIONS
and
BRIEF CALCULUS & ITS APPLICATIONS
THIRTEENTH EDITION

Larry J. Goldstein
Goldstein Educational Technologies

David C. Lay
University of Maryland

David I. Schneider
University of Maryland

Nakhlé Asmar
University of Missouri

PEARSON

Boston Columbus Indianapolis New York San Francisco Upper Saddle River
Amsterdam Cape Town Dubai London Madrid Milan Munich Paris Montreal Toronto
Delhi Mexico City São Paulo Sydney Hong Kong Seoul Singapore Taipei Tokyo

Copyright © 2014, 2010, 2007 Pearson Education, Inc.
Publishing as Pearson, 75 Arlington Street, Boston, MA 02116.

ISBN-13: 978-0-321-87857-1
ISBN-10: 0-321-87857-4

www.pearsonhighered.com

PEARSON

CONTENTS

Chapter 0
Functions

0.1 Functions and Their Graphs

You should read Sections 0.1 and 0.2 even if you don't plan to work the exercises. The concept of a function is a fundamental idea, and the notation for functions is used in nearly every section of the text.

1. The notation [−1, 4] is equivalent to $-1 \le x \le 4$. Both −1 and 4 belong to the interval.

7. The inequality $2 \le x < 3$ describes the half-open interval [2, 3). The right parenthesis indicates that 3 is not in the set.

13. If $f(x) = x^2 - 3x,$ then

$$f(0) = (0)^2 - 3(0) = 0,$$
$$f(5) = (5)^2 - 3(5) = 25 - 15 = 10,$$
$$f(3) = (3)^2 - 3(3) = 9 - 9 = 0,$$
$$f(-7) = (-7)^2 - 3(-7) = 49 + 21 = 70.$$

19. (a) $f(0)$ represents the number of laptops sold in 2010.

(b) Substitute 6 for x in $f(x) = 150 + 2x + x^2$:

$$f(6) = 150 + 2(6) + 6^2$$
$$= 150 + 12 + 36$$
$$= 198$$

25. The domain of $f(x) = \dfrac{3x-5}{x^2+x-6}$ consists of those x for which the denominator does not equal zero since division by zero is not permissible. We must solve the equation $x^2+x-6=0$ to find those values of x.

$$x^2+x-6=0$$
$$(x+3)(x-2)=0 \qquad \text{Factor.}$$
$$x+3=0 \ \Big| \ x-2=0 \quad \text{Use the zero-factor property.}$$
$$x=-3 \ \Big| \quad x=2 \quad \text{Solve for } x.$$

The domain of f is all real numbers such that $x \ne -3$ or $x \ne 2$. Written using interval notation, this is $(-\infty,\,-3)\cup(-3,\,2)\cup(2,\infty)$.

31. This is not the graph of a function by the vertical line test.

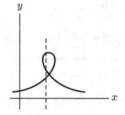

37. By referring to the graph, we find that $f(4)$ is positive since the graph of the function is above the x-axis when $x=4$.

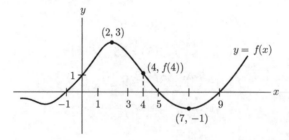

43. The concentration of the drug when $t=1$ is about .03 unit because the graph tells us that $f(1) \approx .03$. The concentration of the drug when $t=5$ is about .04 unit because the graph tells us that $f(5)\approx.04$.

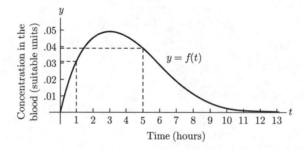

49. If the point $\left(\dfrac{1}{2}, \dfrac{2}{5}\right)$ is on the graph of the function $g(x) = \dfrac{3x-1}{x^2+1}$, then $g\left(\dfrac{1}{2}\right)$ must equal $\dfrac{2}{5}$. This is the case since

$$g\left(\frac{1}{2}\right) = \frac{3\left(\frac{1}{2}\right)-1}{\left(\frac{1}{2}\right)^2+1} = \frac{\frac{3}{2}-1}{\frac{1}{4}+1} = \frac{\frac{1}{2}}{\frac{5}{4}} = \frac{1}{2} \cdot \frac{4}{5} = \frac{2}{5}$$

so $\left(\dfrac{1}{2}, \dfrac{2}{5}\right)$ is on the graph.

55. Since $f(x) = \pi x^2$ for $x < 2$, $f(1) = \pi(1)^2 = \pi$.
Since $f(x) = 1 + x$ for $2 \leq x \leq 2.5$, $f(2) = 1 + 2 = 3$.
Since $f(x) = 4x$ for $2.5 < x$, $f(3) = 4(3) = 12$.

61. Entering $\mathbf{Y_1 = X \wedge 3 / 4}$ will graph the function $f(x) = \dfrac{x^3}{4}$. In order to graph the function $y = x^{3/4}$, you need to include parentheses in the exponent: $\mathbf{Y_1 = X \wedge (3/4)}$.

Help for Technology Exercises: The examples in the text use the family of TI-83/84 calculators. Helpful information about the use of calculators appears at the end of most sections in subsections titled **Incorporating Technology**. You might wish to place a paper or plastic tab in your text, to help you reference the appropriate information.

0.2 Some Important Functions

The most important functions here are the linear functions, the quadratic functions, and the power functions. Absolute values appear only briefly in Sections 4.5 and 6.1, and then more frequently in Chapters 9 and 10.

1. The equation $y = 2x - 1$ is linear. Since a line is determined by any two of its points, we may choose any two points on the graph of $f(x)$ and draw the line through them. For instance, when $x = 0$,

$$y = 2(0) - 1 = 0 - 1 = -1$$

and when $y = 0$,

$$0 = 2x - 1 \Rightarrow 2x = 1 \Rightarrow x = \frac{1}{2}$$

So $(0, -1)$ and $\left(\frac{1}{2}, 0\right)$ are two points on the graph of $f(x)$.

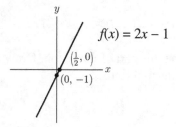

7. Since the equation is linear, its graph is a straight line. To simplify finding points on the line, solve for *y* first.

$$x - y = 0$$
$$x = y$$

Even though only two points are needed to identify and graph the line, we will compute three points to get a better sketch of the line and verify that it is correct.

x	y	(x, y)
1	1	(1, 1)
0	0	(0, 0)
–1	–1	(–1, –1)

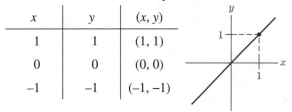

13. To find the *y*-intercept, evaluate $f(x)$ at $x = 0$.

$$f(0) = 5$$

So the *y*-intercept is (0, 5). To find the *x*-intercept, set $f(x) = 0$ and solve for *x*.

$$5 = 0$$

which is not true, so there is no *x*-intercept.

19. **(a)** The cost of renting the car for one day and driving 200 miles is

$$24 + 200(.25) = \$74.$$

(b) The total rental expense for one day is given by

$$f(x) = .25x + 24.$$

25. Write $y = 3x^2 - 4x$ in the form $y = ax^2 + bx + c$. That is, $y = 3x^2 + (-4)x + 0$, so $a = 3, b = -4$, and $c = 0$.

31. This function is defined by two distinct linear functions. For $0 \le x \le 1$, $f(x) = 3x$. This graph is determined by two points, say at $x = 0$ and $x = 1$.

$$f(0) = 3 \cdot 0 = 0, \qquad f(1) = 3 \cdot 1 = 3.$$

So (0, 0) and (1, 3) determine this part of the graph. Draw a line segment between these points. For $x > 1$, we have $f(x) = \frac{9}{2} - \frac{3}{2}x$. This graph is also determined by two points, say at $x = 1$ and $x = 3$.

$$f(1) = \frac{9}{2} - \frac{3}{2}(1) = 3, \qquad f(3) = \frac{9}{2} - \frac{3}{2}(3) = 0.$$

So (1, 3) and (3, 0) determine this part of the graph. Draw a line segment between these points and extend it to the right as *x* tends to infinity.

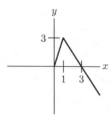

37. $f(-1) = (-1)^{100} = 1$ since a negative number raised to an even power is positive.

43. Using a TI-84, set $Y_1 = 3X^3 + 8$. Then find $f(-11)$ and $f(10)$ by pressing VARS, then Y-VARS, then choice 1, followed by choice 1 again. Then enter "(" followed by the value and ")" as shown below.

```
Plot1  Plot2  Plot3        Y₁(-11)
\Y₁∎3X³+8                            -3985
\Y₂=                        Y₁(10)
\Y₃=                                  3008
\Y₄=
\Y₅=
\Y₆=
\Y₇=
```

0.3   The Algebra of Functions

The algebraic skills in this section will be used frequently throughout the course. You should review the material here on composition of functions before you read Section 3.2.

1. $f(x) + g(x) = (x^2 + 1) + 9x = x^2 + 9x + 1.$

7. $f(x) + g(x) = \dfrac{2}{x-3} + \dfrac{1}{x+2}.$

In order to add two fractions, the denominators must be the same. A common denominator for

$$\frac{2}{x-3} \quad \text{and} \quad \frac{1}{x+2}$$

is $(x-3)(x+2)$. If we multiply

$$\frac{2}{x-3} \quad \text{by} \quad \frac{x+2}{x+2}$$

we obtain an equivalent expression whose denominator is $(x-3)(x+2)$. Similarly, if we multiply

$$\frac{1}{x+2} \quad \text{by} \quad \frac{x-3}{x-3}$$

we obtain an equivalent expression whose denominator is $(x-3)(x+2)$. Thus

$$f(x) + g(x) = \frac{2}{x-3} + \frac{1}{x+2}$$

$$= \frac{2}{x-3} \cdot \frac{x+2}{x+2} + \frac{1}{x+2} \cdot \frac{x-3}{x-3}$$

$$= \frac{2(x+2)}{(x-3)(x+2)} + \frac{x-3}{(x-3)(x+2)}$$

$$= \frac{2x+4+x-3}{(x-3)(x+2)}$$

$$= \frac{3x+1}{(x-3)(x+2)} = \frac{3x+1}{x^2-x-6}.$$

13. $f(x) - g(x) = \dfrac{x}{x-2} - \dfrac{5-x}{5+x}$. A common denominator for

$$\frac{x}{x-2} \quad \text{and} \quad \frac{5-x}{5+x}$$

is $(x-2)(5+x)$.

Proceeding as in Exercise 7:

$$f(x) - g(x) = \frac{x}{x-2} \cdot \frac{5+x}{5+x} - \frac{5-x}{5+x} \cdot \frac{x-2}{x-2}$$

$$= \frac{x(5+x)}{(x-2)(5+x)} - \frac{(5-x)(x-2)}{(x-2)(5+x)}$$

$$= \frac{5x+x^2 - (-x^2+7x-10)}{(x-2)(5+x)}$$

$$= \frac{2x^2-2x+10}{x^2+3x-10}.$$

19. Substitute $x+1$ for x in the expressions for $f(x)$ and $g(x)$, then simplify.

$$f(x+1)g(x+1) = \left[\frac{(x+1)}{(x+1)-2}\right]\left[\frac{5-(x+1)}{5+(x+1)}\right]$$

$$= \left[\frac{x+1}{x-1}\right]\left[\frac{4-x}{6+x}\right] = \frac{(x+1)(4-x)}{(x-1)(6+x)}$$

$$= \frac{-x^2+3x+4}{x^2+5x-6}.$$

25. To find $f(g(x))$, substitute $g(x)$ in place of each x in $f(x) = x^6$. Thus

$$f(g(x)) = (g(x))^6 = \left(\frac{x}{1-x}\right)^6.$$

31. If $f(x) = x^2$ then,

$$f(x+h) - f(x) = (x+h)^2 - x^2$$
$$= x^2 + 2hx + h^2 - x^2$$
$$= 2hx + h^2.$$

37. $f(x) = \frac{1}{8}x$, $g(x) = 8x + 1$. Substitute $g(x)$ in place of each x in $f(x)$. Thus

$$h(x) = f(g(x)) = \frac{1}{8}g(x)$$

$$= \frac{1}{8}[8x+1]$$

$$= x + \frac{1}{8}.$$

$g(x)$ converts British sizes to French sizes, while $f(x)$ converts French sizes to U.S. sizes. Thus, $h(x) = f(g(x))$ converts British sizes to U.S. sizes. For example, consider the British hat size of $6\frac{3}{4}$. Since $h(x) = x + \frac{1}{8}$, the U.S. size is $h\left(6\frac{3}{4}\right) = 6\frac{3}{4} + \frac{1}{8} = 6\frac{7}{8}$, which agrees with the table.

43. On a TI-84, if $Y_1 = X/(X-1)$, set $Y_2 = Y_1(Y_1)$. Before graphing Y_2, you should "deselect" Y_1, so its graph will not appear also. Use the TRACE feature to get an impression of what the formula for Y_2 might be.

To actually determine the formula for $f(f(x))$, substitute $x/(x-1)$ for each occurrence of x in the formula for $f(x)$. (Note that $f(x)$ is not defined for $x = 1$.)

$$f(f(x)) = \frac{x/(x-1)}{[x/(x-1)] - 1}, \quad x \neq 1.$$

The denominator can be simplified:

$$\frac{x}{x-1} - 1 = \frac{x}{x-1} - \frac{x-1}{x-1} = \frac{x - (x-1)}{x-1} = \frac{1}{x-1}.$$

Multiply numerator and denominator of the formula for $f(f(x))$ by $x - 1$ and obtain

$$f(f(x)) = \frac{x/(x-1)}{1/(x-1)} = \frac{x}{1} = x, \quad x \neq 1.$$

$[-15, 15]$ by $[-10, 10]$

0.4   Zeros of Functions—The Quadratic Formula and Factoring

The material in this section will be used routinely throughout the text, beginning in Section 2.3. (Factoring skill is also needed for a few exercises in Section 1.4.) By Section 2.3 you must have mastered the ability to factor quadratic polynomials and to solve problems such as Exercises 39–44 (in Section 0.4). You should also be able to factor cubic polynomials where every term contains a power of x. (See (b) and (c) of Example 7.) Problems such as those in Exercises 31–38 will appear in Section 6.4 and in later sections. The quadratic formula should be memorized although you will not need to use it as often as factoring skills.

1. In the equation $2x^2 - 7x + 6 = 0$, we have $a = 2$, $b = -7$, and $c = 6$. Substitute these values into the quadratic formula and obtain

$$x = \frac{-(-7) \pm \sqrt{(-7)^2 - 4(2)(6)}}{2(2)} = \frac{7 \pm \sqrt{49 - 48}}{4} = \frac{7 \pm 1}{4}.$$

Thus,
$$x = \frac{7 + 1}{4} = 2 \quad \text{or} \quad x = \frac{7 - 1}{4} = \frac{3}{2}.$$

So the zeros of the function $2x^2 - 7x + 6$ are 2 and $\frac{3}{2}$.

7. In the equation $5x^2 - 4x - 1 = 0$, we have $a = 5$, $b = -4$, and $c = -1$. By the quadratic formula,

$$x = \frac{-(-4) \pm \sqrt{(-4)^2 - 4(5)(-1)}}{2(5)}$$

$$= \frac{4 \pm \sqrt{16 + 20}}{10} = \frac{4 \pm \sqrt{36}}{10}$$

$$= \frac{4 \pm 6}{10}.$$

So the solutions of the equation $5x^2 - 4x - 1 = 0$ are

$$x = \frac{4 + 6}{10} = \frac{10}{10} = 1, \quad \text{and} \quad x = \frac{4 - 6}{10} = \frac{-2}{10} = -\frac{1}{5}.$$

13. To factor $x^2 + 8x + 15$, find values for c and d such that $cd = 15$ and $c + d = 8$. The solution is $c = 3, d = 5$, and

$$x^2 + 8x + 15 = (x + 3)(x + 5).$$

19. To factor $30 - 4x - 2x^2$, first factor out the coefficient -2 in front of x^2. That is,

$$30 - 4x - 2x^2 = -2(x^2 + 2x - 15).$$

Then to factor $x^2 + 2x - 15$, find values for c and d such that $cd = -15$ and $c + d = 2$. The solution is $c = 5, d = -3$, and $x^2 + 2x - 15 = (x + 5)(x - 3)$. Thus

$$30 - 4x - 2x^2 = -2(x + 5)(x - 3).$$

25. The expression $x^3 - 1$ is a difference of cubes. Thus,

$$x^3 - 1 = (x-1)(x^2 + x + 1)$$

31. If a point (x, y) is on both graphs, then its coordinates must satisfy both equations. That is, x and y must satisfy $y = 2x^2 - 5x - 6$ and $y = 3x + 4$. Equate the two expressions for y:

$$2x^2 - 5x - 6 = 3x + 4.$$

To use the quadratic formula, rewrite the equation in the form

$$2x^2 - 8x - 10 = 0,$$

and divide both sides by 2:

$$x^2 - 4x - 5 = 0.$$

By the quadratic formula,

$$x = \frac{-(-4) \pm \sqrt{(-4)^2 - 4(1)(-5)}}{2(1)} = \frac{4 \pm \sqrt{16 + 20}}{2}$$

$$= \frac{4 \pm \sqrt{36}}{2} = \frac{4 \pm 6}{2}.$$

So $x = \dfrac{4+6}{2} = 5$ or $x = \dfrac{4-6}{2} = -1$. Thus the x-coordinates of the points of intersection are 5 and -1.

To find the y-coordinates, substitute these values of x into either equation, $y = 2x^2 - 5x - 6$ or $y = 3x + 4$. Since $y = 3x + 4$ is simpler, use this equation to find that at $x = 5$, $y = 3(5) + 4 = 15 + 4 = 19$. Also, at $x = -1$, $y = 3(-1) + 4 = -3 + 4 = 1$. Thus the points of intersection are $(5, 19)$ and $(-1, 1)$.

37. As in Exercise 31, equate the two expressions for y to get

$$\frac{1}{2}x^3 + x^2 + 5 = 3x^2 - \frac{1}{2}x + 5.$$

Then rewrite this as

$$\frac{1}{2}x^3 - 2x^2 + \frac{1}{2}x = 0,$$

and factor out a common factor of x to get

$$x\left(\frac{1}{2}x^2 - 2x + \frac{1}{2}\right) = 0.$$

This shows that one of the points of intersection has x-coordinate 0. To find the x-coordinates of the remaining points of intersection, apply the quadratic formula to $\dfrac{1}{2}x^2 - 2x + \dfrac{1}{2} = 0$:

$$x = \frac{-(-2) \pm \sqrt{(-2)^2 - 4\left(\frac{1}{2}\right)\left(\frac{1}{2}\right)}}{2\left(\frac{1}{2}\right)}$$

$$= \frac{2 \pm \sqrt{4-1}}{1}$$

$$= 2 \pm \sqrt{3}.$$

Substitute these values into $y = 3x^2 - \frac{1}{2}x + 5$. At $x = 0$, $y = 3(0)^2 - \frac{1}{2}(0) + 5 = 5$.

At $x = 2 + \sqrt{3}$,

$$y = 3(2 + \sqrt{3})^2 - \frac{1}{2}(2 + \sqrt{3}) + 5$$

$$= 3(7 + 4\sqrt{3}) - \frac{1}{2}(2 + \sqrt{3}) + 5$$

$$= 21 + 12\sqrt{3} - 1 - \frac{1}{2}\sqrt{3} + 5$$

$$= 25 + \frac{23}{2}\sqrt{3}.$$

At $x = 2 - \sqrt{3}$,

$$y = 3(2 - \sqrt{3})^2 - \frac{1}{2}(2 - \sqrt{3}) + 5$$

$$= 3(7 - 4\sqrt{3}) - 1 + \frac{1}{2}\sqrt{3} + 5$$

$$= 21 - 12\sqrt{3} + 4 + \frac{1}{2}\sqrt{3}$$

$$= 25 - \frac{23}{2}\sqrt{3}.$$

Thus the points of intersection are:

$$(0, 5), \quad \left(2 + \sqrt{3}, 25 + \frac{23}{2}\sqrt{3}\right), \text{ and } \left(2 - \sqrt{3}, 25 - \frac{23}{2}\sqrt{3}\right).$$

43. A rational function will be zero only if the numerator is zero. Solve

$$x^2 + 14x + 49 = 0,$$

$$(x + 7)^2 = 0,$$

$$x + 7 = 0.$$

That is, $x = -7$. Since the denominator is not 0 at $x = -7$, the solution is $x = -7$.

49. You can graph the function and use **TRACE** to estimate the zero as approximately 4.56. (Your estimate is influenced by your choice of viewing window.) Another procedure, which is usually more accurate, is described on page 32 in the text.

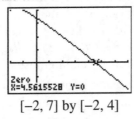

Zero
X=4.5615528   Y=0

[−2, 7] by [−2, 4]

Helpful Hint: Exercises 55–58 are somewhat difficult because you do not yet have the tools of calculus available. To find an interval that contains the zeros of $f(x)$, use the comment that follows the Incorporating Technology box on page 32 in the text. In exercise 57, for example, write

$$f(x) = 3\left(x^3 + \frac{52}{3}x^2 - 4x - 4\right)$$

and let M be the number that is one more than the largest magnitude of the coefficients 52/3, −4, and −4. That is, let $M = 55/3$. Then the zeros of $f(x)$ lie between $-M$ and M. For a first graphing attempt, use x values in the interval $-55/3 \le x \le 55/3$, or perhaps $-19 \le x \le 19$.

55. Since $f(x) = x^3 - 22x^2 + 17x + 19$, and the coefficient of the highest power of x is 1, you can let M be the number that is one more that the largest magnitude of the coefficients –22, 17, and 19. That is, let $M = 23$. Then the zeros of the polynomial lie between $-M$ and M, in the interval $-23 \le x \le 23$. Sketch the graph in the window $[-23, 23]$ *by* $[-10, 10]$

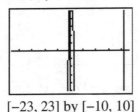

[−23, 23] by [−10, 10]

In this first window, you see three nearly vertical lines. The "lines" connect somewhere off the screen. So enlarge the range of y-values, say, $-100 \le y \le 100$. Also, since two of the "lines" in the graph are close to $x = 0$ and one is close to $x = 23$, you can chop off most of the negative x-axis. Try the window $[-5, 25]$ by $[-100, 100]$.

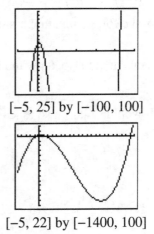

[−5, 25] by [−100, 100]

[−5, 22] by [−1400, 100]

This time, you should observe that the y-values seem to get very negative. The **TRACE** command might be helpful now, to find out about how negative the y-values become. Move the cursor along the curve until it disappears at the bottom of the screen. The coordinates of points on the graph will continue to appear on the screen even when the cursor is not visible. Watching the y-coordinates, you should see them go down to about −1300. So, try the window $[-5, 22]$ *by* $[-1400, 100]$. (When you use such a large y-range, set the y-scale to, say, 100.)

0.5   Exponents and Power Functions

We have found that operations with exponents cause our students more difficulty than any other algebraic skill, so we have included lots of drill exercises. Try some of each group of problems. If you cannot work them accurately with relative ease and confidence, keep working more problems. If you need more practice, get a college algebra text or Schaum's *Outline of College Algebra*. Do this immediately, because operations with exponents are used in Section 1.3 and in most sections thereafter.

Helpful Hint: We are all so accustomed to reading formulas from left to right that it is more difficult to use a formula such as $b^r b^s = b^{r+s}$ in "reverse," that is, in the form $b^{r+s} = b^r b^s$. Yet the laws of exponents are sometimes needed in reverse from the way they are written in the text. So *memorize* the following list, which incorporates the list on page 34 of the text.

Laws of Exponents

1. $b^r \cdot b^s = b^{r+s}$ 4. $(b^r)^s = b^{rs}$

1′. $b^{r+s} = b^r \cdot b^s$ 4′. $b^{rs} = (b^r)^s$

2. $b^{-r} = \dfrac{1}{b^r}$ 5. $(ab)^r = a^r b^r$

2′. $\dfrac{1}{b^r} = b^{-r}$ 5′. $a^r b^r = (ab)^r$

3. $\dfrac{b^r}{b^s} = b^r \cdot b^{-s} = b^{r-s}$ 6. $\left(\dfrac{a}{b}\right)^r = \dfrac{a^r}{b^r}$

3′. $b^{r-s} = b^r \cdot b^{-s} = \dfrac{b^r}{b^s}$ 6′. $\dfrac{a^r}{b^r} = \left(\dfrac{a}{b}\right)^r$

A common use of Law 4′ is in the form $b^{m/n} = (b^{1/n})^m$. For instance, $9^{3/2} = (9^{1/2})^3 = (3)^3 = 27$, and $27^{4/3} = (27^{1/3})^4 = (3)^4 = 81$. Most instructors will assume that you can compute the square roots of the following numbers without using a calculator:

$$4, 9, 16, 25, 36, 49, 64, 81, 100.$$

You should also know the following cube roots:

$$8^{1/3} = 2, \quad 27^{1/3} = 3, \quad 64^{1/3} = 4, \quad \text{and possibly} \quad 125^{1/3} = 5.$$

It wouldn't hurt also to learn the following fourth roots:

$$16^{1/4} = 2,$$
$$81^{1/4} = 3.$$

1. $3^3 = 3 \cdot 3 \cdot 3 = 9 \cdot 3 = 27$.

7. $-4^2 = -16$. Note that the exponent 2 does not act on the negative sign. That is, -4^2 is not the same as $(-4)^2 = (-4) \cdot (-4) = +16$.

13. $6^{-1} = \dfrac{1}{6}$ (Law 2)

19. $(25)^{3/2} = (25^{1/2})^3$ (Law 4')
$$= 5^3$$
$$= 125$$

25. $4^{-1/2} = \dfrac{1}{4^{1/2}}$ (Law 2)
$$= \dfrac{1}{2}$$

31. $6^{1/3} \cdot 6^{2/3} = 6^{1/3 + 2/3}$ (Law 1)
$$= 6^1$$
$$= 6$$

37. $\left(\dfrac{8}{27}\right)^{2/3} = \dfrac{8^{2/3}}{27^{2/3}}$ (Law 6)
$$= \dfrac{(8^{1/3})^2}{(27^{1/3})^2}$$ (Law 4')
$$= \dfrac{2^2}{3^2} = \dfrac{4}{9}$$

43. $\dfrac{x^4 \cdot y^5}{xy^2} = \left(\dfrac{x^4}{x}\right)\left(\dfrac{y^5}{y^2}\right)$
$$= (x^{4-1})(y^{5-2})$$ (Law 3)
$$= x^3 y^3$$

49. $(x^3 y^5)^4 = (x^3)^4 (y^5)^4$ (Law 5)
$$= x^{12} y^{20}$$ (Law 4)

55. $\dfrac{-x^3 y}{-xy} = \dfrac{x^3 y}{xy} = \dfrac{x^3}{x} \cdot \dfrac{y}{y}$
$$= x^{3-1} y^{1-1}$$ (Law 3)
$$= x^2 y^0 = x^2$$

61. $\left(\dfrac{3x^2}{2y}\right)^3 = \dfrac{(3x^2)^3}{(2y)^3}$ (Law 6)   [Don't forget parentheses]

$= \dfrac{3^3(x^2)^3}{2^3 y^3}$ (Law 5)   [Careful: This step is where mistakes usually happen.]

$= \dfrac{27(x^2)^3}{8y^3} = \dfrac{27x^6}{8y^3}$ (Law 4)

67. $\sqrt{x}\left(\dfrac{1}{4x}\right)^{5/2} = x^{1/2}\left(\dfrac{1}{4x}\right)^{5/2}$

$= x^{1/2} \cdot \dfrac{1^{5/2}}{(4x)^{5/2}}$ (Law 6)   [Careful: Don't forget parentheses around $4x$]

$= x^{1/2} \cdot \dfrac{1}{4^{5/2}x^{5/2}}$ (Law 5)

$= \dfrac{x^{1/2}}{(4^{1/2})^5 x^{5/2}}$ (Law 4)

$= \dfrac{x^{1/2}}{2^5 x^{5/2}} = \dfrac{1}{32} \cdot x^{1/2-5/2}$ (Law 3)

$= \dfrac{1}{32} \cdot x^{-2} = \dfrac{1}{32x^2}$ (Law 2)

73. If $f(x) = \sqrt[3]{x}$ and $g(x) = \dfrac{1}{x^2}$, then

$\dfrac{g(x)}{f(x)} = \dfrac{\frac{1}{x^2}}{\sqrt[3]{x}} = \dfrac{\frac{1}{x^2}}{x^{1/3}}$ Substitute; rewrite denominator in exponential form.

$= x^{-2} \cdot x^{-1/3}$ Law 2′

$= x^{-2} \cdot x^{-1/3} = x^{-7/3}$ Law 1

$= \dfrac{1}{x^{7/3}}$ Law 2

79. If $f(x) = \sqrt[3]{x}$ and $g(x) = \dfrac{1}{x^2}$, then

$$f(g(x)) = f\left(\frac{1}{x^2}\right) \qquad \text{Substitute.}$$

$$= f\left(x^{-2}\right) \qquad \text{Law 2}'$$

$$= \sqrt[3]{x^{-2}}$$

$$= \left(x^{-2}\right)^{1/3} \qquad \text{Write in exponential form}$$

$$= x^{-2(1/3)} = x^{-2/3} \qquad \text{Law 4}$$

$$= \frac{1}{x^{2/3}} \qquad \text{Law 2}$$

85. $x^{-1/4} + 6x^{1/4} = x^{-1/4}\left(1 + 6x^{1/2}\right)$ or $x^{-1/4}\left(1 + 6\sqrt{x}\right)$

91. $f(4) = (4)^{-1} = \dfrac{1}{4}$

97. $A = P\left(1 + \dfrac{r}{m}\right)^{mt}$, where $P = 500$, $r = .06$, $m = 1$, $t = 6$

$$= 500\left(1 + \frac{.06}{1}\right)^{1 \cdot 6}$$

$$= 500(1.06)^6$$

$$\approx \$709.26$$

103. $A = P\left(1 + \dfrac{r}{m}\right)^{mt}$, where $P = 1500$, $r = .06$, $m = 360$, $t = 1$

$$= 1500\left(1 + \frac{.06}{360}\right)^{360 \cdot 1}$$

$$= 1500(1.0001667)^{360}$$

$$\approx \$1592.75$$

109. $[\text{new stopping distance}] = \dfrac{1}{20}(2x)^2$

$$= \frac{1}{20} \cdot 4x^2$$

$$= 4\left(\frac{1}{20}x^2\right)$$

$$= 4 \cdot [\text{old stopping distance}]$$

0.6   Functions and Graphs in Applications

The crucial step in many of these problems is to express the function in one variable using the information given. Exercise 7 is a typical example. Both the height and width are expressed in terms of x. The function, here perimeter, is now in terms of x.

1. If x = width, then the height is $3(\text{width}) = 3x$.

7. $\text{Perimeter} = 2 \cdot \text{height} + 2 \cdot \text{width}$
    $$= 2(3x) + 2(x) = 6x + 2x = 8x.$$
    $\text{Area} = \text{height} \times \text{width}$
    $$= (3x)(x) = 3x^2.$$
    But Area = 25 square feet, so $3x^2 = 25$.

13. Volume $= \pi r^2 h$, where r is the radius of the circular ends and h is the height of the cylinder. Since the volume is supposed to be 100 cubic inches,
    $$\pi r^2 h = 100.$$
    Using the solutions to Practice Problems 0.6,

    Area of the left end: πr^2,

    Area of the right end: πr^2,

    Area of the side (a "rolled-up rectangle"): $2\pi rh$.

    Cost for the left end: $5(\pi r^2) = 5\pi r^2$,

    Cost for the right end: $6(\pi r^2) = 6\pi r^2$,

    Cost for the side: $7(2\pi rh) = 14\pi rh$.

    So the total cost is $11\pi r^2 + 14\pi rh$ (dollars).

19. From Exercise 7, perimeter $= 8x$ (for the rectangle of Exercise 1), hence
    $$8x = 40$$
    $$x = 5.$$
    Also from Exercise 7, the area is $3x^2$ (for the rectangle of Exercise 1). Therefore, since $x = 5$,
    $$\text{Area} = 3(5)^2 = 75 \text{ cm}^2.$$

25. (a) $P(x) = R(x) - C(x)$
    $$= 21x - (9x + 800)$$
    $$= 12x - 800$$

    (b) x = number of sales, and here $x = 120$, so
    $$P(x) = P(120)$$
    $$= 12(120) - 800$$
    $$= 640 \text{ dollars in profit.}$$

(c) The weekly profit function is $P(x) = 12x - 800$, from (a). A weekly profit of $1000 yields

$$12x - 800 = 1000,$$
$$12x = 1800,$$
$$x = 150.$$

The weekly revenue is $R(x) = 21x$, so revenue $= 21(150) = \$3150$.

31. $(3, 162)$, $(6, 270)$ are points on the graph $y = f(r)$. The cost of constructing a cylinder of radius 3 inches is 162 cents, and similarly, for a radius of 6 inches the cost is 270 cents. Thus the additional cost of increasing the radius from 3 inches to 6 inches is $270 - 162 = 108$ cents $= \$1.08$.

37. $C(1000) = 4000$.

43. "Solve $P(x) = 30,000$" translates to "find the x-coordinates of the points on the graph whose y-coordinate is 30,000."

49. The phrase "determine when" means "find the time or times." In this case, find the values of t that make $h(t) = 100$ feet. Graphically, the task is to find the t-coordinates of the points on the graph of $h(t)$ whose y-coordinate is 100. See Example 7(d).

Chapter 0 Review Exercises

Study the Review of Fundamental Concepts. Write out your own answers. Can you handle the algebra of functions as in Section 0.3? Can you factor quadratic (and simple cubic) polynomials? Have you memorized the quadratic formula? Have you thoroughly (and successfully) practiced using the laws of exponents? If you cannot answer yes to all these questions by the time you finish Chapter 1, you are not seriously interested in succeeding in this calculus course, or you need to take a refresher course in college algebra before you study calculus.

The supplementary exercises provide a brief review of the main skills of the chapter. While solutions for all review exercises are included, expanded explanations are included for every sixth exercise.

1. If $f(x) = x^3 + \dfrac{1}{x}$, then

$$f(1) = 1^3 + \frac{1}{1} = 2,$$

$$f(3) = 3^3 + \frac{1}{3} = 27 + \frac{1}{3} = 27\frac{1}{3},$$

$$f(-1) = (-1)^3 + \frac{1}{-1} = -1 - 1 = -2,$$

$$f\left(-\frac{1}{2}\right) = \left(-\frac{1}{2}\right)^3 + \frac{1}{-\frac{1}{2}} = -\frac{1}{8} - 2 = -2\frac{1}{8},$$

$$f(\sqrt{2}) = (\sqrt{2})^3 + \frac{1}{\sqrt{2}} = 2\sqrt{2} + \frac{1}{\sqrt{2}} = 2\sqrt{2} + \frac{\sqrt{2}}{2} = \frac{5\sqrt{2}}{2}.$$

2. $f(x) = 2x + 3x^2$

$f(0) = 2(0) + 3(0)^2 = 0$

$f\left(-\dfrac{1}{4}\right) = 2\left(-\dfrac{1}{4}\right) + 3\left(-\dfrac{1}{4}\right)^2 = -\dfrac{5}{16}$

$f\left(\dfrac{1}{\sqrt{2}}\right) = 2\left(\dfrac{1}{\sqrt{2}}\right) + 3\left(\dfrac{1}{\sqrt{2}}\right)^2 = \dfrac{3 + 2\sqrt{2}}{2}$

3. $f(x) = x^2 - 2$

$f(a-2) = (a-2)^2 - 2 = a^2 - 4a + 2$

4. $f(x) = \dfrac{1}{x+1} - x^2$

$f(a+1) = \dfrac{1}{(a+1)+1} - (a+1)^2 = \dfrac{1}{a+2} - (a^2 + 2a + 1) = -\dfrac{a^3 + 4a^2 + 5a + 1}{a+2}$

5. $f(x) = \dfrac{1}{x(x+3)} \Rightarrow x \neq 0, -3$

6. $f(x) = \sqrt{x-1} \Rightarrow x \geq 1$

7.   The domain of $f(x) = \sqrt{x^2 + 1}$ consists of all values of x since $x^2 + 1$ is never negative.

8. $f(x) = \dfrac{1}{\sqrt{3x}}, \; x > 0$

9. $h(x) = \dfrac{x^2 - 1}{x^2 + 1}$

$h\left(\dfrac{1}{2}\right) = \dfrac{\left(\frac{1}{2}\right)^2 - 1}{\left(\frac{1}{2}\right)^2 + 1} = -\dfrac{3}{5}$

So the point $\left(\dfrac{1}{2}, -\dfrac{3}{5}\right)$ is on the graph.

10. $k(x) = x^2 + \dfrac{2}{x}$

$k(1) = 1^2 + \dfrac{2}{1} = 3$

So the point $(1, -2)$ is not on the graph.

11. $5x^3 + 15x^2 - 20x = 5x(x^2 + 3x - 4) = 5x(x-1)(x+4)$

12. $3x^2 - 3x - 60 = 3(x^2 - x - 20) = 3(x - 5)(x + 4)$

13. To factor $18 + 3x - x^2$, first factor out the coefficient -1 of x^2 to get $-1(x^2 - 3x - 18)$. To factor $x^2 - 3x - 18$, you need to find c and d such that $cd = -18$ and $c + d = -3$. The solution is $c = -6$ and $d = 3$, so

$$x^2 - 3x - 18 = (x - 6)(x + 3).$$

Thus, $$18 + 3x - x^2 = -1(x^2 - 3x - 18) = -1(x - 6)(x + 3).$$

14. $x^5 - x^4 - 2x^3 = x^3(x^2 - x - 2) = x^3(x - 2)(x + 1)$

15. $y = 5x^2 - 3x - 2 \Rightarrow 5x^2 - 3x - 2 = 0.$

$$x = \frac{-b \pm \sqrt{b^2 - 4ac}}{2a} = \frac{3 \pm \sqrt{(-3)^2 - 4(5)(-2)}}{2(5)} = \frac{3 \pm 7}{10} \Rightarrow x = 1 \text{ or } x = -\frac{2}{5}$$

16. $y = -2x^2 - x + 2 \Rightarrow -2x^2 - x + 2 = 0.$

$$x = \frac{-b \pm \sqrt{b^2 - 4ac}}{2a} = \frac{1 \pm \sqrt{(-1)^2 - 4(-2)(2)}}{2(-2)} = \frac{1 \pm \sqrt{17}}{-4} \Rightarrow x = \frac{-1 + \sqrt{17}}{4} \text{ or } x = \frac{-1 - \sqrt{17}}{4}$$

17. Substitute $2x - 1$ for y in the quadratic equation, then find the zeros:
$5x^2 - 3x - 2 = 2x - 1 \Rightarrow 5x^2 - 5x - 1 = 0.$

$$x = \frac{-b \pm \sqrt{b^2 - 4ac}}{2a} = \frac{5 \pm \sqrt{(-5)^2 - 4(5)(-1)}}{2(5)} = \frac{5 \pm 3\sqrt{5}}{10}$$

Now find the y-values for each x value:

$$y = 2x - 1 = 2\left(\frac{5 + 3\sqrt{5}}{10}\right) - 1 = \frac{3\sqrt{5}}{5}$$

$$y = 2x - 1 = 2\left(\frac{5 - 3\sqrt{5}}{10}\right) - 1 = \frac{-3\sqrt{5}}{5}$$

Points of intersection: $\left(\dfrac{5 + 3\sqrt{5}}{10}, \dfrac{3\sqrt{5}}{5}\right), \left(\dfrac{5 - 3\sqrt{5}}{10}, -\dfrac{3\sqrt{5}}{5}\right)$

18. Substitute $x - 5$ for y in the quadratic equation, then find the zeros:
$-x^2 + x + 1 = x - 5 \Rightarrow x^2 - 6 = 0 \Rightarrow x = \pm\sqrt{6}$

Now find the y-values for each x value: $y = x - 5 = \sqrt{6} - 5$

$y = -\sqrt{6} - 5$

Points of intersection: $\left(\sqrt{6}, \sqrt{6} - 5\right), \left(-\sqrt{6}, -\sqrt{6} - 5\right)$

19. $f(x) + g(x) = (x^2 - 2x) + (3x - 1) = x^2 + x - 1.$

20. $f(x) - g(x) = (x^2 - 2x) - (3x - 1) = x^2 - 5x + 1$

21. $f(x)h(x) = (x^2 - 2x)(\sqrt{x}) = x^2 \cdot x^{1/2} - 2x \cdot x^{1/2} = x^{5/2} - 2x^{3/2}$

22. $f(x)g(x) = (x^2 - 2x)(3x - 1) = 3x^3 - x^2 - 6x^2 + 2x = 3x^3 - 7x^2 + 2x$

23. $\dfrac{f(x)}{h(x)} = \dfrac{x^2 - 2x}{\sqrt{x}} = x^{3/2} - 2x^{1/2}$

24. $g(x)h(x) = (3x - 1)\sqrt{x} = 3x \cdot x^{1/2} - x^{1/2} = 3x^{3/2} - x^{1/2}$

25. $f(x) - g(x) = \dfrac{x}{(x^2 - 1)} - \dfrac{(1 - x)}{(1 + x)}.$

In order to add two fractions, their denominators must be the same. Observe that $x^2 - 1 = (x + 1)(x - 1)$. Thus a common denominator for

$$\frac{x}{(x^2 - 1)} \quad \text{and} \quad \frac{(1 - x)}{(1 + x)} \quad \text{is} \quad x^2 - 1.$$

If we multiply

$$\frac{1 - x}{1 + x} \quad \text{by} \quad \frac{x - 1}{x - 1},$$

we get an equivalent expression whose denominator is $x^2 - 1$. Thus

$$\frac{x}{x^2 - 1} - \frac{1 - x}{1 + x} = \frac{x}{x^2 - 1} - \frac{1 - x}{1 + x} \cdot \frac{x - 1}{x - 1}$$

$$= \frac{x}{x^2 - 1} - \frac{(1 - x)(x - 1)}{x^2 - 1}$$

$$= \frac{x - (-x^2 + 2x - 1)}{x^2 - 1} = \frac{x^2 - x + 1}{x^2 - 1}.$$

26. $f(x) - g(x + 1) = \dfrac{x}{x^2 - 1} - \dfrac{1 - (x + 1)}{1 + (x + 1)} = \dfrac{x(x + 2) - (-x)(x^2 - 1)}{(x^2 - 1)(x + 2)} = \dfrac{x^3 + x^2 + x}{(x^2 - 1)(x + 2)}$

27. $g(x) - h(x) = \dfrac{1 - x}{1 + x} - \dfrac{2}{3x + 1} = \dfrac{(1 - x)(3x + 1) - 2(1 + x)}{(1 + x)(3x + 1)} = -\dfrac{3x^2 + 1}{(1 + x)(3x + 1)} = -\dfrac{3x^2 + 1}{3x^2 + 4x + 1}$

28. $f(x) + h(x) = \dfrac{x}{x^2 - 1} + \dfrac{2}{3x + 1} = \dfrac{x(3x + 1) + 2(x^2 - 1)}{(x^2 - 1)(3x + 1)} = \dfrac{5x^2 + x - 2}{(x^2 - 1)(3x + 1)}$

29. $g(x) - h(x-3) = \dfrac{1-x}{1+x} - \dfrac{2}{3(x-3)+1} = \dfrac{(1-x)(3x-8) - 2(1+x)}{(1+x)(3x-8)}$

$\qquad = \dfrac{-3x^2 + 9x - 10}{(1+x)(3x-8)} = \dfrac{-3x^2 + 9x - 10}{3x^2 - 5x - 8}$

30. $f(x) + g(x) = \dfrac{x}{x^2-1} + \dfrac{1-x}{1+x} = \dfrac{x + (1-x)(x-1)}{x^2-1} = \dfrac{-x^2 + 3x - 1}{x^2-1}$

31. Substitute $g(x)$ for each occurrence of x in $f(x)$ to obtain $f(g(x))$. Thus

$$f(g(x)) = \left(\dfrac{1}{x^2}\right)^2 - 2\left(\dfrac{1}{x^2}\right) + 4 = \dfrac{1}{x^4} - \dfrac{2}{x^2} + 4.$$

32. $g\left(f(x)\right) = g(x^2 - 2x + 4) = \dfrac{1}{\left(x^2 - 2x + 4\right)^2} = \dfrac{1}{x^4 - 4x^3 + 12x^2 - 16x + 16}$

33. $g\left(h(x)\right) = g\left(\dfrac{1}{\sqrt{x}-1}\right) = \dfrac{1}{\left(\frac{1}{\sqrt{x}-1}\right)^2} = \dfrac{1}{\frac{1}{x - 2\sqrt{x}+1}} = x - 2\sqrt{x} + 1 = \left(\sqrt{x}-1\right)^2$

34. $h\left(g(x)\right) = h\left(\dfrac{1}{x^2}\right) = \dfrac{1}{\sqrt{\frac{1}{x^2}} - 1} = \dfrac{1}{\frac{1}{|x|} - 1} = \dfrac{|x|}{1 - |x|}$

35. $f\left(h(x)\right) = f\left(\dfrac{1}{\sqrt{x}-1}\right) = \left(\dfrac{1}{\sqrt{x}-1}\right)^2 - 2\left(\dfrac{1}{\sqrt{x}-1}\right) + 4 = \dfrac{1}{\left(\sqrt{x}-1\right)^2} - \dfrac{2}{\sqrt{x}-1} + 4$

36. $h\left(f(x)\right) = h\left(x^2 - 2x + 4\right) = \dfrac{1}{\sqrt{x^2 - 2x + 4} - 1} = \left(\sqrt{x^2 - 2x + 4} - 1\right)^{-1}$

37. $(81)^{3/4} = (81^{1/4})^3 = 3^3 = 27,$ (Law 4′)

$\qquad 8^{5/3} = (8^{1/3})^5 = 2^5 = 32,$ (Law 4′)

$\qquad (.25)^{-1} = \left(\dfrac{1}{4}\right)^{-1} = 4$

38. $(100)^{3/2} = \left(\sqrt{100}\right)^3 = 1000$

$\qquad (.001)^{1/3} = \left(\sqrt[3]{.001}\right) = .1$

39. $C(x)$ = carbon monoxide level corresponding to population x
$P(t)$ = population of the city in t years
$C(x) = 1 + .4x$
$P(t) = 750 + 25t + .1t^2$
$C(P(t)) = 1 + .4(750 + 25t + .1t^2) = 1 + 300 + 10t + .04t^2 = .04t^2 + 10t + 301$

40. $R(x) = 5x - x^2$

$$f(d) = 6\left(1 - \frac{200}{d+200}\right)$$

$$R(f(d)) = 5 \cdot 6\left(1 - \frac{200}{d+200}\right) - \left[6\left(1 - \frac{200}{d+200}\right)\right]^2 = 30\left(1 - \frac{200}{d+200}\right) - 36\left(1 - \frac{200}{d+200}\right)^2$$

41. $\left(\sqrt{x+1}\right)^4 = (x+1)^{4/2} = (x+1)^2 = x^2 + 2x + 1$

42. $\dfrac{xy^3}{x^{-5}y^6} = x \cdot x^5 \cdot y^3 \cdot y^{-6} = \dfrac{x^6}{y^3}$

43. $\dfrac{x^{3/2}}{\sqrt{x}} = \dfrac{x^{3/2}}{x^{1/2}} = x^{(3/2-1/2)} = x^1 = x$ \quad (Law 3)

44. $\sqrt[3]{x}(8x^{2/3}) = x^{1/3} \cdot 8x^{2/3} = 8x$

Chapter 1
The Derivative

1.1 The Slope of a Straight Line

Slope Property 1 will help you understand the concept of the slope of a line. Property 2 is needed when you have to find the slope of a line between two points. Property 3 is the most useful, and you must memorize the point-slope form of the equation of a line. Exercises 7–10 are simple but very important and will give you practice using the point-slope form. Check with your instructor to see how much attention you should give to Slope Properties 4 and 5. They are seldom needed later in the text.

1. Write the equation $y = 3 - 7x$ in the form $y = mx + b$. That is,

$$y = -7x + 3.$$

Hence the slope is –7, and the y-intercept is (0, 3).

7. Let $(x_1, y_1) = (7, 1)$ and $m = -1$, then use Slope Property 3. The equation of the line is $y - 1 = -1(x - 7)$, or, equivalently, $y = -x + 8$.

13. By Slope Property 2, the slope of the line is

$$\frac{0 - 0}{1 - 0} = 0.$$

Since (0, 0) is on the line, the point-slope equation of the line is $y - 0 = 0(x - 0)$, or $y = 0$.

19. The x-intercept of the line is –2, so the point $(-2, 0)$ is on the line. Let $(x_1, y_1) = (-2, 0)$ and $m = -2$, then use Slope Property 3. The equation of the line is $y - 0 = -2\big(x - (-2)\big)$, or, equivalently, $y = -2x - 4$.

25. The line whose equation is sought is perpendicular to $x + y = 0$, or $y = -x$. The slope of this line is –1, so by Slope Property 5 the slope m of the line whose equation is sought satisfies the equation

$$-1 \cdot m = -1$$

or $m = 1$. Now let $(x_1, y_1) = (2, 0)$ and $m = 1$, then use Slope Property 3. The equation of the line is $y - 0 = 1(x - 2)$, or, equivalently, $y = x - 2$.

31. No units are shown on the graphs in the figure, but you can see whether a graph has positive or negative slope, and you can see whether the y-intercept is on the positive or negative y-axis.

(a) Rewrite the equation as $y = 1 - x = (-1)x + 1$. The slope is negative and the y-intercept (0, 1) is on the positive y-axis. The only graph with these two properties is (C).

(b) Rewrite the equation as $y = x - 1$. The slope is positive and the y-intercept $(0, -1)$ is on the negative y-axis. The graph is (B).

(c) Rewrite the equation as $y = (-1)x - 1$. The slope is negative, the y-intercept is on the negative y-axis, and so the graph is (D).

(d) This must be (A) because the other graphs are already chosen. The fact that (A) works is also easy to see from the equation $y = x + 1$ (positive slope and positive y-intercept).

37. Since the slope of this line is 2, if we start at a point on the line and move 1 unit to the right and then 2 units up (in the positive y-direction) we will reach another point on the line. If we start at $(1, 3)$ and move 1 unit to the right and 2 units up, we arrive at $(2, 5)$. So $(2, 5)$ is on the line. A similar move from $(2, 5)$ takes us to $(3, 7)$, so $(3, 7)$ is on the line. Finally, suppose that $(0, y)$ is on the line. This point is one unit to the left of $(1, 3)$ in the x-direction. Starting at $(0, y)$ and moving one unit to the right and 2 units up, we arrive at $(1, y + 2)$. This point is on the line and so is $(1, 3)$. Hence we must have $y + 2 = 3$, and $y = 1$. Thus $(0, 1)$ is on the line.

Alternatively, we can determine the equation of the line using the point-slope form:

$$y - 3 = 2(x - 1) \text{ or } y = 2x + 1.$$

If $x = 2$, then $y = 2(2) + 1 = 5$. If $x = 3$, then $y = 2(3) + 1 = 7$. If $x = 0$, then $y = 2(0) + 1 = 1$. Thus, the points $(2, 5)$, $(3, 7)$, and $(0, 1)$ lie on the line.

43. For slope -2 and y-intercept $(0, -1)$, the slope-intercept equation $y = mx + b$ is $y = -2x + (-1)$, or

$$y = -2x - 1.$$

To graph this line, first find the x-intercept by letting $y = 0$ and solving for x.

$$0 = -2x - 1$$
$$1 = -2x$$
$$-\frac{1}{2} = x$$

Thus, the x-intercept is $\left(-\frac{1}{2}, 0\right)$. Plot both intercepts and draw the line.

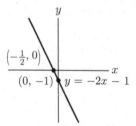

49. Let x be the number of months since January 1, 2012, and $P(x)$ be the price of gasoline per gallon x months since January 1, 2012. Since the price of gasoline on January 1, 2012 was \$4.12/gallon, the point $(0, 4.12)$ is on our line. Also, since the price is rising at a rate of 6 cents per gallon, the slope of our line in .06. By Slope Property 3, the equation of the line through point $(0, 4.12)$ with slope .06 is

$$y - 4.12 = .06(x - 0)$$
$$y - 4.12 = .06x$$
$$y = .06x + 4.12.$$

Therefore, the price of gasoline per gallon x months after January 1, 2012 is

$$P(x) = .06x + 4.12.$$

April, 2012 is 3 months after Jan. 1, 2012, thus the price for a gallon of gasoline on April 1, 2012 is

$$P(3) = .06(3) + 4.12 = \$4.30/\text{gallon}$$

and 15 gallons of gasoline on April 1, 2012 would cost $15 \times \$4.30/\text{gallon} = \$64.50$.

Sept., 2012 is 8 months after Jan. 1, 2012, thus the price for a gallon of gasoline on Sept. 1, 2012 is

$$P(9) = .06(8) + 4.12 = \$4.60/\text{gallon}$$

and 15 gallons of gasoline on Sept. 1, 2012 would cost $15 \times \$4.60/\text{gallon} = \$69.00$.

55. Assuming the total cost $C(x)$ is linearly related to the daily production level x, total cost can be expressed as

$$C(x) = mx + b$$

(a) Fixed costs are $1500, therefore $b = 1500$. Also, we know the total cost is $2200 when 100 rods are produced per day, thus

$$C(100) = m(100) + 1500 = 2200.$$

Solving for the slope m, we find $m = 7$. Therefore, the total cost can be expressed as a function of daily production levels as follows:

$$C(x) = 7x + 1500.$$

(b) In Example 1, we saw that the marginal cost which is the additional cost incurred when the production level is increased by 1 unit, is the same as the slope of the line. Thus, the marginal cost at $x = 100$ is $7 per rod.

(c) The additional cost of raising the daily production level from 100 to 101 rods is the marginal cost, $7 per rod. Also, the additional cost of raising production from 100 to 101 rods can be expressed by

$$C(101) - C(100) = 7(101) + 1500 - (7(100) + 1500)$$
$$= 707 + 1500 - 700 - 1500$$
$$= \$7 \text{ per rod.}$$

61. The diver starts at a depth of 212 ft, which is represented as -212. Thus, the function is

$$y(t) = 2t - 212.$$

67. Using $\dfrac{f(x_2) - f(x_1)}{x_2 - x_1} = m$ and the hint, we have

$$\frac{f(x) - f(x_1)}{x - x_1} = m$$
$$f(x) - f(x_1) = m(x - x_1)$$
$$f(x) = m(x - x_1) + f(x_1)$$
$$= mx + (-mx_1 + f(x_1))$$

Let $b = -mx_1 + f(x_1)$. Then $f(x) = mx + b$.

Help for Technology Exercises: You may wish to review the **Incorporating Technology** section in chapter 0 to help you with the technology exercises in chapter 1.

1.2   The Slope of a Curve at a Point

This brief section should be read carefully. Example 2 and Exercises 15, 16, and 32 are very important, and so we have included a solution of Exercise 15. Of the students who do *not* use this *Manual,* at least 20% will miss an exam problem on the equation of a tangent line. If you carefully study the solution of Exercise 15 here and the solution of Exercise 43 in Section 1.6, you should have no difficulty on an exam.

1.

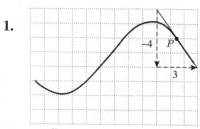

The slope of the curve at the point P is, by definition, the slope of the tangent line at P. If you move four units in the negative y-direction from P, you can return to the line by moving three units in the positive x-direction. Therefore the slope is $-\frac{4}{3}$.

7.

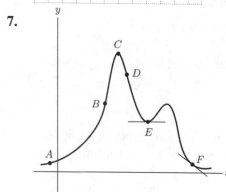

At the point E, the tangent is a horizontal line. Therefore, the slope is 0.

At the point F, the tangent has a small negative slope.

13.   Use the *slope formula* $2x$ to find the slope of the graph at $x = -\frac{1}{4}$.

$$m = 2\left(-\frac{1}{4}\right) = -\frac{1}{2}$$

Thus, the slope of the tangent line to the graph of $y = x^2$ at the point where $x = -\frac{1}{4}$ is $-\frac{1}{2}$.

15.   Use the point-slope equation of the tangent line to the graph of $y = x^2$ at $x = 2.5$. Find a point (x_1, y_1) and a slope m. There are two basic principles to keep in mind.

   (a)   Use the *original equation* $y = x^2$ to find a *point* on the graph. For $x = 2.5$, compute

   $y = (2.5)^2 = 6.25$. Thus $(2.5, 6.25)$ is on the graph of $y = x^2$.

   (b)   Use the *slope formula* $2x$ to find the *slope* of the graph at a point. When $x = 2.5$, the slope of the graph is $2(2.5) = 5$. By definition, this slope is the slope of the tangent line to the graph at the point where $x = 2.5$.

   The desired tangent line equation has the form $y - y_1 = m(x - x_1)$, where $x_1 = 2.5$, $y_1 = 6.25$, and $m = 5$. That is,

   $$y - 6.25 = 5(x - 2.5).$$

   Leave the answer in this form unless you are specifically asked to rewrite the answer in some equivalent form such as the slope-intercept form.

Warning: A common mistake in Exercise 15 is to replace m in the equation $y - y_1 = m(x - x_1)$ by the slope formula $2x$ instead of a *specific value* of m. But the equation $y - 6.25 = 2x(x - 2.5)$ is *not* the equation of a line. (In fact, this equation simplifies to $y - 6.25 = 2x^2 - 5x$, or $y = 2x^2 - 5x + 6.25$, which is a quadratic equation.) The equation of a tangent *line* must involve a specific *number m* that gives the slope of the desired tangent line.

19. Rewrite the equation $2x + 3y = 4$ in slope-intercept form to find the slope of the line.

$$2x + 3y = 4$$
$$3y = -2x + 4$$
$$y = -\frac{2}{3}x + \frac{4}{3}$$

So, the slope of the line is $-\frac{2}{3}$. Since the tangent line is parallel to this line, its slope is also $-\frac{2}{3}$ by Slope Property 4. We know that the slope of the graph of $y = x^2$ at the point (x, y) is given by the formula $2x$, so solve $-\frac{2}{3} = 2x$ to find the x-coordinate of the point where the tangent line touches the graph.

$$-\frac{2}{3} = 2x$$
$$-\frac{1}{3} = x$$

If $x = -\frac{1}{3}$, then

$$y = x^2 = \left(-\frac{1}{3}\right)^2 = \frac{1}{9}.$$

Thus, the point on the graph of $y = x^2$ where the tangent line is parallel to the line $2x + 3y = 4$ is $\left(-\frac{1}{3}, \frac{1}{9}\right)$.

25.

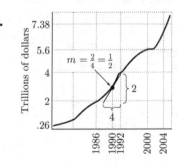

It appears the points (1988, 2) and (1990, 3) are on the tangent line. The slope is

$$m = \frac{3 - 2}{1990 - 1988} = .5$$

Therefore, the annual rate of increase of the federal debt in 1990 is approximately \$.5 trillion/year.

31. The slope formula for the curve $y = x^3$ is given by the formula $3x^2$. The point $\left(-\frac{1}{2}, -\frac{1}{8}\right)$ corresponds to $x = -\frac{1}{2}$, so the slope at $\left(-\frac{1}{2}, -\frac{1}{8}\right)$ is $3x^2 = 3\left(-\frac{1}{2}\right)^2 = 3\left(\frac{1}{4}\right) = \frac{3}{4}$.

37. **(a)** By Slope Property 2, the slope of the line *l* is

$$\frac{13-4}{5-2} = \frac{9}{3} = 3.$$

The length of line segment *d* is the absolute value of the difference between the *y*-coordinates of the Points *P* and *Q*, i.e.,

$$d = |13-4| = 9.$$

(b) As the Point *Q* is moved along the curve toward *P*, the slope of the line *l* through the Points *P* and *Q* increases. For example, if the Point *Q′* lies on the curve midway between *P* and *Q*, then the slope of the line through *P* and *Q′* is greater than the slope of the line *l* through *P* and *Q*.

39. Review the material in the **Incorporating Technology** section on page 69 of the text to learn how to zoom in on a graph.

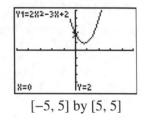
[−5, 5] by [5, 5]

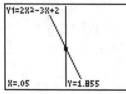
[−.078125, .078125] by [1.927923, 2.084173]

When *x* = 0, *y* = 2. Find a second point on the line using **value:** *x* = .05, *y* = 1.855

$$m = \frac{1.855-2}{.05-0} = -2.9$$

The actual value of $f'(0)$ is −3.

1.3   The Derivative and Limits

Your short-term goal for this section should be to learn the derivative formulas that appear in boxes on page 74 and to become familiar with the notation introduced on pages 77–78. This will enable you to do the homework for this section. Your long-term goal should be to have some understanding of the secant-line calculation of the derivative, as described on pages 78–79. This material is not grasped easily. Reading it *out loud* will help you to go over the ideas slowly and carefully. Plan to review this section before your first exam and again when you reach Chapter 3.

1. If $f(x) = mx + b$, then $f'(x) = m$. Hence if $f(x) = 3x + 7$, then $f'(x) = 3$.

7. Apply the power rule with $r = \frac{2}{3}$:

$$f(x) = x^{2/3},$$

$$f'(x) = \frac{2}{3}x^{(2/3)-1} = \frac{2}{3}x^{-1/3}$$

$$= \frac{2}{3}\left(\frac{1}{x^{1/3}}\right) = \frac{2}{3\sqrt[3]{x}}$$

13. Rewrite $f(x) = \dfrac{1}{x^{-2}}$ as $f(x) = x^2$. Then apply the power rule with $r = 2$.

$$f'(x) = 2x.$$

19. If $f(x) = \dfrac{1}{x}$, then $f'(x) = -\dfrac{1}{x^2} \, (x \neq 0)$, by formula (5) on Page 75. Hence at $x = \dfrac{2}{3}$,

$$f'\left(\frac{2}{3}\right) = -\frac{1}{(2/3)^2} = -\frac{1}{4/9} = -\frac{9}{4}.$$

Helpful Hint: Formula (5) for the derivative of $f(x) = \frac{1}{x}$ is used so frequently that you should memorize it even though it is only a special case of the power rule.

25. Use the power rule with $r = 4$:

$$\frac{d}{dx}(x^4) = 4x^{4-1} = 4x^3.$$

Setting $x = 2$ in this derivative, we find that the slope of the curve $y = x^4$ at $x = 2$ is $4(2)^3 = 4 \cdot 8 = 32$.

31. If $f(x) = \dfrac{1}{x^5}$, then $f(-2) = \dfrac{1}{(-2)^5} = -\dfrac{1}{32}$. To compute $f'(2)$, first determine $f'(x)$ and then substitute -2 for x in the expression for $f'(x)$. Since $\dfrac{1}{x^5} = x^{-5}$, apply the power rule with $r = -5$.

$$f(x) = x^{-5},$$
$$f'(x) = -5x^{-5-1} = -5x^{-6} = -\frac{5}{x^6},$$
$$f'(2) = -\frac{5}{2^6} = -\frac{5}{64}.$$

37. If $f(x) = \sqrt{x} = x^{1/2}$, then $f\left(\dfrac{1}{9}\right) = \dfrac{1}{3}$. Applying the power rule with $r = \dfrac{1}{2}$, $f'(x) = \dfrac{1}{2}x^{-1/2} = \dfrac{1}{2\sqrt{x}}$.

Then the slope of the line tangent to $f(x)$ at $x = \dfrac{1}{9}$ is $f'\left(\dfrac{1}{9}\right) = \dfrac{1}{2}\left(\dfrac{1}{9}\right)^{-1/2} = \dfrac{1}{2}(9)^{1/2} = \dfrac{3}{2}$. Since the tangent line passes through the point $\left(\dfrac{1}{9}, \dfrac{1}{3}\right)$, its equation is

$$y - \frac{1}{3} = \frac{3}{2}\left(x - \frac{1}{9}\right).$$

43. The slope of the graph $y = \sqrt{x}$ at the point $x = a$ is found by using y' evaluated at $x = a$.

$$y' = \frac{1}{2}x^{-1/2} = \frac{1}{2\sqrt{x}}.$$

At $x = a$,

$$y' = \frac{1}{2\sqrt{a}}.$$

The slope of the tangent line $y = 2x + b$ at the point $(a, \sqrt{a})$ is $m = 2$. So we have

$$2 = \frac{1}{2\sqrt{a}}$$

$$\sqrt{a} = \frac{1}{4}$$

$$(\sqrt{a})^2 = \left(\frac{1}{4}\right)^2 \Rightarrow a = \frac{1}{16}.$$

To find P: at $a = \frac{1}{16}$, $y = \sqrt{\frac{1}{16}} = \frac{1}{4}$. Therefore, $P = \left(\frac{1}{16}, \frac{1}{4}\right)$. To find b:

$$y = 2x + b$$

$$\frac{1}{4} = 2\left(\frac{1}{16}\right) + b$$

$$b = \frac{1}{4} - 2\left(\frac{1}{16}\right)$$

$$b = \frac{1}{8}.$$

49. Use the power rule with $r = 8$:

$$\frac{d}{dx}(x^8) = 8x^{8-1} = 8x^7.$$

55. Apply the power rule with $r = \frac{1}{5}$:

$$y = x^{1/5},$$

$$\frac{dy}{dx} = \frac{1}{5}x^{(1/5)-1}$$

$$= \frac{1}{5}x^{-4/5}.$$

61.

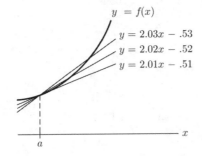

The three lines and $y = f(x)$ intersect at the point $(a, f(a))$. Take the line $y = 2.01x - .51$ at this point. Then

$$f(a) = 2.01a - .51.$$

Similarly using $y = 2.02x - .52,$

$$f(a) = 2.02a - .52.$$

So,
$$2.01a - .51 = 2.02a - .52,$$
$$0.01 = 0.01a,$$
$$a = 1,$$
$$f(a) = 2.02(1) - .52 = 2.02 - .52 = 1.5.$$

To estimate $f'(a)$ use the secant-line calculation. The secant line $y = 2.01x - .51$ is "nearly" a tangent line, therefore the slope of this line is nearly $f'(a)$. The slope of the tangent line is slightly less than the secant line's slope of 2.01. Hence

$$f'(a) \approx 2.$$

67. $f(x) = -x^2 + 2x \Rightarrow$

$$\frac{f(x+h) - f(x)}{h} = \frac{\left[-(x+h)^2 + 2(x+h) \right] - \left(-x^2 + 2x \right)}{h}$$

$$= \frac{-x^2 - 2xh - h^2 + 2x + 2h + x^2 - 2x}{h}$$

$$= \frac{h(-2x + 2 - h)}{h}$$

$$= -2x + 2 - h$$

73. We must apply the three-step method to find the derivative of $f(x) = 7x^2 + x - 1$.

Step 1:

$$\frac{f(x+h) - f(x)}{h} = \frac{\left[7(x+h)^2 + (x+h) - 1\right] - \left(7x^2 + x - 1\right)}{h}$$

$$= \frac{7x^2 + 14xh + 7h^2 + x + h - 1 - 7x^2 - x + 1}{h}$$

$$= \frac{14xh + 7h^2 + h}{h}$$

$$= \frac{h(14x + 7h + 1)}{h}$$

$$= 14x + 1 + 7h$$

Steps 2 and 3: As h approaches 0, the quantity $14x + 1 + 7h$ approaches $14x + 1$. Thus,

$$f'(x) = 14x + 1.$$

For exercises 79–91, refer to the **Incorporating Technology** section on page 79.

79. $f'(0)$, where $f(x) = 2^x$

85.

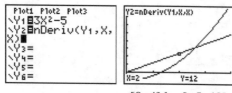

[0, 4] by [−5, 40]

Using **TRACE**, we find that the value of the derivative of Y_1 at $x = 2$ is 12.

91. $f(x) = \dfrac{5}{x}$, $g(x) = 5 - 1.25x$

To solve graphically, graph the functions and use the **intersect** command to find where $g(x)$ is tangent to $f(x)$.

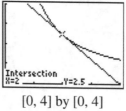

[0, 4] by [0, 4]

Thus, $a = 2$.

Alternatively, we can solve $\dfrac{5}{x} = 5 - 1.25x$ to find the x-value of the intersection of $f(x)$ and $g(x)$.

$$\frac{5}{x} = 5 - 1.25x$$

$$5 = 5x - 1.25x^2$$

$$1.25x^2 - 5x + 5 = 0$$

$$1.25\left(x^2 - 4x + 4\right) = 0$$

$$1.25\left(x - 2\right)^2 = 0$$

$$x = 2$$

1.4   Limits and the Derivative

Instructors differ widely about how much theoretical material to include in a course using our text. We recommend that at a minimum, you read pages 82–84, even if no exercises are assigned. The skills practice in this section are not needed for later work, but the exercises will give you valuable experience with simple limits and will improve your understanding of derivatives.

1. $\lim_{x \to 3} g(x)$ does not exist. As x approaches 3 from the right, $g(x)$ approaches 2, but as x approaches 3 from the left, the values for $g(x)$ do not approach 2. In order for a limit to exist, the values of the function must approach the same number as x approaches 2 from each direction.

7. Since $1 - 6x$ is a polynomial, the limit exists and
$$\lim_{x \to 1} (1 - 6x) = 1 - 6(1)$$
$$= 1 - 6 = -5.$$

13. Using the Limit Theorems
$$\lim_{x \to 7} (x + \sqrt{x - 6})(x^2 - 2x + 1)$$

$$= \left[\lim_{x \to 7} (x + \sqrt{x - 6}) \right]\left[\lim_{x \to 7} (x^2 - 2x + 1) \right] \qquad \text{(Thm. V)}$$

$$= \left[\lim_{x \to 7} x + \lim_{x \to 7} (x - 6)^{1/2} \right]\left[\lim_{x \to 7} (x^2 - 2x + 1) \right] \qquad \text{(Thm. III)}$$

$$= \left[\lim_{x \to 7} x + \left(\lim_{x \to 7} (x - 6) \right)^{1/2} \right]\left[\lim_{x \to 7} (x^2 - 2x + 1) \right] \qquad \text{(Thm. II)}$$

$$= [7 + (1)^{1/2}][49 - 14 + 1] \qquad \text{(Limits of polynomial functions)}$$

$$= (8)(36) = 288.$$

19. Since $\dfrac{-2x^2 + 4x}{x - 2} = \dfrac{-2x(x - 2)}{x - 2} = -2x$ for $x \neq 2$,

$$\lim_{x \to 2} \frac{-2x^2 + 4x}{x - 2} = \lim_{x \to 2} (-2x)$$
$$= -2(2) = -4.$$

25. No limit exists. Observe that

$$\lim_{x\to 8}\left(x^2+64\right)=128 \quad \text{and} \quad \lim_{x\to 8}\left(x-8\right)=0,$$

as x approaches 8. So the denominator gets very small and the numerator approaches 128. For example, if $x=8.00001$, then the numerator is 128.00016 and the denominator is .00001. The quotient is 12,800,016. As x approaches 8 even more closely, the quotient gets arbitrarily large and cannot possibly approach a limit.

31. Since

$$f'(a)=\lim_{h\to 0}\frac{f(a+h)-f(a)}{h},$$

we must calculate

$$\lim_{h\to 0}\frac{f(0+h)-f(0)}{h}.$$

If $f(x)=x^3+3x+1,$ then

$$\frac{f(0+h)-f(0)}{h}=\frac{[(0+h)^3+3(0+h)+1]-[0^3+3(0)+1]}{h}$$

$$=\frac{h^3+3h+1-1}{h}$$

$$=\frac{h^3+3h}{h}$$

$$=h^2+3.$$

Therefore, $f'(0)=\lim_{h\to 0}\left(h^2+3\right)=0^2+3=3.$

37. We must calculate

$$f'(x)=\lim_{h\to 0}\frac{f(x+h)-f(x)}{h}.$$

If $f(x)=3x+1,$ then

$$\lim_{h\to 0}\frac{f(x+h)-f(x)}{h}=\lim_{h\to 0}\frac{3(x+h)+1-(3x+1)}{h}$$

$$=\lim_{h\to 0}\frac{3x+3h+1-3x-1}{h}$$

$$=\lim_{h\to 0}\frac{3h}{h}$$

$$=\lim_{h\to 0}3$$

$$=3.$$

We conclude $f'(x)=3.$

43. We must calculate

$$f'(x) = \lim_{h \to 0} \frac{f(x+h) - f(x)}{h}.$$

If $f(x) = \dfrac{1}{x^2+1}$, then

$$\lim_{h \to 0} \frac{f(x+h) - f(x)}{h} = \lim_{h \to 0} \frac{\frac{1}{(x+h)^2+1} - \frac{1}{x^2+1}}{h}$$

$$= \lim_{h \to 0} \frac{\frac{(x^2+1)-((x+h)^2+1)}{((x+h)^2+1)(x^2+1)}}{h} = \lim_{h \to 0} \frac{\frac{x^2+1-x^2-2xh-h^2-1}{((x+h)^2+1)(x^2+1)}}{h}$$

$$= \lim_{h \to 0} \frac{\frac{-h(2x+h)}{((x+h)^2+1)(x^2+1)}}{h} = \lim_{h \to 0} \frac{-h(2x+h)}{((x+h)^2+1)(x^2+1)}\left(\frac{1}{h}\right)$$

$$= \lim_{h \to 0} \frac{-(2x+h)}{((x+h)^2+1)(x^2+1)} = \frac{-2x}{(x^2+1)(x^2+1)}$$

$$= \frac{-2x}{(x^2+1)^2}.$$

We conclude $f'(x) = \dfrac{-2x}{\left(x^2+1\right)^2}.$

49. We want to find $f(x)$ so that $\lim\limits_{h \to 0} \dfrac{(1+h)^2 - 1}{h}$ has the same form as

$$f'(a) = \lim_{h \to 0} \frac{f(a+h) - f(a)}{h}.$$

So we want $f(a+h) = (1+h)^2$; and $f(a) = 1$. From this we see that $f(x) = x^2$ and $a = 1$. Note that $f(a+h) = f(1+h) = (1+h)^2$ as needed.

55. We want to find $f(x)$ so that $\lim\limits_{h \to 0} \dfrac{(2+h)^2 - 4}{h}$ has the same form as

$$f'(a) = \lim_{h \to 0} \frac{f(a+h) - f(a)}{h}.$$

So we want $f(a+h) = (2+h)^2$; and $f(a) = 4$. From this we see that $f(x) = x^2$ and $a = 2$. Therefore, the given limit is $f'(2)$ where $f(x) = x^2$. Also, since we know $f'(x) = 2x \Rightarrow f'(2) = 2(2) = 4$, we conclude

$$f'(2) = \lim_{h \to 0} \frac{(2+h)^2 - 4}{h} = 4$$

61. As x increases without bound so does x^2. Therefore, $\dfrac{1}{x^2}$ approaches zero as x approaches ∞. That is

$$\lim_{x \to \infty} \frac{1}{x^2} = 0.$$

67. Referring to the figure in the text, we see that as x approaches 0 from the left, $f(x)$ approaches $\frac{3}{4}$. Likewise, as x approaches 0 from the right, $f(x)$ approaches $\frac{3}{4}$. That is, $f(x)$ approaches $\frac{3}{4}$ from both the left and right as x approaches 0, so we have established that

$$\lim_{x \to 0} f(x) = \frac{3}{4}.$$

73. Examining the graph of the function $f(x) = \sqrt{25+x} - \sqrt{x}$, it appears as if the values of the function approach 0 as $x \to \infty$.

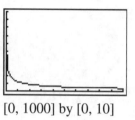

[0, 1000] by [0, 10]

To verify this, lets examine $f(x)$ for large values of x:

x	$f(x) = \sqrt{25+x} - \sqrt{x}$
10,000	$f(10,000) = \sqrt{25+10,000} - \sqrt{10,000} \approx .12492197250$
100,000	$f(100,000) = \sqrt{25+100,000} - \sqrt{100,000} \approx .0395260005$
1,000,000	$f(1,000,000) = \sqrt{25+1,000,000} - \sqrt{1,000,000} \approx .0124999219$
10,000,000	$f(10,000,000) = \sqrt{25+10,000,000} - \sqrt{10,000,000} \approx .0039528446$
100,000,000	$f(100,000,000) = \sqrt{25+100,000,000} - \sqrt{100,000,000} \approx .00125$
1,000,000,000	$f(1,000,000,000) = \sqrt{25+1,000,000,000} - \sqrt{1,000,000,000} \approx .00039528$

It also appears from the table that the function approaches 0 as x approaches ∞. We conclude

$$\lim_{x \to \infty} f(x) = 0.$$

1.5 Differentiability and Continuity

We want you to be aware that real applications sometimes involve functions that may not be differentiable at one or more points in their domains. So this section gives you rare opportunity to see functions whose graphs are not as "nice" as the ones we usually consider.

Exercises 1 and 7 refer to the following figure.

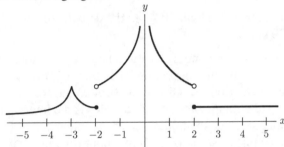

1. No, the graph in the figure is not continuous at $x = 0$ since the limit as x approaches zero does not exist.

7. No, the graph in the figure is not differentiable at $x = 0$ because the function is not even *defined* at $x = 0$. Even if the function were given some value at $x = 0$, say $f(0) = 1$, the graph would have a vertical tangent line at $x = 0$.

13. Since $f(x) = x^2$, the power rule gives a slope-formula, namely, $2x$, that is valid for all x. This slope-formula was verified in Section 1.3 using the limit of the slopes of secant lines. Thus $f(x)$ has a derivative (in the formal sense of Section 1.4) for all x. In particular, $f(x)$ is differentiable at $x = 1$. By Theorem 1 on page 94, $f(x)$ is necessarily continuous at $x = 1$.

19. At $x = 1$ the function $f(x)$ is defined, namely $f(1) = 0$. When computing $\lim_{x \to 1} f(x)$, we exclude consideration of the value $x = 1$; therefore,

$$\lim_{x \to 1} f(x) = \lim_{x \to 1} \frac{1}{x - 1}$$

which does not exist. Hence $\lim_{x \to 1} f(x) \neq f(1)$, so $f(x)$ is not continuous at $x = 1$. By Theorem 1, since $f(x)$ is not continuous at $x = 1$, it cannot be differentiable at $x = 1$.

25. $f(x)$ is continuous at $x = a$, if $\lim_{x \to a} f(x) = f(a)$. Here

$$f(x) = \frac{(6 + x)^2 - 36}{x}, \quad x \neq 0.$$

Where $a = 0$, we have:

$$\lim_{x \to 0} \frac{(6 + x)^2 - 36}{x} = \lim_{x \to 0} \frac{36 + 12x + x^2 - 36}{x}$$
$$= \lim_{x \to 0} (12 + x) = 12.$$

So define $f(0) = 12$, and this definition will make $f(x)$ continuous for all x.

31. (a) Referring to the graph in the text, we see that at 8 A.M., a total of $4000 in sales have already been made. The graph also shows that by 10 A.M., $10,000 worth of sales have been made. Thus the rate of sales during the period between 8 A.M. and 10 A.M. is given by the slope formula given in Slope Property 2:

$$\frac{y_2 - y_1}{x_2 - x_1} = \frac{10,000 - 4,000}{10 - 8} = \frac{6,000}{2} = 3,000.$$

In the time period between 8 A.M. and 10 A.M., the department store is selling goods at an average rate of $3000 per hour.

(b) Each portion of the graph corresponding to a two hour time period starting at an even hour is a straight line segment. Visually inspecting these line segments reveals that the segment corresponding to the time period from 8 A.M. to 10 A.M. has the largest slope (it's the steepest). Thus, the 2-hour interval with the highest rate of sales is the interval between 8 A.M. and 10 A.M. The rate of sales in this interval is $3000 per hour, as calculated in Part (a). Note that it is possible to compute the rate of sales for each two hour time interval starting at an even hour as in Part (a), but this is not necessary for this problem.

1.6 Some Rules for Differentiation

The rules in this section are mastered by working lots of problems. Make sure you learn to distinguish between a constant that is *added* to a function and a constant that *multiplies* a function. For instance, $x^3 + 5$ is the sum of the cube function $f(x) = x^3$ and the constant function $g(x) = 5$. Thus

$$\frac{d}{dx}(x^3 + 5) = \frac{d}{dx}(x^3) + \frac{d}{dx}5 = 3x^2 + 0, \quad \text{(Sum rule)}$$

because the derivative of a constant function is zero. However, the constant 5 in the formula $5x^3$ *multiplies* the function x^3. Hence, by the constant-multiple rule,

$$\frac{d}{dx}5x^3 = 5 \cdot \frac{d}{dx}(x^3)$$
$$= 5 \cdot 3x^2$$
$$= 15x^2.$$

Exercise 43 is very important. But before you look at the solution here, go back and read the solution to Exercise 15 in Section 1.2. Then try Exercise 43 in Section 1.6 by yourself. Peak at the solution only if you are stuck.

1. To differentiate $y = 6x^3$, apply the constant-multiple rule. $k = 6$ and $f(x) = x^3$.

$$\frac{dy}{dx} = \frac{d}{dx}(6x^3)$$
$$= 6 \cdot \frac{d}{dx}(x^3)$$
$$= 6(3x^2)$$
$$= 18x^2$$

7. $\dfrac{d}{dx}(x^4 + x^3 + x) = \dfrac{d}{dx}(x^4) + \dfrac{d}{dx}(x^3 + x)$ (Sum rule)

$\qquad\qquad = \dfrac{d}{dx}(x^4) + \dfrac{d}{dx}(x^3) + \dfrac{d}{dx}(x)$ (Sum rule)

$\qquad\qquad = 4x^3 + 3x^2 + 1.$

13. Write $\dfrac{4}{x^2}$ in the form $4 \cdot x^{-2}$. Then

$$\dfrac{dy}{dx} = \dfrac{d}{dx}(4 \cdot x^{-2})$$

$$\qquad = 4 \cdot \dfrac{d}{dx}(x^{-2}) \qquad\qquad \text{(Constant-multiple rule)}$$

$$\qquad = 4(-2)x^{-3} \qquad\qquad \text{(Power rule)}$$

$$\qquad = -8x^{-3}, \quad \text{or} \quad -\dfrac{8}{x^3}.$$

19. Since $\dfrac{1}{5x^5} = \dfrac{1}{5} \cdot \dfrac{1}{x^5} = \left(\dfrac{1}{5}\right) \cdot x^{-5}$, use the constant-multiple rule and the power rule.

$$\dfrac{dy}{dx} = \dfrac{d}{dx}\left(\dfrac{1}{5} \cdot x^{-5}\right) = \dfrac{1}{5} \cdot \dfrac{d}{dx}(x^{-5}) = \dfrac{1}{5}(-5)x^{-6} = -x^{-6}, \quad \text{or} \quad -\dfrac{1}{x^6}.$$

Warning: Problems like those in Exercises 13 and 19 tend to be missed on exams by many students. The difficulty lies in the first step—recognizing how to write the function as a constant times a power of x. Be sure to review this before the exam.

25. $\dfrac{d}{dx} 5\sqrt{3x^3 + x} = \dfrac{d}{dx} 5(3x^3 + x)^{1/2}$

$$\qquad = \dfrac{5}{2}(3x^3 + x)^{-1/2} \cdot \dfrac{d}{dx}(3x^3 + x) \qquad\qquad \text{(General power rule)}$$

$$\qquad = \dfrac{5}{2}(3x^3 + x)^{-1/2}\left[\dfrac{d}{dx}(3x^3) + \dfrac{d}{dx}(x)\right] \qquad\qquad \text{(Sum rule)}$$

$$\qquad = \dfrac{5}{2}(3x^3 + x)^{-1/2}\left[3 \cdot \dfrac{d}{dx}(x^3) + \dfrac{d}{dx}(x)\right] \qquad\qquad \text{(Constant-multiple rule)}$$

$$\qquad = \dfrac{5}{2}(3x^3 + x)^{-1/2}[3(3x^2) + 1] = \dfrac{45x^2 + 5}{2\sqrt{3x^3 + x}}.$$

Warning: Don't forget to use parentheses (or brackets) when appropriate in the general power rule. The last line in the solution above is incorrect if written in the form:

$$= \dfrac{5}{2}(3x^3 + x)^{-1/2} \cdot \underbrace{3(3x^2) + 1}_{\text{missing brackets}}.$$

31. Note that

$$y = \frac{2}{1-5x} = 2 \cdot \frac{1}{1-5x} = 2 \cdot (1-5x)^{-1}.$$

Hence

$$\frac{dy}{dx} = \frac{d}{dx}(2 \cdot (1-5x)^{-1})$$

$$= 2 \cdot \frac{d}{dx}(1-5x)^{-1} \qquad \text{(Constant-multiple rule)}$$

$$= 2(-1)(1-5x)^{-2} \cdot \frac{d}{dx}(1-5x) \qquad \text{(General power rule)}$$

$$= -2(1-5x)^{-2}(-5) = 10(1-5x)^{-2}.$$

Warning: Forgetting the (–5) in the last line is a common mistake. Another common error is to carelessly forget the parentheses and write the derivative as

$$-2(1-5x)^{-2} - 5.$$

This is definitely incorrect, and many instructors will give no part credit for such an answer. Your ability to find a derivative is of no value to anyone if your algebra is careless and incorrect.

In this *Manual*, we'll point where algebra errors are likely to occur so you can guard against them. One key to avoiding such errors on exams is to practice working carefully on your homework.

37. If $f(x) = \left(\frac{\sqrt{x}}{2}+1\right)^{3/2}$ then

$$f'(x) = \frac{d}{dx}\left(\frac{\sqrt{x}}{2}+1\right)^{3/2}$$

$$= \frac{3}{2}\left(\frac{\sqrt{x}}{2}+1\right)^{1/2} \cdot \frac{d}{dx}\left(\frac{\sqrt{x}}{2}+1\right) \qquad \text{(General power rule)}$$

$$= \frac{3}{2}\left(\frac{\sqrt{x}}{2}+1\right)^{1/2}\left[\frac{d}{dx}\left(\frac{\sqrt{x}}{2}\right)+\frac{d}{dx}(1)\right] \qquad \text{(Sum rule)}$$

$$= \frac{3}{2}\left(\frac{\sqrt{x}}{2}+1\right)^{1/2}\left[\frac{1}{2}\cdot\frac{d}{dx}(x^{1/2})+\frac{d}{dx}(1)\right] \qquad \text{(Constant-multiple rule)}$$

$$= \frac{3}{2}\left(\frac{\sqrt{x}}{2}+1\right)^{1/2}\left[\frac{1}{2}\left(\frac{1}{2}\right)x^{-1/2}+0\right]$$

$$= \frac{3}{2}\left(\frac{\sqrt{x}}{2}+1\right)^{1/2}\left(\frac{1}{4}x^{-1/2}\right).$$

43. The general slope-formula for the curve $y = (x^2 - 15)^6$ is given by the derivative.

$$\frac{dy}{dx} = 6(x^2 - 15)^5 \cdot \frac{d}{dx}(x^2 - 15) \qquad \text{(General power rule)}$$
$$= 6(x^2 - 15)^5 (2x)$$
$$= 12x(x^2 - 15)^5.$$

To find the *slope m* of the tangent line at the particular point where $x = 4$, substitute 4 for x in the *derivative* formula to get

$$m = 12(4)[(4)^2 - 15]^5 = 48(1)^5 = 48.$$

For the equation of the tangent line at $x = 4$, use the point-slope form with slope $m = 48$. To find a *point* (x_1, y_1) on the curve when $x_1 = 4$, substitute 4 for x in the original equation to get

$$y_1 = [(4)^2 - 15]^6 = (1)^6 = 1.$$

The equation of the tangent line is

$$y - 1 = 48(x - 4).$$

Warning: Exercise 43 helped you to get the equation of the tangent line by asking you first for the slope of the line. Exam questions tend to be more like Exercise 44. You are expected to know that you need to find the slope of the line and a point on the line. [Try Exercise 44; the answer is $y - 1 = -\frac{5}{8}(x - 2)$.] Also, see the solution to Exercise 15 in Section 1.2.

49. $f(5) = 2, g(5) = 4$, and $f'(5) = 3, g'(5) = 1$. Then

$$h(x) = 3f(x) + 2g(x),$$
$$h(5) = 3f(5) + 2g(5) = 3(2) + 2(4) = 14,$$
$$h'(x) = 3f'(x) + 2g'(x),$$
$$h'(5) = 3f'(5) + 2g'(5) = 3(3) + 2(1) = 11.$$

55. The point of intersection of the tangent line and $f(x)$ is $(4, 5)$, which alternatively can be stated as $f(4) = 5$. The tangent line goes through $(0, 3)$ and $(4, 5)$, which means its slope is

$$m = \frac{5 - 3}{4 - 0} = \frac{2}{4} = \frac{1}{2}. \text{ That is, } f'(4) = \frac{1}{2}.$$

1.7   More About Derivatives

Your ability to differentiate functions easily and correctly depends upon how much you practice. Use the exercises in this section to gain the experience you need. Don't be annoyed that you have to learn so much notation. We want you to be able to read technical articles in your field and be familiar with some of the notation you will find there.

1. $f(t) = (t^2 + 1)^5$

$$f'(t) = 5(t^2+1)^4 \cdot \frac{d}{dt}(t^2+1)$$
$$= 5(t^2+1)^4 2t = 10t(t^2+1)^4.$$

7. $\frac{d}{dP}\left(3P^2 - \frac{1}{2}P + 1\right) = \frac{d}{dP}(3P^2) + \frac{d}{dP}\left(-\frac{1}{2}P\right) + \frac{d}{dP}(1)$

$$= 3 \cdot \frac{d}{dP}(P^2) + -\frac{1}{2} \cdot \frac{d}{dP}(P) + \frac{d}{dP}(1)$$
$$= 3(2P) - \frac{1}{2}(1) + 0$$
$$= 6P - \frac{1}{2}.$$

Helpful Hint: The notation $\frac{d}{dt}$ in Exercise 9 indicates that t is the independent variable when you differentiate. Any other letters appearing in the function represent constants, even though the values of these constants are not specified.

13. $y = \sqrt{x} = x^{1/2}, \quad \frac{dy}{dx} = \frac{1}{2}x^{-1/2}, \quad$ and

$$\frac{d^2y}{dx^2} = \frac{d}{dx}\left(\frac{dy}{dx}\right) = \frac{d}{dx}\left(\frac{1}{2}x^{-1/2}\right) = \frac{1}{2}\left(-\frac{1}{2}\right)x^{-3/2} = -\frac{1}{4}x^{-3/2}.$$

Helpful Hint: Problems involving radical signs (mainly square roots) are handled more easily when you switch to exponential notation, as in the solution to Exercise 13, above. Your instructor will probably permit you to write your answers in this form, too.

19. $f(P) = (3P+1)^5,$

$$f'(P) = 5(3P+1)^4 \cdot \frac{d}{dP}(3P+1) = 5(3P+1)^4 \cdot 3$$
$$= 15(3P+1)^4. \qquad \text{[Don't forget to multiply by 3 above.]}$$
$$f''(P) = \frac{d}{dP}f'(P) = \frac{d}{dP}[15(3P+1)^4] = 15 \cdot \frac{d}{dP}(3P+1)^4$$
$$= 15 \cdot 4(3P+1)^3 \cdot \frac{d}{dP}(3P+1) = 60(3P+1)^3(3)$$
$$= 180(3P+1)^3.$$

25. $\dfrac{d}{dx}(3x^3 - x^2 + 7x - 1) = \dfrac{d}{dx}(3x^3) + \dfrac{d}{dx}(-x^2) + \dfrac{d}{dx}(7x) + \dfrac{d}{dx}(-1)$

$$= 9x^2 - 2x + 7 + 0$$

$$= 9x^2 - 2x + 7$$

$$\dfrac{d^2}{dx^2}(3x^2 - x^2 + 7x - 1) = \dfrac{d}{dx}(9x^2 - 2x + 7)$$

$$= 18x - 2 + 0 = 18x - 2$$

$$\dfrac{d^2}{dx^2}(3x^2 - x^2 + 7x - 1)\Big|_{x=2} = 18(2) - 2 = 36 - 2 = 34.$$

31. $\dfrac{dR}{dx} = \dfrac{d}{dx}(1000 + 80x - .02x^2)$

$$= 0 + 80 - (.02)(2x) = 80 - .04x.$$

$$\dfrac{dR}{dx}\Big|_{x=1500} = 80 - .04(1500) = 80 - 60 = 20.$$

37. $C(x)$ is the cost in dollars of manufacturing x bicycles per day. Given $C(50) = 5000$, $x = 50$ so we interpret this as the cost of manufacturing 50 bicycles in one day is $5000. Given $C'(50) = 45$ this tells us it costs an additional $45 dollars to manufacture the 51st bicycle in a given day.

43. **(a)** The sales at the end of January reached $120,560, so $S(1) = \$120,560$. Sales rising at a rate of $1500/month means that $S'(1) = \$1500$.

(b) At the end of March, the sales for the month dropped to $80,000, so $S(3) = \$80,000$. Sales falling by about $200/day means that $S'(30) = -\$200(30) = -\$6000$.

49. The federal debt for the years 1995 to 2004 is given by

$$D(x) = 4.95 + .402x - .1067x^2 + .0124x^3 - .00024x^4,$$

where x is the number of years elapsed since the end of 1995. To estimate the federal debt at the end of 1999, compute $D(4)$.

$$D(4) = 4.95 + .402(4) - .1067(4)^2 + .0124(4)^3 - .00024(4)^4$$

$$= \$5.58296 \text{ trillion}$$

We estimate the rate at which it was increasing at that time by computing $D'(4)$. First find $D'(x)$.

$$D'(x) = .402 - 2(.1067x) + 3(.0124x^2) - 4(.00024x^3)$$

$$= .402 - .2134x + .0372x^2 - .00096x^3$$

$$D'(4) = .402 - .2134(4) + .0372(4)^2 - .00096(4)^3$$

$$= \$.08216 \text{ trillion/year}$$

The federal debt was increasing at the rate of $.08216 trillion/year at the end of 1999.

1.8   The Derivative as a Rate of Change

This section is crucial for later work. There are three main categories of problems: (1) the rate of change of some function—either an abstract function or a function in an application where some quantity is changing with respect to time; (2) the rate of change of one economic quantity with respect to another—problems involving the adjective "marginal"; and (3) velocity and acceleration problems.

Problems in one category may seem quite different from those in another category. But this difference is superficial and is due to the terminology involved. Try to discover similarities in the problems. The key to this lies in the Practice Problems. Remember to use the practice problems correctly. Try to answer all six problems yourself before you look at the solutions.

1. (a) For $f(x) = 4x^2$, the average rate of change of $f(x)$ over the interval 1 to 2 is

$$\frac{f(2) - f(1)}{2 - 1} = \frac{4(2)^2 - 4(1)^2}{2 - 1} = \frac{16 - 4}{1} = 12.$$

On the interval 1 to 1.5, the average rate of change is

$$\frac{f(1.5) - f(1)}{1.5 - 1} = \frac{4(1.5)^2 - 4(1)^2}{1.5 - 1} = \frac{9 - 4}{.5} = 10.$$

On the interval 1 to 1.1:

$$\frac{f(1.1) - f(1)}{1.1 - 1} = \frac{4(1.1)^2 - 4(1)^2}{1.1 - 1} = \frac{4.84 - 4}{.1} = 8.4.$$

(b) The (instantaneous) rate of change of $4x^2$ at $x = 1$ is

$$\frac{d}{dx} 4x^2 \bigg|_{x=1} = 8x \bigg|_{x=1} = 8(1) = 8.$$

7. (a) If $s(t)$ represents the position function of an object moving in a straight line, then the velocity $v(t)$ of the object at time t is given by $s'(t)$. Since $s(t) = 2t^2 + 4t$, we know that

$$v(t) = s'(t) = 2(2t) + 4 = 4t + 4.$$
$$v(6) = s'(6) = 4(6) + 4 = 28 \text{ km/hr}$$

(b) In 6 hours, the object has traveled $s(6) = 2(6)^2 + 4(6) = 72 + 24 = 96$ km.

(c) We are asked to determine when $v(t) = s'(t) = 6$.

$$s'(t) = 4t + 4$$
$$6 = 4t + 4 \Rightarrow t = \frac{1}{2}$$

The object is traveling at the rate of 6 km/hr when $t = \frac{1}{2}$ hr.

13. This exercise, along with Exercise 17, is a key problem. It is essential that you try all five parts without looking at the text or the solutions below. If you must have help, look again at the practice problems. After you have tried Exercise 13, compare your answers with those below. Also look again at the practice problems.

(a) Since velocity is the rate of change in position, the initial velocity of the rocket when $t = 0$ is $s'(0)$.

$$s(t) = 160t - 16t^2,$$
$$s'(t) = 160 - 32t,$$
$$s'(0) = 160 - 32(0) = 160.$$

So the velocity at $t = 0$ is 160 ft/sec.

(b) "Velocity after 2 seconds" means velocity when $t = 2$.

$$v(2) = s'(2) = 160 - 32(2) = 96 \text{ ft/sec.}$$

(c) Acceleration involves the rate of change of velocity, that is, the derivative of $v(t) = 160 - 32t$.

$$a(t) = v'(t) = -32.$$

Since the distance is in feet and time is in seconds, the acceleration is -32 feet per second per second or -32 ft/sec^2. If the units were kilometers and hours, for example, the acceleration would be measured in kilometers per hour per hour.

(d) In this part the time is unknown. The rocket hits the ground when the distance above the ground is zero. So set $s(t) = 0$ and solve for t:

$$160t - 16t^2 = 0,$$
$$16t(10 - t) = 0,$$
$$t = 0, \quad \text{and} \quad t = 10.$$

The rocket hits the ground when $t = 10$ seconds.

(e) We know the time from part (d). The velocity at $t = 10$ seconds is
$$v(t) = 160 - 32(10) = -160 \text{ ft/sec.}$$

A negative sign on the velocity indicates that the distance function is decreasing, that is, the rocket is falling down.

Helpful Hint: Exercise 13(e) is a good exam question. If 13(d) is not on the exam, too, here is how to analyze the problem:

1. "At what velocity" means you must find some value of the velocity. To do this you need the velocity function and some specific time t.
2. You can get the velocity function by computing $s'(t)$.
3. Since you aren't given the time, you must determine the time from the fact that the rocket has just smashed into the ground. That is, you must solve a question like #13(d), even though it may not be listed specifically on the exam.

Helpful Hint: Check with your instructor about whether your test answers for velocity and acceleration problems must include the correct units. Almost all instructors require your answer to include the correct units.

19. Suppose $f(100) = 5000$ and $f'(100) = 10$. Then when x is close to 100, the values of $f(x)$ change at the rate of approximately 10 units for each unit change in x. A change in x of h units produces a change of about $f'(100)h = 10h$ units in the values of $f(x)$.

    (a) For $f(101)$, x changes by 1 unit, so the values change by about 10 units. Thus
    $$f(101) \approx f(100) + 10 = 5010.$$

    (b) For $f(100.5)$, x increases by $h = .5$, so the values of $f(x)$ change by about $10(.5)$ units. Thus $f(100.5) \approx f(100) + 5 = 5005$.

    (c) For $f(99)$, x changes by $h = -1$ unit, so the values of $f(x)$ change by about $10(-1)$ units. That is, $f(99) \approx f(100) + (-10) = 4990$.

    (d) For $f(98)$, $h = -2$ units and
    $$f(98) \approx f(100) + 10(-2)$$
    $$= 5000 - 20 = 4980.$$

    (e) For $f(99.75)$, $h = -.25$ unit and
    $$f(99.75) \approx f(100) + 10(-.25)$$
    $$= 5000 - 2.5 = 4997.5.$$

25. $f(x)$ is the number (in thousands) of computers sold when the price is x hundred dollars per computer. Given $f(12) = 60$, this tells us that when the price of a computer is \$1200 ($x = 12$), 60,000 computers will be sold. Given $f'(12) = -2$, this tells us that at that price (\$1200), the number of computers sold decreases by 2000 for every \$100 increase in the price of the computer.
    $$f(12.5) \approx f(12) + .5 f'(12)$$
    $$= 60 + .5(-2) = 59$$

    About 59,000 computers will be sold if the price increases to \$1250.

31. (a) 1987 corresponds to $t = 7$ years after 1980. The y-coordinate of the point on the graph where $t = 7$ is 500, so about 500 billion dollars were spent in 1987.

    (b) The *rate* of expenditures is given by the graph of $f'(t)$. The y-coordinate of the point on this graph where $t = 7$ is about 50. So expenditures were rising at the rate of about 50 billion dollars per year.

    (c) A question that asks "when" requires you to find an appropriate value of t. Since the question involves expenditures, look at the graph of $f(t)$. An expenditure of one trillion (which is one thousand billion) dollars corresponds to the point on the graph of $f(t)$ whose y-coordinate is 1000. This appears to be $(14, 1000)$. That is, the expenditures reached one thousand billion in 1994, about 14 years after 1980.

    (d) This question involves the rate of expenditures, so look at the graph of $f'(t)$. Rates of expenditures are the y-coordinates on the graph. The rate of \$100 billion per year occurs at the point on the graph of $f(t)$ whose y-coordinate is 100, which seems to be the point $(14, 100)$. Thus, the rate of 100 billion per year occurred in 1994, 14 years after 1980.

Review of Chapter 1

The derivative is presented in three different ways. The derivative is defined *geometrically* in Section 1.3.

The derivative of $f(x)$ is a function $f'(x)$ whose value at $x = a$ gives the slope of the graph of $f(x)$ at $x = a$.

The secant-line approximation describes how derivative formulas are obtained. This is made more precise in Section 1.4 when the derivative is defined *analytically* as the limit of difference quotients.

$$f'(a) = \lim_{h \to 0} \frac{f(a+h) - f(a)}{h}$$

Finally, in Section 1.8 the derivative is described *operationally* as the rate of change of a function. Your review of Chapter 1 should include an attempt to see how these three aspects of the derivative concept are related. At the same time, of course, you should review the basic techniques for *calculating* derivatives (Sections 1.6 and 1.7). The supplementary exercises will help here. Some instructors tend to look at them when they prepare exams. Finally, don't forget to review Section 1.1.

Chapter 1 Review Exercises

1. Use the point-slope equation $y - y_1 = m(x - x_1)$, with $(x_1, y_1) = (0, 3)$ and $m = -2$. That is,

    $$y - 3 = -2(x - 0), \quad \text{or} \quad y = -2x + 3.$$

    To graph this line, first find the *x*-intercept by letting $y = 0$ and solving for *x*.

    $$0 = -2x + 3$$
    $$-3 = -2x$$
    $$\frac{3}{2} = x$$

    Thus the *x*-intercept is $\left(\frac{3}{2}, 0\right)$. Plot both intercepts and draw the line.

    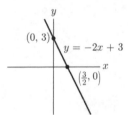

2. Let $(x_1, y_1) = (0, -1)$.

    $$y - (-1) = \frac{3}{4}(x - 0) \Rightarrow y = \frac{3}{4}x - 1$$

    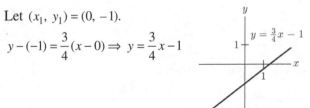

3. Let $(x_1, y_1) = (2, 0)$.

$$y - 0 = 5(x - 2)$$
$$y = 5x - 10$$

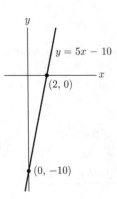

4. Let $(x_1, y_1) = (1, 4)$.

$$y - 4 = -\frac{1}{3}(x - 1)$$

$$y = -\frac{x}{3} + \frac{13}{3}$$

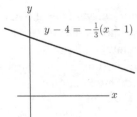

5. $y = -2x$, slope $= -2$

Let $(x_1, y_1) = (3, 5)$.

$$y - 5 = -2(x - 3)$$
$$y = 11 - 2x \text{ or}$$
$$y = -2x + 11$$

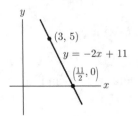

6. $-2x + 3y = 6$

$$y = 2 + \frac{2}{3}x, \text{ slope} = \frac{2}{3}$$

Let $(x_1, y_1) = (0, 1)$.

$$y - 1 = \frac{2}{3}(x - 0)$$

$$y = \frac{2}{3}x + 1$$

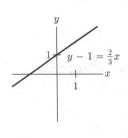

7. Apply Slope Property 2 with $(x_1, y_1) = (-1, 4)$ and $(x_2, y_2) = (3, 7)$. The slope of the line is

$$\frac{y_2 - y_1}{x_2 - x_1} = \frac{7 - 4}{3 - (-1)} = \frac{3}{4}.$$

Set $(x_1, y_1) = (-1, 4)$ and $m = \frac{3}{4}$ in the point-slope equation of a line:

$$y - 4 = \frac{3}{4}(x - (-1)), \quad \text{or} \quad y = \frac{3}{4}x + \frac{19}{4}.$$

To graph this line plot the points $(-1, 4)$ and $(3, 7)$ and draw a straight line through them.

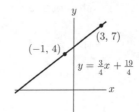

8. slope $= \dfrac{1-1}{5-2} = 0$

Let $(x_1, y_1) = (2, 1)$.

$y - 1 = 0(x - 2)$ or $y = 1$

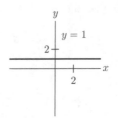

9. Slope of $y = 3x + 4$ is 3, thus a perpendicular line has slope of $-\dfrac{1}{3}$. The perpendicular line through $(1, 2)$ is

$$y - 2 = \left(-\dfrac{1}{3}\right)(x-1) \text{ or } y = -\dfrac{1}{3}x + \dfrac{7}{3}$$

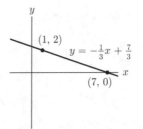

10. Slope of $3x + 4y = 5$ is $-\dfrac{3}{4}$ since $y = -\dfrac{3}{4}x + \dfrac{5}{4}$, thus a perpendicular line has slope of $\dfrac{4}{3}$. The perpendicular line through $(6, 7)$ is $y - 7 = \dfrac{4}{3}(x-6)$ or $y = \dfrac{4}{3}x - 1$.

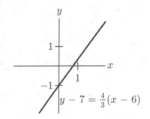

11. The equation of the x-axis is $y = 0$, so the equation of this line is $y = 3$.

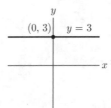

12. The equation of the y-axis is $x = 0$, so 4 units to the right is $x = 4$.

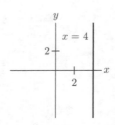

13. The *y*-axis has the equation $x = 0$.

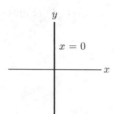

14. The *x*-axis has the equation $y = 0$.

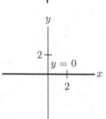

15. $y = x^7 + x^3$; $y' = 7x^6 + 3x^2$

16. $y = 5x^8$; $y' = 40x^7$

17. $y = 6\sqrt{x} = 6x^{1/2}$; $y' = 3x^{-1/2} = \dfrac{3}{\sqrt{x}}$

18. $y = x^7 + 3x^5 + 1$; $y' = 7x^6 + 15x^4$

19. $\dfrac{d}{dx}\left(\dfrac{3}{x}\right) = \dfrac{d}{dx}(3 \cdot x^{-1}) = 3 \cdot \dfrac{d}{dx}(x^{-1})$

   (Constant-multiple rule)

   $= 3(-1)x^{-2} = -3x^{-2}$,   or $-\dfrac{3}{x^2}$.

20. $y = x^4 - \dfrac{4}{x} = x^4 - 4x^{-1}$

   $y' = 4x^3 + 4x^{-2} = 4x^3 + \dfrac{4}{x^2}$

21. $y = (3x^2 - 1)^8$

   $y' = 8(3x^2 - 1)^7 (6x) = 48x(3x^2 - 1)^7$

22. $y = \dfrac{3}{4}x^{4/3} + \dfrac{4}{3}x^{3/4}$

   $y' = x^{1/3} + x^{-1/4}$

23. $y = \dfrac{1}{5x - 1} = (5x - 1)^{-1}$

   $\dfrac{dy}{dx} = -(5x - 1)^{-2}(5) = -\dfrac{5}{(5x - 1)^2}$

24. $y = (x^3 + x^2 + 1)^5$

   $y' = 5(x^3 + x^2 + 1)^4(3x^2 + 2x)$

25. $\dfrac{d}{dx}\sqrt{x^2 + 1} = \dfrac{d}{dx}(x^2 + 1)^{1/2}$

   $= \dfrac{1}{2}(x^2 + 1)^{-1/2} \cdot \dfrac{d}{dx}(x^2 + 1)$

   (General power rule)

   $= \dfrac{1}{2}(x^2 + 1)^{-1/2}(2x) = x(x^2 + 1)^{-1/2}$

   $= \dfrac{x}{\sqrt{x^2 + 1}}$.

26. $y = \dfrac{5}{7x^2 + 1} = 5(7x^2 + 1)^{-1}$

   $\dfrac{dy}{dx} = -5(7x^2 + 1)^{-2}(14x) = -\dfrac{70x}{(7x^2 + 1)^2}$

27. $f(x) = \dfrac{1}{\sqrt[4]{x}} = x^{-1/4}$

   $f'(x) = -\dfrac{1}{4}x^{-5/4} = -\dfrac{1}{4x^{5/4}}$

28. $f(x) = (2x+1)^3$

$$f'(x) = 3(2x+1)^2(2) = 6(2x+1)^2$$

29. $f(x) = 5; \ f'(x) = 0$

30. $f(x) = \dfrac{5x}{2} - \dfrac{2}{5x} = \dfrac{5}{2}x - \dfrac{2}{5}x^{-1}$

$$f'(x) = \dfrac{5}{2} + \dfrac{2}{5}x^{-2} = \dfrac{5}{2} + \dfrac{2}{5x^2}$$

31. If $f(x) = [x^5 - (x-1)^5]^{10}$, then

$$f'(x) = 10[x^5 - (x-1)^5]^9 \cdot \dfrac{d}{dx}[x^5 - (x-1)^5] \qquad \text{(General power rule)}$$

$$= 10[x^5 - (x-1)^5]^9 \left[\dfrac{d}{dx}(x^5) - \dfrac{d}{dx}(x-1)^5 \right] \qquad \text{(Sum rule)}$$

$$= 10[x^5 - (x-1)^5]^9 \left[5x^4 - 5(x-1)^4 \cdot \dfrac{d}{dx}(x-1) \right] \qquad \text{(General power rule)}$$

$$= 10[x^5 - (x-1)^5]^9 [5x^4 - 5(x-1)^4].$$

32. $f(t) = t^{10} - 10t^9; \ f'(t) = 10t^9 - 90t^8$

33. $g(t) = 3\sqrt{t} - \dfrac{3}{\sqrt{t}} = 3t^{1/2} - 3t^{-1/2}$

$$g'(t) = \dfrac{3}{2}t^{-1/2} + \dfrac{3}{2}t^{-3/2}$$

34. $h(t) = 3\sqrt{2}; \ h'(t) = 0$

35. $f(t) = \dfrac{2}{t - 3t^3} = 2(t - 3t^3)^{-1}$

$$f'(t) = -2(t - 3t^3)^{-2}(1 - 9t^2) = \dfrac{-2(1 - 9t^2)}{(t - 3t^3)^2} = \dfrac{2(9t^2 - 1)}{(t - 3t^3)^2}$$

36. $g(P) = 4P^{.7}; \ g'(P) = 2.8P^{-.3}$

37. If $h(x) = \dfrac{3}{2}x^{3/2} - 6x^{2/3}$, then

$$h'(x) = \dfrac{3}{2}\dfrac{d}{dx}(x^{3/2}) - 6 \cdot \dfrac{d}{dx}(x^{2/3})$$

$$= \dfrac{3}{2}\left(\dfrac{3}{2}\right)x^{1/2} - 6\left(\dfrac{2}{3}\right)x^{-1/3}$$

$$= \dfrac{9}{4}x^{1/2} - 4x^{-1/3}.$$

38. $f(x) = \sqrt{x + \sqrt{x}} = (x + x^{1/2})^{1/2}$

$$f'(x) = \frac{1}{2}(x + x^{1/2})^{-1/2}\left(1 + \frac{1}{2}x^{-1/2}\right) = \frac{1}{2\sqrt{x + \sqrt{x}}}\left(1 + \frac{1}{2\sqrt{x}}\right)$$

39. $f(t) = 3t^3 - 2t^2$

$f'(t) = 9t^2 - 4t$

$f'(2) = 36 - 8 = 28$

40. $V(r) = 15\pi r^2$

$V'(r) = 30\pi r$

$V'\left(\frac{1}{3}\right) = 10\pi$

41. $g(u) = 3u - 1$

$g(5) = 15 - 1 = 14$

$g'(u) = 3$

$g'(5) = 3$

42. $h(x) = -\frac{1}{2}; \quad h(-2) = -\frac{1}{2}$

$h'(x) = 0; \quad h'(-2) = 0$

43. If $f(x) = x^{5/2}$, then

$$f'(x) = \frac{5}{2}x^{3/2} \quad \text{and} \quad f''(x) = \left(\frac{5}{2}\right)\left(\frac{3}{2}\right)x^{1/2} = \frac{15}{4}\sqrt{x}.$$

Hence $\qquad f''(4) = \frac{15}{4}\sqrt{4} = \frac{15}{4}(2) = \frac{15}{2}.$

44. $g(t) = \frac{1}{4}(2t - 7)^4$

$g'(t) = (2t - 7)^3(2) = 2(2t - 7)^3$

$g''(t) = 6(2t - 7)^2(2) = 12(2t - 7)^2$

$g''(3) = 12[2(3) - 7]^2 = 12$

45. $y = (3x - 1)^3 - 4(3x - 1)^2$

slope $= y' = 3(3x - 1)^2(3) - 8(3x - 1)(3) = 9(3x - 1)^2 - 24(3x - 1)$

When $x = 0$, slope $= 9 + 24 = 33$.

46. $y = (4 - x)^5$

slope $= y' = 5(4 - x)^4(-1) = -5(4 - x)^4$

When $x = 5$, slope $= -5$.

47. $\dfrac{d}{dx}(x^4 - 2x^2) = 4x^3 - 4x$

48. $\dfrac{d}{dt}(t^{5/2} + 2t^{3/2} - t^{1/2}) \quad = \dfrac{5}{2}t^{3/2} + 3t^{1/2} - \dfrac{1}{2}t^{-1/2}$

49. $\dfrac{d}{dP}\sqrt{(1-3P)} = \dfrac{d}{dP}(1-3P)^{1/2}$

$\qquad\qquad = \dfrac{1}{2}(1-3P)^{-1/2} \cdot \dfrac{d}{dP}(1-3P)$ \qquad (General power rule)

$\qquad\qquad = \dfrac{1}{2}(1-3P)^{-1/2}(-3)$

$\qquad\qquad = -\dfrac{3}{2}(1-3P)^{-1/2}.$

50. $\dfrac{d}{dn}\left(n^{-5}\right) = -5n^{-6}$

51. $\dfrac{d}{dz}(z^3 - 4z^2 + z - 3)\Big|_{z=-2} = (3z^2 - 8z + 1)\Big|_{z=-2} = 12 + 16 + 1 = 29$

52. $\dfrac{d}{dx}(4x-10)^5\Big|_{x=3} = [5(4x-10)^4(4)]\Big|_{x=3} = [20(4x-10)^4]\Big|_{x=3} = 320$

53. $\dfrac{d^2}{dx^2}(5x+1)^4 = \dfrac{d}{dx}[4(5x+1)^3(5)] = 60(5x+1)^2(5) = 300(5x+1)^2$

54. $\dfrac{d^2}{dt^2}\left(2\sqrt{t}\right) = \dfrac{d^2}{dt^2}2t^{1/2} = \dfrac{d}{dt}t^{-1/2} = -\dfrac{1}{2}t^{-3/2}$

55. $\dfrac{d}{dt}(t^3 + 2t^2 - t) = \dfrac{d}{dt}(t^3) + 2 \cdot \dfrac{d}{dt}(t^2) - \dfrac{d}{dt}(t) = 3t^2 + 4t - 1$

$\qquad \dfrac{d^2}{dt^2}(t^3 + 2t - t) = \dfrac{d}{dt}(3t^2 + 4t - 1) = 6t + 4$

$\qquad \dfrac{d^2}{dt^2}(t^3 + 2t - t)\Big|_{t=-1} = 6(-1) + 4 = -2$

56. $\dfrac{d^2}{dP^2}(3P+2)\Big|_{P=4} = \dfrac{d}{dP}3\Big|_{P=4} = 0\Big|_{P=4} = 0$

57. $\dfrac{d^2y}{dx^2}(4x^{3/2}) = \dfrac{dy}{dx}(6x^{1/2}) = 3x^{-1/2}$

58. $\dfrac{d}{dt}\left(\dfrac{1}{3t}\right) = \dfrac{d}{dt}\left(\dfrac{1}{3}t^{-1}\right) = -\dfrac{1}{3}t^{-2}$ or $-\dfrac{1}{3t^2}$

$\dfrac{d}{dt}\left(-\dfrac{1}{3}t^{-2}\right) = \dfrac{2}{3}t^{-3}$ or $\dfrac{2}{3t^3}$

59. $f(x) = x^3 - 4x^2 + 6$

slope $= f'(x) = 3x^2 - 8x$

When $x = 2$, slope $= 3(2)^2 - 8(2) = -4$.

When $x = 2$, $y = 2^3 - 4(2)^2 + 6 = -2$.

Let $(x_1,\, y_1) = (2,\, -2)$.

$y - (-2) = -4(x - 2) \Rightarrow y = -4x + 6$

60. $y = \dfrac{1}{3x - 5} = (3x - 5)^{-1}$

$y' = -(3x - 5)^{-2}(3) = -\dfrac{3}{(3x - 5)^2}$

When $x = 1$, slope $= -\dfrac{3}{\left(3(1) - 5\right)^2} = -\dfrac{3}{4}$.

When $x = 1$, $y = \dfrac{1}{3(1) - 5} = -\dfrac{1}{2}$.

Let $(x_1,\, y_1) = \left(1,\, -\dfrac{1}{2}\right)$.

$y - \left(-\dfrac{1}{2}\right) = -\dfrac{3}{4}(x - 1) \Rightarrow y = -\dfrac{3}{4}x + \dfrac{1}{4}$

61. The slope of the graph of $y = x^2$ at the point $(x,\, y)$ is $2x$, hence the slope of the tangent line

at $\left(\dfrac{3}{2}, \dfrac{9}{4}\right)$ is $2\left(\dfrac{3}{2}\right) = 3$. Let $(x_1,\, y_1) = \left(\dfrac{3}{2}, \dfrac{9}{4}\right)$ and $m = 3$, and use the point-slope equation for the

tangent line to get

$$y - \dfrac{9}{4} = 3\left(x - \dfrac{3}{2}\right), \quad \text{or} \quad y = 3x - \dfrac{9}{4}.$$

Now sketch the curve $y = x^2$ by plotting a few points, including the point $\left(\dfrac{3}{2}, \dfrac{9}{4}\right)$. Move one unit to

the right of $\left(\dfrac{3}{2}, \dfrac{9}{4}\right)$ and then 3 units in the positive y-direction to reach another point on the tangent

line. Draw the straight line through these two points.)

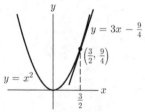

Helpful Hint: Make sure you can work Exercises 61 and 63 without any help. Make a note to work Exercise 63 later when you have not just finished reading this *Manual*. If you need help, see the solutions for Exercise 15 in Section 1.2 and Exercise 43 in Section 1.6.

62. $y = x^2$

slope $= y' = 2x$

When $x = -2$, slope $= 2(-2) = -4$.

Let $(x_1, y_1) = (-2, 4)$.

Then, $y - 4 = -4(x + 2) \Rightarrow y = -4x - 4$

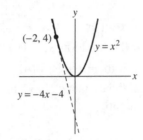

63. $y = 3x^3 - 5x^2 + x + 3$

slope $= y' = 9x^2 - 10x + 1$

When $x = 1$, slope $= 9(1)^2 - 10(1) + 1 = 0$.

When $x = 1$, $y = 3(1)^3 - 5(1)^2 + 1 + 3 = 2$.

Let $(x_1, y_1) = (1, 2)$. Then, $y - 2 = 0(x - 1) \Rightarrow y = 2$

64. $y = (2x^2 - 3x)^3$

slope $= y' = 3(2x^2 - 3x)^2(4x - 3)$

When $x = 2$, slope $= 3\left(2(2)^2 - 3(2)\right)^2 \left(4(2) - 3\right) = 60$.

When $x = 2$, $y = \left(2(2)^2 - 3(2)\right)^3 = 8$.

Let $(x_1, y_1) = (2, 8)$.

$y - 8 = 60(x - 2) \Rightarrow y = 60x - 112$

65. The line has slope –1 and contains the point (5, 0).

$y - 0 = -1(x - 5) \Rightarrow y = -x + 5$

$f(2) = -2 + 5 = 3$

$f'(2) = -1$

66. The tangent line contains the points $(0, 2)$ and (a, a^3) and has slope $= 3a^2$. Thus,

$$\frac{a^3 - 2}{a} = 3a^2 \Rightarrow a^3 - 2 = 3a^3 \Rightarrow -2 = 2a^3 \Rightarrow$$

$a = -1$

67. $s(t) = -16t^2 + 32t + 128$,   (height of binoculars)

$s'(t) = -32t + 32$.   (velocity of binoculars)

To answer the question "How fast…" you must give the value of the velocity, $s'(t)$. But at what time? The time is identified only by the phrase, "when they hit the ground." Can you describe this time using $s(t)$ or $s'(t)$? Yes, since $s(t) = 0$ when the binoculars are on the ground. So set $s(t)$ equal to zero and solve to find the time when this happens:

$$-16t^2 + 32t + 128 = 0,$$
$$-16(t^2 - 2t - 8) = 0,$$
$$-16(t - 4)(t + 2) = 0.$$

Thus $t - 4 = 0$ or $t + 2 = 0$, so that $t = 4$ or $t = -2$. You may discard the possibility $t = -2$ because the appropriate domain for the function $s(t)$ is $t \geq 0$. The binoculars hit the ground when $t = 4$ seconds, and their velocity at that time is $s'(4) = -32(4) + 32 = -128 + 32 = -96$ feet per second. The negative sign indicates that the distance above the ground is decreasing. So the binoculars are *falling* at the rate of 96 feet per second.

Helpful Hint: Exercise 67 is a typical exam question. The solutions require two steps—finding $t = 4$ and computing $s'(4)$. Students who miss this problem usually try to work the problem in one step and don't know where to begin.

68. $40t + t^2 - \dfrac{1}{15}t^3$ tons is the total output of a coal mine after t hours. The rate of output is

$40 + 2t - \dfrac{1}{5}t^2$ tons per hour. At $t = 5$, the rate of output is $40 + 2(5) - \dfrac{1}{5}(5)^2 = 45$ tons/hour.

69. 11 feet

70. $\dfrac{s(4) - s(1)}{4 - 1} = \dfrac{6 - 1}{4 - 1} = \dfrac{5}{3}$ ft/sec

71. Slope of the tangent line is $\dfrac{5}{3}$ so $\dfrac{5}{3}$ ft/sec.

72. $t = 6$, since $s(t)$ is steeper at $t = 6$ than at $t = 5$.

73. (a)
$$C(x) = .1x^3 - 6x^2 + 136x + 200,$$
$$C(21) = .1(21)^3 - 6(21)^2 + 136(21) + 200 = 1336.1,$$
$$C(20) = .1(20)^3 - 6(20)^2 + 136(20) + 200 = 1320.0,$$
$$C(21) - C(20) = 1336.1 - 1320.0 = 16.1 = \$16.10$$

This is the extra cost of raising the production from 20 to 21 units.

(b) The true marginal cost involves the derivative:
$$C'(x) = .3x^2 - 12x + 136,$$
$$C'(20) = .3(20)^2 - 12(20) + 136$$
$$= 16.0 \text{ dollars per unit.}$$

74. $f(235) = 4600$
$f'(235) = -100$
$f(a + h) \approx f'(a) \cdot h + f(a)$
(a) $237 = 235 + 2$
$f(235 + 2) \approx f'(235) \cdot 2 + f(235) \approx -100 \cdot 2 + 4600 \approx 4400$ riders

(b) $234 = 235 + (-1)$

$f(235 + (-1)) \approx f'(235) \cdot (-1) + f(235) \approx -100 \cdot (-1) + 4600 \approx 4700$ riders

(c) $240 = 235 + 5$

$f(235 + 5) \approx f'(235) \cdot 5 + f(235) \approx -100 \cdot 5 + 4600 \approx 4100$ riders

(d) $232 = 235 + (-3)$

$f(235 + (-3)) \approx f'(235) \cdot (-3) + f(235) \approx -100 \cdot (-3) + 4600 \approx 4900$ riders

75. $h(12.5) - h(12) \approx h'(12)(.5) = (1.5)(.5) = .75$ in.

76. $f\left(7 + \dfrac{1}{2}\right) - f(7) \approx f'(7) \cdot \dfrac{1}{2} = (25.06) \cdot \dfrac{1}{2} = 12.53$

$12.53 is the additional money earned if the bank paid $7\dfrac{1}{2}\%$ interest.

77. $\lim_{x \to 2} \dfrac{x^2 - 4}{x - 2} = \lim_{x \to 2} \dfrac{(x+2)(x-2)}{x-2} = \lim_{x \to 2}(x+2) = 2 + 2 = 4$

78. The limit does not exist.

79. $x^2 - 8x + 16 = (x-4)^2$, so

$$\lim_{x \to 4} \dfrac{x-4}{x^2 - 8x + 16} = \lim_{x \to 4} \dfrac{x-4}{(x-4)^2}$$
$$= \lim_{x \to 4} \dfrac{1}{x-4},$$

which does not exist since the denominator is 0 at $x = 4$ and $\dfrac{1}{0}$ is not defined.

80. $\lim_{x \to 5} \dfrac{x-5}{x^2 - 7x + 2} = \dfrac{5-5}{25 - 35 + 2} = 0$

81. $f'(5) = \lim_{h \to 0} \dfrac{f(5+h) - f(5)}{h}$

If $f(x) = \dfrac{1}{2x}$, then

$f(5+h) - f(5) = \dfrac{1}{2(5+h)} - \dfrac{1}{2(5)} = \dfrac{1}{2(5+h)} \cdot \dfrac{5}{5} - \dfrac{1}{2(5)} \cdot \left(\dfrac{5+h}{5+h}\right) = \dfrac{5 - (5+h)}{10(5+h)} = \dfrac{-h}{10(5+h)}$

Thus,

$f'(5) = \lim_{h \to 0}[f(5+h) - f(5)] \cdot \dfrac{1}{h} = \lim_{h \to 0} \dfrac{-h}{10(5+h)} \cdot \dfrac{1}{h} = \lim_{h \to 0} \dfrac{-1}{10(5+h)} = -\dfrac{1}{50}$

82. $f'(3) = \lim\limits_{h \to 0} \dfrac{f(3+h) - f(3)}{h}$

If $f(x) = x^2 - 2x + 1,$ then

$f(3+h) - f(3) = (3+h)^2 - 2(3+h) + 1 - (9 - 6 + 1) \qquad = h^2 + 4h.$

Thus, $f'(3) = \lim\limits_{h \to 0} \dfrac{f(3+h) - f(3)}{h} = \lim\limits_{h \to 0} \dfrac{h^2 + 4h}{h} = \lim\limits_{h \to 0} (h + 4) = 4.$

83. The slope of a secant line at $(3, 9)$

84. $\dfrac{\frac{1}{2+h} - \frac{1}{2}}{h} = \dfrac{\frac{2-2-h}{2(2+h)}}{h} = \dfrac{-1}{2(2+h)}$

As $h \to 0, \dfrac{-1}{2(2+h)} \to -\dfrac{1}{4}.$

Chapter 2
Applications of the Derivative

This chapter culminates with one of the most powerful applications of the derivative—the solution of optimization problems. Although graphs play a key role in these problems, graphs are important in their own right. Sections 2.1 and 2.2 provide the background for the key topics covered in the remaining sections of the chapter.

2.1   Describing Graphs of Functions

The terms defined in this section are used throughout the text. The words *maximum* and *minimum* are used to refer to both points and values of the function. When referring to a point, these words are usually preceded with the word "relative." When referring to a value, "maximum value" means the largest value that the function assumes on its domain, and "minimum value" means the smallest value the function assumes on its domain.

1. Functions a, e, and f are increasing for all x since the graphs rise as we move along them from left to right.

7. Since the graph is superimposed on graph paper, we see that $(0, 2)$ is a relative minimum point, $(1, 3)$ is an inflection point, and $(2, 4)$ is a relative maximum point. Therefore, the key features of the graph are as follows:

    1. Decreasing for $x < 0$.
    2. Relative minimum point at $x = 0$.
    3. Increasing for $0 < x < 2$.
    4. Relative maximum point at $x = 2$.
    5. Decreasing for $x > 2$.
    6. Concave up for $x < 1$, concave down for $x > 1$.
    7. Inflection point at $(1, 3)$.
    8. y-intercept, $(0, 2)$, x-intercept at about $(3.6, 0)$.

13. Take an arbitrary point on the left-hand side of the graph, say $(0, 0)$, and gradually increase its value. Initially the slope of the graph is positive. As x increases from 0 to $x = .5$, this slope becomes less positive, i.e., it decreases until finally at the point $(.5, 1)$ the slope is zero. This point is a relative maximum. Continuing, as x goes from .5 to $x = 1$, the slope is negative and decreases. The graph is concave down for all x in the domain, hence there are no inflection points.

19.

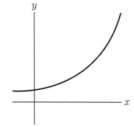

It is clear that both $f(x)$ and its slope increase as x increases, because the curve is rising and it lies above the tangent line at each point.

25.

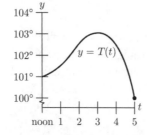

31. Since the curve has the line $y = -15t + 10$ as an asymptote, we can use the equation of the line as an approximation for $v(t)$, the velocity of the parachutist, as t increases. Keeping in mind that $d = rt$, we see that the parachutist's velocity levels off to 15 ft/sec.

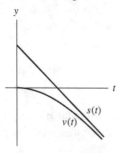

37. (a) Yes, a curve with two relative maximum points must have a relative minimum point between them.

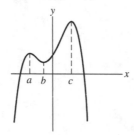

Notice that the graph changes from increasing to decreasing at $x = a$ and again at $x = c$. There must be some point between $x = a$ and $x = c$ where the graph changes from decreasing to increasing. In this case, that occurs at $x = b$. By definition, the point where a graph changes from decreasing to increasing is a relative minimum point.

(b) Yes, if a curve has two relative extreme points, it must have an inflection point. In the figure above, $x = b$ is a relative minimum and the curve is concave up. There is a relative maximum at $x = c$ and the curve is concave down. Thus, there must be a point between $x = b$ and $x = c$ at which the graph changes from concave up to concave down. By definition, this is an inflection point.

43. Set $Y_1 = 1/X + X$ and set $Y_2 = X$ specify the window $[-6, 6]$ *by* $[-6, 6]$, and graph the functions.

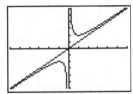

When $x = 6$, $Y_1 = \dfrac{37}{6}$ and $Y_2 = 6$, so the two graphs are $\dfrac{37}{6} - 6 = \dfrac{1}{6}$ unit apart.

2.2   The First and Second Derivative Rules

The exercises in this section provide a warm-up to the problem of sketching the graphs of functions. Here we gain insights into the types of questions to ask in order to determine the main features of the graph of a given function.

Examples 3, 4, and 5 discuss important relations between the graphs of $f(x)$ and $f'(x)$. Study the examples carefully. The discussion in Example 5(e) shows that if the graph of $f'(x)$ is decreasing and crosses the *x*-axis at, say, $x = 3$, then $f(x)$ has a relative maximum at $x = 3$. Similarly, if the graph of $f'(x)$ is increasing and crosses the *x*-axis at, say, $x = a$, then $f(x)$ has a relative minimum at $x = a$ (because the value of $f'(x)$ changes from negative to positive at $x = a$ and so the graph of $f(x)$ changes from decreasing to increasing at $x = a$).

 1. If a function has a positive first derivative for all *x*, the First Derivative Rule tells us that the function is increasing for all value of *x*. This is only true of graph *e*.

 7. The only specific point on the graph is (2, 1). Plot this point and then use the fact that $f'(2) = 0$ to sketch the tangent line at $x = 2$. Since the graph is concave up, the point (2, 1) must be a minimum point. Since it is concave up for all *x*, there are no other relative extreme points or inflection points.

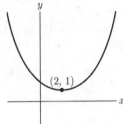

13. $f(3) = 4 \Rightarrow (3, 4)$ on the graph.

   $f'(3) = -1/2,$ so the slope of the graph at (3, 4) is $-1/2$.

   $f''(3) = 5 > 0 \Rightarrow$ the graph is concave up at $x = 3$.

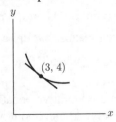

19.

	f	f'	f''
A	POS	POS	NEG
B	0	NEG	0
C	NEG	0	POS

25. Look at the point (6, 2) on the graph of $f'(x)$. Since the y-coordinate of that point is positive, the *slope* of the graph of $f(x)$ is positive at $x = 6$. So $f(x)$ is increasing at $x = 6$, by the first derivative rule.

31. At $x = 1$, the graph of $f'(x)$ appears to have a relative maximum. That is, the graph of $f'(x)$ changes from increasing to decreasing at $x = 1$. An increasing derivative means that the graph of the original function $f(x)$ is concave up; a decreasing derivative means that the graph of $f(x)$ is concave down. So the graph of $f(x)$ changes at $x = 1$ from concave up to concave down, which means that $f(x)$ has an inflection point at $x = 1$.

Helpful Hint: To find values of x at which $f(x)$ has a relative extreme point, look for places at which the graph of $f'(x)$ *crosses* the x-axis. To find values of x at which $f(x)$ has an inflection point, look for places at which the graph of $f'(x)$ has a relative extreme point. If you are asked to do this on an exam, be sure that you can justify your answers, as in the discussions of Exercises 25 and 31 (above).

37. **(a)** If $h'(100) = \dfrac{1}{3}$, then at time $t = 100$, the water level is rising at the rate of 1/3 inch per hour. In a half-hour, the water will rise about half that much, namely, 1/6 inch.

  (b) The conditions $h'(100) = 2$, and $h''(100) = -5$ say that the water is rising, but the *rate* of rise is declining since the derivative of the rate function is negative. That's good news, because the river may not rise much higher. However, condition (ii), $h'(100) = -2$, says that water level is falling, which is even better news.

Helpful Hint: Exercises 43 and 44 will help you learn how to extract information from graphs of a function and its derivatives. When you look at the graph of $f(t)$, you see how it increases or decreases and how the graph "bends." When you look at a graph of $f'(t)$ or $f''(t)$, focus mainly on the y-coordinates of various points, not on the shapes of the graphs.

43. **(a)** The number of farms is described by the graph of $f(t)$. The year 1990 corresponds to $t = 65$ (years after 1925), and the point (65, 2) appears to be on the graph of $f(t)$. So there were about 2 million farms in 1990.

  (b) The *rate* of change in the number of farms is described by $f'(t)$. When $t = 65$, $f'(65)$ is approximately $-.03$. Thus in 1990, the number of farms is decreasing (because $f'(65)$ is negative) at the rate of about .03 million per year, that is, at the rate of about 30,000 farms per year.

  (c) 6 million farms corresponds to the point on the graph of $f(t)$ whose y-coordinate is 6. This point is approximately (15, 6), which means that there were about 6 million farms in 1940.

  (d) Here a rate is given and the question is "when?" The answer is found on the graph of $f'(t)$, looking for the value of t that corresponds to the given rate of $-60,000$ farms per year, that is, $-.06$ million farms per year. The rate is negative because the number of farms is declining. There are two such values of t, namely, 20 and about 53, which correspond to 1945 and 1978.

(e) The number of farms is declining the fastest when the rate is the most negative. This corresponds to the inflection point on the graph of $f(t)$, but that point is hard to find. Looking at the graph of

$f'(t)$, you might guess that the minimum rate occurs at about $t = 35$. To be certain, look at

$f''(t)$, because the minimum of $f'(t)$ occurs where its derivative crosses the t-axis. This happens at $t = 35$ (that is, 35 years after 1925). Thus, the number of farms was declining the fastest in 1960.

2.3   The First- and Second-Derivative Tests and Curve Sketching

The curves sketched in this section will look like one of the curves shown below or one of these curves turned upside down. The first type of curve is called a *parabola* and is the graph of a quadratic function. The second curve is called a *cubic* and is the graph of a certain type of cubic polynomial. The purpose of this section is to apply the first and second derivative rules to find the graphs of functions. Therefore, you should not automatically assume that the graph will be one of those below, but should reason its shape from the two derivative rules. For instance, every time that you claim that a point is an extreme point, you should determine the value of the second derivative at that point and draw your conclusion from the second derivative test. Later in the text, we will graph functions about which we have no prior information.

1. First we find the critical values and critical points of f:

$$f'(x) = 3x^2 - 27 = 3(x^2 - 9)$$
$$= 3(x + 3)(x - 3).$$

The first derivative $f'(x) = 0$ if $x + 3 = 0$ or $x - 3 = 0$. Thus the critical values are

$$x = -3 \quad \text{and} \quad x = 3.$$

Substituting the critical values into the expression of f:

$$f(-3) = (-3)^3 - 27(-3) = 54$$
$$f(3) = (3)^3 - 27(3) = -54$$

The critical points are $(-3, 54)$ and $(3, -54)$. To determine whether these are relative maximums, minimums, or neither, we will apply the first derivative test. We will use the following chart to study the sign of $f'(x)$.

Critical Values		-3		3	
	$x < -3$		$-3 < x < 3$		$3 < x$
$3(x + 3)$	$-$	0	$+$		$+$
$x - 3$	$-$		$-$	0	$+$
$f'(x)$	$+$	0	$-$	0	$+$
$f(x)$	Increasing on $(-\infty, -3)$		Decreasing on $(-3, 3)$		Increasing on $(3, \infty)$
		54		-54	

We can see from the chart that the sign of $f'(x)$ changes from positive to negative at $x = -3$. Therefore, according to the first derivative test, f has a local maximum at $x = -3$. Also, the sign of $f'(x)$ changes from negative to positive at $x = 3$. Therefore, according to the first derivative test, f has a local minimum at $x = 3$. We conclude f has a local maximum at $(-3, 54)$ and a local minimum at $(3, -54)$. We can verify this by examining the graph.

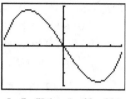

[-5, 5] by [-60, 60]

7. First we find the critical values and critical points of f:

$$f'(x) = -3x^2 - 24x = -3x(x+8)$$

The first derivative $f'(x) = 0$ if $-3x = 0$ or $x + 8 = 0$. Thus the critical values are

$$x = 0 \quad \text{and} \quad x = -8.$$

Substituting the critical values into the expression of f:

$$f(-8) = -(-8)^3 - 12(-8)^2 - 2 = -258$$
$$f(0) = -(0)^3 - 12(0)^2 - 2 = -2$$

The critical points are $(-8, -258)$ and $(0, -2)$. To determine whether these are relative maximums, minimums, or neither, we will apply the first derivative test. We will use the following chart to study the sign of $f'(x)$.

Critical Values		-8		0	
	$x < -8$		$-8 < x < 0$		$0 < x$
$x + 8$	$-$	0	$+$		$+$
$-3x$	$+$		$+$	0	$-$
$f'(x)$	$-$	0	$+$	0	$-$
$f(x)$	Decreasing on $(-\infty, -8)$		Increasing on $(-8, 0)$		Decreasing on $(0, \infty)$
		-258		258	

We can see from the chart that the sign of $f'(x)$ changes from negative to positive at $x = -8$. Therefore, according to the first derivative test, f has a local minimum at $x = -8$. Also, the sign of $f'(x)$ changes from positive to negative at $x = 0$. Therefore, according to the first derivative test, f has a local maximum at $x = 0$. We conclude f has a local minimum at $(-8, -258)$ and a local maximum at $(0, -2)$. We can verify this by examining the graph.

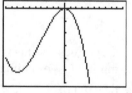

[-10, 10] by [-300, 25]

13. $f(x) = 1 + 6x - x^2$, $f'(x) = 6 - 2x$, $f''(x) = -2$

Set $f'(x) = 0$ and solve for x.

$$6 - 2x = 0 \Rightarrow -2x = -6 \Rightarrow x = 3$$

Substitute this value for x back into $f(x)$ to find the y-coordinate of this possible extreme point.

$$f(3) = 1 + 6(3) - (3)^2$$
$$= 1 + 18 - 9 = 10.$$

The positive extreme point is (3, 10). Since $f''(x) = -2$, which is negative, the graph of $f(x)$ is concave down at $x = 3$, and (3, 10) is a relative maximum point.

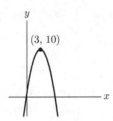

19. $f(x) = x^3 - 12x$, $f'(x) = 3x^2 - 12$, $f''(x) = 6x$

Set $f'(x) = 0$ and solve for x.

$$3x^2 - 12 = 0$$
$$3(x^2 - 4) = 0$$
$$3(x+2)(x-2) = 0$$
$$x = -2 \quad \text{or} \quad x = 2.$$

Substituting these values of x back into $f(x)$ to find the y-coordinates of these possible relative extreme points:

$$f(-2) = (-2)^3 - 12(-2) = -8 + 24 = 16$$
$$f(2) = (2)^3 - 12(2) = 8 - 24 = -16$$

The possible extreme points are (−2, 16) and (2, −16). Now,

$$f''(-2) = 6(-2) = -12 < 0,$$
$$f''(2) = 6(2) = 12 > 0.$$

The graph of $f(x)$ is concave down at $x = -2$ and concave up at $x = 2$. Therefore, (−2, 16) is a relative maximum point and (2, −16) is a relative minimum point.

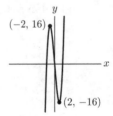

25. $y = x^3 - 3x + 2$, $y' = 3x^2 - 3$, $y'' = 6x$

Set $y' = 0$ and solve for x.

$$3x^2 - 3 = 0$$
$$3(x^2 - 1) = 0$$
$$3(x+1)(x-1) = 0$$
$$x = -1 \quad \text{or} \quad x = 1.$$

Use these values to find the y-coordinates of the possible relative extreme points:

$$\text{At } x = -1, \quad y = (-1)^3 - 3(-1) + 2 = -1 + 3 + 2 = 4$$
$$\text{At } x = 1, \quad y = (1)^3 - 3(1) + 2 = 1 - 3 + 2 = 0.$$

The possible relative extreme points are $(-1, 4)$ and $(1, 0)$. Now,

$$\text{At } x = -1, \quad y'' = 6(-1) = -6 < 0 \Rightarrow \text{ graph is concave down at } x = -1.$$

$$\text{At } x = 1, \quad y'' = 6(1) = 6 > 0 \Rightarrow \quad \text{ graph is concave up at } x = 1.$$

So the function has a relative maximum at $(-1, 4)$ and a relative minimum at $(1, 0)$. Since the concavity changes somewhere between $x = -1$ and $x = 1$, there must be at least one inflection point. To find this, set $y'' = 0$ and solve for x:

$$6x = 0 \Rightarrow x = 0$$

Substitute $x = 0$ into the equation to find the y-coordinate of this inflection point:

$$y = (0)^3 - 3(0) + 2 = 0 - 0 + 2 = 2.$$

The inflection point is $(0, 2)$.

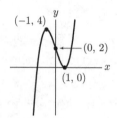

31. $y = 2x^3 - 3x^2 - 36x + 20,$

$y' = 6x^2 - 6x - 36,$

$y'' = 12x - 6,$

Set $y' = 0$ and solve for x.

$$6x^2 - 6x - 36 = 0$$
$$6(x^2 - x - 6) = 0$$
$$3(x + 2)(x - 3) = 0$$
$$x = -2 \quad \text{or} \quad x = 3.$$

Use these values the find the y-coordinates of the possible relative extreme points:

$$\text{At } x = -2, \quad y = 2(-2)^3 - 3(-2)^2 - 36(-2) + 20 = -16 - 12 + 72 + 20 = 64$$
$$\text{At } x = 3, \quad y = 2(3)^3 - 3(3)^2 - 36(3) + 20 = 54 - 27 - 108 + 20 = -61.$$

The possible relative extreme points are $(-2, 64)$ and $(3, -61)$. Now,

$$\text{At } x = -2, \quad y'' = 12(-2) - 6 = -30 < 0 \Rightarrow \text{ graph is concave down at } x = -2.$$

$$\text{At } x = 3, \quad y'' = 12(3) - 6 = 30 > 0 \Rightarrow \quad \text{ graph is concave up at } x = 3.$$

So the function has a relative maximum at $(-2, 64)$ and a relative minimum at $(3, -61)$. Since the concavity changes somewhere between $x = -2$ and $x = 3$, there must be at least one inflection point. To find this, set $y'' = 0$ and solve for x:

$$12x - 6 = 0 \Rightarrow 12x = 6 \Rightarrow x = \frac{1}{2}$$

Substitute $x = \dfrac{1}{2}$ into the equation to find the y-coordinate of this inflection point:

$$y = 2\left(\frac{1}{2}\right)^3 - 3\left(\frac{1}{2}\right)^2 - 36\left(\frac{1}{2}\right) + 20 = \frac{2}{8} - \frac{3}{4} - \frac{36}{2} + 20 = \frac{3}{2}.$$

The inflection point is $\left(\dfrac{1}{2}, \dfrac{3}{2}\right)$.

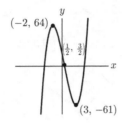

37. $g(x) = 3 + 4x - 2x^2,$

$g'(x) = 4 - 4x,$

$g''(x) = -4,$

Set $g'(x) = 0$ and solve for x:

$$4 - 4x = 0 \Rightarrow 4 = 4x \Rightarrow x = 1$$

Substitute this value back into $g(x)$ to find the y-coordinate of this possible relative extreme point.

$$g(1) = 3 + 4(1) - 2(1)^2 = 3 + 4 - 2 = 5$$

Since $g''(x) = -4,$ which is negative, the graph is concave down at $x = 1,$ and thus, $(1, 5)$ is a relative maximum point.

43.

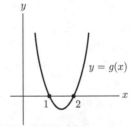

(a) Refer to the figure above. Assume $g(x)$ is the first derivative of $f(x)$. Then, when $x = 2$, $g(2) = 0$, $g(x)$ is negative for $1 < x < 2$ and positive for $x > 2$ (or equivalently; $f'(2) = 0$, $f'(x)$ is negative for $1 < x < 2$ and positive for $x > 2$). This tells us that $f(x)$ is decreasing for $1 < x < 2$ and increasing for $x > 2$ and $f(x)$ has a relative minimum at $x = 2$.

(b) Refer to the figure above. Assume $g(x)$ is the second derivative of $f(x)$. Then, when $x = 2$, $g(2) = 0$, $g(x)$ is negative for $1 < x < 2$ and positive for $x > 2$ (or equivalently; $f''(2) = 0$, $f''(x)$ is negative for $1 < x < 2$ and positive for $x > 2$). This tells us that $f(x)$ is concave down for $1 < x < 2$ and concave up for $x > 2$ and $f(x)$ has an inflection point at $x = 2$.

2.4   Curve Sketching (Conclusion)

The two most important types of curves that are graphed in this section are the cubics without relative extreme points (such as the first curve or the upside down version of this curve) and the curves with asymptotes. These curves are stressed since, along with the curves considered in Section 2.3, they occur frequently in applications. However, there are a few other types of curves that occur in the exercises. As previously, all reasoning should be based on the first and second derivative tests.

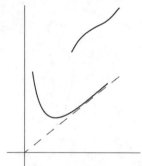

1. To find the x-intercepts of the function $y = x^2 - 3x + 1$, use the quadratic formula to find the values of x for which $f(x) = 0$:

$$y = x^2 - 3x + 1 \quad (a = 1, b = -3, c = 1),$$

$$b^2 - 4ac = (-3)^2 - 4 \cdot 1 \cdot 1 = 5,$$

$$x = \frac{-(-3) \pm \sqrt{5}}{2 \cdot 1} = \frac{3 \pm \sqrt{5}}{2}.$$

Therefore, the x-intercepts are $\left(\dfrac{3 + \sqrt{5}}{2}, 0 \right)$ and $\left(\dfrac{3 - \sqrt{5}}{2}, 0 \right)$.

7. If $f(x) = \frac{1}{3}x^3 - 2x^2 + 5x$, then

$$f'(x) = x^2 - 4x + 5.$$

If we apply the quadratic formula to $f'(x)$ with $a = 1$, $b = -4$, and $c = 5$, we find that $b^2 - 4ac = (-4)^2 - 4 \cdot 1 \cdot 5 = -4$, a negative number. Since we cannot take the square root of a negative number, there are no values of x for which $f'(x) = 0$. Hence $f(x)$ has no relative extreme points.

13. $f(x) = 5 - 13x + 6x^2 - x^3$,

$f'(x) = -13 + 12x - 3x^2$,

$f''(x) = 12 - 6x$

If we apply the quadratic formula to $f'(x)$, we find that

$$b^2 - 4ac = (12)^2 - 4(-3)(-13) = 144 - 156 = -12,$$

which is negative.

Since we cannot take the square root of a negative number, there are no values of x for which $f'(x) = 0$. So $f(x)$ has no relative extreme points. If we evaluate $f'(x)$ at some x, say $x = 0$, we see that the first derivative is negative, and so $f(x)$ is decreasing there. Since the graph of $f(x)$ is a smooth curve with no relative extreme points and no breaks, $f(x)$ must be decreasing for all x. Now, to check the concavity of $f(x)$, we must find where $f''(x)$ is negative, positive, or zero.

$f''(x) = 12 - 6x = 6(2 - x)$ is:

$$\begin{cases} \text{positive if } x < 2, & \text{(graph is concave up)} \\ \text{negative if } x > 2, & \text{(graph is concave down)} \\ \text{zero if } x = 2. & \text{(concavity reverses)} \end{cases}$$

The inflection point is $(2, f(2)) = (2, -5)$. The y-intercept is $(0, f(0)) = (0, 5)$. To further improve the sketch of the graph, first sketch the tangent line at the inflection point. To do this compute the slope of the graph at $(2, -5)$.

$$f'(2) = -13 + 12(2) - 3(2)^2 = -1.$$

To sketch the graph, plot the inflection point and the y-intercept, draw the tangent line at $(2, -5)$, and then draw a curve that has this tangent line and is decreasing for all x, is concave down for $x > 2$, and concave up for $x < 2$.

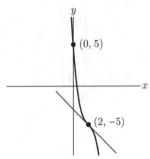

19. $f(x) = x^4 - 6x^2,$

$f'(x) = 4x^3 - 12x,$

$f''(x) = 12x^2 - 12.$

To find possible relative extreme points, set $f'(x) = 0$ and solve for x:

$$4x^3 - 12x = 0$$
$$4x(x^2 - 3) = 0$$
$$4x(x - \sqrt{3})(x + \sqrt{3}) = 0$$
$$x = 0 \quad \text{or} \quad x = \pm\sqrt{3}.$$

Substitute these values into $f(x)$ to find the y-coordinates of these possible relative extreme points:

$$f(0) = (0)^4 - 6(0)^2 = 0,$$
$$f(\sqrt{3}) = (\sqrt{3})^4 - 6(\sqrt{3})^2 = -9,$$
$$f(-\sqrt{3}) = (-\sqrt{3})^4 - 6(-\sqrt{3})^2 = -9$$

The points are $(0, 0)$, $(\sqrt{3}, -9)$, and $(-\sqrt{3}, -9)$.

Now $f''(0) = 12(0)^2 - 12 = -12$. Therefore, $f(x)$ is concave down at $x = 0$ and $(0, 0)$ is a relative maximum point. For the other two points, compute

$$f''(-\sqrt{3}) = 12(-\sqrt{3})^2 - 12 = 12(3) - 12 = 24,$$
$$f''(\sqrt{3}) = 12(\sqrt{3})^2 - 12 = 12(3) - 12 = 24$$

Hence $f(x)$ is concave up at $x = -\sqrt{3}$ and $x = \sqrt{3}$, and so $(\sqrt{3}, -9)$ and $(-\sqrt{3}, -9)$ are relative minimum points. The concavity of this function reverses twice, so there must be at least two inflection points. To find these, set $f''(x) = 0$ and solve for x:

$$12x^2 - 12 = 0,$$
$$x^2 - 1 = 0,$$
$$(x-1)(x+1) = 0, \quad \text{or} \quad x = \pm 1.$$

The corresponding y-coordinates are given by

$$f(1) = (1)^4 - 6(1)^2 = -5,$$
$$f(-1) = (-1)^4 - 6(-1)^2 = -5.$$

The inflection points are $(1, -5)$.

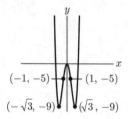

25. $y = \dfrac{9}{x} + x + 1, \quad x > 0$

$$y' = -\frac{9}{x^2} + 1,$$

$$y'' = \frac{18}{x^3}$$

To find possible extrema, set $y' = 0$ and solve for x:

$$-\frac{9}{x^2} + 1 = 0,$$

$$1 = \frac{9}{x^2} \qquad \text{(Multiply both sides by } x^2.\text{)}$$

$$x^2 = 9,$$

$$x = 3. \qquad (\textit{Note}: \text{ Only consider } x > 0.)$$

When $x = 3$, $y = \dfrac{9}{3} + 3 + 1 = 7$, and

$$y'' = \frac{18}{3^3} > 0 \quad \text{(Concave up).}$$

Therefore, (3, 7) is a relative minimum point. Since y'' can never be 0, there are no inflection points. The term $\dfrac{9}{x}$ in the function y tells us that the y-axis is an asymptote. Also, as x gets large, the graph of y gets arbitrarily close to the straight line $y = x + 1$. Therefore, $y = x + 1$ is an asymptote of the graph.

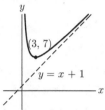

Helpful Hint: In general $f'(a) = 0$ does not necessarily mean that $f(x)$ has an extreme point at $x = a$. Also, $f''(a) = 0$ does not necessarily mean that $f(x)$ has an inflection point at $x = a$. However, if $f(x)$ is quadratic or of the form $hx + \dfrac{k}{x} + c$, then $f'(a) = 0$ guarantees an extreme point at $x = a$. (Neither of these two types of functions has inflection points.) If $f(x)$ is a cubic polynomial, then there is one inflection point and it occurs where the second derivative is zero. (*Note:* These observations will help you in checking your work. On exams, you must still show that extreme and inflection points have the properties you claim.)

31. $g(x) = f'(x)$, because $f(x)$ has 2 relative minima and a relative maximum and $g(x) = 0$ for these values of x.

$f(x) \neq g'(x)$. To see this, look at the relative maximum of $g(x)$. Since $f(x)$ is not zero for this value of x, $f(x) \neq g'(x)$.

37. (a)

[0, 20] by [−12, 50]

(b) The weight of the rat in grams is given by $f(t) = 4.96 + .48t + .17t^2 - .00048t^3$, after t days. After 7 days, the rat weighed

$$f(7) = 4.96 + .48(7) + .17(7)^2 - .00048(7)^3$$
$$= 15.0036 \text{ grams.}$$

(c) To find when the rat weighed 27 grams, solve the following for t:

$$4.96 + .48t + .17t^2 - .00048t^3 = 27.$$

Use graphing calculator techniques to obtain $t \approx 12.0380$ (one method is to graph both $f(t)$ and $y = 27$ and use the intersect command to find where $f(t) = 27$). Therefore, the rat's weight reached 27 grams after approximately 12 days.

(d) Note that $f'(t) = .48 + .34t - .0144t^2$. Therefore, the rate at which the rat was gaining weight after 4 days is found by

$$f'(4) = .48 + .34(4) - .0144(4)^2 = 1.6096.$$

After 4 days, the rat was gaining weight at a rate of 1.6096 grams per day.

(e) To find when the rat was gaining weight at a rate of 2 grams per day, solve the following for t:

$$.48 + .34t - .0144t^2 = 2$$

Use graphing calculator techniques to obtain $t \approx 5.990$ or $t \approx 17.6207$. (Graph both $f'(t)$ and $y = 2$ and use the intersect command to find where $f'(t) = 2$). The rat was gaining weight at the rate of 2 grams per day after about 6 days and after about 17.6 days.

(f) To find when the rat was gaining wait at the fastest rate, we want to find the maximum value of $f'(t)$. In order to do this, solve $f''(t) = 0$, or

$$f''(t) = .34 - .0288t = 0$$

to obtain $t \approx 11.8056$. Confirm this is a maximum by examining the graph of $f'(t)$, or notice $f'''(t) = -.0288 < 0$. The rat was growing at the fastest rate after about 11.8 days.

2.5   Optimization Problems

The procedure for solving an optimization problem can be thought of as consisting of the following two primary parts:

(a) Find the function to be optimized.

(b) Make a rough sketch of the graph of the function.

Part (a) requires careful reading of the problem. Part (b) relies on techniques presented in Sections 2.3 and 2.4. However, certain shortcuts can be taken when sketching curves in this section. For instance, if you only want to find the x-coordinate of a relative extreme point, you can make a very rough estimate of the y-coordinate when sketching the curve.

One of the common mistakes that students make when working optimization problems on exams is just to set the first derivative equal to 0 and solve for x. They forget to apply the second derivative test at the value of x found.

1. $g(x) = 10 + 40x - x^2$,

$g'(x) = 40 - 2x$,

$g''(x) = -2$.

The graph is obtained by the curve-sketching technique of Section 2.3. Therefore, the maximum value of $g(x)$ occurs at $x = 20$.

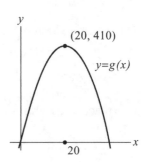

7. The function to be minimized is $Q = x^2 + y^2$. By solving the equation $x + y = 6$ for y, we can write y in terms of x:

$$y = 6 - x,$$

and then substitute in order to express Q as a function of the single variable x:

$$Q = x^2 + (6 - x)^2$$

Simplifying, we have

$$Q = x^2 + (36 - 12x + x^2), \quad \text{or} \quad Q = 2x^2 - 12x + 36.$$

Differentiating,

$$Q' = 4x - 12,$$
$$Q'' = 4$$

Now, solve $Q' = 0$:

$$4x - 12 = 0,$$
$$4x = 12,$$
$$x = 3$$

$Q' = 0$ and Q'' is positive at $x = 3$. Therefore Q has a relative minimum at $x = 3$. Plug $x = 3$ into the equation for Q in order to find that when $x = 3$, $Q = 18$. Since $x = 3$ is the only critical point, 18 is the absolute minimum.

13. (a)

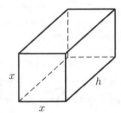

(b) The girth of the box is given by $x + x + x + x = 4x$. The formula for the length plus the girth is

$$h + 4x$$

(c) The equation for volume, which must be maximized, is the objective equation,

$$V = x^2 h.$$

The constraint equation is

$$h + 4x = 84,$$

or

$$h = 84 - 4x.$$

(d) Substitute $h = 84 - 4x$ into the objective equation:

$$V = x^2(84 - 4x)$$
$$= -4x^3 + 84x^2.$$

(e) Make a rough sketch of the graph of $V = -4x^3 + 84x^2$ to find the value of x corresponding to the greatest volume.

$$V' = -12x^2 + 168x,$$
$$V'' = -24x + 168.$$

Solve $V' = 0$:

$$-12x^2 + 168x = 0,$$
$$x^2 - 14x = 0,$$
$$x(x - 14) = 0,$$
$$x = 0, \quad \text{or} \quad x = 14.$$

V'' is positive at $x = 0$ and negative at $x = 14$. Therefore V has a relative minimum at $x = 0$ and a relative maximum at $x = 14$. When $x = 0$, $V = 0$. When $x = 14$, $V = -4(14)^3 + 84(14)^2 = 5488$. The sketch of the graph shows that the maximum value of V occurs when $x = 14$. From the constraint equation, $h = 84 - 4(14) = 28$.

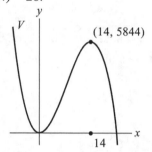

19. The problem asks that total area be maximized. Let A be the area, let x be the length of the fence parallel to the river, and let y be the length of each section perpendicular to the river. The objective equation gives an expression for A in terms of the other variables:

$$A = xy \quad \text{(objective equation).}$$

The cost of the fence parallel to the river is $6x$ dollars (x feet at \$6 per foot) and the cost of the three sections perpendicular to the river is $3 \cdot 5y$ or $15y$ dollars. Since \$1500 is available to build the fence, we must have

$$6x + 15y = 1500 \quad \text{(constraint equation).}$$

Solving for y, we have

$$15y = -6x + 1500,$$
$$y = -\frac{2}{5}x + 100.$$

Substituting this expression for y into the objective equation, we obtain

$$A = x\left(-\frac{2}{5}x + 100\right) = -\frac{2}{5}x^2 + 100x.$$

The graph of $A = -\frac{2}{5}x^2 + 100x$ is easily obtained by our curve sketching techniques. The maximum value occurs when $x = 125$. Substituting into $y = -\frac{2}{5}x + 100$, this gives us

$y = -\frac{2}{5}(125) + 100 = 50$. Therefore, the optimum dimensions are $x = 125$ ft, $y = 50$ ft.

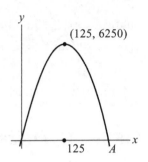

25. $x = 20 - \dfrac{w}{2}$ (from the constraint equation).

$A = wx$ (the objective function).

Therefore:

$$A = w\left(20 - \frac{w}{2}\right) = -\frac{w^2}{2} + 20w \qquad \text{(has a parabola shape)},$$

$$\frac{dA}{dw} = -\frac{2w}{2} + 20 = -w + 20.$$

A relative minimum or maximum occurs when $\dfrac{dA}{dw} = 0$. Hence $-w + 20 = 0 \Rightarrow w = 20$. (Since

$x = 20 - \dfrac{w}{2}$, $x = 10$.$\Big)$ The next question is whether this is a relative maximum. Look at the second derivative:

$$\frac{d^2 A}{dw^2} = -1 < 0,$$

which shows that $w = 20$ is a relative maximum.

To sketch the graph of A, observe that $A = 0$ when

$w = 0$ and when $x = 0$: $20 - \dfrac{w}{2} = 0 \Rightarrow \dfrac{w}{2} = 20$, or $w = 40$.

At $w = 20$, $A = 20\left(20 - \dfrac{20}{2}\right) = 200$.

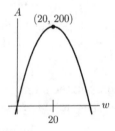

31. The distance from any point (x, y) on the line $y = -2x + 5$ to the origin is given by

$$\text{Distance} = \sqrt{x^2 + y^2} = \sqrt{x^2 + (-2x + 5)^2} = \sqrt{5x^2 - 20x + 25}.$$

The distance has its smallest value when $5x^2 - 20x + 25$ does, so we need to minimize $D(x) = 5x^2 - 20x + 25$:

$$D(x) = 5x^2 - 20x + 25$$
$$D'(x) = 10x - 20$$
$$D'(x) = 0 \Rightarrow 10x - 20 = 0$$
$$x = 2$$

To confirm $x = 2$ is a minimum, we check $D''(x)$: $D''(x) = 10 > 0$, so $x = 2$ is in fact a minimum of $D(x)$. Substituting $x = 2$ into $y = -2x + 5$, we find

$$y = -2(2) + 5 = 1$$

Therefore, the point on the line $y = -2x + 5$ closest to the origin is $(2, 1)$.

2.6   Further Optimization Problems

The exercises in this section are of four types.

1. Variations of the types of exercises considered in Section 2.5.
2. Inventory problems. These problems all have the same constraint equation, $r \cdot x =$ [number of items used or manufactured during the year], and similar objective equations, [costs] = ...
3. Inventory control problems where the constraint equations are found by reading the problem to see if [amount of money] depends on [quantity] or vice versa, and expressing the dependent variable in terms of the other. These problems have the same type of objective equation, [profit or revenue] = [amount of money] · [quantity]. See practice problem 1 or Example 2.
4. Original problems. Situations that use the same type of machinery as the examples worked in the text but require a fresh approach or area/volume problems.

1. **(a)** In each order-reorder period, the amount of cherries in inventory decreases linearly from 180 pounds at the beginning of the period to 0 pounds at the end of the period. Thus the average amount of cherries in inventory in one order-reorder period is the average of 0 and 180, i.e., 90 pounds.

    (b) The maximum amount of cherries in inventory during a given order-reorder period is 180 pounds. This occurs at the beginning of the period.

    (c) 6 orders were placed during the year; each order-reorder period is represented by one of the 6 line segments on the graph.

    (d) Since 180 pounds of cherries were sold during each order-reorder period, and there were 6 order-reorder periods during the year, a total of $180 \cdot 6 = 1080$ pounds of cherries were sold during the year.

7. Let r represent the number of production runs and x the number of microscopes manufactured per run. Hence the total number of microscopes manufactured is

    $$\begin{bmatrix} \text{number of microscopes per} \\ \text{production run} \end{bmatrix} \cdot [\text{number of runs}] = xr.$$

    The constraint equation is

    $$xr = 1600, \quad \text{or} \quad x = \frac{1600}{r}.$$

    We wish to minimize inventory expenses. There are three expenses which make up the total cost.

The storage costs (SC), based on the maximum number of microscopes in the warehouse, will be
$$SC = 15x,$$
since the warehouse is the most full just after a run, which will produce x microscopes. Since the insurance costs are based on the average number of microscopes in the warehouse, and the average number of microscopes is given by $\dfrac{x}{2}$, the insurance costs (IC) are

$$IC = 20\left(\frac{x}{2}\right).$$

Each production run costs $2500, so the total production costs (PC) are given by
$$PC = 2500r.$$

Hence the objective equation is

$$C = PC + SC + IC$$

$$= 2500r + 15x + 20\left(\frac{x}{2}\right)$$

$$= 2500r + 25x.$$

Substituting $x = 1600/r$ into the objective equation we have,

$$C = 2500r + 25\left(\frac{1600}{r}\right)$$

$$= 2500r + \frac{40,000}{r}.$$

Use our curve sketching techniques to make a rough sketch of this function. Only positive values of r are relevant; that is, $r > 0$.

$$C' = 2500 - \frac{40,000}{r^2},$$

$$C'' = \frac{80,000}{r^3}.$$

Set $C' = 0$ and solve for r:

$$2500 - \frac{40,000}{r^2} = 0,$$

$$2500 = \frac{40,000}{r^2},$$

$$r^2 = \frac{40,000}{2,500} = 16, \quad \text{or} \quad r = 4.$$

When $r = 4$, $C'' > 0$. There is a relative minimum at $r = 4$. Therefore, there should be 4 production runs. *Note:* The value of C when $r = 4$ need not be calculated. After establishing the existence of relative minimum point when $r = 4$, use your familiarity with graphs of functions of the form $y = \dfrac{a}{x} + bx$ to make a rough sketch.

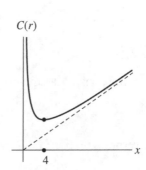

13.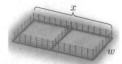

Let w and x be the dimensions of the corral. The corral is to have an area of 54 square meters, hence the constraint equation is

$$xw = 54 \quad \text{or} \quad w = \frac{54}{x} \quad \text{(Constraint)}.$$

The total amount of fencing needed is two pieces of length x and three pieces of length w. Hence the objective equation is

$$F = 2x + 3w \quad \text{(Objective)}.$$

Substituting $w = \frac{54}{x}$ into the objective equation we have,

$$F = 2x + 3\left(\frac{54}{x}\right) = 2x + \frac{162}{x}.$$

Next, use curve sketching techniques to make a rough sketch of the graph.

$$F = 2x + \frac{162}{x}, \quad x > 0,$$

$$F' = 2 - \frac{162}{x^2},$$

$$F'' = \frac{324}{x^3}.$$

Set $F' = 0$ and solve for x:

$$2 - \frac{162}{x^2} = 0 \Rightarrow 2 = \frac{162}{x^2} \Rightarrow x^2 = \frac{162}{2} = 81, \quad \text{or} \quad x = 9.$$

Notice that $F'' > 0$ for all positive x, so the graph of F is concave up, and F has a relative minimum at $x = 9$. In fact the formula for F has the form $y = \frac{a}{x} + bx$, and so the graph has the basic shape shown below.

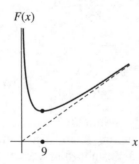

From the constraint equation,

$$w = \frac{54}{x} = \frac{54}{9} = 6.$$

Therefore, $w = 6$ m, $x = 9$ m are the optimum dimensions.

19. First consider the diagram. Let x be the length of each edge of the square ends and h the other dimension.

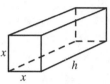

Since the sum of the three dimensions can be at most 120 centimeters, the constraint equation is

$$2x + h = 120, \quad h = 120 - 2x.$$

Since we wish to maximize the volume of the package, the objective equation is

$$V = x^2 h.$$

Substitute $h = 120 - 2x$ into the objective equation and compute

$$V = x^2(120 - 2x) = 120x^2 - 2x^3,$$
$$V' = 240x - 6x^2,$$
$$V'' = 240 - 12x.$$

Set $V' = 0$ and solve for x:

$$240x - 6x^2 = 0,$$
$$6x(40 - x) = 0,$$
$$x = 0, \quad \text{and} \quad x = 40.$$

Check concavity: At $x = 0$,

$$V'' = 240 - 12(0) = 240 > 0 \qquad \text{(graph concave up)}.$$

At $x = 40$,

$$V'' = 240 - 12(40) = -240 < 0 \qquad \text{(graph concave down)}.$$

Therefore, the graph has a relative minimum at $x = 0$ and a relative maximum at $x = 40$. When $x = 0$, $V = 0$; at $x = 40$, $V = 120(40)^2 - 2(40)^3 = 64,000$. A rough sketch of the graph

$$V = 120x^2 - 2x^3, \quad x \geq 0,$$

reveals a maximum value at $x = 40$.

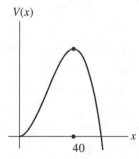

The value of h corresponding to $x = 40$ is found from the constraint equation

$$h = 120 - 2x = 120 - 2(40) = 40.$$

Thus to achieve maximum volume, the package should be 40 cm × 40 cm × 40 cm.

25. First, consider the diagram below. The base of the window has length $2x$. The area of the window is given by

$$A = 2xy = 2x(9 - x^2)$$
$$= 18x - 2x^3.$$

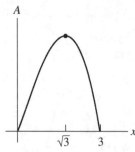

Next, make a rough sketch of the graph of $A = 18x - 2x^3$, for $x > 0$:

$$A' = 18 - 6x^2,$$
$$A'' = -12x.$$

Set $A' = 0$:

$$18 - 6x^2 = 0,$$
$$18 = 6x^2,$$
$$3 = x^2, \quad \text{or} \quad x = \sqrt{3}.$$

When $x = \sqrt{3}$, $A'' = -12(\sqrt{3})$ is negative. Therefore the graph of $A = 18x - 2x^3$ has a relative maximum value at $x = \sqrt{3}$. When $x = \sqrt{3}$,

$$A = 18\sqrt{3} - 2(\sqrt{3})^3$$
$$= 18\sqrt{3} - 2 \cdot 3 \cdot \sqrt{3} \quad \left[\text{since } (\sqrt{3})^3 = (\sqrt{3} \cdot \sqrt{3}) \cdot \sqrt{3} = 3\sqrt{3} \right]$$
$$= 12\sqrt{3}.$$

A rough sketch of the graph reveals a maximum value at $x = \sqrt{3}$.

The height y, corresponding to $x = \sqrt{3}$ is found from the equation of the parabola

$$y = 9 = x^2$$
$$= 9(\sqrt{3})^2 = 9 - 3 = 6.$$

The window of maximum area should be 6 units high and $2\sqrt{3}$ units wide. (The value of x gives only half the width of the base of the window.)

2.7   Applications of Derivatives to Business and Economics

This section explores cost, revenue, and profit functions, and applies calculus to the optimization of profits of a firm.

1. If the cost function is

$$C(x) = x^3 - 6x^2 + 13x + 15,$$

the marginal cost function is

$$M(x) = C'(x) = 3x^2 - 12x + 13.$$

We find the minimum value of $M(x)$ by making a rough graph, a simple matter since $M(x)$ is quadratic function.

$$M'(x) = 6x - 12,$$
$$M''(x) = 6.$$

Set $M'(x) = 0$ to locate the x where the marginal cost is minimized.

$$6x - 12 = 0$$
$$6x = 12, \quad \text{or} \quad x = 2.$$

Now, $M(2) = 3(2)^2 - 12(2) + 13 = 1$ and $M''(2)$ is positive. Therefore, $M(x)$ has a relative minimum at $x = 2$. A sketch of the graph reveals a minimum value of 1. That is, the minimum marginal cost is \$1.

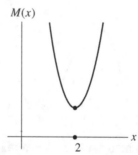

Note: Sketching the graph is not absolutely necessary. Once you realize that it is a parabola opening upward, you should know that its minimum value will occur where the first derivative is zero.

7. If the demand equation for the commodity is given by $p = \frac{1}{12}x^2 - 10x + 300, \ 0 \le x \le 60,$ then the revenue function is

$$R(x) = x \cdot p = x\left(\frac{1}{12}x^2 - 10x + 300\right) = \frac{1}{12}x^3 - 10x^2 + 300x.$$

To find the maximum value of $R(x)$, first make a rough sketch of its graph.

$$R(x) = \frac{1}{12}x^3 - 10x^2 + 300x,$$

$$R'(x) = \frac{1}{4}x^2 - 20x + 300,$$

$$R''(x) = \frac{1}{2}x - 20.$$

Set $R'(x) = 0$ and solve for x:

$$\frac{1}{4}x^2 - 20x + 300 = 0,$$

$$x^2 - 80x + 1200 = 0, \qquad \text{(multiply by 4)}$$

$$(x - 20)(x - 60) = 0.$$

$$x = 20 \quad \text{or} \quad x = 60.$$

[*Note:* To see how to factor $x^2 - 80x + 1200$, first try to factor $x^2 - 8x + 12$.] Now,

$$R(20) = \frac{1}{12}(20)^3 - 10(20)^2 + 300(20) = 2666\frac{2}{3},$$

$$R(60) = \frac{1}{12}(60)^3 - 10(60)^2 + 300(60) = 0,$$

$$R''(20) = \frac{1}{2}(20) - 20 = -10 < 0,$$

$$R''(60) = \frac{1}{2}(60) - 20 = 10 > 0.$$

A rough sketch of the graph reveals that revenue is maximized when $x = 20$. From the demand equation,

$$p = \frac{1}{12}(20)^2 - 10(20) + 300 = 133\frac{1}{3} \approx \$133.33.$$

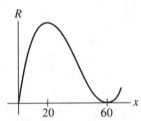

13. Let x represent the number of prints that the artist makes and let $R(x)$ represent the artist's total revenue. Since she can charge \$400 per print if she offers 50 prints for sale, and this price will decrease by \$5 for each print offered in excess of 50, the price per print if she offers x prints for sale is:

$$p(x) = 400 - 5(x - 50).$$

Thus

$$R(x) = x \cdot p(x) = x \cdot (400 - 5(x - 50)),$$

i.e.,

$$R(x) = 650x - 5x^2.$$

Differentiating,

$$R'(x) = 650 - 10x,$$

$$R''(x) = -10$$

Setting $R'(x) = 0$ and solving for x yields $x = 65$.

Since R'' is negative, $x = 65$ is a maximum. When $x = 65$, the artist's revenue is

$$R(x) = 650(65) - 5(65)^2 = 21{,}125.$$

The point (65, 21125) is the maximum point of $R(x)$. In order to attain the maximum revenue of $21,125, the artist should make 65 prints.

19. The savings and loan association spends money by paying interest to customers with savings accounts, and earns money by charging interest to customers taking loans. The association's profit is determined by the formula $P = R - C$, where R is the revenue from interest charged on loans and C is the cost of interest paid to savings accounts. Let

$$i_d = \text{interest rate on deposits, and}$$
$$i_l = \text{interest rate on loans.}$$

Then

$$R = (\text{amount loaned}) \cdot (i_l), \text{ and}$$
$$C = (\text{amount deposited in savings}) \cdot (i_d).$$

From the information given in the problem, we have:

$$i_l = .1, \text{ and}$$
$$(\text{amount loaned}) = (\text{amount deposited})$$
$$= (1{,}000{,}000) \cdot (i_d).$$

Thus

$$P = (1{,}000{,}000)(i_d) \cdot (.1) - (1{,}000{,}000) \cdot (i_d) \cdot (i_d),$$

i.e.,

$$P = (100{,}000)(i_d) - (1{,}000{,}000)(i_d)^2.$$

Differentiating,

$$P' = 100{,}000 - (2{,}000{,}000)(i_d).$$

Setting P' equal to 0 to find the critical point yields $i_d = \frac{1}{20}$, or 5%. Thus the savings and loan association should offer a 5% interest rate on deposits in order to generate the most profit.

Chapter 2 Review Exercises

The most difficult problems in this chapter are the optimization problems. These problems cannot be solved all at once, but require that you break them down into small manageable pieces. Although some of them fit into neat categories, others require patient analysis.

While solutions for all review exercises are included, expanded explanations are included for every sixth exercise.

1. (a) The graph of $f(x)$ is increasing for values of x at which $f'(x)$ is positive. The graph of $f'(x)$ shows that this happens for $-3 < x < 1$ and $x > 5$. The graph of $f(x)$ is decreasing for $x < -3$ and $1 < x < 5$, because the values of $f'(x)$ are negative.

(b) The graph of $f(x)$ is concave up for values of x at which the slopes of the graph are increasing, that is, for values of x at which the graph of $f'(x)$ is *increasing*. Be careful to distinguish between "increasing" and "positive". The graph of $f'(x)$ is increasing for $x < -1$ and for $x > 3$, even though the values of $f'(x)$ are not always positive there.

The graph of $f(x)$ is concave down for values of x at which the slopes of the graph are decreasing, that is, for values of x at which the graph of $f'(x)$ is *decreasing*. This happens for $-1 < x < 3$.

2. **(a)** $f(3) = 2$

   (b) The tangent line has slope $\dfrac{1}{2}$, so $f'(3) = \dfrac{1}{2}$.

   (c) Since the point $(3, 2)$ appears to be an inflection point, $f''(3) = 0$.

3.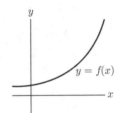

4.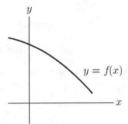

5.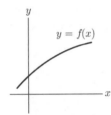

6.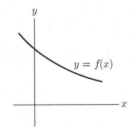

7. $f'(x)$ is positive at d and e.

8. b   9. c, d   10. a   11. e   12. b

13. Since $f(1) = 2$, the graph goes through the point $(1, 2)$. Since $f'(1) > 0$, the graph is increasing at $x = 1$.

14. Graph goes through $(1, 5)$, decreasing at $x = 1$.

15. Increasing and concave up at $x = 3$.

16. Decreasing and concave down at $x = 2$.

17. $(10, 2)$ is a relative minimum point.

18. Graph goes through $(4, -2)$, increasing and concave down at $x = 4$.

19. Since $g(5) = -1$, the graph goes through the point $(5, -1)$. Since $g'(5) = -2 < 0$, the graph is decreasing at $x = 5$. The fact that $g''(5) = 0$ is inconclusive since it is possible for $g''(x)$ to be 0 without x being an inflection point. For example, if $g(x)$ is a straight line, $g''(x) = 0$ for all x but $g(x)$ has no inflection points.

20. (0, 0) is a relative minimum point.

21. (a) $f(t) = 1$ at $t = 2$, after 2 hours.

    (b) $f(5) = .8$

    (c) $f'(t) = -.08$ at $t = 3$, after 3 hours.

    (d) Since $f'(8) = -.02$, the rate of change is $-.02$ unit per hour.

22. (a) Since $f(50) = 400$, the amount of energy produced was 400 trillion kilowatt-hours.

    (b) Since $f'(50) = 35$, the rate of change was 35 trillion kilowatt-hours per year.

    (c) Since $f(t) = 3000$ at $t = 95$, the production level reached 300 trillion kilowatt-hours in 1995.

    (d) Since $f'(t) = 10$ at $t = 35$, the production level was rising at the rate of 10 trillion kilowatt-hours per year in 1935.

    (e) Looking at the graph of $y = f'(t)$, the value of $f'(t)$ appears to be greatest at $t = 70$. To confirm, observe that the graph of $y = f''(t)$ crosses the t-axis at $t = 70$. Energy production was growing at the greatest rate in 1970. Since $f(70) = 1600$, the production level at that time was 1600 trillion kilowatt-hours.

23.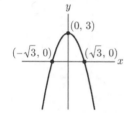

24.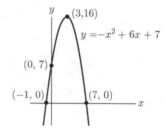

25. If $y = x^2 + 3x - 10$, then

$$y' = 2x + 3,$$
$$y'' = 2.$$

Set $y' = 0$ and solve for x to find the relative extreme point:

$$2x + 3 = 0,$$
$$x = -\frac{3}{2}.$$

If $x = -\frac{3}{2}$, then $y = \left(-\frac{3}{2}\right)^2 + 3\left(-\frac{3}{2}\right) - 10 = \frac{9}{4} - \frac{9}{2} - 10 = -\frac{49}{4}$. Since $y'' = 2 > 0$, the parabola is concave up and the point $\left(-\frac{3}{2}, -\frac{49}{4}\right)$ is a minimum point. When $x = 0$, $y = 0^2 + 3(0) - 10 = -10$. Hence the y-intercept is $(0, -10)$. To find the x-intercepts, set $y = 0$ and solve for x:

$$x^2 + 3x - 10 = 0,$$
$$(x + 5)(x - 2) = 0,$$
$$x = -5 \quad \text{and} \quad x = 2.$$

So the x-intercepts are $(-5, 0)$ and $(2, 0)$.

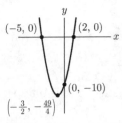

26.

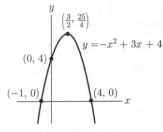

27.

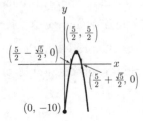

28.

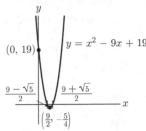

29.

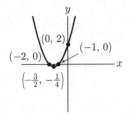

30.

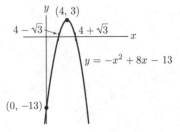

31. If $y = -x^2 + 20x - 90$, then

$$y' = -2x + 20,$$
$$y'' = -2.$$

Set $y' = 0$ and solve for x to find the relative extreme point:

$$-2x + 20 = 0,$$
$$x = 10.$$

If $x = 10$, $y = -(10)^2 + 20(10) - 90 = -100 + 200 - 90 = 10$. Since $y'' = -2 < 0$, the parabola is concave down and $(10, 10)$ is the relative maximum point. When $x = 0$, $y = -0^2 + 20(0) - 90 = -90$. Hence the y-intercept is $(0, -90)$. To find the x-intercept, use the quadratic formula:

$$b^2 - 4ac = (20)^2 - 4(-1)(-90) = 40,$$

$$\sqrt{b^2 - 4ac} = \sqrt{40} = \sqrt{4 \cdot 10} = \sqrt{4} \cdot \sqrt{10} = 2\sqrt{10}.$$

$$x = \frac{-20 \pm 2\sqrt{10}}{-2} = 10 \pm \sqrt{10}.$$

So the x-intercepts are $(10 + \sqrt{10},\, 0)$ and $(10 - \sqrt{10},\, 0)$.

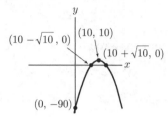

32.

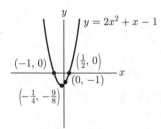

33.

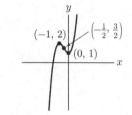

34.

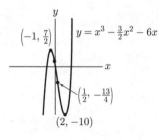

35.

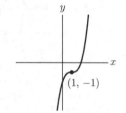

36.

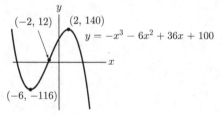

37. $\quad y = \dfrac{11}{3} + 3x - x^2 - \dfrac{1}{3}x^3,$

$\quad y' = 3 - 2x - x^2,$

$\quad y'' = -2 - 2x.$

Set $y' = 0$ and solve for x to find the possible relative extreme points:

$$3 - 2x - x^2 = 0.$$

Multiple by −1 and rearrange terms.

$$x^2 + 2x - 3 = 0,$$
$$(x-1)(x+3) = 0,$$
$$x = 1, \quad \text{or} \quad x = -3.$$

If $x = 1$, $y = \frac{11}{3} + 3(1) - (1)^2 - \frac{1}{3}(1)^3 = \frac{16}{3}$, and

$$y'' = -2 - 2 = -4 < 0.$$

Hence the graph is concave down and has a relative maximum at $\left(1, \frac{16}{3}\right)$.

At $x = -3$, $y = \frac{11}{3} + 3(-3) - (-3)^2 - \frac{1}{3}(-3)^3 = -\frac{16}{3}$ and $y'' = -2 - 2(-3) = 4 > 0$. Hence the graph is concave up and has a relative minimum at $\left(-3, -\frac{16}{3}\right)$. To find the inflection point, set $y'' = 0$ and solve for x:

$$-2 - 2x = 0,$$
$$-2x = 2,$$
$$x = -1.$$

If $x = -1$, $y = \frac{11}{3} + 3(-1) - (-1)^2 - \frac{1}{3}(-1)^3 = 0$. Hence the inflection point is $(-1, 0)$.

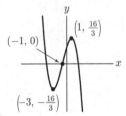

38.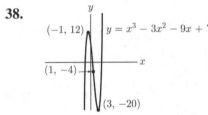

39.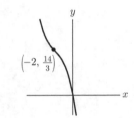

40.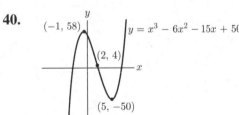

41.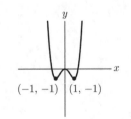

42.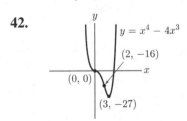

43. This function has the form $y = g(x) + mx + b,$ where

$$g(x) = \frac{20}{x} \quad \text{and} \quad mx+b = \frac{x}{5}+3 \quad \left(=\frac{1}{5}x+3,\text{ a straight line of slope } \frac{1}{5}\right).$$

$$y = \frac{x}{5}+\frac{20}{x}+3, \quad y' = \frac{1}{5}-\frac{20}{x^2}, \quad y'' = \frac{40}{x^3}.$$

Set $y'=0$ and solve for x:

$$\frac{1}{5}-\frac{20}{x^2}=0 \Rightarrow \frac{1}{5}=\frac{20}{x^2} \Rightarrow x^2=100$$

(The last equation was obtained by cross multiplication). Thus $x=10,$ since only $x>0$ is being considered. When $x=10,$ $y=\frac{10}{5}+\frac{20}{10}+3=2+2+3=7,$ and y'' is obviously positive. Therefore the graph is concave up at $x=10,$ and so $(10,.7)$ is a relative minimum point. Since y'' can never be 0, there are no inflection points. The graph has the y-axis as a vertical asymptote and approaches the straight line $y=\frac{x}{5}+3$ as x gets large.

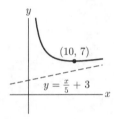

44.

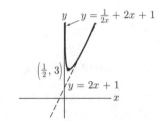

45. $f'(x)=\frac{3}{2}\left(x^2+2\right)^{1/2}(2x)=3x\left(x^2+2\right)^{1/2}$

Since $f'(0)=0,$ f has a possible extreme value at $x=0.$

46. $f'(x)=\frac{3}{2}\left(2x^2+3\right)^{1/2}(4x)$

$2x^2+3>0$ for all $x,$ so the sign of $f'(x)$ is determined by the sign of $4x.$ Therefore, $f'(x)>0$ if $x>0,$ $f'(x)<0$ if $x<0.$ This means that $f(x)$ is decreasing for $x<0$ and increasing for $x>0.$

47. $f''(x)=\frac{-2x}{(1+x^2)^2},$ so $f''(0)=0.$ Since $f'(x)>0$ for all $x,$ it follows that 0 must be an inflection point.

48. $f''(x)=\frac{1}{2}\left(5x^2+1\right)^{-1/2}(10x),$ so $f''(0)=0.$ Since $f'(x)>0$ for all $x,$ $f''(x)$ is negative for $x<0$ and positive for $x>0,$ and it follows that 0 must be an inflection point.

49. *A* and *c*, because a constant function corresponds to no change in the position of the car.

B and *e*, because a positive derivative corresponds to an increase of the distance from the reference point, and the fact that $s'(t)$ is constant means that the velocity is a "steady rate."

C and *f*, because information about $s'(a)$ gives information about $f(t)$ for *t* close to $t = a$, and a positive derivative corresponds to "moving forward."

D and *b*, for the same reason that *C* matches *f*, except that a negative derivative goes with "backing up."

E and *a*. A positive **first** derivative corresponds to "moving forward" and a positive **second** derivative goes with increasing velocity ("speeding up").

F and *d*, for the same reason that *E* matches *a*, except that a negative **second** derivative corresponds to decreasing velocity.

50. A – c, B – e, C – f, D – b, E – a, F – d

51. **a.** The number of people living between $10 + h$ and 10 miles from the center of the city.

b. If so, $f(x)$ would be decreasing at $x = 10$, which is not possible.

52. $f(x) = \dfrac{1}{4}x^2 - x + 2 \ (0 \le x \le 8)$

$f'(x) = \dfrac{1}{2}x - 1$

$f'(x) = 0 \Rightarrow \dfrac{1}{2}x - 1 = 0 \Rightarrow x = 2$

$f''(x) = \dfrac{1}{2}$

Since $f'(2)$ is a relative minimum, the maximum value of $f(x)$ must occur at one of the endpoints. $f(0) = 2, f(8) = 10$. 10 is the maximum value, attained $x = 8$.

53. $f(x) = 2 - 6x - x^2 \ (0 \le x \le 5)$

$f'(x) = -6 - 2x$

Since $f'(x) < 0$ for all $x > 0$, $f(x)$ is decreasing on the interval [0, 5]. Thus, the maximum value occurs at $x = 0$. The maximum value is $f(0) = 2$.

54. $g(t) = t^2 - 6t + 9 \ (1 \le t \le 6)$

$g'(t) = 2t - 6$

$g'(t) = 0 \Rightarrow 2t - 6 = 0 \Rightarrow t = 3$

$g''(t) = 2$

The minimum value of $g(t)$ is $g(3) = 0$.

55. First consider the figure. Let x be the width and h be the height of the box.

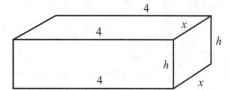

The box is to have a volume of 200 cubic feet, hence the constraint equation is

$$4xh = 200 \quad \text{or} \quad h = \frac{50}{x}.$$

We minimize the amount of material by minimizing the amount of surface area. The objective equation is

$$A = \underset{\substack{\text{area of} \\ \text{base}}}{4x} + \underset{\substack{\text{area of 2} \\ \text{ends}}}{2xh} + \underset{\substack{\text{area of 2} \\ \text{sides}}}{8h}$$

Substitute the value $h = \dfrac{50}{x}$ into the objective equation:

$$A = 4x + 2x\left(\frac{50}{x}\right) + 8\left(\frac{50}{x}\right)$$

$$= 4x + 100 + \frac{400}{x} \quad (x > 0).$$

Next, make a rough sketch of the graph of this function:

$$A = 4x + 100 + \frac{400}{x}, \quad x > 0,$$

$$A' = 4 - \frac{400}{x^2},$$

$$A'' = \frac{800}{x^3}.$$

Set $A' = 0$ and solve for x:

$$4 - \frac{400}{x^2} = 0,$$

$$4 = \frac{400}{x^2},$$

$$x^2 = \frac{400}{4} = 100, \quad \text{or} \quad x = 10.$$

When $x = 10$, $A = 4(10) + 100 + \frac{400}{10} = 180$. Since the second derivative $\dfrac{800}{x^3}$ is positive for all $x > 0$, the graph is concave up for $x > 0$; also, the point $(10, 180)$ is a minimum point.

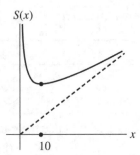

So the amount of material of the box is minimized when x is 10 feet. The other dimension is $h = \dfrac{50}{x} = \dfrac{50}{10} = 5$ feet. The dimensions of the box are $4 \times 5 \times 10$ feet.

56. Let x be the length of the base of the box and let y be the other dimension. The objective is $V = x^2 y$

and the constraint is $3x^2 + x^2 + 4xy = 48$

$$y = \frac{48 - 4x^2}{4x} = \frac{12 - x^2}{x}$$

$$V(x) = x^2 \cdot \frac{12 - x^2}{x} = 12x - x^3$$

$$V'(x) = 12 - 3x^2$$

$$V'(x) = 0 \Rightarrow 12 - 3x^2 = 0 \Rightarrow x = 2$$

$$V''(x) = -6x;\ V''(2) < 0.$$

The maximum value for x for $x > 0$ occurs at $x = 2$. The optimal dimensions are 2 ft $\times$ 2 ft $\times$ 4 ft.

57. Let x be the number of inches turned up on each side of the gutter. The objective is $A(x) = (30 - 2x)x$. (A is the cross-sectional area of the gutter—maximizing this will maximize the volume).
$A'(x) = 30 - 4x$

$$A'(x) = 0 \Rightarrow 30 - 4x = 0 \Rightarrow x = \frac{15}{2}$$

$$A''(x) = -4,\ A''\left(\frac{15}{2}\right) < 0$$

$x = \dfrac{15}{2}$ inches gives the maximum value for A.

58. Let x be the number of trees planted. The objective is $f(x) = \left(25 - \dfrac{1}{2}(x - 40)\right)x\ (x \geq 40)$.

$$f(x) = 45x - \frac{1}{2}x^2$$

$$f'(x) = 45 - x$$

$$f'(x) = 0 \Rightarrow 45 - x = 0 \Rightarrow x = 45$$

$$f''(x) = -1;\ f''(45) < 0.$$

The maximum value of $f(x)$ occurs at $x = 45$. Thus, 45 trees should be planted.

59. Let r be the number of production runs and let x be the lot size. Then the objective is

$$C = 1000r + .5\left(\frac{x}{2}\right) \text{ and the constraint is } rx = 400,000 \Rightarrow r = \frac{400,000}{x} \text{ so } C(x) = \frac{4 \cdot 10^8}{x} + \frac{x}{4}.$$

$$C'(x) = \frac{-4 \cdot 10^8}{x^2} + \frac{1}{4}$$

$$C'(x) = 0 \Rightarrow \frac{-4 \cdot 10^8}{x^2} + \frac{1}{4} = 0 \Rightarrow x = 4 \cdot 10^4$$

$$C''(x) = \frac{8 \cdot 10^8}{x^3}; C''(4 \cdot 10^4) > 0$$

The minimum value of $C(x)$ for $x > 0$ occurs at $x = 4 \cdot 10^4 = 40,000$. Thus the economic lot size is 40,000 books/run.

60. Let x be the width of the poster and let y be its height. Then the objective is $A = (x-4)(y-5)$ and the constraint is $xy = 125 \Rightarrow y = \frac{125}{x}$.

$$A(x) = (x-4)\left(\frac{125}{x} - 5\right)$$

$$= 125 - 5x - \frac{500}{x} + 20 = 145 - 5x - \frac{500}{x}$$

$$A'(x) = -5 + \frac{500}{x^2}$$

$$A'(x) = 0 \Rightarrow -5 + \frac{500}{x^2} = 0 \Rightarrow x = 10$$

$$A''(x) = -\frac{1000}{x^3}; A''(10) = -\frac{1000}{10^3} = -1 < 0.$$

The maximum value of $A(x)$ for $x > 0$ occurs at $x = 10$. Thus the optimal dimensions are $x = 10$, $y = 12.5$ inches.

61. Since the profit function is

$$P(x) = R(x) - C(x) \qquad \text{(revenue minus cost)},$$

and since

$$R(x) = x \cdot p = x(150 - .02x) = 150x - .02x^2$$

the profit function is

$$P(x) = (150x - .02x^2) - (10x + 300)$$

$$= 140x - .02x^2 - 300$$

$$= -.02x^2 + 140x - 300.$$

The graph of $P(x)$ is a parabola opening downward (since the coefficient of x^2 is negative) and therefore assumes its maximum value where the first derivative is zero.

$$P'(x) = -.04x + 140.$$

Setting $P'(x) = 0$ and solving for x

$$-.04x + 140 = 0$$
$$-.04x = -140$$
$$x = \frac{-140}{-.04} = 3500.$$

[*Note:* One way to evaluate $\frac{140}{.04}$ is to first multiply the numerator and denominator by 100 and then divide. That is, $\frac{140}{.04} = \frac{14,000}{4} = 3500.$] Thus the profit is maximized when the sales level x is 3500 units.

62. 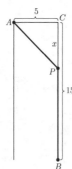 The distance from point A to point P is $\sqrt{25 + x^2}$ and the distance from point P to point B is $15 - x$. The time it takes to travel from point A to point P is $\frac{\sqrt{25 + x^2}}{8}$ and the time it takes to travel from point P to point B is $\frac{15 - x}{17}$.

Therefore, the total trip takes $T(x) = \frac{1}{8}(25 + x^2)^{1/2} + \frac{1}{17}(15 - x)$ hours.

$$T'(x) = \frac{1}{16}(25 + x^2)^{-1/2}(2x) - \frac{1}{17}$$

$$T'(x) = 0 \Rightarrow \frac{1}{16}(25 + x^2)^{-1/2}(2x) - \frac{1}{17} = 0 \Rightarrow x = \frac{8}{3}$$

$$T''(x) = \frac{1}{8}(25 + x^2)^{-1/2} - \frac{1}{8}x^2(25 + x^2)^{-3/2}$$

$$T''\left(\frac{8}{3}\right) = \frac{1}{8}\left(25 + \left(\frac{8}{3}\right)^2\right)^{-1/2} - \frac{1}{8}\left(\frac{8}{3}\right)^2\left(25 + \left(\frac{8}{3}\right)^2\right)^{-3/2} = \frac{675}{39304} > 0$$

The minimum value for $T(x)$ occurs at $x = \frac{8}{3}$. Thus, Jane should drive from point A to point P, $\frac{8}{3}$ miles from point C, then down to point B.

63. Let $12 \leq x \leq 25$ be the size of the tour group. Then, the revenue generated from a group of x people, $R(x)$, is $R(x) = [800 - 20(x - 12)]x$. To maximize revenue:

$R'(x) = 1040 - 40x$

$R'(x) = 0 \Rightarrow 1040 - 40x = 0 \Rightarrow R = 26.$

$R''(x) = -40;\ R''(26) < 0$.

Revenue is maximized for a group of 26 people, which exceeds the maximum allowed. Although, $R(x)$ is an increasing function on [12, 25], therefore $R(x)$ reaches its maximum at $x = 25$ on the interval $12 \leq x \leq 25$. The tour group that produces the greatest revenue is size 25.

Chapter 3
Techniques of Differentiation

3.1 The Product and Quotient Rules

The first twenty-eight exercises are routine drill. You will be able to tell from the answers if you are mastering the use of the product and quotient rules. In each case, try to obtain the simplified version of the answer shown in the answer section. Most students find this difficult, but now is a good time to practice a skill that may be important for the next exam!

1. To differentiate the product $(x+1)(x^3+5x+2)$, let $f(x)=(x+1)$ and $g(x)=(x^3+5x+2)$ and apply the product rule. Then

$$\frac{d}{dx}[(x+1)(x^3+5x+2)] = (x+1) \cdot \frac{d}{dx}(x^3+5x+2)$$

$$+ (x^3+5x+2) \cdot \frac{d}{dx}(x+1)$$

$$= (x+1)(3x^2+5) + (x^3+5x+2)(1).$$

Carry out the multiplication in the first product above and combine like terms:

$$\frac{d}{dx}[(x+1)(x^3+5x+2)] = (3x^3+3x^2+5x+5)+(x^3+5x+2)$$

$$= 4x^3+3x^2+10x+7.$$

7. To differentiate $(x^2+3)(x^2-3)^{10}$, let $f(x)=(x^2+3)$ and $g(x)=(x^2-3)^{10}$ and apply the product rule. To compute $g'(x)$, use the general power rule.

$$\frac{d}{dx}[(x^2+3)(x^2-3)^{10}] = (x^2+3) \cdot \frac{d}{dx}(x^2-3)^{10} + (x^2-3)^{10} \cdot \frac{d}{dx}(x^2+3)$$

$$= (x^2+3)10(x^2-3)^9 2x + (x^2-3)^{10} 2x.$$

To simplify the answer, factor $2x(x^2-3)^9$ out of each term of the derivative:

$$\frac{d}{dx}[(x^2+3)(x^2-3)^{10}] = 2x(x^2-3)^9[(x^2+3)10+(x^2-3)]$$

$$= 2x(x^2-3)^9(11x^2+27).$$

Helpful Hint: Here is a good way to remember the order of the terms in the quotient rule:

1. Draw a long fraction bar and write $[g(x)]^2$ in the denominator:

$$\frac{}{[g(x)]^2}$$

2. While you are still thinking about $g(x)$, write it in the numerator:

$$\frac{g(x)}{[g(x)]^2}$$

3. The rest is easy because each term in the numerator of the derivative involves one function and one derivative. Since $g(x)$ is already written, it must go with $f'(x)$. Next comes the minus sign and then the function and derivative that have not yet been used.

$$\frac{g(x)f'(x)-f(x)g'(x)}{[g(x)]^2}$$

Think of $g(x)$ and $g'(x)$ as being "around the outside" of this formula.

13. To differentiate $\dfrac{x^2-1}{x^2+1}$, let $f(x)=x^2-1$ and $g(x)=x^2+1$ and apply the quotient rule:

$$\frac{d}{dx}\left(\frac{x^2-1}{x^2+1}\right)=\frac{(x^2+1)\cdot\frac{d}{dx}(x^2-1)-(x^2-1)\cdot\frac{d}{dx}(x^2+1)}{(x^2+1)^2}$$

$$=\frac{(x^2+1)2x-(x^2-1)2x}{(x^2+1)^2}.$$

If the products in the numerator are expanded, this derivative simplifies:

$$\frac{d}{dx}\left(\frac{x^2-1}{x^2+1}\right)=\frac{2x^3+2x-2x^3+2x}{(x^2+1)^2}$$

$$=\frac{4x}{(x^2+1)^2}.$$

19. To differentiate $\dfrac{x^2}{\left(x^2+1\right)^2}$, let $f(x)=x^2$ and $g(x)=\left(x^2+1\right)^2$ and apply the quotient rule.

$$\frac{d}{dx}\left(\frac{x^2}{\left(x^2+1\right)^2}\right)=\frac{\left(x^2+1\right)^2\frac{d}{dx}\left(x^2\right)-\left(x^2\right)\frac{d}{dx}\left(\left(x^2+1\right)^2\right)}{\left(\left(x^2+1\right)^2\right)^2}$$

$$=\frac{\left(x^2+1\right)^2(2x)-\left(x^2\right)(2)\left(x^2+1\right)(2x)}{\left(x^2+1\right)^4}\qquad\text{Use the chain rule to find }\frac{d}{dx}\left(\left(x^2+1\right)^2\right).$$

If the products in the numerator are expanded, we have

$$\frac{d}{dx}\left(\frac{x^2}{\left(x^2+1\right)^2}\right) = \frac{\left(x^2+1\right)^2(2x)-\left(x^2\right)(2)\left(x^2+1\right)(2x)}{\left(x^2+1\right)^4}$$

$$= \frac{\left(x^4+2x^2+1\right)(2x)-\left(4x^5+4x^3\right)}{\left(x^2+1\right)^4}$$

$$= \frac{2x^5+4x^3+2x-4x^5-4x^3}{\left(x^2+1\right)^4}$$

$$= \frac{-2x^5+2x}{\left(x^2+1\right)^4}$$

Alternatively, we can factor $2x\left(x^2+1\right)$ in the numerator to obtain

$$\frac{d}{dx}\left(\frac{x^2}{\left(x^2+1\right)^2}\right) = \frac{\left(x^2+1\right)^2(2x)-x^2(2)\left(x^2+1\right)(2x)}{\left(\left(x^2+1\right)^2\right)^2}$$

$$= \frac{2x\left(x^2+1\right)\left[\left(x^2+1\right)-2x^2\right]}{\left(x^2+1\right)^4}$$

$$= \frac{2x\left(1-x^2\right)}{\left(x^2+1\right)^3} = \frac{2x-2x^3}{\left(x^2+1\right)^3}$$

Warning: Some students prefer to work #19 using the product rule, but this may cause them more work than they realize. It is possible to write

$$y = \frac{x^2}{\left(x^2+1\right)^2} = x^2\left(x^2+1\right)^{-2},$$

and

$$\frac{dy}{dx} = x^2 \cdot \frac{d}{dx}\left(x^2+1\right)^{-2} + \left(x^2+1\right)^{-2} \cdot \frac{d}{dx}\left(x^2\right)$$

$$= \left(x^2\right)\cdot(-2)\left(x^2+1\right)^{-3}(2x) + \left(x^2+1\right)^{-2}\cdot(2x).$$

This method is not really shorter, but it does avoid the quotient rule. The main difficulty arises when you try to simplify the derivative, for you must factor out terms that involve negative exponents. However, if you really prefer this method, the next paragraph explains how to handle negative exponents.

How to simplify a sum involving negative exponents: Suppose that various powers of some quantity Q occur in each term of the sum. Find the exponent on Q that lies farthest to the left on the number line and factor out this power of Q from each term in the sum.

In the solution above, notice that $\frac{dy}{dx}$ involves $\left(x^2+1\right)^{-2}$ and $\left(x^2+1\right)^{-3}$. Since -3 is to the left of -2 on the number line, factor out $2x \cdot \left(x^2+1\right)^{-3}$. In the second term in the third line below we need $\left(x^2+1\right)^{1}$, because $\left(x^2+1\right)^{-2} = \left(x^2+1\right)^{-3}\left(x^2+1\right)^{+1}$.

$$\frac{dy}{dx} = x^2 \cdot \frac{d}{dx}\left(x^2+1\right)^{-2} + \left(x^2+1\right)^{-2} \cdot \frac{d}{dx}\left(x^2\right)$$

$$= \left(x^2\right) \cdot (-2)\left(x^2+1\right)^{-3}(2x) + \left(x^2+1\right)^{-2} \cdot (2x).$$

$$= (2x)\left(x^2+1\right)^{-3}\left(-2x^2 + \left(x^2+1\right)\right)$$

$$= (2x)\left(x^2+1\right)^{-3}\left(-x^2+1\right)$$

Moral of this story: If you have to simplify the derivative of a quotient, use the quotient rule for the differentiation step.

Helpful Hint: A problem such as #22 can be worked with the quotient rule, but since the numerator is a constant, the differentiation is easier if you rewrite the quotient with the notation $(\cdots)^{-1}$. See Practice Problem #2.

25. $\dfrac{d}{dx}\left(\dfrac{x+11}{x-3}\right)^3 = 3\left(\dfrac{x+11}{x-3}\right)^2 \dfrac{d}{dx}\left(\dfrac{x+11}{x-3}\right)$ Use the general power rule.

$$= 3\left(\frac{x+11}{x-3}\right)^2\left(\frac{(x-3)(1)-(x+11)(1)}{(x-3)^2}\right) \quad \text{Use the quotient rule.}$$

$$= 3\left(\frac{x+11}{x-3}\right)^2\left(\frac{-14}{(x-3)^2}\right)$$

$$= -\frac{42(x+11)^2}{(x-3)^4}$$

31. The tangent line is horizontal when $\dfrac{dy}{dx}=0.$ Using the quotient rule, we find

$$\frac{dy}{dx} = \frac{(x-4)^3(5)(x-2)^4 - (x-2)^5(3)(x-4)^2}{[(x-4)^3]^2}$$

$$= \frac{(x-4)^2(x-2)^4(5x-20-3x+6)}{(x-4)^6}$$

$$= \frac{(x-2)^4(2x-14)}{(x-4)^4}.$$

$\frac{dy}{dx}=0$ when $(x-2)^4(2x-14)=0,$ or when $x=2$ or $x=7.$ Thus, the tangent line is horizontal for $x=2$ or $x=7.$

Helpful Hint: Do you try the Practice Problems before starting your homework? You should. If you worked Practice Problem #1 in this section, you may have already learned to think about simplifying a function before starting to differentiate. Often there is no need to simplify first, but in Exercise 35 this really helps. Notice that

$$y = \frac{x^2 + 3x - 1}{x} = x + 3 - \frac{1}{x},$$

so that

$$\frac{dy}{dx} = 1 + \frac{1}{x^2}.$$

37. Here we are asked to determine $\frac{d^2 y}{dx^2}$, the second derivative. So, we differentiate the function, and then differentiate the result.

$$\frac{dy}{dx} = 4\left(x^2 + 1\right)^3 (2x) \quad \text{Use the general power rule. (See section 1.6.)}$$

$$= 8x\left(x^2 + 1\right)^3 \qquad \text{Simplify.}$$

Now use the product rule to differentiate $8x\left(x^2 + 1\right)^3$.

$$\frac{d^2 y}{dx^2} = 8x(3)\left(x^2 + 1\right)^2 (2x) + \left(x^2 + 1\right)^3 (8)$$

$$= 8\left(x^2 + 1\right)^2 \left(3x(2x) + \left(x^2 + 1\right)\right)$$

$$= 8\left(x^2 + 1\right)^2 \left(7x^2 + 1\right)$$

43. Applying the quotient rule, we have

$$h'(x) = \frac{d}{dx}\left[\frac{f(x)}{x^2 + 1}\right] = \frac{\left(x^2 + 1\right) f'(x) - f(x)(2x)}{\left(x^2 + 1\right)^2}$$

49. Recall that the marginal revenue, *MR*, is defined by

$$MR = R'(x).$$

The average revenue is maximized when $\frac{d}{dx}(AR) = 0$.

$$\frac{d}{dx}(AR) = \frac{d}{dx}\left(\frac{R(x)}{x}\right) = \frac{xR'(x) - R(x)(1)}{x^2}$$

If $\frac{xR'(x) - R(x)}{x^2} = 0$, then $xR'(x) - R(x) = 0 \Rightarrow xR'(x) = R(x) \Rightarrow R'(x) = \frac{R(x)}{x} \Rightarrow MR = AR$.

Helpful Hint: Remember that when you have to determine when the derivative of a quotient is zero, the algebra is usually easier if you use the quotient rule.

55. We use the first derivative rule to determine the *x*-coordinate of the maximum point.

$$y = \frac{10x}{1+.25x^2}$$

$$\frac{dy}{dx} = \frac{\left(1+.25x^2\right)10 - 10x(.5x)}{\left(1+.25x^2\right)^2}$$

$$= \frac{10 - 2.5x^2}{\left(1+.25x^2\right)^2}$$

Now solve $\dfrac{dy}{dx} = 0.$

$$\frac{dy}{dx} = \frac{10 - 2.5x^2}{\left(1+.25x^2\right)^2} = 0$$

$$10 = 2.5x^2$$

$$x = \pm 2$$

The function is defined only for $x \geq 0$, so the *x*-coordinate of the maximum point is $x = 2$. Now determine $f(2)$ to find the *y*-coordinate of the maximum point.

$$y = \frac{10(2)}{1+.25(2)^2} = \frac{20}{2} = 10$$

Thus, the coordinates of the maximum point are (2, 10).

61. In exercise 60, we are given $f(1) = 2,\ f'(1) = 3,\ g(1) = 4,\ $ and $\ g'(1) = 5.$

$$\frac{d}{dx}\left[\frac{f(x)}{g(x)}\right]\Bigg|_{x=1} = \frac{g(1)f'(1) - f(1)g'(1)}{g(1)^2}$$

$$= \frac{(4)(3) - (2)(5)}{4^2} = \frac{1}{8}$$

67. $b(t) = \dfrac{w(t)}{\left[h(t)\right]^2}$

Using the quotient rule, we have

$$b'(t) = \frac{d}{dt}\left[\frac{w(t)}{\left[h(t)\right]^2}\right]$$

$$= \frac{\left[h(t)\right]^2 w'(t) - w(t)(2)h(t)h'(t)}{\left[h(t)\right]^4}$$

$$= \frac{h(t)w'(t) - 2h'(t)w(t)}{\left[h(t)\right]^3}$$

3.2 The Chain Rule and the General Power Rule

The chain rule is used in subsequent chapters to derive formulas for the derivatives of composite functions where the outer function is either a logarithm function, an exponential function, or one of the trigonometric functions. In each case there will be a formula that has the same feel as the general power rule. These formulas will become so automatic that you will sometimes forget that they are special cases of the chain rule.

Many of the exercises in this section may be solved with the general power rule, but you should approach the exercises with the abstract chain rule in mind. The practice you gain here will make it easier to use and remember the later versions of the chain rule.

1. If $f(x) = \dfrac{x}{x+1}$, $g(x) = x^3$, then $f(g(x)) = \dfrac{x^3}{x^3+1}$.

7. If $f(g(x)) = \sqrt{4-x^2}$, then $f(x) = \sqrt{x}$ and $g(x) = 4 - x^2$.

13. If $y = 6x^2(x-1)^3$,

$$\frac{dy}{dx} = \left[6x^2 \cdot \frac{d}{dx}(x-1)^3 \right] + (x-1)^3 \cdot \frac{d}{dx}(6x^2) \quad \text{[Product Rule]}$$

$$= 6x^2(3)(x-1)^2(1) + (x-1)^3(12x) \quad \text{[General Power Rule]}$$

Factor $6x(x-1)^2$ out of each term to get

$$\frac{dy}{dx} = 6x(x-1)^2[3x + 2(x-1)]$$

$$= 6x(x-1)^2(5x-2).$$

19. To differentiate $\left(\dfrac{4x-1}{3x+1} \right)^3$, let $f(x) = x^3$, $g(x) = \dfrac{4x-1}{3x+1}$, and use the chain rule. Now $f'(x) = 3x^2$ and

$$g'(x) = \frac{(3x+1)(4) - (4x-1)(3)}{(3x+1)^2} \quad \text{[Quotient Rule]}$$

Hence

$$\frac{d}{dx} f(g(x)) = f'(g(x))g'(x)$$

$$= 3\left(\frac{4x-1}{3x+1} \right)^2 \cdot \frac{(3x+1)4 - (4x-1)3}{(3x+1)^2}.$$

This simplifies to

$$\frac{d}{dx} f(g(x)) = \frac{3(4x-1)^2(12x+4-12x+3)}{(3x+1)^4}$$

$$= \frac{21(4x-1)^2}{(3x+1)^4}.$$

25. Using the quotient rule:

$$h'(x) = \left[\frac{f(x^2)}{x}\right]$$

$$= \frac{x \cdot \frac{d}{dx}[f(x^2)] - f(x^2) \cdot \frac{d}{dx}(x)}{x^2}$$

$$= \frac{x \cdot \frac{d}{dx}[f(x^2)] - f(x^2) \cdot (1)}{x^2}$$

Now apply the chain rule to in order to find $\frac{d}{dx}[f(x^2)]$:

$$\frac{d}{dx}[f(x^2)] = f'(x^2) \cdot \frac{d}{dx}(x^2) = f'(x^2) \cdot (2x).$$

Putting these results together, we have

$$h'(x) = \frac{x \cdot f'(x^2) \cdot (2x) - f(x^2) \cdot (1)}{x^2} = \frac{2x^2 \cdot f'(x^2) - f(x^2)}{x^2}.$$

31. If $f(x) = \dfrac{1}{x} = x^{-1}$ and $g(x) = 1 - x^2$, then $f'(x) = -x^{-2} = \frac{-1}{x^2}$ and $g'(x) = -2x$. Hence

$$\frac{d}{dx} f(g(x)) = f'(g(x)) \cdot g'(x)$$

$$= \frac{-1}{(1-x^2)^2} \cdot (-2x) = \frac{2x}{(1-x^2)^2}.$$

37. If $y = u^{3/2}$ and $u = 4x + 1$, then

$$\frac{dy}{du} = \frac{3}{2} u^{1/2}, \quad \text{and} \quad \frac{du}{dx} = 4.$$

Hence

$$\frac{dy}{dx} = \frac{dy}{du} \cdot \frac{du}{dx} = \frac{3}{2} u^{1/2} \cdot (4) = 6u^{1/2}.$$

To express $\dfrac{dy}{dx}$ as a function of x alone, substitute $4x + 1$ for u to obtain

$$\frac{dy}{dx} = 6 \cdot (4x + 1)^{1/2}.$$

43. If $y = \dfrac{x+1}{x-1}$ and $x = \dfrac{t^2}{4} = \dfrac{1}{4}t^2$, then

$$\frac{dy}{dx} = \frac{(x-1)(1) - (x+1)(1)}{(x-1)^2} \qquad \text{[Quotient Rule]}$$

$$= \frac{-2}{(x-1)^2}$$

and $$\frac{dx}{dt} = \frac{1}{4}(2t) = \frac{t}{2}.$$

Therefore, $$\frac{dy}{dx} = \frac{dy}{dx} \cdot \frac{dx}{dt} = \frac{-2}{(x-1)^2} \cdot \frac{t}{2}.$$

To express $\frac{dy}{dt}$ as a function of t alone, substitute $\frac{t^2}{4}$ for x to obtain

$$\frac{dx}{dt} = \frac{-2}{\left(\frac{t^2}{4}-1\right)^2} \cdot \frac{t}{2}.$$

Now, $$\left.\frac{dy}{dt}\right|_{t_0=3} = \left.\frac{-2}{\left(\frac{t^2}{4}-1\right)^2} \cdot \frac{t}{2}\right|_{t_0=3} = \frac{-2}{\left(\frac{9}{4}-1\right)^2} \cdot \frac{3}{2}$$

$$= \frac{-32}{25} \cdot \frac{3}{2} = -\frac{48}{25}.$$

49. (a) The volume of the cube is a function of the length, x, of the edge of the cube, and the length x is a function of time. Thus by the chain rule we have:

$$\frac{dV}{dt} = \frac{dV}{dx} \cdot \frac{dx}{dt}$$

While the formula for the function describing x as a function of time is not given, the function that describes V as a function of x is $V = x^3$. Thus $\frac{dV}{dx} = 3x^2$. Substituting, we have

$$\frac{dV}{dt} = 3x^2 \cdot \frac{dx}{dt}.$$

(b) When $\frac{dV}{dx} = 12,$ we have

$$\frac{dV}{dt} = 12 \cdot \frac{dx}{dt},$$

which says that the volume of the cube is increasing twelve times as quickly as the length of the edge of the cube. To find the value of x which makes this the case, we must solve the equation $3x^2 = 12$ for x. Choosing the positive solution since x represents a length, we arrive at $x = 2$.

55. (a) Since $L = 10 + .4x + .0001x^2$, differentiating gives

$$\frac{dL}{dx} = .4 + .0002x.$$

(b) Similarly, since $x = 752 + 23t + .5t^2$, differentiating gives

$$\frac{dx}{dt} = 23 + t.$$

Evaluating at time $t = 2$ years,

$$\left.\frac{dx}{dt}\right|_{t=2} = 23 + 2 = 25.$$

The population is increasing at the rate of 25 thousand people per year.

(c) Use the chain rule to find $\dfrac{dL}{dt}$:

$$\frac{dL}{dt} = \frac{dL}{dx} \cdot \frac{dx}{dt} = (.4 + .0002x) \cdot (23 + t).$$

At time $t = 2$ years, $x = 752 + 23(2) + .5 \cdot (2)^2 = 800$. Thus

$$\left.\frac{dL}{dt}\right|_{t=2} = (.4 + .0002(800)) \cdot (25) = (.56)(25) = 14.$$

The carbon monoxide level is increasing at the rate of 14 parts per million per year.

61. **(a)** Referring to Figure 1(a) in the text, we see that the value of a share of the company's stock at time $t = 1.5$ months is \$40. Next, referring to Figure 1(b), we see that the total value of the company when the value of a single share is \$40 is approximately 78 million dollars. Using the formula, we have

$$W(40) = 10\left(\frac{12 + 8(40)}{3 + 40}\right) \approx \$77.209 \text{ million}$$

Similarly, at $t = 3.5$ months, the value of one share of the stock is \$30. At this share price, the company is valued at approximately 76 million dollars. Using the formula, we have

$$W(30) = 10\left(\frac{12 + 8(30)}{3 + 30}\right) \approx \$76.364 \text{ million}$$

(b) The graph in Figure 1(a) in the text shows a straight line segment between the points $(0, 10)$ and $(2, 50)$. Thus, for each value of t between 0 and 2, the value of $\frac{dx}{dt}$ is the slope of this line segment.

In particular, the value of $\frac{dx}{dt}$ at time $t = 1.5$ months is

$$\left.\frac{dx}{dt}\right|_{t=1.5} = \frac{50 - 10}{2 - 0} = \frac{40}{2} = 20.$$

This means that 1.5 months after the company went public, the value of a share of the company's stock was increasing at a rate of \$20 per month.

At time $t = 3.5$ months,

$$\left.\frac{dx}{dt}\right|_{t=3.5} = \frac{30 - 30}{5 - 3} = \frac{0}{2} = 0.$$

(Also note that the graph is horizontal; therefore, the slope is 0.) This means that 3.5 months after the company went public, the value of a share of the company's stock was holding constant (neither increasing nor decreasing) at \$30.

3.3   Implicit Differentiation and Related Rates

The material in this section is not used in the rest of the text. However, the two techniques introduced here are fundamental concepts of calculus and are commonly found in applications. Both techniques involve using the chain rule in situations where we don't know the specific formula for the inner part of a composite function.

1. Differentiate $x^2 - y^2 = 1$ term by term. The first term x^2 has derivative $2x$. Think of the second term y^2 as having the form $[g(x)]^2$. Use the chain rule to differentiate.

$$\frac{d}{dx}[g(x)]^2 = 2[g(x)]g'(x).$$

Hence,

$$\frac{d}{dx}y^2 = 2y \cdot \frac{dy}{dx}.$$

On the right side of the original equation, the derivative of the constant function 1 is zero. Thus implicit differentiation of $x^2 - y^2 = 1$ yields

$$2x - 2y \cdot \frac{dy}{dx} = 0.$$

Solving for $\dfrac{dy}{dx}$,

$$-2y \cdot \frac{dy}{dx} = -2x.$$

If $y \neq 0$,

$$\frac{dy}{dx} = \frac{x}{y}.$$

7. Differentiate $2x^3 + y = 2y^3 + x$ term by term. The derivative of $2x^3$ is $6x^2$. Write $\dfrac{dy}{dx}$ for the derivative of y. On the right side of the equation, $2y^3$ has derivative $6y^2 \cdot \dfrac{dy}{dx}$ and x has derivative 1.

Thus implicit differentiation of $2x^3 + y = 2y^3 + x$ yields

$$6x^2 + \frac{dy}{dx} = 6y^2 \cdot \frac{dy}{dx} + 1.$$

Solving for $\dfrac{dy}{dx}$,

$$(1 - 6y^2)\frac{dy}{dx} = 1 - 6x^2,$$

$$\frac{dy}{dx} = \frac{1 - 6x^2}{1 - 6y^2}.$$

13. To differentiate $x^3 y^2$, use the product rule, treating y as a function of x:

$$\frac{d}{dx} x^3 y^2 = x^3 \cdot \frac{d}{dx}(y^2) + y^2 \cdot \frac{d}{dx}(x^3)$$

$$= x^3 2y \cdot \frac{dy}{dx} + y^2 (3x^2)$$

$$= 2x^3 y \frac{dy}{dx} + 3x^2 y^2.$$

Hence implicit differentiation of $x^3 y^2 - 4x^2 = 1$ yields

$$2x^3 y \frac{dy}{dx} + 3x^2 y^2 - 8x = 0.$$

Solving for $\dfrac{dy}{dx}$,

$$2x^3 y \frac{dy}{dx} = 8x - 3x^2 y^2,$$

$$\frac{dy}{dx} = \frac{8x - 3x^2 y^2}{2x^3 y} = \frac{8 - 3xy^2}{2x^2 y}.$$

19. Implicit differentiation of $4y^3 - x^2 = -5$ yields

$$12y^2 \cdot \frac{dy}{dx} - 2x = 0.$$

Solving for $\dfrac{dy}{dx}$,

$$12y^2 \frac{dy}{dx} = 2x,$$

$$\frac{dy}{dx} = \frac{2x}{12y^2} = \frac{x}{6y^2}.$$

Hence,

$$\left. \frac{dy}{dx} \right|_{\substack{x=3 \\ y=1}} = \frac{3}{6(1)^2} = \frac{1}{2}.$$

25. A two-step procedure is required.
   (a) Find the slope at each point.
   (b) Use the point-slope formula to obtain the equation of the tangent line.

For (a), note that

$$\frac{d}{dx} x^2 y^4 = x^2 \cdot \frac{d}{dx} y^4 + y^4 \cdot \frac{d}{dx} x^2 = x^2 \cdot 4y^3 \frac{dy}{dx} + y^4 \cdot 2x.$$

Thus implicit differentiation of $x^2 y^4 = 1$ yields

$$4y^3 x^2 \frac{dy}{dx} + 2y^4 x = 0,$$

$$4y^3 x^2 \frac{dy}{dx} = -2y^4 x,$$

$$\frac{dy}{dx} = \frac{-2y^4 x}{4y^3 x^2} = \frac{-y}{2x}.$$

At the point $\left(4, \frac{1}{2}\right)$, the slope of the curve is given by

$$\left.\frac{dy}{dx}\right|_{\substack{x=4 \\ y=1/2}} = \frac{-\frac{1}{2}}{2(4)} = -\frac{1}{16}.$$

Hence the equation of the tangent line at $\left(4, \frac{1}{2}\right)$ is

$$y - \frac{1}{2} = -\frac{1}{16}(x-4).$$

At the point $\left(4, -\frac{1}{2}\right)$, the slope of the curve is given by

$$\left.\frac{dy}{dx}\right|_{\substack{x=4 \\ y=-1/2}} = \frac{-\left(-\frac{1}{2}\right)}{2(4)} = \frac{1}{16}.$$

Hence the equation of the tangent line at $\left(4, -\frac{1}{2}\right)$ is

$$y + \frac{1}{2} = \frac{1}{16}(x-4).$$

31. Differentiate $x^4 + y^4 = 1$ term by term. Since x is a function of t, the general power rule gives

$$\frac{d}{dx}x^4 = 4x^3 \cdot \frac{dx}{dt}.$$

Similarly,

$$\frac{d}{dt}y^4 = 4y^3 \cdot \frac{dy}{dt}.$$

And

$$\frac{d}{dx}(1) = 0.$$

Hence

$$4x^3 \frac{dx}{dt} + 4y^3 \frac{dy}{dt} = 0.$$

Solving for $\dfrac{dy}{dt}$,

$$4y^3 \frac{dy}{dt} = -4x^3 \frac{dx}{dt},$$

$$\frac{dy}{dt} = -\frac{x^3}{y^3}\frac{dx}{dt}.$$

37. First compute $\dfrac{dy}{dt}$. Differentiating $x^2 - 4y^2 = 9$ term by term,

$$2x\frac{dx}{dt} - 8y\frac{dy}{dt} = 0.$$

Solving for $\dfrac{dy}{dt}$,

$$-8y\frac{dy}{dt} = -2x\frac{dx}{dt},$$

$$\frac{dy}{dt} = \frac{x}{4y}\frac{dx}{dt}.$$

The problem says that at the point $(5, -2)$, the x-coordinate is increasing at the rate of 3 units per second, that is $\dfrac{dx}{dt} = 3$. Hence,

$$\frac{dy}{dt} = \frac{5}{4(-2)}(3) = -\frac{15}{8}$$

The y-coordinate is decreasing at $\dfrac{15}{8}$ units per second.

43. $\dfrac{d}{dt}P^5V^7 = P^5 \cdot \dfrac{d}{dt}V^7 + V^7 \cdot \dfrac{d}{dt}P^5$

$\qquad = P^5 7V^6 \dfrac{dV}{dt} + V^7 5P^4 \dfrac{dP}{dt}.$

$\dfrac{d}{dx}(k) = 0$ (since k is a constant).

Hence $7P^5V^6 \dfrac{dV}{dt} + 5P^4V^7 \dfrac{dP}{dt} = 0$. Solving for $\dfrac{dV}{dt}$,

$$7P^5V^6\frac{dV}{dt} = -5P^4V^7\frac{dP}{dt},$$

$$\frac{dV}{dt} = \frac{-5P^4V^7}{7P^5V^6}\frac{dP}{dt},$$

$$= \frac{-5V}{7P}\frac{dP}{dt}.$$

The problem says that when $V = 4$ liters, $P = 200$ units and $\dfrac{dP}{dt} = 5$ units per second. Hence

$$\frac{dV}{dt} = \frac{-5(4)}{7(200)}(5) = -\frac{1}{14}.$$

Therefore the volume is decreasing at the rate of $\frac{1}{14}$ liter per second.

Chapter 3 Review Exercises

The techniques of differentiation presented in the first two sections of this chapter must be mastered. They will be used throughout the text.

You may need to review curve sketching and optimization problems because these problems can be more difficult when they involve the product rule or quotient rule. On an exam that covers both Chapters 2 and 3 (or on the final exam) you may see a problem similar to Exercises 45 and 46 in Section 3.1 or Exercises 27 and 28 in Section 3.2.

While solutions for all review exercises are included, expanded explanations are included for every sixth exercise.

1. $\dfrac{d}{dx}[(4x-1)(3x+1)^4] = (4x-1)\cdot\dfrac{d}{dx}(3x+1)^4 + (3x+1)^4\cdot\dfrac{d}{dx}(4x-1)$ [Product Rule]

$$= (4x-1)4(3x+1)^3\cdot(3) + (3x+1)^4\cdot(4).$$

Factor $4(3x+1)^3$ from each term to obtain

$$\frac{d}{dx}[(4x-1)(3x+1)^4] = 4(3x+1)^3[3(4x-1)+(3x+1)]$$
$$= 4(3x+1)^3[12x-3+3x+1]$$
$$= 4(3x+1)^3(15x-2).$$

2. $\dfrac{d}{dx}\left[2(5-x)^3(6x-1)\right] = 2\left[(5-x)^3(6)+(6x-1)3(5-x)^2(-1)\right] = 2\left[-3(5-x)^2(6x-1)+6(5-x)^3\right]$

$$= 6\left[2(5-x)^3-(5-x)^2(6x-1)\right] = 6(5-x)^2\left[10-2x-6x+1\right] = 6(5-x)^2(11-8x)$$

3. $\dfrac{d}{dx}\left[x(x^5-1)^3\right] = (x)(3)(x^5-1)^2(5x^4)+(x^5-1)^3(1) = (x^5-1)^2(16x^5-1)$

4. $\dfrac{d}{dx}\left[(2x+1)^{5/2}(4x-1)^{3/2}\right] = (2x+1)^{5/2}\left(\dfrac{3}{2}\right)(4x-1)^{1/2}(4)+(4x-1)^{3/2}\left(\dfrac{5}{2}\right)(2x+1)^{3/2}(2)$

$$= 5(2x+1)^{3/2}(4x-1)^{3/2}+6(4x-1)^{1/2}(2x+1)^{5/2}$$
$$= (2x+1)^{3/2}(4x-1)^{1/2}\left[5(4x-1)+6(2x+1)\right]$$
$$= (2x+1)^{3/2}(4x-1)^{1/2}(32x+1)$$

5. $\dfrac{d}{dx}\left[5\left(\sqrt{x}-1\right)^4\left(\sqrt{x}-2\right)^2\right]=5\left[\left(\sqrt{x}-1\right)^4(2)\left(\sqrt{x}-2\right)\left(\dfrac{1}{2}x^{-1/2}\right)+\left(\sqrt{x}-2\right)^2(4)\left(\sqrt{x}-1\right)^3\left(\dfrac{1}{2}x^{-1/2}\right)\right]$

$$=5\left[2x^{-1/2}\left(\sqrt{x}-1\right)^3\left(\sqrt{x}-2\right)^2+x^{-1/2}\left(\sqrt{x}-2\right)\left(\sqrt{x}-1\right)^4\right]$$

$$=5x^{-1/2}\left(\sqrt{x}-1\right)^3\left(\sqrt{x}-2\right)\left[2\left(\sqrt{x}-2\right)+\left(\sqrt{x}-1\right)\right]$$

$$=\dfrac{5}{\sqrt{x}}\left(\sqrt{x}-1\right)^3\left(\sqrt{x}-2\right)\left(3\sqrt{x}-5\right)$$

6. $\dfrac{d}{dx}\left[\dfrac{\sqrt{x}}{\sqrt{x}+4}\right]=\dfrac{\left(\sqrt{x}+4\right)\left(\frac{1}{2\sqrt{x}}\right)-\sqrt{x}\left(\frac{1}{2\sqrt{x}}\right)}{\left(\sqrt{x}+4\right)^2}=\dfrac{\frac{1}{2\sqrt{x}}\left(\sqrt{x}+4-\sqrt{x}\right)}{\left(\sqrt{x}+4\right)^2}=\dfrac{2}{\sqrt{x}\left(\sqrt{x}+4\right)^2}$

7. If $y=3(x^2-1)^3(x^2+1)^5$, then

$$\dfrac{dy}{dx}=3(x^2-1)^3\cdot\dfrac{d}{dx}(x^2+1)^5+(x^2+1)^5\cdot\dfrac{d}{dx}3(x^2-1)^3 \qquad \text{[Product Rule]}$$

$$=3(x^2-1)^3\cdot5(x^2+1)^4(2x)+(x^2+1)^5\cdot9(x^2-1)^2(2x).$$

Factor $6x(x^2-1)^2(x^2+1)^4$ from each term to obtain

$$\dfrac{dy}{dx}=6x(x^2-1)^2(x^2+1)^4[5(x^2-1)+3(x^2+1)]$$

$$=6x(x^2-1)^2(x^2+1)^4(8x^2-2)$$

$$=12x(x^2-1)^2(x^2+1)^4(4x^2-1).$$

8. $\dfrac{d}{dx}\left[\dfrac{1}{(x^2+5x+1)^6}\right]=\dfrac{(x^2+5x+1)^6(0)-(1)(6)(x^2+5x+1)^5(2x+5)}{(x^2+5x+1)^{12}}=\dfrac{-12x-30}{(x^2+5x+1)^7}$

9. $\dfrac{d}{dx}\left[\dfrac{x^2-6x}{x-2}\right]=\dfrac{(x-2)(2x-6)-(x^2-6x)(1)}{(x-2)^2}=\dfrac{2x^2-10x+12-x^2+6x}{(x-2)^2}=\dfrac{x^2-4x+12}{(x-2)^2}$

10. $\dfrac{d}{dx}\left[\dfrac{2x}{2-3x}\right]=\dfrac{(2-3x)(2)-(2x)(-3)}{(2-3x)^2}=\dfrac{4-6x+6x}{(2-3x)^2}=\dfrac{4}{(2-3x)^2}$

11. $\dfrac{d}{dx}\left[\left(\dfrac{3-x^2}{x^3}\right)^2\right]=2\left(\dfrac{3-x^2}{x^3}\right)\left(\dfrac{(x^3)(-2x)-(3-x^2)(3x^2)}{x^6}\right)=2\left(\dfrac{3-x^2}{x^3}\right)\left(\dfrac{-2x^4-9x^2+3x^4}{x^6}\right)$

$$=2\left(\dfrac{3-x^2}{x^3}\right)\left(\dfrac{x^4-9x^2}{x^6}\right)=2\left(\dfrac{3-x^2}{x^3}\right)\left(\dfrac{x^2-9}{x^4}\right)=\dfrac{2(3-x^2)(x^2-9)}{x^7}$

12. $\dfrac{d}{dx}\left[\dfrac{x^3+x}{x^2-x}\right]=\dfrac{(x^2-x)(3x^2+1)-(x^3+x)(2x-1)}{(x^2-x)^2}=\dfrac{3x^4-3x^3+x^2-x-2x^4-2x^2+x^3+x}{(x^2-x)^2}$

$$=\dfrac{x^4-2x^3-x^2}{x^4-2x^3+x^2}=\dfrac{x^2-2x-1}{x^2-2x+1}=\dfrac{x^2-2x-1}{(x-1)^2}$$

13. If $f(x)=(3x+1)^4(3-x)^5$, then

$$f'(x)=(3x+1)^4\cdot\frac{d}{dx}(3-x)^5+(3-x)^5\cdot\frac{d}{dx}(3x+1)^4 \qquad \text{[Product Rule]}$$

$$=(3x+1)^4\cdot 5(3-x)^4(-1)+(3-x)^5\cdot 4(3x+1)^3(3)$$

$$=(3x+1)^3(3-x)^4[-5(3x+1)+12(3-x)]$$

$$=(3x+1)^3(3-x)^4(-27x+31).$$

Set $f'(x)=0$ and solve for x:

$$(3x+1)^3(3-x)^4(-27x+31)=0.$$

$$x=-\frac{1}{3}, \quad x=3, \quad \text{or} \quad x=\frac{31}{27}.$$

14. $f(x)=\dfrac{x^2+1}{x^2+5}$

$f'(x)=\dfrac{(x^2+5)(2x)-(x^2+1)(2x)}{(x^2+5)^2}$

$\quad=\dfrac{2x^3+10x-2x^3-2x}{(x^2+5)^2}=\dfrac{8x}{(x^2+5)^2}$

Let $f'(x)=0$ and solve for x.

$\dfrac{8x}{(x^2+5)^2}=0\Rightarrow x=0$

15. $y=(x^3-1)(x^2+1)^4$

slope $=y'$

$y'=(x^3-1)(4)(x^2+1)^3(2x)+(x^2+1)^4(3x^2)$

$\quad=3x^2(x^2+1)^4+8x(x^2+1)^3(x^3-1)$

When $x=-1$,

slope $=3(-1)^2(1+1)^4+8(-1)(1+1)^3(-1-1)$

$\quad=48+128=176$

When $x=-1$, $y=(-1-1)(1+1)^4=-32$.

Let $(x_1,\,y_1)=(-1,\,-32)$. Then

$y+32=176(x+1)$ or $y=176x+144$.

16. $y=\dfrac{x-3}{\sqrt{4+x^2}}$

slope $=y'$

$y'=\dfrac{(4+x^2)^{1/2}(1)-(x-3)(\frac{1}{2})(4+x^2)^{-1/2}(2x)}{4+x^2}$

$\quad=\dfrac{(4+x^2)^{1/2}}{4+x^2}-\dfrac{(x^2-3x)}{(4+x^2)^{3/2}}$

$\quad=\dfrac{4+x^2-x^2+3x}{(4+x^2)^{3/2}}=\dfrac{3x+4}{(4+x^2)^{3/2}}$

When $x=0$, $y=-\dfrac{3}{2}$. When $x=0$,

slope $=\dfrac{4}{8}=\dfrac{1}{2}$. Let $(x_1,\,y_1)=\left(0,\,-\dfrac{3}{2}\right)$.

Then

$y+\dfrac{3}{2}=\dfrac{1}{2}(x-0)$ or $y=\dfrac{1}{2}x-\dfrac{3}{2}$.

17. The objective is
$A = 2y(2) + (x-4)(2) = 4y + 2x - 8.$ The constraint is $(x-4)(y-2) = 800$ or
$$y = \frac{800}{x-4} + 2.$$
Thus,
$$A(x) = 4\left(\frac{800}{x-4} + 2\right) + 2x - 8 = \frac{3200}{x-4} + 2x$$
$$A'(x) = -\frac{3200}{(x-4)^2} + 2 \Rightarrow A'(44) = 0$$
The minimum value of $A(x)$ for $x > 0$ occurs at $x = 44$. Thus, the optimal values of x and y are $x = 44$ m,
$$y = \frac{800}{44-4} + 2 = 22 \text{ m}.$$

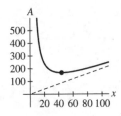

18. The objective is
$A = 2(x-4)(2) + 2(y)(2) = 4x - 16 + 4y$ and the constraint is $(x-4)(y-4) = 800$ or
$$y = \frac{800}{x-4} + 4.$$
$$A(x) = 4x + 4\left(\frac{800}{x-4} + 4\right) - 16 = 4x + \frac{3200}{x-4}$$
$$A'(x) = 4 - \frac{3200}{(x-4)^2} \Rightarrow A'\left(4 + 20\sqrt{2}\right) = 0$$
The minimum value of $A(x)$ occurs at $x = 4 + 20\sqrt{2}$. The optimal dimensions are $x = y = 4 + 20\sqrt{2} \approx 32.3$ meters.

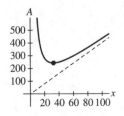

19. We want to evaluate $\dfrac{dC}{dt}$. The problem tells us that daily sales are rising at the rate of three lamps per day. This means $\dfrac{dx}{dt} = 3$.

Now if $C = 40x + 30$, then $\dfrac{dC}{dx} = 40$.

Hence by the chain rule,
$$\frac{dC}{dt} = \frac{dC}{dx} \cdot \frac{dx}{dt} = 40(3) = 120.$$
Costs are rising by \$120 per day.

20. $\dfrac{dy}{dt} = \dfrac{dy}{dP} \cdot \dfrac{dP}{dt}$

21. $f(1) = 3,\; f'(1) = \dfrac{1}{2},\; g(1) = 2,\; g'(1) = \dfrac{3}{2}$
$$h(1) = 2f(1) - 3g(1) = 2(3) - 3(2) = 0$$
$$h'(1) = 2f'(1) - 3g'(1) = 2\left(\frac{1}{2}\right) - 3\left(\frac{3}{2}\right) = -\frac{7}{2}$$

22. $h(x) = f(x)g(x),$
$$h'(x) = f(x)g'(x) + g(x)f'(x)$$
$$h(1) = 3(2) = 6;$$
$$h'(1) = 3\left(\frac{3}{2}\right) + (2)\left(\frac{1}{2}\right) = \frac{11}{2}$$

23. $h(x) = \dfrac{f(x)}{g(x)},$
$$h'(x) = \frac{g(x)f'(x) - g'(x)f(x)}{[g(x)]^2},$$
$$h(1) = \frac{3}{2},\; h'(x) = \frac{2\left(\frac{1}{2}\right) - 3\left(\frac{3}{2}\right)}{2^2} = -\frac{7}{8}$$

24. $h(x) = [f(x)]^2,\; h'(x) = 2f(x)f'(x),$
$$h(1) = 3^2 = 9,\; h'(1) = 2(3)\left(\frac{1}{2}\right) = 3$$

25. $h(x) = f(g(x))$, $h(1) = f(g(1))$. From the graph of $g(x)$, you can see that $g(1) = 2$, so $h(1) = f(2)$. From the graph of $f(x)$, you can see that $f(2) = 1$, so $h(1) = f(2) = 1$. Next, by the Chain Rule,

$$h'(x) = f'(g(x))g'(x)$$
$$h'(1) = f'(g(1))g'(1)$$
$$= f'(2)g'(1).$$

The slope of the tangent line at $x = 2$ on the graph of $f(x)$ is -1. This is clear from an examination of the grid background in Figure 2. Therefore, $f'(2) = -1$. Similarly, Figure 2 shows that the slope of the tangent line at $x = 1$ on the graph of $g(x)$ is $\frac{3}{2}$, so

$$g'(1) = \frac{3}{2},$$

$$h'(1) = (-1)\frac{3}{2} = -\frac{3}{2}.$$

26. $h(x) = g(f(x))$, $h(x) = g'(f(x))f'(x)$

$$f(1) = 3, \ f'(1) = \frac{1}{2}, \ g(3) = 1, \ g'(3) = -\frac{1}{2}$$

$$h(1) = 1, \ h'(1) = g'(3)f'(1) = \left(-\frac{1}{2}\right)\left(\frac{1}{2}\right) = -\frac{1}{4}$$

27. $g(x) = x^3$, $g'(x) = 3x^2$, $\dfrac{d}{dx}f(g(x)) = \dfrac{1}{x^6 + 1}(3x^2) = \dfrac{3x^2}{x^6 + 1}$

28. $g(x) = \dfrac{1}{x}$, $g'(x) = -\dfrac{1}{x^2}$

$$\frac{d}{dx}f(g(x)) = \frac{1}{\left(\frac{1}{x}\right)^2 + 1}\left(-\frac{1}{x^2}\right) = \left(\frac{x^2}{1 + x^2}\right)\left(-\frac{1}{x^2}\right) = -\frac{1}{1 + x^2}, \ x \neq 0$$

29. $g(x) = x^2 + 1$, $g'(x) = 2x$

$$\frac{d}{dx}f(g(x)) = \frac{1}{(x^2 + 1)^2 + 1}(2x) = \frac{2x}{(x^2 + 1)^2 + 1}$$

30. $g(x) = x^2$, $g'(x) = 2x$

$$\frac{d}{dx}f(g(x)) = x^2\sqrt{1 - x^4}(2x) = 2x^3\sqrt{1 - x^4}$$

31. If $g(x) = \sqrt{x} = x^{1/2}$, then $g'(x) = \frac{1}{2}x^{-1/2}$. Since $f'(x) = x\sqrt{1 - x^2}$,

$$\frac{d}{dx}f(g(x)) = f'(g(x))g'(x) = (x^{1/2})\sqrt{1 - x}\left(\frac{1}{2}x^{-1/2}\right)$$

$$= \frac{1}{2}\sqrt{1 - x}.$$

32. $g(x) = x^{3/2}, \; g'(x) = \dfrac{3}{2}x^{1/2}$

$$\frac{d}{dx}f(g(x)) = x^{3/2}\sqrt{1-x^3}\left(\frac{3}{2}x^{1/2}\right)$$

$$= \frac{3}{2}x^2\sqrt{1-x^3}$$

33. $\dfrac{dy}{du} = \dfrac{u}{u^2+1}, \; u = x^{3/2}, \; \dfrac{du}{dx} = \dfrac{3}{2}x^{1/2},$

$$\frac{dy}{dx} = \frac{dy}{du}\frac{du}{dx} = \left(\frac{u}{u^2+1}\right)\frac{3}{2}x^{1/2}$$

Substitute $x^{3/2}$ for u.

$$\frac{dy}{dx} = \frac{x^{3/2}}{x^3+1}\left(\frac{3}{2}x^{1/2}\right) = \frac{3x^2}{2(x^3+1)}$$

34. $\dfrac{dy}{du} = \dfrac{u}{u^2+1}, \; u = x^2+1, \; \dfrac{du}{dx} = 2x,$

$$\frac{dy}{dx} = \left(\frac{u}{u^2+1}\right)(2x)$$

Substitute (x^2+1) for u.

$$\frac{dy}{dx} = \frac{(x^2+1)}{(x^2+1)^2+1}(2x) = \frac{2x(x^2+1)}{(x^2+1)^2+1}$$

35. $\dfrac{dy}{du} = \dfrac{u}{u^2+1}, \; u = \dfrac{5}{x}, \; \dfrac{du}{dx} = -\dfrac{5}{x^2}$

$$\frac{dy}{dx} = \frac{\frac{5}{x}}{\frac{25}{x^2}+1}\left(-\frac{5}{x^2}\right) = -\frac{25}{x(25+x^2)}$$

36. $\dfrac{dy}{du} = \dfrac{u}{\sqrt{1+u^4}}, \; u = x^2, \; \dfrac{du}{dx} = 2x$

$$\frac{dy}{dx} = \frac{x^2}{\sqrt{1+x^8}}(2x) = \frac{2x^3}{\sqrt{1+x^8}}$$

37. If $u = \sqrt{x} = x^{1/2},$ then

$$\frac{du}{dx} = \frac{1}{2}x^{-1/2}.$$

Since

$$\frac{dy}{du} = \frac{u}{\sqrt{1+u^4}},$$

$$\frac{du}{dx} = \frac{dy}{du}\cdot\frac{du}{dx} = \frac{u}{\sqrt{1+u^4}}\left(\frac{1}{2}x^{-1/2}\right).$$

To express $\dfrac{dy}{dx}$ as a function of x alone, substitute $x^{1/2}$ for u to obtain

$$\frac{dy}{dx} = \frac{x^{1/2}}{(1+x^2)^{1/2}}\left(\frac{1}{2}x^{-1/2}\right) = \frac{1}{2\sqrt{1+x^2}}.$$

38. $\dfrac{dy}{du} = \dfrac{u}{\sqrt{1+u^4}}, \; u = \dfrac{2}{x}, \; \dfrac{du}{dx} = \dfrac{-2}{x^2}$

$$\frac{dy}{dx} = \frac{\frac{2}{x}}{\sqrt{1+\frac{16}{x^4}}}\left(-\frac{2}{x^2}\right) = -\frac{4}{x^3\sqrt{1+\frac{16}{x^4}}}$$

39. (a) $\dfrac{dR}{dA}, \dfrac{dA}{dt}, \dfrac{dR}{dx}, \dfrac{dx}{dA}$

(b) $\dfrac{dR}{dt} = \dfrac{dR}{dx}\dfrac{dx}{dA}\dfrac{dA}{dt}$

40. (a) $\dfrac{dP}{dt}, \dfrac{dA}{dP}, \dfrac{dS}{dP}, \dfrac{dA}{dS}$

(b) $\dfrac{dA}{dt} = \dfrac{dA}{dS}\cdot\dfrac{dS}{dP}\cdot\dfrac{dP}{dt}$

41. $x^{2/3} + y^{2/3} = 8$

(a) $\dfrac{2}{3}x^{-1/3} + \dfrac{2}{3}y^{-1/3}\dfrac{dy}{dx} = 0$

$$\frac{dy}{dx} = -\frac{2}{3}x^{-1/3}\cdot\frac{3}{2}y^{1/3} = -\frac{y^{1/3}}{x^{1/3}}$$

(b) $\text{slope} = \dfrac{dy}{dx}$

When $x = 8, y = -8, \text{slope} = 1.$

42. $x^3 + y^3 = 9xy$

(a) $3x^2 + 3y^2\dfrac{dy}{dx} = 9x\dfrac{dy}{dx} + 9y$

$$\frac{dy}{dx} = \frac{9y-3x^2}{3y^2-9x} = \frac{3y-x^2}{y^2-3x} \text{ or } \frac{x^2-3y}{3x-y^2}$$

(b) $\text{slope} = \dfrac{dy}{dx}$

When $x = 2, y = 4,$

$$\text{slope} = \frac{3(4)-4}{16-6} = \frac{4}{5}.$$

43. By the product rule,

$$\frac{d}{dx} x^2 y^2 = x^2 \cdot \frac{d}{dx} y^2 + y^2 \cdot \frac{d}{dx} x^2$$

$$= x^2 2y \frac{dy}{dx} + y^2 2x.$$

Hence, implicit differentiation of $x^2 y^2 = 9$ yields

$$2x^2 y \frac{dy}{dx} + 2xy^2 = 0.$$

Solving for $\frac{dy}{dx}$,

$$2x^2 y \frac{dy}{dx} = -2xy^2$$

$$\frac{dy}{dx} = \frac{-2xy^2}{2x^2 y} = -\frac{y}{x}.$$

If $x = 1$ and $y = 3$, then $\frac{dy}{dx} = -\frac{3}{1} = -3.$

44. $xy^4 = 48$

$$y^4 + 4y^3 x \frac{dy}{dx} = 0 \Rightarrow \frac{dy}{dx} = -\frac{y^4}{4y^3 x} = -\frac{y}{4x}$$

When $x = 3$, $y = 2$, $\frac{dy}{dx} = \frac{-2}{4(3)} = -\frac{1}{6}.$

45. $x^2 - xy^3 = 20$

$$2x - \left(y^3 + 3y^2 x \frac{dy}{dx} \right) = 0 \Rightarrow \frac{dy}{dx} = \frac{2x - y^3}{3y^2 x}$$

When $x = 5$, $y = 1$, $\frac{dy}{dx} = \frac{10 - 1}{3(1)(5)} = \frac{3}{5}.$

46. $xy^2 - x^3 = 10$

$$\left(y^2 + 2yx \frac{dy}{dx} \right) - 3x^2 = 0 \Rightarrow \frac{dy}{dx} = \frac{3x^2 - y^2}{2yx}$$

When $x = 2$, $y = 3$, $\frac{dy}{dx} = \frac{3(4) - 9}{2(3)(2)} = \frac{1}{4}.$

47. $y^2 - 5x^3 = 4$

(a) $2y \frac{dy}{dx} - 15x^2 = 0 \Rightarrow \frac{dy}{dx} = \frac{15x^2}{2y}$

(b) When $x = 4$ and $y = 18$,
$$\frac{dy}{dx} = \frac{15(16)}{2(18)} = \frac{20}{3} \text{ (thousand dollars per thousand-unit increase in production)}.$$

(c) $\dfrac{dy}{dt} = \dfrac{dy}{dx}\dfrac{dx}{dt} = \dfrac{15x^2}{2y}\dfrac{dx}{dt}$

(d) When $x = 4$, $y = 18$, and $\dfrac{dx}{dt} = .3$, $\dfrac{dy}{dt} = \dfrac{15(16)}{2(18)}(.3) = 2$ (thousand dollars per week).

48. $y^3 - 8000x^2 = 0$

(a) $3y^2\dfrac{dy}{dx} - 16{,}000x = 0 \Rightarrow \dfrac{dy}{dx} = \dfrac{16{,}000x}{3y^2}$

(b) When $x = 27$, $y = 180$,
$$\frac{dy}{dx} = \frac{16{,}000(27)}{3(32{,}400)} = \frac{40}{9} \approx 4.44 \text{ (thousand books per person)}.$$

(c) $\dfrac{dy}{dt} = \dfrac{dy}{dx}\dfrac{dx}{dt} = \dfrac{16{,}000x}{3y^2}\dfrac{dx}{dt}$

(d) When $x = 27$, $y = 180$, and $\dfrac{dx}{dt} = 1.8$,

$$\frac{dy}{dt} = \frac{16{,}000(27)}{3(32{,}400)}(1.8) = 8 \text{ (thousand books per year)}.$$

49. First, to find $\dfrac{dx}{dt}$, differentiate $6p + 5x + xp = 50$ term by term, with respect to t, using the product rule on xp.

$$6\frac{dp}{dt} + 5\frac{dx}{dt} + x\frac{dp}{dt} + p\frac{dx}{dt} = 0.$$

Solving for $\dfrac{dx}{dt}$,

$$5\frac{dx}{dt} + p\frac{dx}{dt} = -6\frac{dp}{dt} - x\frac{dp}{dt},$$
$$(5 + p)\frac{dx}{dt} = -(6 + x)\frac{dp}{dt},$$
$$\frac{dx}{dt} = -\left(\frac{6 + x}{5 + p}\right)\frac{dp}{dt}.$$

If $x = 4$, $p = 3$, and $\dfrac{dp}{dt} = -2$, then

$$\frac{dx}{dt} = -\left(\frac{6 + 4}{5 + 3}\right)(-2) = \frac{10}{4} = \frac{5}{2} = 2.5.$$

Therefore, the quantity x is increasing at the rate of 2.5 units per unit time.

50. $V = .005\pi r^2$, so $\dfrac{dV}{dt} = .005\pi(2)r\dfrac{dr}{dt} = .01\pi r\dfrac{dr}{dt}$.

Now, $\dfrac{dV}{dt} = 20$, so $20 = .01\pi r\dfrac{dr}{dt} \Rightarrow \dfrac{dr}{dt} = \dfrac{2000}{\pi r}$.

When $r = 50$, $\dfrac{dr}{dt} = \dfrac{2000}{50\pi} = \dfrac{40}{\pi}$ m/hr.

51. $S = .1W^{2/3}$, so $\dfrac{dS}{dt} = \dfrac{2}{3}(.1)W^{-1/3}\dfrac{dW}{dt} = \dfrac{.2}{3}W^{-1/3}\dfrac{dW}{dt}$.

When $W = 350$ and $\dfrac{dW}{dt} = 200$, $\dfrac{dS}{dt} = \dfrac{.2}{3\sqrt[3]{350}}(200) = \dfrac{40}{3\sqrt[3]{350}} \approx 1.89$ m^2/yr.

52. $xy - 6x + 20y = 0$, $\dfrac{dx}{dt}y + \dfrac{dy}{dt}x - 6\dfrac{dx}{dt} + 20\dfrac{dy}{dt} = 0$

Currently, $x = 10$, $y = 2$, $\dfrac{dx}{dt} = 1.5$. Thus, $1.5(2) + \dfrac{dy}{dt}(10) - 6(1.5) + 20\dfrac{dy}{dt} = 0$, so

$30\dfrac{dy}{dt} = 6 \Rightarrow \dfrac{dy}{dt} = .2$ or 200 dishwashers/month.

Chapter 4
The Exponential and Natural Logarithm Functions

4.1 Exponential Functions

The exercises in Section 4.1 focus on the specific skills you will use later in Chapter 4. If you need more practice, go back to Section 0.5.

It will be helpful to make a table that lists the laws of exponents shown on page 227 and to add to it the "reverse" laws listed in the *Manual* notes for Section 0.5. (Use x and y in the exponents now instead of r and s.) You definitely need to memorize this enlarged list of laws. Working scores of problems will help, of course, because you will have to refer to your list. Here is an additional method. Write only the left side of each law in the table on page 227. Maybe mix up the order of the laws. On a separate sheet of paper, write the left side of the laws listed in the *Manual* for Section 0.5 (with x, y in place of r, s). Put these partial lists aside for a day or two while you work exercises from Section 4.1. Later, try to fill in the lists of laws without any help. You may need to do this more than once.

1. $4^x = (2^2)^x = 2^{2x}$,

   $(\sqrt{3})^x = (3^{1/2})^x = 3^{(1/2)x}$,

   $\left(\dfrac{1}{9}\right)^x = (3^{-2})^x = 3^{-2x}$.

7. $6^x \cdot 3^{-x} = (2 \cdot 3)^x \cdot 3^{-x} = 2^x \cdot 3^x \cdot 3^{-x} = 2^x \cdot 3^0 = 2^x \cdot 1 = 2^x$,

   $\dfrac{15^x}{5^x} = \left(\dfrac{15}{5}\right)^x = 3^x$,   or $\dfrac{15^x}{5^x} = (3 \cdot 5)^x \cdot 5^{-x} = 3^x \cdot 5^x \cdot 5^{-x} = 3^x$,

   $\dfrac{12^x}{2^{2x}} = 12x \cdot 2^{-2x} = (3 \cdot 2^2)^x \cdot 2^{-2x} = 3^x \cdot 2^{2x} \cdot 2^{-2x} = 3^x \cdot 2^0 = 3^x$,   or $\dfrac{12^x}{2^{2x}} = \dfrac{12^x}{4^x} = \left(\dfrac{12}{4}\right)^x = 3^x$.

13. $(2^{-3x} \cdot 2^{-2x})^{2/5} = (2^{-5x})^{2/5} = 2^{(-5x)(2/5)} = 2^{-2x}$,

   $(9^{1/2} \cdot 9^4)^{x/9} = 9^{(9/2)(x/9)} = 9^{x/2} = (3^2)^{x/2} = 3^x$,

   or $(9^{1/2} \cdot 9^4)^{x/9} = (3 \cdot (3^2)^4)^{x/9} = (3^9)^{x/9} = 3^x$.

19. If $(2.5)^{2x+1} = (2.5)^5$, then equating exponents,

$$2x + 1 = 5 \Rightarrow 2x = 4 \Rightarrow x = 2.$$

25. If $(2^{x+1} \cdot 2^{-3})^2 = 2$, then

$$(2^{(x+1)-3})^2 = 2, \quad (2^{x-2})^2 = 2, \quad 2^{2x-4} = 2^1.$$

Equating exponents, we have $2x - 4 = 1$ and hence $x = \dfrac{5}{2}$.

31. If $2^x - \dfrac{8}{2^{2x}} = 0$, then

$$2^x = \frac{8}{2^{2x}}$$
$$2^x \cdot 2^{2x} = 8$$
$$2^{x+2x} = 8$$
$$2^{3x} = 8.$$

But $8 = 2^3$, thus $2^{3x} = 2^3$. Equating exponents,

$$3x = 3, \quad x = 1.$$

37. Since $2^{3+h} = 2^3(2^h)$, the missing factor is 2^h.

43. Set $Y_1 = 2\text{\textasciicircum}X$. Review the **INCORPORATING TECHNOLOGY** material in section 1.2 (page 70) to approximate the slope of a graph at a point.

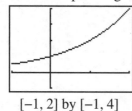
[−1, 2] by [−1, 4]

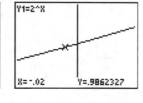

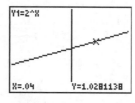

Alternatively, if Y_4 is the derivative of Y_1, then evaluate $Y_4(0)$, as described in **INCORPORATING TECHNOLOGY** in section 0.2 (and apply the steps to Y_4 in place of Y_1). You should obtain .69314724.

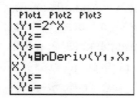

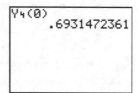

45. Apply the methods for Exercise 43 to $b\text{\textasciicircum}X$ in place of $2\text{\textasciicircum}X$. Try different numbers for b, such as 2.5 and 3.0. By trial and error you should obtain $b = 2.7$.

4.2   The Exponential Function e^x

The purpose of this brief section and its exercise is to introduce the function e^x and to help you get used to working with it. Here are the main facts about e^x. The number e is between 2 and 3; the graph of the function e^x is similar to the graphs of 2^x and 3^x and lies between them. The number e is chosen so that the slope of $y = e^x$ is 1 at $x = 0$. The slope of $y = e^x$ at an arbitrary point (x, e^x) on the graph has the same numerical value as the y-coordinate of the point. That is

$$[\text{slope at } (x, e^x)] = [\,y\text{-coordinate of}\,(x, e^x)]$$

$$\frac{d}{dx}e^x = e^x.$$

See Figure 3 on page 232.

1. If $y = 3^x$, then the slope of the secant line through $(0, 1)$ and $(h, 3^h)$ is

$$\frac{3^h - 1}{h}.$$

As h approaches zero, the slope of the secant line approaches the slope of $y = 3^x$ at $x = 0$; that is,

$$\left.\frac{d}{dx}3^x\right|_{x=0}.$$

$$\text{If } h = .1, \quad\quad \text{then} \quad\quad \frac{3^h - 1}{h} = \frac{1.11612 - 1}{.1} = 1.1612.$$

$$\text{If } h = .01, \quad\quad \text{then} \quad\quad \frac{3^h - 1}{h} = \frac{1.01105 - 1}{.01} = 1.105.$$

$$\text{If } h = .001, \quad\quad \text{then} \quad\quad \frac{3^h - 1}{h} = \frac{1.00110 - 1}{.001} = 1.10.$$

$$\text{Therefore, } \left.\frac{d}{dx}3^x\right|_{x=0} = \lim_{x \to 0} \frac{3^h - 1}{h} \approx 1.1.$$

Helpful Hint: The calculations in Exercise 1 can be automated in various ways on a graphing calculator. Even if such calculators are not required in your course, you should consider learning how to perform simple computations. For instance, set $Y_1 = (3\,^\wedge\,X - 1)/X$ and then evaluate $Y_1(.1)$, $Y_1(.01)$, and $Y_1(.001)$. Or, in the home screen, compute $(3\,^\wedge\,.1 - 1)/.1$. Then use the key that recalls the last computation and edit the command, replacing .1 with .01. Repeat, with .001 in place of .01

7. $(e^2)^x = e^{2x} \Rightarrow k = 2.$ $\quad\left(\dfrac{1}{e}\right)^x = (e^{-1}) = e^{-x} \Rightarrow k = -1.$

13. To solve $e^{5x} = e^{20}$, equate exponents:

$$5x = 20, \quad x = 4.$$

19. The tangent passes through the point $\left(-1, e^{-1}\right) = \left(-1, \dfrac{1}{e}\right)$, or $(-1, .37)$. The slope of the tangent is given by

$$\frac{dy}{dx}\bigg|_{x=-1} = e^x\big|_{x=-1} = \frac{1}{e} \approx .37.$$

Thus, the equation of the tangent in point-slope form is

$$y - \frac{1}{e} = \frac{1}{e}(x+1) \quad \text{or} \quad y - .37 = .37(x+1).$$

25. $\dfrac{d}{dx}\left(3e^x - 7x\right) = \dfrac{d}{dx}\left(3e^x\right) - \dfrac{d}{dx}(7x) = 3e^x - 7$

31. By the quotient rule,

$$\frac{d}{dx}\frac{e^x}{x+1} = \frac{(x+1)e^x - e^x \cdot 1}{(x+1)^2} = \frac{xe^x}{(x+1)^2}.$$

37. The tangent line is horizontal when $\dfrac{dy}{dx} = 0$. Using the product rule we find

$$\frac{dy}{dx} = (1+x^2)e^x + e^x(2x)$$

$$= e^x(x^2 + 2x + 1).$$

$\dfrac{dy}{dx} = 0$ when $x^2 + 2x + 1 = (x+1)^2 = 0$, or when $x = -1$. To find the y-coordinate, substitute $x = -1$ into the equation:

$$y = (1 + (-1)^2)e^{-1} = 2e^{-1}$$

Thus, the tangent line is horizontal at the point $(-1, 2e^{-1})$.

43. $f(x) = e^x(1+x)^2$. Use the product rule along with the chain rule to find $f'(x)$:

$$f'(x) = e^x 2(1+x)(1) + (1+x)^2 e^x$$

$$= e^x(2 + 2x + 1 + 2x + x^2)$$

$$= e^x(x^2 + 4x + 3).$$

Use the product rule to find $f''(x)$:

$$f''(x) = e^x(2x+4) + (x^2 + 4x + 3)e^x$$

$$= e^x(2x + 4 + x^2 + 4x + 3)$$

$$= e^x(x^2 + 6x + 7).$$

49.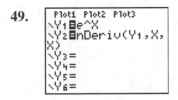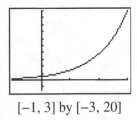

$[-1, 3]$ by $[-3, 20]$

4.3 Differentiation of Exponential Functions

Everything in this section is important. It is essential that you not fall behind the class at this point. Make every effort to completely finish the work in Sections 4.1–4.3 before the class begins Section 4.4. You may have difficulty with the material on logarithms in Sections 4.4–4.6 if you try to learn it while you are still uncertain about exponential functions. Be sure to read the boxes on pages 236–237. The importance of the differential equation $y' = ky$ will not be apparent until you reach Chapter 5, but it is desirable to think a little bit now about questions such as Exercises 41–44.

1. $\dfrac{d}{dx} 4e^{2x} = 4 \cdot \dfrac{d}{dx} e^{2x} = 4(2e^{2x}) = 8e^{2x}.$

7. Begin with the general power rule.

$$\frac{d}{dx}(e^x + e^{-x})^3 = 3(e^x + e^{-x})^2 \cdot \frac{d}{dx}(e^x + e^{-x}) = 3(e^x + e^{-x})^2(e^x - e^{-x}).$$

13. Use the chain rule for $e^{g(x)}$:

$$\frac{d}{dx}(e^{x^2 - 5x + 4}) = e^{x^2 - 5x + 4} \cdot \frac{d}{dx}(x^2 - 5x + 4)$$
$$= e^{x^2 - 5x + 4} \cdot (2x - 5).$$

19. Use the product rule, then the chain rule for e^{2t} and $(t+1)^2$:

$$\frac{d}{dt}(t+1)^2 e^{2t} = (t+1)^2 \cdot \frac{d}{dx} e^{2t} + e^{2t} \cdot \frac{d}{dx}(t+1)^2$$
$$= (t+1)^2 \cdot 2e^{2t} + e^{2t} \cdot 2 \cdot (t+1) \cdot 1$$
$$= 2e^{2t}\left(t^2 + 2t + 1\right) + 2e^{2t}\left(t + 1\right)$$
$$= 2e^{2t}\left(t^2 + 3t + 2\right)$$
$$= 2e^{2t}(t+1)(t+2)$$

25. Write $\dfrac{1}{x}$ as x^{-1} and use the product rule:

$$\frac{d}{dx}(x^{-1} + 3)e^{2x} = (x^{-1} + 3) \cdot \frac{d}{dx} e^{2x} + e^{2x} \cdot \frac{d}{dx}(x^{-1} + 3)$$
$$= (x^{-1} + 3) \cdot 2e^{2x} + e^{2x} \cdot (-x^{-2})$$
$$= (6 + 2x^{-1} - x^{-2})e^{2x} \text{ or } \left(-\frac{1}{x^2} + \frac{2}{x} + 6\right)e^{2x}$$

31. $f(x) = (5x - 2)e^{1-2x}$. Use the product rule:

$$f'(x) = (5x - 2) \cdot \frac{d}{dx} e^{1-2x} + e^{1-2x} \cdot \frac{d}{dx}(5x - 2)$$

$$= (5x - 2) \cdot e^{1-2x} \cdot (-2) + e^{1-2x} \cdot 5.$$

Simplify $f'(x)$ first, and then find $f''(x)$:

$$f'(x) = (-10x + 4)e^{1-2x} + 5e^{1-2x}$$

$$= (-10x + 4 + 5)e^{1-2x}$$

$$= (9 - 10x)e^{1-2x}$$

$$f''(x) = (9 - 10x) \cdot e^{1-2x}(-2) + e^{1-2x} \cdot (-10)$$

To find values of x at which $f(x)$ has a possible relative max or min, set the derivative equal to zero, and solve for x:

$$f'(x) = (9 - 10x)e^{1-2x} = 0 \Rightarrow 9 - 10x = 0 \quad \text{and} \quad x = \frac{9}{10}$$

Here we used the fact that e^{1-2x} is never zero. To check the concavity of the graph of $f(x)$ at $x = .9$, compute $f''(.9)$. Looking at $f''(x)$, notice that the first term is zero when $9 - 10x = 0$, so

$$f''(.9) = 0 + e^{1-2(.9)} \cdot (-10) < 0,$$

because every value of e^x is positive. The graph is concave down at $x = .9$, and so $f(x)$ has relative maximum point at $x = .9$ or $9/10$.

Helpful Hint: Simplify $f'(x)$ before computing $f''(x)$, but don't simplify $f''(x)$. When you find an x that makes $f'(x) = 0$, this value of x may satisfy some equation that will make part of the formula for $f''(x)$ equal 0. In Exercise 31 above, the equation was $9 - 10x = 0$. This fact simplifies calculations when you check $f''(x)$ for concavity.

37. The phrase "after 7 weeks" means that 7 weeks have passed since t was zero; that is, $t = 7$. The rate of growth involves the derivative, so compute:

$$f(t) = (.05 + e^{-.4t})^{-1}$$

$$f'(t) = (-1)(.05 + e^{-.4t})^{-2} \cdot \frac{d}{dt}(.05 + e^{-.4t})$$

$$= -(.05 + e^{-.4t})^{-2} \cdot e^{-.4t}(-.4).$$

There is no need to simplify $f'(t)$, because further differentiation is not required.

$$f'(7) = -(.05 + e^{-2.8})^{-2} \cdot e^{-2.8}(-.4) \approx 1.98.$$

The plant was growing at the rate of nearly 2 inches per week after 7 weeks.

43. This question relies on the basic fact discussed just before Example 5. If $y' = -.5y$, then y must be a function of the form $f(t) = Ce^{-.5t}$, for some constant C. If, in addition, $f(0) = 1$, then $Ce^{-.5(0)} = 1$, which shows that $C = 1$ (because $e^0 = 1$). Thus, $f(x) = e^{-.5x}$.

49. Set $Y_1 = 1.825 \wedge 3*(1-1.6*e \wedge (-.4196X)) \wedge 3$, set Y_4 to be the derivative (possibly a numerical derivative) of Y_1, and set Y_5 to be the derivative of Y_4.

(a) A suitable window for graphing is [1, 15] *by* [0, 8]. The graph seems to have a horizontal asymptote close to $y = 6$ milliliters. As X grows, the negative exponential goes to zero, and values of Y_1 approach $1.825 \wedge 3$, which equals 6.0784, to four decimal places. This means that the tumor's volume appears to stabilize around 6 ml.

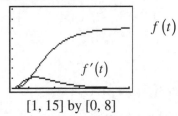

[1, 15] by [0, 8]

(b) Use **TRACE** or function evaluation to find $Y_1(5) \approx 3.16$ milliliters at $x = 5$ weeks.

(c) Set $Y_2 = 5$ (a constant function), and use **intersect** to find the X at which the graphs of Y_1 and Y_2 cross. You should obtain $X \approx 7.71$ weeks.

(d) Graph Y_4 and use **TRACE**, or evaluate $Y_4(5)$, to find the rate of growth when X = 5. You should the find the $Y_4(5)$ is about .97 milliliters per week.

(e) The growth is fastest when the derivative has a maximum, which can happen only if the second derivative is zero. So use a **Solver** command to find when $Y_5 = 0$. You should obtain $X \approx 3.74$ weeks.

(f) Evaluate Y_4 at the value of X found in Part (e). You should obtain $Y_4(3.74) \approx 1.13$ ml/wk.

4.4 The Natural Logarithm Function

The natural logarithm function is defined by its graph, and this is what you should think of when you are trying to understand what the ln x function really is. Study Figure 2 in Section 4.4. The natural logarithm is a function whose graph is defined only for $x > 0$; the graph is increasing, passes through (1, 0), and is concave downward. The graph of ln x is obtained by interchanging the x- and y-coordinates of points on the graph of $y = e^x$. This fact leads immediately to the fundamental relations (numbered below as in the text):

$$e^{\ln x} = x \qquad \text{(all positive } x\text{)} \qquad (2)$$

$$\ln e^x = x \qquad \text{(all } x\text{)} \qquad (3)$$

The main purpose of Section 4.4 is to teach you how to use these relations to solve equations involving ln x and e^x.

1. $\ln(\sqrt{e}) = \ln e^{1/2} = \dfrac{1}{2}$ [Relation (3)]

7. $\ln e^{-3} = -3$ [Relation (3)]

13. $e^{2\ln x} = (e^{\ln x})^2 = (x)^2 = x^2$ [Relation (2)]

19. If $e^{2x} = 5$, take the (natural) logarithm of each side to obtain:

$$\ln e^{2x} = \ln 5,$$
$$2x = \ln 5, \qquad \text{[R elation (3)]}$$
$$x = \frac{1}{2} \ln 5.$$

25. The first step is to isolate the term $e^{-.00012x}$ in order to use to the logarithm function to undo the effect of the exponentiation. Given $6e^{-.00012x} = 3$, divide by 6 to obtain:

$$e^{-.00012x} = .5.$$

Take the natural logarithm of each side and solve:

$$\ln e^{-.00012x} = \ln .5,$$
$$-.00012x = \ln .5,$$
$$x = \frac{\ln .5}{-.00012}.$$

31. If $2e^{x/3} - 9 = 0$ then

$$2e^{x/3} = 9,$$
$$e^{x/3} = \frac{9}{2},$$
$$\ln(e^{x/3}) = \ln \frac{9}{2},$$
$$\frac{x}{3} = \ln \frac{9}{2}, \quad \text{or} \quad x = 3 \ln \frac{9}{2}.$$

37. Given $4e^x \cdot e^{-2x} = 6$, observe that x occurs in two places. Use a property of exponents to write

$$4e^{-x} = 6.$$

Now proceed as in Exercise 25.

$$e^{-x} = \frac{6}{4} = \frac{3}{2},$$
$$\ln e^{-x} = \ln \frac{3}{2},$$
$$-x = \ln \frac{3}{2}, \quad \text{or} \quad x = -\ln \frac{3}{2}.$$

Warning: A slight change in Exercise 37 can make it seem harder. Consider the equation

$$e^x = \frac{3}{2} e^{2x}.$$

Taking logarithms at this point is of no help because the equation $x = \ln\left(\frac{3}{2}e^{2x}\right)$ expresses x in terms of something involving x. The correct step is to get x on only one side of the equation. There are two ways to do this:

$$\text{(i)} \qquad e^x = \frac{3}{2}e^{2x}, \quad \text{or} \quad \text{(ii)} \qquad e^x = \frac{3}{2}e^{2x},$$

$$e^x - \frac{3}{2}e^{2x} = 0. \qquad\qquad e^x \cdot e^{-2x} = \frac{3}{2}e^{2x}e^{-2x}.$$

Equation (i) leads nowhere because you can't take the logarithm of each side, since $\ln 0$ is not defined. Also, $\ln\left(e^x - \frac{3}{2}e^{2x}\right)$ cannot be simplified. Equation (ii) is the correct approach, since a property of exponents leads to $e^{-x} = \frac{3}{2}e^0$, that is, $e^{-x} = \frac{3}{2}$. This can be solved for x as in Exercise 37, yielding $x = -\ln\frac{3}{2}$.

43. $f(x) = e^{-x} + 3x; \quad f'(x) = -e^{-x} + 3; \quad f''(x) = e^{-x}$

Set $f'(x) = 0$ and solve for x.

$$-e^{-x} + 3 = 0$$
$$-e^{-x} = -3$$
$$e^{-x} = 3$$
$$\ln e^{-x} = \ln 3$$
$$-x = \ln 3$$
$$x = -\ln 3.$$

Substitute this value for x back into $f(x)$ to find the y-coordinate of this possible extreme point.

$$f(3) = e^{-(-\ln 3)} + 3(-\ln 3)$$
$$= e^{\ln 3} - 3\ln 3$$
$$= 3 - 3\ln 3.$$

The possible extreme point is $(-\ln 3,\ 3 - 3\ln 3)$. Since $f''(-\ln 3) = e^{-(-\ln 3)} = e^{\ln 3} = 3$, which is positive, the graph of $f(x)$ is concave up at $x = -\ln 3$ and $(-\ln 3,\ 3 - 3\ln 3)$ is a relative minimum point.

49.   Graph of $y = \ln(e^x)$:   Graph of $y = e^{\ln x}$:

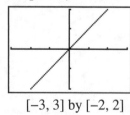

   [−3, 3] by [−2, 2]   [−3, 3] by [−2, 2]

The graph of $y = e^{\ln x}$ is the same as the graph of $y = x$ for $x > 0$.

4.5   The Derivative of ln *x*

The exercises in this section combine the derivative formula for ln x with a review of the product rule, the quotient rule, and the chain rule. You should memorize the chain rule for the logarithm function:

$$\frac{d}{dx}[\ln g(x)] = \frac{1}{g(x)} \cdot g'(x) = \frac{g'(x)}{g(x)}$$

This rule is used in Exercises 1, 7, 13, and 19, discussed below.

1. $\dfrac{d}{dx}\ln(2x) = \dfrac{1}{2x} \cdot \dfrac{d}{dx}(2x) = \dfrac{1}{2x}(2) = \dfrac{1}{x}$

7. $\dfrac{d}{dx}e^{\ln x + x} = e^{\ln x + x} \cdot \dfrac{d}{dx}(\ln x + x) = \left(\dfrac{1}{x} + 1\right)e^{\ln x + x}$

13. $\dfrac{d}{dx}\ln(kx) = \dfrac{1}{kx} \cdot \dfrac{d}{dx}(kx) = \dfrac{1}{kx}(k) = \dfrac{1}{x}$

19. $\dfrac{d}{dx}\ln(e^{5x} + 1) = \dfrac{1}{e^{5x}+1} \cdot \dfrac{d}{dx}(e^{5x} + 1) = \dfrac{1}{e^{5x}+1} \cdot e^{5x}(5) = \dfrac{5e^{5x}}{e^{5x}+1}$

25. For the equation of the tangent line, you need a point on the line and the slope of the line. Use the original equation $y = \ln(x^2 + e)$ to find a point. If $x = 0$, then $y = \ln(0^2 + e) = \ln e = 1$, and hence (0, 1) is on the line. For the slope of the tangent line, first find the general slope formula:

$$\frac{dy}{dx} = \frac{1}{x^2 + e} \cdot \frac{d}{dx}(x^2 + e)$$

$$= \frac{1}{x^2 + e}(2x) \quad [e \text{ is a constant}]$$

$$= \frac{2x}{x^2 + e}.$$

The slope of the tangent line when $x = 0$ is $\frac{2(0)}{0^2 + e} = 0$. Therefore the tangent line is the horizontal line passing through the point (0, 1). The equation of this line is $y - 1 = 0(x - 0)$; that is, $y = 1$.

Warning: Exercises 25–30 make good exam questions because they review important concepts and skills from Chapters 1–3 as well as testing your ability to differentiate ln $g(x)$.

31. The marginal cost at $x = 10$ is $C'(10)$. Using the quotient, first find

$$C'(x) = \frac{d}{dx}\left(\frac{100 \ln x}{40 - 3x}\right)$$

$$= \frac{(40 - 3x) \cdot \frac{d}{dx}(100 \ln x) - (100 \ln x) \cdot \frac{d}{dx}(40 - 3x)}{(40 - 3x)^2}$$

$$= \frac{(40 - 3x)\left(\frac{100}{x}\right) - (100 \ln x)(-3)}{(40 - 3x)^2}.$$

Now

$$C'(10) = \frac{(40 - 3(10))\frac{100}{10} + 300\ln 10}{(40 - 3(10))^2} = \frac{100 + 300\ln 10}{(10)^2} = 1 + 3\ln 10.$$

37. Set $Y_1 = \ln(\text{abs}(X))$ and set Y_4 to be the derivative of Y_1. The "abs" command is located in the MATH NUM menu. Alternatively, use the **CATALOG** command to find "abs" in the list of all commands. Graph both functions in the **ZDecimal** window (found in the **ZOOM** menu). Use **TRACE** on the graph of Y_4 to check the values of Y_4 at $-4, -2, -1, 1, 2,$ and 4. (You should obtain $-.25, -.5, -1, 1, .5,$ and $.25$.) Using the graph verifies the derivative formula. What happens at $x = 0$?

$$f(x) = \ln|x|$$

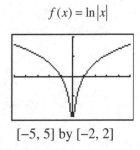

$[-5, 5]$ by $[-2, 2]$

$$f'(x) = \frac{1}{x}$$

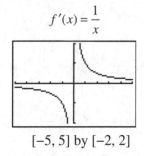

$[-5, 5]$ by $[-2, 2]$

4.6 Properties of the Natural Logarithm Function

Just as with laws of exponents, it is essential to know the properties of logarithms "backwards and forwards." You will find it helpful to add the following properties to the list on page 247.

LI′	$\ln x + \ln y = \ln xy$
LII′	$-\ln x = \ln\left(\dfrac{1}{x}\right)$
LIII′	$\ln x - \ln y = \ln\left(\dfrac{x}{y}\right)$
LIV′	$b\ln x = \ln(x^b)$

Unfortunately, many students make up additional "laws" that are not true. A study of the following facts will help you avoid the most common incorrect "laws."

(a) $\ln(x + y)$ is <u>not</u> equal to $\ln x + \ln y$.

(b) $\ln(e + e^y)$ is <u>not</u> equal to $1 + y$

(c) $(\ln x)(\ln y)$ is <u>not</u> equal to $\ln x + \ln y$.

(d) $\dfrac{\ln x}{\ln y}$ is <u>not</u> equal to $\ln x - \ln y$.

1. $\ln 5 + \ln x = \ln 5x.$ (LI′)

7. $e^{2\ln x} = e^{\ln x^2} = x^2.$ (LIV′)

13. Property (LIV′) shows that $2\ln 5 = \ln 5^2 = \ln 25$ and $3\ln 3 = \ln 3^3 = \ln 27$. Since $25 < 27$, and since the graph of the natural logarithm function is increasing, $\ln 25 < \ln 27$. Therefore, $3\ln 3$ is larger.

19. Using property (LIV'), we see $4 \ln 2x = \ln(2x)^4 = \ln(2^4 x^4) = \ln(16x^4)$. The answer is (d).

25. $\ln x^4 - 2 \ln x = 1 \Rightarrow 4 \ln x - 2 \ln x = 1 \Rightarrow 2 \ln x = 1 \Rightarrow \ln x = \dfrac{1}{2} \Rightarrow x = e^{1/2} = \sqrt{e}.$

31. $\ln(x+1) - \ln(x-2) = 1 \Rightarrow \ln\left(\dfrac{x+1}{x-2}\right) = 1$. Exponentiate each side to obtain

$$\frac{x+1}{x-2} = e^1 = e$$
$$x+1 = e(x-2)$$
$$x+1 = ex - 2e$$
$$1 + 2e = ex - x$$
$$1 + 2e = x(e-1)$$
$$x = \frac{1+2e}{e-1}.$$

Helpful Hint: Whenever you have a function to differentiate, you should pause and check if the form of the function can be simplified *before* you begin to differentiate. Exercises 33–40 illustrate how much this will simplify your work.

37. $y = \ln\left[\sqrt{xe^{x^2+1}}\right] = \ln\left(xe^{x^2+1}\right)^{1/2} = \dfrac{1}{2} \ln\left(xe^{x^2+1}\right) = \dfrac{1}{2} \ln x + \dfrac{1}{2} \ln e^{x^2+1} = \dfrac{1}{2} \ln x + \dfrac{1}{2}\left(x^2+1\right)$

Therefore,

$$\frac{dy}{dx} = \frac{d}{dx}\left(\frac{1}{2} \ln x\right) + \frac{d}{dx} \frac{1}{2}\left(x^2+1\right)$$
$$= \frac{1}{2} \cdot \frac{1}{x} + \frac{1}{2} \cdot 2x = \frac{1}{2x} + x$$

Warning: Remember that the natural logarithm converts products and quotients into sums and differences. But a *product* of logarithms *cannot* be simplified by a standard logarithm rule. Be on guard when you work Exercises 41 and 42.

43. Let $f(x) = (x+1)^4(4x-1)^2$. Then,

$$\ln f(x) = \ln\left[(x+1)^4(4x-1)^2\right]$$
$$= \ln(x+1)^4 + \ln(4x-1)^2$$
$$= 4 \ln(x+1) + 2 \ln(4x-1).$$

Differentiate both sides with respect to x:

$$\frac{f'(x)}{f(x)} = 4 \cdot \frac{1}{x+1} + 2 \cdot \frac{1}{4x-1}(4)$$
$$= \frac{4}{x+1} + \frac{8}{4x-1}.$$

So,

$$f'(x) = f(x)\left(\frac{4}{x+1} + \frac{8}{4x-1}\right) = (x+1)^4(4x-1)^2\left(\frac{4}{x+1} + \frac{8}{4x-1}\right).$$

49. Let $f(x) = x^x$. Then,

$$\ln f(x) = \ln x^x = x \ln x.$$

Differentiate both sides with respect to x:

$$\frac{f'(x)}{f(x)} = x \cdot \frac{1}{x} + \ln x \cdot (1) = 1 + \ln x.$$

So,

$$f'(x) = f(x)(1 + \ln x) = x^x(1 + \ln x).$$

Review of Chapter 4

There are many facts in this chapter to remember and keep straight. The learning process requires time and lots of practice. Don't wait until the last minute to review. Your efforts to master this chapter will be rewarded later, since the exponential and natural logarithm functions appear in nearly every section in the rest of the text.

By now you should have constructed the expanded lists of properties of exponents and logarithms. To these lists add notes about common mistakes—yours and the ones mentioned in the *Manual* notes for this chapter. Here is another common error: if you take logarithms of each side of an equation of the form

$$A = B + C$$

you *cannot* write $\ln A = \ln B + \ln C$. The correct form is

$$\ln A = \ln(B + C).$$

Similarly, if you exponentiate each side of $A = B + C$, you *cannot* write $e^A = e^B + e^C$. The correct form is

$$e^A = e^{(B+C)}.$$

Here are four problems that contain "traps" for the unwary student. Try them now. You will find answers later in the manual as you work through the supplementary exercises. *Please* don't look for the answers until you have done your best to work the problems.

(A) Solve for y in terms of x: $\ln y - \ln x^2 = \ln 5$.

(B) Solve for y in terms of x: $e^y - e^{-3} = e^{2x}$.

(C) Solve for x: $\dfrac{\ln 10x^3}{\ln 2x^2} = 1$.

(D) Simplify, if possible: $\ln(x^3 - x^2)$.

Chapter 4 Review Exercises

In Exercises 9–14 and 39–44, "*simplify*" means to use the laws of exponents and logarithms to write the expression in another form that either is less complicated or at least is more useful for some purposes.

While solutions for all review exercises are included, expanded explanations are included for every sixth exercise.

1. $27^{4/3} = (27^{1/3})^4 = 3^4 = 81.$

2. $4^{1.5} = (2^2)^{3/2} = 2^3 = 8$

3. $5^{-2} = \dfrac{1}{5^2} = \dfrac{1}{25}$

4. $16^{-.25} = 16^{-1/4} = \dfrac{1}{(2^4)^{1/4}} = \dfrac{1}{2}$

5. $(2^{5/7})^{14/5} = 2^{14/7} = 2^2 = 4$

6. $8^{1/2} \cdot 2^{1/2} = (2^3)^{1/2} \cdot 2^{1/2} = 2^{3/2} \cdot 2^{1/2}$
$\qquad = 2^{4/2} = 4$

7. $\dfrac{9^{5/2}}{9^{3/2}} = 9^{5/2-3/2} = 9^{2/2} = 9.$

8. $4^{.2} \cdot 4^{.3} = 4^{.5} = 4^{1/2} = 2$

9. $(e^{x^2})^3 = e^{3x^2}$

10. $e^{5x} \cdot e^{2x} = e^{7x}$

11. $\dfrac{e^{3x}}{e^x} = e^{3x-x} = e^{2x}$

12. $2^x \cdot 3^x = (2 \cdot 3)^x = 6^x$

13. $(e^{8x} + 7e^{-2x})e^{3x} = e^{8x} \cdot e^{3x} + 7e^{-2x} \cdot e^{3x}$
$\qquad\qquad = e^{11x} + 7e^x$

14. $\dfrac{e^{5x/2} - e^{3x}}{\sqrt{e^x}} = (e^{5x/2} - e^{3x})e^{(-1/2)x}$
$\qquad\qquad = e^{4x/2} - e^{5x/2} = e^{2x} - e^{5x/2}$

15. $e^{-3x} = e^{-12} \Rightarrow \ln e^{-3x} = \ln e^{-12} \Rightarrow$
$\qquad -3x = -12 \Rightarrow x = 4$

16. $\qquad e^{x^2-x} = e^2$
$\qquad \ln e^{x^2-x} = \ln e^2$
$\qquad\qquad x^2 - x = 2$
$\qquad\quad x^2 - x - 2 = 0$
$\qquad (x-2)(x+1) = 0 \Rightarrow x = 2 \text{ or } x = -1$

17. $(e^x \cdot e^2)^3 = e^{-9} \Rightarrow e^{3x+6} = e^{-9} \Rightarrow$
$\qquad \ln e^{3x+6} = \ln e^{-9} \Rightarrow 3x + 6 = -9 \Rightarrow x = -5$

18. $e^{-5x} \cdot e^4 = e \Rightarrow e^{-5x+4} = e \Rightarrow$
$\qquad \ln e^{-5x+4} = \ln e \Rightarrow -5x + 4 = 1 \Rightarrow x = \dfrac{3}{5}$

19. $\dfrac{d}{dx} 10e^{7x} = 10\dfrac{d}{dx}e^{7x} = 10e^{7x}(7) = 70e^{7x}.$

20. $\dfrac{d}{dx} e^{\sqrt{x}} = \dfrac{d}{dx}e^{x^{1/2}} = e^{x^{1/2}} \cdot \dfrac{1}{2}x^{-1/2} = \dfrac{e^{\sqrt{x}}}{2\sqrt{x}}$

21. $\dfrac{d}{dx}[xe^{x^2}] = xe^{x^2}(2x) + e^{x^2}(1) = e^{x^2}(2x^2+1)$

22. $\dfrac{d}{dx}\left[\dfrac{e^x+1}{x-1}\right] = \dfrac{(x-1)e^x - (e^x+1)(1)}{(x-1)^2}$
$\qquad\qquad = \dfrac{-2e^x + xe^x - 1}{(x-1)^2} = \dfrac{(x-2)e^x - 1}{(x-1)^2}$

23. $\dfrac{d}{dx}[e^{e^x}] = e^{e^x}(e^x) = e^{x+e^x}$

24. $\dfrac{d}{dx}\left[\left(\sqrt{x}+1\right)e^{-2x}\right]$
$\qquad = \left(\sqrt{x}+1\right)e^{-2x}(-2) + e^{-2x}\left(\dfrac{1}{2}x^{-1/2}\right)$
$\qquad = e^{-2x}\left(\dfrac{1}{2\sqrt{x}} - 2\sqrt{x} - 2\right)$

25. $\dfrac{d}{dx}\left(\dfrac{x^2-x+5}{e^{3x}+3}\right)=\dfrac{(e^{3x}+3)(2x-1)-(x^2-x+5)\cdot 3e^{3x}}{(e^{3x}+3)^2}$ [Quotient Rule]

26. $\dfrac{d}{dx}x^e=ex^{e-1}$

Helpful Hint: In Exercises 27–30, the solution is a *function,* not a number. An equation in which both the unknown function and its (unknown) derivative appear is called a *differential equation.* The differential equations that appear here are solved with the boxed result (3) of Section 4.3 (page 237).

27. $y'=-y$; $y=Ce^{-x}$ for some C.

28. $y'=-1.5y$; $f(x)=y=Ce^{-1.5x}$ for some C. Since $f(0)=2000$, $2000=Ce^{(-1.5)(0)}=C$. Thus, $y=2000e^{-1.5x}$.

29. $y'=1.5y$; $f(x)=y=Ce^{1.5x}$ for some C. Since $f(0)=2$, $2=Ce^{(1.5)(0)}=C$. Thus, $y=2e^{1.5x}$.

30. $y'=\dfrac{1}{3}y$; $y=Ce^{(1/3)x}$ for some C.

31. $y=e^{-x}+x$, $y'=-e^{-x}+1$, $y''=e^{-x}$.

Now set $y'=0$ and solve for x:

$$-e^{-x}+1=0,$$
$$e^{-x}=1,$$
$$\ln e^{-x}=\ln(1),$$
$$-x=0, \quad \text{or} \quad x=0$$

If $x=0$, then $y=e^{-0}+0=1$ and $y''=e^{-0}=1>0$, hence the graph is concave up at $(0,1)$. Thus the graph has a relative minimum at $(0,1)$. Since $y''=e^{-x}>0$ for all x, the graph is concave up for all x, and there are no inflection points. As x becomes large, e^{-x} approaches 0, and for this case, $e^{-x}+x$ is only slightly larger than x. Hence for large positive values of x, the graph of $y=e^{-x}+x$ has $y=x$ as an asymptote.

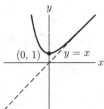

32.

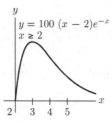

33.

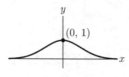

34.

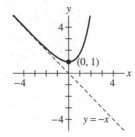

35.

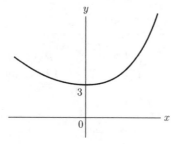

36.

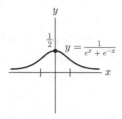

37. The slope of the tangent line to the graph of $y = \dfrac{e^x}{1+e^x}$ at $(0, .5)$ is the value of the derivative $\dfrac{dy}{dx}$ when $x = 0$. By the quotient rule,

$$\frac{d}{dx}\left(\frac{e^x}{1+e^x}\right)\bigg|_{x=0} = \frac{(1+e^x)\cdot e^x - (e^x)(e^x)}{(1+e^x)^2}\bigg|_{x=0}$$

$$= \frac{(1+e^0)\cdot e^0 - (e^0)(e^0)}{(1+e^0)^2}$$

$$= \frac{(1+1)\cdot 1 - (1)(1)}{(1+1)^2}$$

$$= \frac{2-1}{2^2}$$

$$= \frac{1}{4}.$$

The tangent line is thus a line with slope $\frac{1}{4}$ passing through the point $(0, .5)$. Since the line passes through $(0, .5)$, the y-intercept is $\frac{1}{2}$ and the equation of the line is:

$$y = \frac{1}{4}x + \frac{1}{2}.$$

38. $y = \dfrac{e^x - e^{-x}}{e^x + e^{-x}}$

$\dfrac{dy}{dx} = \dfrac{(e^x + e^{-x})(e^x + e^{-x}) - (e^x - e^{-x})(e^x - e^{-x})}{\left(e^x + e^{-x}\right)^2} = \dfrac{4}{\left(e^x + e^{-x}\right)^2}$

$\left.\dfrac{dy}{dx}\right|_{x=1} = \dfrac{4}{\left(e^1 + e^{-1}\right)^2}$

$\left.\dfrac{dy}{dx}\right|_{x=-1} = \dfrac{4}{\left(e^{-1} + e^{-(-1)}\right)^2} = \dfrac{4}{\left(e^1 + e^{-1}\right)^2}$

The tangent lines at $x = 1$ and $x = -1$ have the same slope, so they are parallel.

39. $e^{(\ln 5)/2} = e^{\ln \sqrt{5}} = \sqrt{5}$

40. $e^{\ln(x^2)} = x^2$

Warning: Be careful with Exercise 41. Don't use an incorrect property of logarithms, and don't look at the answer until you have tried the problem!

41. $\dfrac{\ln x^2}{\ln x^3} = \dfrac{2\ln x}{3\ln x} = \dfrac{2}{3}$

42. $e^{2\ln 2} = e^{\ln 2^2} = 2^2 = 4$

43. $e^{-5\ln 1} = e^{-5(0)} = e^0 = 1.$

44. $[e^{\ln x}]^2 = x^2$

45. $t^{\ln t} = e \Rightarrow \ln t^{\ln t} = \ln e \Rightarrow \ln t(\ln t) = 1 \Rightarrow (\ln t)^2 = 1$

Taking the square root of both sides, $|\ln t| = 1 \Rightarrow t = e$ or $t = \dfrac{1}{e}$.

46. $\ln(\ln 3t) = 0 \Rightarrow e^{\ln(\ln 3t)} = e^0 \Rightarrow \ln 3t = 1 \Rightarrow e^{\ln 3t} = e \Rightarrow 3t = e \Rightarrow t = \dfrac{e}{3}$

47. $3e^{2t} = 15 \Rightarrow e^{2t} = 5 \Rightarrow \ln e^{2t} = \ln 5 \Rightarrow 2t = \ln 5 \Rightarrow t = \dfrac{1}{2}\ln 5$

48. $3e^{t/2} - 12 = 0 \Rightarrow 3(e^{t/2} - 4) = 0 \Rightarrow e^{t/2} = 4 \Rightarrow \ln e^{t/2} = \ln 4 \Rightarrow t = 2\ln 4 \Rightarrow t = \ln 16$

49. The equation $2\ln t = 5$ leads to $\ln t = \frac{5}{2}$ and $t = e^{5/2}$.

50. $2e^{-.3t} = 1 \Rightarrow e^{-.3t} = \dfrac{1}{2} \Rightarrow \ln e^{-.3t} = \ln\dfrac{1}{2} \Rightarrow -.3t = \ln\dfrac{1}{2} \Rightarrow t = -\dfrac{1}{.3}\ln\dfrac{1}{2} = \dfrac{\ln 2}{.3}$

51. $\dfrac{d}{dx}\ln(x^6+3x^4+1) = \dfrac{1}{x^6+3x^4+1}(6x^5+12x^3) = \dfrac{6x^5+12x^3}{x^6+3x^4+1}$

52. $\dfrac{d}{dx}\left[\dfrac{x}{\ln x}\right] = \dfrac{\ln x - x\left(\frac{1}{x}\right)}{(\ln x)^2} = \dfrac{\ln x - 1}{(\ln x)^2}$

53. $\dfrac{d}{dx}\ln(5x-7) = \dfrac{1}{5x-7}(5) = \dfrac{5}{5x-7}$

54. $\dfrac{d}{dx}\ln(9x) = \dfrac{1}{9x}(9) = \dfrac{1}{x}$

55. $\dfrac{d}{dx}(\ln x)^2 = 2(\ln x)\cdot\dfrac{d}{dx}(\ln x) = 2(\ln x)\left(\dfrac{1}{x}\right) = \dfrac{2\ln x}{x}$

56. $\dfrac{d}{dx}\left[(x\ln x)^3\right] = 3(x\ln x)^2\left(x\cdot\dfrac{1}{x}+\ln x\right) = 3(x\ln x)^2(1+\ln x)$

57. $\dfrac{d}{dx}\ln\left[\dfrac{xe^x}{\sqrt{1+x}}\right] = \dfrac{d}{dx}\left[\ln\left(xe^x\right)-\ln\sqrt{1+x}\right] = \dfrac{1}{xe^x}(xe^x+e^x)-\dfrac{1}{\sqrt{1+x}}\cdot\dfrac{1}{2}(1+x)^{-1/2} = 1+\dfrac{1}{x}-\dfrac{1}{2(1+x)}$

58. $\dfrac{d}{dx}\ln\left[e^{6x}(x^2+3)^5(x^3+1)^{-4}\right] = \dfrac{d}{dx}\left[6x+5\ln(x^2+3)-4\ln(x^3+1)\right]$

$$= 6+\dfrac{5}{x^2+3}(2x)-\dfrac{4}{x^3+1}(3x^2) = 6+\dfrac{10x}{x^2+3}-\dfrac{12x^2}{x^3+1}$$

59. $\dfrac{d}{dx}[x\ln x - x] = x\left(\dfrac{1}{x}\right)+\ln x - 1 = \ln x$

60. $\dfrac{d}{dx}\left[e^{2\ln(x+1)}\right] = \dfrac{d}{dx}\left[e^{\ln(x+1)^2}\right] = \dfrac{d}{dx}\left[(x+1)^2\right] = 2(x+1)$

61. $\dfrac{d}{dx}\ln(\ln\sqrt{x}) = \dfrac{1}{\ln\sqrt{x}}\cdot\dfrac{d}{dx}\ln\sqrt{x} = \dfrac{1}{\ln x^{1/2}}\cdot\dfrac{d}{dx}\ln x^{1/2} = \dfrac{1}{(1/2)\ln x}\cdot\dfrac{d}{dx}\left[\dfrac{1}{2}\ln x\right]$

$$= \dfrac{2}{\ln x}\cdot\dfrac{1}{2}\cdot\dfrac{d}{dx}\ln x = \dfrac{1}{\ln x}\cdot\dfrac{1}{x} = \dfrac{1}{x\ln x}.$$

Helpful Hint: Always be alert to the possibility of simplifying a function before you differentiate it. This is particularly important when the function involves $\ln x$ or e^x. Observe that

$$\ln(\ln \sqrt{x}) = \ln(\ln x^{1/2})$$

$$= \ln\left(\frac{1}{2}\ln x\right)$$

$$= \ln\frac{1}{2} + \ln(\ln x).$$

Thus

$$\frac{d}{dx}\ln(\ln \sqrt{x}) = \frac{d}{dx}\ln\frac{1}{2} + \frac{d}{dx}\ln(\ln x)$$

$$= 0 + \frac{1}{\ln x} \cdot \frac{d}{dx}\ln x = \frac{1}{\ln x} \cdot \frac{1}{x} = \frac{1}{x\ln x}.$$

62. $\dfrac{d}{dx}\left[\dfrac{1}{\ln x}\right] = \dfrac{d}{dx}\left[(\ln x)^{-1}\right] = -1(\ln x)^{-2}\left(\dfrac{1}{x}\right) = -\dfrac{1}{x(\ln x)^2}$

63. $\dfrac{d}{dx}[e^x \ln x] = e^x\left(\dfrac{1}{x}\right) + e^x \ln x = \dfrac{e^x}{x} + e^x \ln x$

64. $\dfrac{d}{dx}\ln(x^2 + e^x) = \dfrac{1}{x^2 + e^x}(2x + e^x) = \dfrac{2x + e^x}{x^2 + e^x}$

65. $\dfrac{d}{dx}\ln\sqrt{\dfrac{x^2+1}{2x+3}} = \dfrac{d}{dx}\ln\left(\dfrac{x^2+1}{2x+3}\right)^{1/2} = \dfrac{d}{dx}\dfrac{1}{2}\left[\ln\left(x^2+1\right) - \ln(2x+3)\right]$

$$= \dfrac{1}{2}\left[\dfrac{1}{x^2+1}(2x) - \dfrac{1}{2x+3}(2)\right] = \dfrac{x}{x^2+1} - \dfrac{1}{2x+3}$$

66. $-2x + 1 > 0$ gives us $x < \dfrac{1}{2}$. $-2x + 1 < 0$ gives us $x > \dfrac{1}{2}$.

For $x < \dfrac{1}{2}$, we have $\dfrac{d}{dx}\ln|-2x+1| = \dfrac{d}{dx}\ln(-2x+1) = \dfrac{1}{-2x+1}(-2) = \dfrac{2}{2x-1}$.

For $x > \dfrac{1}{2}$, we have $\dfrac{d}{dx}\ln|-2x+1| = \dfrac{d}{dx}\ln(-(-2x+1)) = \dfrac{d}{dx}\ln(2x-1) = \dfrac{1}{2x-1}(2) = \dfrac{2}{2x-1}$.

So, for $x \neq \dfrac{1}{2}, \dfrac{d}{dx}\ln|-2x+1| = \dfrac{2}{2x-1}$.

67. First, simplify the original function using the properties of the logarithm from Chapter 4.

$$\ln\left(\frac{e^{x^2}}{x}\right) = \ln(e^{x^2}) - \ln(x) \qquad \text{(LIII)}$$

$$= x^2 - \ln(x) \qquad \text{(Equation 3, Section 4.4)}$$

Now take the derivative of the simplified function.

$$\frac{d}{dx}(x^2 - \ln(x)) = 2x - \frac{1}{x}$$

Note that this problem could also be solved using the quotient rule and multiple applications of the chain rule. However, using the properties of the natural logarithm to simplify before differentiating simplifies the solution substantially.

68. $\dfrac{d}{dx}\ln\sqrt[3]{x^3+3x-2} = \dfrac{d}{dx}\left[\dfrac{1}{3}\ln(x^3+3x-2)\right] = \dfrac{1}{3}\dfrac{1}{x^3+3x-2}(3x^2+3) = \dfrac{x^2+1}{x^3+3x-2}$

69. $\dfrac{d}{dx}\ln(2^x) = \dfrac{d}{dx}(x\ln 2) = \ln 2$

70. $\dfrac{d}{dx}\left[\ln(3^{x+1}) - \ln 3\right] = \dfrac{d}{dx}\left[(x+1)\ln 3 - \ln 3\right] = \ln 3$

71. $x - 1 > 0$ gives $x > 1$. $x - 1 < 0$ gives $x < 1$.

For $x > 1$, we have $\dfrac{d}{dx}\ln|x-1| = \dfrac{d}{dx}\ln(x-1) = \dfrac{1}{x-1}$.

For $x < 1$, we have $\dfrac{d}{dx}\ln|x-1| = \dfrac{d}{dx}\ln(-(x-1)) = \dfrac{d}{dx}\ln(-x+1) = \dfrac{1}{-x+1}(-1) = \dfrac{1}{x-1}$.

So, for $x \neq 1$, $\dfrac{d}{dx}\ln|x-1| = \dfrac{1}{x-1}$.

72. $\dfrac{d}{dx}e^{2\ln(2x+1)} = \dfrac{d}{dx}e^{\ln(2x+1)^2} = \dfrac{d}{dx}(2x+1)^2 = 2(2x+1)(2) = 8x+4$

73. First, simplify the original function using the properties of the logarithm from this chapter.

$$\ln\left(\frac{1}{e^{\sqrt{x}}}\right) = -\ln(e^{\sqrt{x}}) \qquad \text{(LII)}$$

$$= -\sqrt{x}. \qquad \text{(Equation 3, Section 4.4)}$$

Now take the derivative of the simplified function.

$$\frac{d}{dx}(-\sqrt{x}) = \frac{d}{dx}(-x^{1/2}) = \frac{-1}{2}x^{-1/2} = \frac{-1}{2\sqrt{x}}$$

Again, this problem could also be solved using the quotient rule and multiple applications of the chain rule, but it is much easier to use the properties of the natural logarithm to simplify before differentiating.

74. $\dfrac{d}{dx}\ln(e^x + 3e^{-x}) = \dfrac{1}{e^x + 3e^{-x}}\left(e^x + 3e^{-x}(-1)\right) = \dfrac{e^x - 3e^{-x}}{e^x + 3e^{-x}}$

75. $\ln f(x) = \ln \sqrt[5]{\dfrac{x^5+1}{x^5+5x+1}} = \dfrac{1}{5}\ln(x^5+1) - \dfrac{1}{5}\ln(x^5+5x+1)$

Differentiating both sides, we have $\dfrac{f'(x)}{f(x)} = \dfrac{1}{5}\dfrac{1}{x^5+1}(5x^4) - \dfrac{1}{5}\dfrac{1}{x^5+5x+1}(5x^4+5) \Rightarrow$

$f'(x) = \sqrt[5]{\dfrac{x^5+1}{x^5+5x+1}}\left(\dfrac{x^4}{x^5+1} - \dfrac{x^4+1}{x^5+5x+1}\right).$

76. $\ln f(x) = \ln 2^x = x \ln 2$

Differentiating both sides, we have $\dfrac{f'(x)}{f(x)} = \ln 2 \Rightarrow f'(x) = 2^x \ln 2.$

77. $\ln f(x) = \ln x^{\sqrt{x}} = \sqrt{x}\ln x$

Differentiating both sides, we have

$\dfrac{f'(x)}{f(x)} = \sqrt{x}\left(\dfrac{1}{x}\right) + \left(\dfrac{1}{2}\right)x^{-1/2}\ln x = \dfrac{1}{\sqrt{x}} + \dfrac{\ln x}{2\sqrt{x}} \Rightarrow f'(x) = x^{\sqrt{x}}\left(\dfrac{1}{\sqrt{x}} + \dfrac{\ln x}{2\sqrt{x}}\right) = x^{\sqrt{x}-1/2}\left(1 + \dfrac{1}{2}\ln x\right)$

78. $\ln f(x) = \ln b^x = x \ln b$

Differentiating both sides, we have $\dfrac{f'(x)}{f(x)} = \ln b \Rightarrow f'(x) = b^x \ln b.$

79. First take the natural logarithm of each side.

$$\begin{aligned}
\ln f(x) &= \ln[(x^2+5)^6(x^3+7)^8(x^4+9)^{10}] \\
&= \ln(x^2+5)^6 + \ln(x^3+7)^8 + \ln(x^4+9)^{10} &&\text{(LI)} \\
&= 6\cdot\ln(x^2+5) + 8\cdot\ln(x^3+7) + 10\cdot\ln(x^4+9). &&\text{(LIV)}
\end{aligned}$$

Now, take the derivative of each side and solve for $f'(x)$.

$$\begin{aligned}
\dfrac{f'(x)}{f(x)} &= 6\left[\dfrac{1}{x^2+5}\cdot 2x\right] + 8\left[\dfrac{1}{x^3+7}\cdot 3x^2\right] + 10\left[\dfrac{1}{x^4+9}\cdot 4x^3\right] \\
&= \dfrac{12x}{x^2+5} + \dfrac{24x^2}{x^3+7} + \dfrac{40x^3}{x^4+9}. \\
f'(x) &= f(x)\left[\dfrac{12x}{x^2+5} + \dfrac{24x^2}{x^3+7} + \dfrac{40x^3}{x^4+9}\right] \\
&= (x^2+5)^6(x^3+7)^8(x^4+9)^{10}\left[\dfrac{12x}{x^2+5} + \dfrac{24x^2}{x^3+7} + \dfrac{40x^3}{x^4+9}\right].
\end{aligned}$$

80. $\ln f(x) = \ln x^{1+x} = (1+x)\ln x = \ln x + x\ln x$

Differentiating both sides, we have $\dfrac{f'(x)}{f(x)} = \dfrac{1}{x} + x\left(\dfrac{1}{x}\right) + \ln x \Rightarrow f'(x) = x^{1+x}\left(\dfrac{1}{x} + 1 + \ln x\right).$

81. $\ln f(x) = \ln 10^x = x \ln 10$

Differentiating both sides, we have $\dfrac{f'(x)}{f(x)} = \ln 10 \Rightarrow f'(x) = 10^x \ln 10.$

82. $\ln f(x) = \ln\left(\sqrt{x^2+5}\,e^{x^2}\right) = \dfrac{1}{2}\ln(x^2+5) + \ln e^{x^2} = \dfrac{1}{2}\ln(x^2+5) + x^2$

Differentiating both sides, we have

$\dfrac{f'(x)}{f(x)} = \dfrac{1}{2}\cdot\dfrac{1}{x^2+5}(2x) + 2x = \dfrac{x}{x^2+5} + 2x \Rightarrow f'(x) = \sqrt{x^2+5}\,e^{x^2}\left[\dfrac{x}{x^2+5} + 2x\right].$

83. $\ln f(x) = \ln\sqrt{\dfrac{xe^x}{x^3+3}} = \dfrac{1}{2}\left[\ln x + \ln e^x - \ln(x^3+3)\right] = \dfrac{1}{2}\left[\ln x + x - \ln(x^3+3)\right]$

Differentiating both sides, we have

$\dfrac{f'(x)}{f(x)} = \dfrac{1}{2}\left[\dfrac{1}{x} + 1 - \dfrac{1}{x^3+3}(3x^2)\right] \Rightarrow f'(x) = \dfrac{1}{2}\sqrt{\dfrac{xe^x}{x^3+3}}\left(\dfrac{1}{x} + 1 - \dfrac{3x^2}{x^3+3}\right).$

84. $\ln f(x) = \ln\left[\dfrac{e^x\sqrt{x+1}(x^2+2x+3)^2}{4x^2}\right] = x + \dfrac{1}{2}\ln(x+1) + 2\ln(x^2+2x+3) - 2\ln(2x)$

Differentiating both sides, we have $\dfrac{f'(x)}{f(x)} = 1 + \dfrac{1}{2(x+1)} + 2\dfrac{1}{x^2+2x+3}(2x+2) - 2\dfrac{1}{2x}(2) \Rightarrow$

$f'(x) = \dfrac{e^x\sqrt{x+1}(x^2+2x+3)^2}{4x^2}\left[1 + \dfrac{1}{2x+2} + \dfrac{4x+4}{x^2+2x+3} - \dfrac{2}{x}\right].$

85. Given

$$f(x) = e^{x+1}(x^2+1)(x)$$
$$= (e^{x+1})(x^3+x),$$

first take the natural logarithm of each side:

$$\ln(f(x)) = \ln\left((e^{x+1})(x^3+x)\right)$$
$$= \ln(e^{x+1}) + \ln(x^3+x) \qquad \text{(LI)}$$
$$= x+1 + \ln(x^3+x). \qquad \text{(Equation 3, Section 4.4)}$$

Now, take the derivative of each side and solve for $f'(x)$.

$$\dfrac{f'(x)}{f(x)} = 1 + \dfrac{3x^2+1}{x^3+x}$$

$$f'(x) = f(x)\left(1 + \dfrac{3x^2+1}{x^3+x}\right) = (e^{x+1})(x^3+x)\left(1 + \dfrac{3x^2+1}{x^3+x}\right)$$

$$= e^{x+1}(x^3+x) + \dfrac{e^{x+1}(x^3+x)(3x^2+1)}{x^3+x}$$

$$= e^{x+1}(x^3+x) + e^{x+1}(3x^2+1)$$

$$= e^{x+1}(x^3+3x^2+x+1).$$

86. $\ln f(x) = \ln(e^x x^2 2^x) = x + 2\ln x + x\ln 2$

Differentiating both sides, we have $\dfrac{f'(x)}{f(x)} = 1 + 2\left(\dfrac{1}{x}\right) + \ln 2 \Rightarrow f'(x) = e^x x^2 2^x \left(1 + \ln 2 + \dfrac{2}{x}\right).$

87.

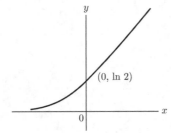

88.

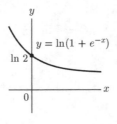

89.

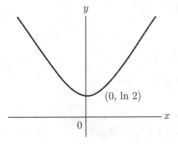

90.

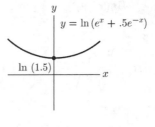

91. $y = (\ln x)^2$,

$$y' = 2(\ln x) \cdot \frac{d}{dx}\ln x = \frac{2\ln x}{x}.$$

$$y'' = \frac{x\frac{d}{dx}(2\ln x) - (2\ln x)\frac{d}{dx}x}{x^2}$$

$$= \frac{x \cdot 2\left(\frac{1}{x}\right) - 2\ln x}{x^2} = \frac{2 - 2\ln x}{x^2}.$$

Set $y' = 0$ and solve for x:

$$\frac{2\ln x}{x} = 0$$

A fraction is zero only when its numerator is zero. Thus $2\ln x = 0$, and $\ln x = 0$. Then $e^{\ln x} = e^0 = 1$, so $x = 1$. If $x = 1$, then $y = (\ln 1)^2 = 0$, and

$$y'' = \frac{2 - 2\ln 1}{1^2} = 2 > 0.$$

Hence, the curve is concave up at $(1, 0)$. Now, set $y'' = 0$ and solve for x to find the possible inflection points.

$$\frac{2 - 2\ln x}{x^2} = 0,$$

$$2 - 2\ln x = 0,$$

$$-2\ln x = -2, \quad \text{or} \quad \ln x = 1.$$

So,

$$e^{\ln x} = e^1, \quad \text{or} \quad x = e.$$

If $x = e$, then $y = (\ln e)^2 = 1$, hence $(e, 1)$ is the only possible inflection point. We have seen that the second derivative is positive at $x = 0$. When x is large, $2 - 2\ln x$ is negative, so the second derivative is negative. Thus the concavity must change somewhere, which shows that $(e, 1)$ is the inflection point.

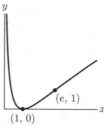

92.

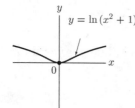

93.

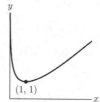

94.

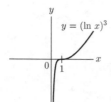

95. $e^y - e^x = 3 \Rightarrow e^y = e^x + 3 \Rightarrow \ln e^y = \ln(e^x + 3) \Rightarrow y = \ln(e^x + 3)$

$$\frac{dy}{dx} = \frac{1}{e^x + 3}(e^x) = \frac{e^x}{e^x + 3}$$

96. $e^{t+y} = t \Rightarrow \ln e^{t+y} = \ln t \Rightarrow t + y = \ln t \Rightarrow y = -t + \ln t \Rightarrow \dfrac{dy}{dt} = -1 + \dfrac{1}{t}$

97. First, take the natural logarithm of both sides in order to solve for y.

$$\ln(e^{xy}) = \ln(x)$$
$$xy = \ln(x) \qquad\qquad \text{(Equation (3), Section 4.4)}$$
$$y = \frac{\ln(x)}{x}.$$

Next, use the quotient rule to find $\dfrac{dy}{dx}$.

$$\frac{dy}{dx} = \frac{x\left(\frac{1}{x}\right) - (1)\ln(x)}{x^2} = \frac{1 - \ln(x)}{x^2}.$$

Finally, evaluate $\dfrac{dy}{dx}$ at $x = 1$.

$$\left.\frac{dy}{dx}\right|_{x=1} = \frac{1 - \ln(1)}{1^2} = \frac{1 - 0}{1} = 1.$$

98. Using implicit differentiation, we have $2e^{2x} + 2e^{2y}\dfrac{dy}{dx} = 0 \Rightarrow \dfrac{dy}{dx} = -\dfrac{e^{2x}}{e^{2y}}$.

99. **(a)** $f(2) \approx 800$ g/cm^2

    (b) $f(x) = 200$ when $x \approx 14$ km

    (c) $f'(8) \approx -50$ g/cm^2 per km

    (d) $f'(x) = -100$ when $x \approx 2$ km

100. **(a)** $f(18) \approx 180$ billion dollars

    (b) $f'(12) \approx 10$ billion dollars per year

    (c) $f(t) = 120$ when $t \approx 14$, so in 2004

    (d) $f'(t) = 20$ when $t \approx 18$, so expenditures were rising at the rate of \$20 billion per year in 2008

Chapter 5
Applications of the Exponential and Natural Logarithm Functions

5.1 Exponential Growth and Decay

The basic differential equation $y' = ky$ was introduced in Section 4.3, just before Example 5. You should review that now. All the exercises in this section refer either to that differential equation or to the general form of functions that satisfy the equation, namely, $f(x) = Ce^{kx}$. In this chapter, we usually use t in place of x, because most of the applications involve functions of time.

The first question to ask when considering an exponential growth and decay problem is, "Does the quantity increase or decrease with time?" For instance, populations increase and radioactive substances decrease. The formula for the quantity present after t units of time will have the form $P(t) = P_0 e^{kt}$ if increasing and $P(t) = P_0 e^{-\lambda t}$ if decreasing. The problems are primarily of the following types:

1. Given P_0 and k (or λ), find the value of $P(t)$ for a specific time t. These problems are solved by just substituting the value of t into the formula.

2. Given P_0 and k (or λ), find the time when $P(t)$ assumes a specific value, call it A. That is, solve $P(t) = A$ for t. Problems of this type are solved by replacing $P(t)$ by its formula, dividing by P_0 and using logarithms. Sometimes A is given as a multiple of P_0 in which case the P_0's cancel each other.

3. Given P_0 and the value of $P(t)$ at some specific time t, solve for k (or λ). This problem reduces to one of the form "solve $P_0 e^{kt} = A$" where A is the value of $P(t)$. This equation is solved by the method in 2 above.

4. The problem gives enough information to solve for k (or λ), but only asks for the value of $P(t)$ at a future time or asks for the time at which $P(t)$ attains a given size. The important point here is that the problem really consists of two parts. First, find k (or λ) and second, answer the question asked. Exercises 9 and 19 are of this type. Students usually have difficulty with such problems on exams since they try to answer the question posed immediately without doing the intermediate step.

5. Determine the rate of change of $P(t)$ at a time when $P(t)$ is a certain size. The actual time is not specified, so you cannot compute $P'(t)$ and evaluate at t. Instead, you must use the differential equation $P'(t) = k \cdot P(t)$. Assuming that k is known, you can determine $P'(t)$ from $P(t)$ (for the same t), and vice versa. You'll find this type of problem in Exercises 1, 13, and elsewhere.

1. **(a)** The differential equation $P'(t) = .02P(t)$ has the form $y' = ky$, with $k = .02$, so the general
    solution $P(t)$ has the form $P(t) = P_0 e^{.02t}$. Also,

    $$P(0) = P_0 e^{.02(0)} = P_0,$$

    so the condition $P(0) = 3$ shows that $P(t) = 3e^{.02t}$.

    (b) *Initial population* is the population when $t = 0$, which in this problem refers to 1990: 3 million
    persons.

    (c) The growth constant is the constant .02 that appears in the differential equation $P'(t) = .02P(t)$.

    (d) 1998 corresponds to $t = 8$. The problem then was $P(8) = 3e^{.02(8)} = 3.52053261298$ million.
    Of course, writing the answer in this form is not realistic. Even the answer 3520533 persons is
    inappropriate. A problem such as this exercise purports to be a model of population growth,
    but it is only an approximation at best. Since the initial population is given with only the
    single digit "3," a rounded answer such as 3.5 million, or perhaps 3.52 million, would be
    more realistic.

    (e) Assume that t_0 is the time when $P(t_0) = 4$ (million). At that value of t, the differential equation
    says that

    $$P'(t_0) = .02P(t_0) = .02 \cdot 4 = .08.$$

    The population is growing at the rate of .08 million (or 80,000) people per year.

    (f) The unit of measurement of the population is millions of people, so convert 70,000 to .07
    million. If t represents the time when $P'(t) = .07$ million people per year, then the differential
    equation

    $$P'(t) = .02P(t)$$

    shows that $.07 = .02P(t)$ and hence $P(t) = .07/.02 = 3.5$. The population is approximately
    3.5 million.

Helpful Hint: In Exercises 5, 6, 9, and 10, the phrase "growing at a rate proportional to its size" describes
the differential equation $y' = ky$. You may assume that the solution of this equation has the form
$y = P_0 e^{kt}$, or $P(t) = P_0 e^{kt}$. This fact was discussed at the beginning of the chapter.

7. See the notes about problem type #2 on page 143 in this *Manual*. The phrase "growing
    exponentially" is another way of saying that the model for the growth is $P(t) = P_0 e^{kt}$. Hence k
    is .05, so $P(t) = P_0 e^{.05t}$. The population will be triple its initial size when $P(3) = 3P_0$. To find the
    time when this happens, set the formula for $P(t)$ equal to $3P_0$ and solve for t:

    $$P_0 e^{.05t} = 3P_0$$
    $$e^{.05t} = 3 \qquad \text{(dividing both sides by } P_0\text{)}.$$

    Apply the natural logarithm to each side:

    $$\ln e^{.05t} = \ln 3$$
    $$.05t = \ln 3$$
    $$t = \frac{\ln 3}{.05} \approx 22 \text{ years} \quad \text{(rounding off 21.97 to 22).}$$

13. (a) From the differential equation $P'(t) = -.021P(t)$, you know that $P(t)$ has the form
$P(t) = P_0 e^{-.021t}$. The initial amount is 8 grams, so $P(t) = 8e^{-.021t}$.

(b) $P(0) = P_0 = 8$ grams.

(c) The decay constant is .021 (not $-.021$)

(d) Compute $P(10) = 8e^{-.021(10)} \approx 6.5$ grams.

(e) $P'(t) = -.021P(t)$. If $P(t) = 1$, then at this same time, $P'(t) = -.021(1) = -.021$ grams/year. The sample is disintegrating at the rate of .021 grams/year.

(f) The time is not specified directly, but it described by the property that P(t) is "disintegrating at the rate of .105 grams per year," that is, $P'(t_0) = -.105$ for some particular time t_0. From the differential equation, you can conclude that

$$P'(t_0) = -.021P(t_0)$$

that is,

$$-.105 = -.021 \cdot P(t_0).$$

Hence, $P(t_0) = \frac{-.105}{-.021} = 5$. Thus 5 grams of material remain at the time when the disintegration rate is .105 grams per year.

(g) A half-life of 33 years means that half of any given amount will remain in 33 years. Of the original 8 grams, 4 will remain in 33 years; of that amount, 2 grams will remain after another 33 years, and 1 gram will remain after yet another 33 years (a total of 99 years).

19. This problem is the problem type #4. The first step is to find the decay constant, using the fact that 5 grams decay to 2 grams in 100 days. You may assume that $P(0) = 5$, and $P(t) = 5e^{-\lambda t}$. Set $P(100) = 2$ and solve for λ:

$$5e^{-\lambda(100)} = 2$$

$$e^{-100\lambda} = \frac{2}{5} = .4$$

$$-100\lambda = \ln(.4)$$

$$\lambda = \frac{\ln(.4)}{-100}$$

$$\approx .00916291 \quad \text{or} \quad .00916.$$

Once λ is determined, write the explicit formula for $P(t)$, that is, $P(t) = 5e^{-.00916t}$. The second step is to find the value of t at which $P(t)$ is 1 gram. Set $P(t) = 1$ and solve for t:

$$5e^{-.00916t} = 1$$

$$e^{-.00916t} = .2$$

$$-.00916t = \ln(.2)$$

$$t = \frac{\ln(.2)}{-.00916}$$

$$\approx 175.70.$$

The material will decay to 1 gram in about 176 days.

Helpful Hint: It is wise to keep at least two or three significant figures in a decay constant or growth constant. Otherwise, the answers to other parts of the problem can vary somewhat. In the solution to Exercise 19, a value of .009 for λ leads to the answer that 1 gram will remain in 179 days.

25. Let P_0 be the original level of ^{14}C in the charcoal. Since the decay constant for ^{14}C is $\lambda = .00012$, the amount of ^{14}C remaining after t years is $P(t) = P_0 e^{-.00012t}$. The discovery in 1947 found that the amount of ^{14}C found in the charcoal is $.20P_0$, assuming that the original ^{14}C level in the charcoal was the same as the level in living organisms today. That is, we assume that the time t satisfies $P(t) = .20P_0$, which says that

$$P_0 e^{-.00012t} = .2P_0.$$

Solving for t,

$$e^{-.00012t} = .2$$

$$-.00012t = \ln(.2)$$

$$t = \frac{(\ln .2)}{(-.00012)}$$

$$\approx 13{,}412 \text{ years.}$$

That was the estimated age of the cave paintings more than 65 years ago. The estimated age now is about 13,500 years.

31. Read through the answers and note that in some cases such as (c) and (d), a function is evaluated at a specified value of t, that is, the time t is known but the value of the function must be computed. In other cases such as (a) and (f), the time t is unknown and an equation is solved to find t.

Next, note that some answers such as (b) and (c) involve the function $P(t)$, the amount of material present, while other answers such as (d) and (f) involve the derivative $P'(t)$, the rate at which the amount of material is changing. Finally, answers (g) and (h) relate to the specific type of function and differential equation that are associated with the model for radioactive decay.

Now, before you read further in this solution, go back to the text, read each question, and try to find the appropriate answer. After you have done this, read the solutions that follow.

A. The only differential equation is in (g).

B. The question "How fast?" indicates an answer involving the derivative (rate of change). The time is known, so the answer is (d): compute $P'(.5)$.

C. The general form of the function $P(t)$ is (h).

D. Half-life is the time at which $P(t)$ is half of the original amount, $P(0)$. The answer is (a).

E. "How many grams?" relates to $P(t)$, not $P'(t)$. The time is known, so the answer is (c): compute $P(.5)$.

F. The phrase "disintegrating at the rate" relates to $P'(t)$, and the question "When?" asks for a value of t. The answer is (f).

G. The question "When?" implies that the answer is either (a), (b) or (f). Rate of change is not involved, so (f) is ruled out. To choose between (a) and (b), you need to realize that the .5 in (b) is the value of $P(t)$ while the .5 in (a) only compares the value of $P(t)$ to $P(0)$. So the answer is (b).

H. "How much" relates to $P(t)$, and "present initially" actually gives the time, namely $t = 0$, although the word "year" is not mentioned. The answer is (e).

Helpful Hint: Exercise 31 provides a good review of basic concepts of exponential decay. Similar questions for exponential growth are in Exercise 25 of Section 5.2 and Exercise 13 of the Review Exercises.

Helpful Hint: Use estimation to check whether your answer to a growth or decay problem is reasonable.

(a) Suppose a population doubles every 24 years and you compute that it will increase tenfold in 100 years. Is this reasonable? Round the doubling time to 25 years and notice that in 100 years, the initial population P_0 will double four times, from P_0 to $2P_0$, then to $4P_0$, $8P_0$ and finally $16P_0$, which is much more than just ten times P_0. Therefore, the answer is not reasonable.

(b) Suppose that a radioactive material has a half-life of 5 years and you compute that 1/10 of the material will remain after 22 years. Is this reasonable? Consider the following table which was constructed solely from the fact that the material will halve every five years.

Number of years	5	10	15	20
Fraction of original remaining	$\dfrac{1}{2}$	$\dfrac{1}{4}$	$\dfrac{1}{8}$	$\dfrac{1}{16}$

Since 1/10 is between 1/8 and 1/16, the time lies between 15 and 20 years. Therefore, the answer of 22 is not reasonable.

5.2 Compound Interest

The exercises in this section are similar to those in Section 5.1. A problem involving continuously compounded interest can be thought of as a problem about a population of money that is growing exponentially. The exercises in this section are of the same five types discussed in the notes for Section 5.1.

1. (a) The initial amount deposited in $A(0) = \$5000$.

(b) The interest rate is .04 or 4% interest per year.

(c) After 10 years, the amount in the account is

$$A(10) = 5000e^{.04(10)} \approx \$7459.12.$$

(d) $A(t)$ satisfies the differential equation

$$A'(t) = .04A(t)$$

which is also written as $y' = .04y$.

(e) From the differential equation and **(c)**,

$$A'(10) = .04 \cdot A(10) = .04 \cdot (7459.12)$$
$$\approx 298.36$$

After 10 years, the savings amount balance is growing at the rate of about $298.36 per year.

(f) Let t_0 be the time at which $A'(t_0) = \$280$ per year. From the differential equation, $A(t_0)$ must satisfy

$$280 = .04 \cdot A(t_0)$$

so that $A(t_0) = 280/.04 = \$7000$. The account balance will be about $7000 when it is growing at the rate of $280 per year.

Helpful Hint: The phrase "after 10 years" in Exercise 1 means "after (exactly) 10 years of time have passed." This is a fairly common way of specifying $t = 10$. (The number 10, of course, could be replaced by any positive number.)

7.   The initial amount and the interest rate are given. So $A(t) = 1000e^{.06t}$. To find when $A(t) = 2500$, set the formula for $A(t)$ equal to 2500, and solve for t:

$$1000e^{.06t} = 2500$$
$$e^{.06t} = 2.5$$

Apply the natural logarithm to both sides:

$$\ln e^{.06t} = \ln 2.5$$
$$.06t = \ln 2.5$$
$$t = \frac{\ln 2.5}{.06} \approx 15.27 \quad \text{(to two decimal places)}.$$

About fifteen and one-quarter years are required for the account balance to reach $2500.

13.   Since $A(t) = Pe^{rt}$, the fact that an initial investment P triples to $3P$ in 15 years means that the interest rate r satisfies

$$Pe^{r(15)} = 3P.$$

Divide by P and take the natural logarithm of each side:

$$e^{15r} = 3$$
$$15r = \ln 3$$
$$r = \frac{(\ln 3)}{15} \approx .07324.$$

Interest rates are usually reported to two or three significant figures. In this case, estimate the rate either as 7.3% or 7.32%.

Helpful Hint: Exercises 15–18 are two-step problems, similar to Exercise 19 in Section 5.1.

Step 1: Find the interest rate r. Keep at least two or preferably three significant figures in r. Otherwise, the answer to step 2 will have less accuracy.

Step 2: Write the formula for $A(t)$ with the value of r filled in, and use $A(t)$ to find the time required for the investment to grow to a certain amount.

19.   The text supplies a formula for present value, but you may find it easier just to remember the equation $A = Pe^{rt}$ in the form:

$$[\text{future } \underline{A}\text{mount}] = [\underline{P}\text{resent value}] \cdot e^{rt}.$$

If P, r, and t are given, you can compute A. If A, r, and t are given, you can compute P (the "present value"). In Exercise 19 you are told the value of the investment at a future time (3 years), so that value is A. Thus $A = \$1000$, $t = 3$ years, r is .08, and

$$1000 = Pe^{.08(3)}$$

$$P = \frac{1000}{e^{.24}} \approx \$786.63 \quad \text{(to the nearest cent)}.$$

25. This exercise is analogous to Exercise 31 in Section 5.1, with the function $A(t) = Pe^{rt}$ in place of $P(t)$. You might review that exercise before continuing here.

For most of the questions in this exercise, you need to decide (1) whether the question involves $A(t)$ or the rate of change of $A(t)$ and (2) whether the time t is known or unknown.

   A. "How fast?" relates to $A'(t)$, the rate of growth of the balance $A(t)$. Since the time is known, the answer is (d): compute $A'(3)$.

   B. Answer: (a), the general form of $A(t)$ is Pe^{rt}.

   C. The question involves the amount $A(t)$ in the account and its relation to the initial amount $A(0)$. The question, "How long?" indicates you must solve to find t. Answer: (h), solve $A(t) = 3A(0)$ for t.

   D. "The balance" refers to $A(t)$. The phrase "after 3 years" means that $t = 3$. Answer: (b), compute $A(3)$.

   E. The question concerns $A(t)$. "When?" means that t must be found. Answer: (f), solve $A(t) = 3$ for t.

   F. The phrase, "the balance . . . growing . . . rate" relates to $A'(t)$. "When?" means that t must be found. Answer: (e), solve $A'(t) = 3$ for t.

   G. Answer: (c), principal amount means initial amount.

   H. The differential equation is in (g): $y' = ry$.

Helpful Hint: Use estimation to check that the answer to a compound interest problem is reasonable.

(a) Suppose $1000 is invested at 5% interest and you compute it will grow to about $1103 in two years. Is this reasonable? Well, the deposit will earn .05($1000) or $50 interest the first year and a little more than that (due to interest on the interest) during the second year. Therefore, it will earn a little more than $100 in interest in two years, and so the result is reasonable.

(b) Suppose an investment doubles every eight years and you compute that it will increase tenfold in 20 years. Is this reasonable? Well, the investment will increase fourfold in 16 years and eightfold in 24 years. Therefore, the answer is not reasonable.

(c) Bankers have a Rule of 70 that can be used to estimate the doubling time of an investment (because 70 is approximately 100 times ln 2). The Rule of 70 says that an investment earning an interest rate of r% will double in about 70/r years. For instance, an investment earning 7 percent interest will double in about 70/7 or 10 years. Similarly, an investment doubling in d years has earned an interest rate of about 70/d percent per year. (In former times, bankers used a rule of 72, because they relied on mental arithmetic, and 72 is easily divisible by many common interest rates.)

31. Set $Y_1 = 1200 * e^{\wedge}(-3X) + 800 * e^{\wedge}(-4X) + 500 * e^{\wedge}(-5X)$. Then use the SOLVER command (on the TI-83/84, $\boxed{\text{MATH}}$ 0) to solve the equation $Y_1 - 2000 = 0$. A good initial guess is $X = .10$ (a 10% interest rate). The answer to six decimal places is $X = .060276$, so the interest rate is about 6.0%.

5.3   Applications of the Natural Logarithm Function to Economics

The material in this section is not used in any other part of the book. Elasticity is one of the most important concepts of economics and provides insights into the pricing of goods and services. In addition to being able to compute the elasticity of demand, you should be able to interpret what it means for the demand to be, say, elastic. The box on page 276 summarizes this. Here is another more informal way to remember it. When demand is elastic, an increase in price causes such a decline in sales that the total revenue falls. When demand is inelastic, an increase in price causes only a relatively small decline in sales, so that the total revenue still increases.

1. $f(t) = t^2$, so $f'(t) = 2t$.

    Thus at $t = 10$, $\dfrac{f'(10)}{f(10)} = \dfrac{2(10)}{10^2} = \dfrac{20}{100} = .2 = 20\%.$

    At $t = 50$, $\dfrac{f'(50)}{f(50)} = \dfrac{2(50)}{50^2} = \dfrac{100}{2500} = .04 = 4\%.$

7. $f(p) = \dfrac{1}{p+2}$, thus $f'(p) = \dfrac{-1}{(p+2)^2}.$

    So at $p = 2$, $\dfrac{f'(2)}{f(2)} = \dfrac{-1/16}{1/4} = -\dfrac{4}{16} = -.25 = -25\%.$

    And at $p = 8$, $\dfrac{f'(8)}{f(8)} = \dfrac{-1/100}{1/10} = -\dfrac{10}{100} = -.1 = -10\%.$

13. $f(p) = q = 700 - 5p$, so $f'(p) = -5$.

    $$E(p) = \frac{-pf'(p)}{f(p)} = \frac{-p(-5)}{700-5p} = \frac{5p}{700-5p} = \frac{p}{140-p}.$$

    So at $p = 80$, $E(80) = \dfrac{80}{140-80} = \dfrac{80}{60} = \dfrac{4}{3}.$

    Since $\frac{4}{3} > 1$, demand is elastic.

19. (a) $f(p) = q = 600(5 - \sqrt{p}) = 3000 - 600p^{1/2}$, and $f'(p) = -300p^{-1/2}$.

    $$E(p) = \frac{-pf'(p)}{f(p)} = \frac{-p(-300p^{-1/2})}{3000-600p^{1/2}} = \frac{300p^{1/2}}{3000-600p^{1/2}} = \frac{p^{1/2}}{10-2p^{1/2}}.$$

    Thus at $p = 4$, $E(4) = \dfrac{4^{1/2}}{10-2(4)^{1/2}} = \dfrac{2}{10-2(2)} = \dfrac{1}{3}.$

    Since $\frac{1}{3} < 1$, demand is inelastic.

    (b) Since demand is inelastic, an increase in price will bring about an increase in revenue. Thus, the price of a ticket should be raised.

25. $E_c(x) = \dfrac{\frac{d}{dx}\ln C(x)}{\frac{d}{dx}\ln x} = \dfrac{\frac{C'(x)}{C(x)}}{\frac{1}{x}} = \dfrac{xC'(x)}{C(x)}.$

29. Use X for the variable p, and set $Y_1 = 60000*e^{\wedge}(-.5*X)$. If Y_4 is the first derivative of Y_1, then use Y_2 for the elasticity of demand function and set $Y_2 = -X*Y_4/Y_1$. To solve $Y_2 = 1$, apply the SOLVER program to the equation $0 = Y_2 - 1$.

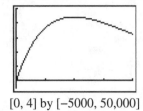

[0, 4] by [−5000, 50,000]

The graph is increasing for $x < 2$, so revenue is increasing for $p < 2$.

5.4   Further Exponential Models

This section illustrates the wide variety of applications in which exponential function appear. The material in this section is not needed for any other part of the book. The logistic curve will be studied further in Chapter 10. However, the discussion there is independent of the discussion in Section 5.4.

1. (a) $f(x) = 5(1-e^{-2x}) = 5 - 5e^{-2x}$, $x \geq 0$. Thus $f'(x) = 10e^{-2x}$. Since $10e^{-2x} > 0$ for every value of x, the derivative is positive for all x, in particular when $x \geq 0$. Thus $f(x)$ is increasing.

$f''(x) = -20e^{-2x}$. Since $-20e^{-2x} < 0$, the second derivative is negative for all x. Thus $f(x)$ is concave down.

(b) $f(x) = 5(1-e^{-2x})$, $x \geq 0$.

Note that as x gets larger, $e^{-2x} = \frac{1}{e^{2x}}$ gets closer and closer to zero. Hence when x is very large the values of $f(x)$ are very close to 5. (The values of $f(x)$ are slightly less than 5 because $1-e^{-2x}$ is slightly less than 1.)

(c)

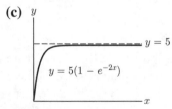

$y = 5$

$y = 5(1 - e^{-2x})$

7. The number of people who have heard about the indictment by time t is given by $f(t) = P(1-e^{-kt})$, where P is the total population. Since after one hour, one quarter of the citizens had heard the news, $f(1) = P(1-e^{-k(1)}) = \frac{1}{4}P = .25P$. Hence

$$1-e^{-k} = .25,$$
$$e^{-k} = .75,$$
$$-k = \ln.75 \approx -.29, \quad \text{or} \quad k = .29.$$

Before continuing with the solution, write out the formula for $f(t)$ for future reference.

$$f(t) = P(1-e^{-.29t}).$$

To find the time t when $f(t) = \frac{3}{4}P = .75P$, set

$$P(1 - e^{-.29t}) = .75P.$$

Thus,

$$1 - e^{-.29t} = .75,$$
$$e^{-.29t} = .25,$$
$$-.29t = \ln .25 \approx -1.4,$$
$$t = \frac{-1.4}{-.29} = \frac{1.4}{.29} \approx 4.8 \text{ hours.}$$

13. Set $Y_1 = 122(e \wedge (-.2X) - e\wedge(-X))$. We assume that Y_4 and Y_5 are the first two derivatives of Y_1.

(a)

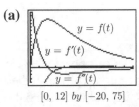

$[0, 12]$ *by* $[-20, 75]$

(b) Evaluate $Y_1(7) \approx 29.97$. About 30 units are present.

(c) Evaluate $Y_4(1) \approx 24.90$. The drug level is increasing at the rate of about 25 units per hour.

(d) Solve $Y_1 - 20 = 0$. If you use the SOLVER command, you'll need to look first at the graph of Y_1 and use TRACE to find an approximate initial guess for X, on the part of the graph where the level is decreasing. Another method is to graph Y_1 and $Y_2 = 20$ in the same window and use the intersect command. You can choose an initial guess for X while in the graphing window. In any case, you should obtain $X \approx 9.038$. So the level of the drug is 20 units at about $t = 9$ hours.

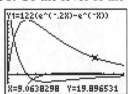

(e) You can use TRACE to estimate the maximum point on the graph of Y_1. On the TI-83, with the suggested window, you should find $(2.04, 65.26)$. A more accurate method is to use SOLVER on the equation $Y_4 = 0$. You should obtain $X \approx 2.0118$. Evaluating $Y_1(X)$, with the exact value of X produced by SOLVER, yields $Y_1 \approx 65.269$. In either case, you might report that after about 2 hours the drug reaches its maximum level of about 65.3 units.

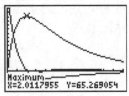

(f) The point of steepest decline on the graph of Y_1 is the inflection point. That's practically impossible to estimate on the graph of Y_1 using TRACE. The two reasonable choices are either to use TRACE on Y_4 and look for the minimum point, or to use SOLVER to find where the second derivative Y_5 is zero. SOLVER will work on Y_5, but only if your initial guess for X is small enough. (If you try X = 12 on the TI-83, the calculator will work for about 2 minutes and give the answer as $X = 9.99...E98$, or about 10^{99}, which is the calculator's view of heaven.)

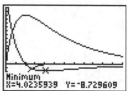

For this graph, the best method is to use TRACE on the graph of Y_4, and find that its minimum point occurs when $X \approx 4.08$. So the drug level is decreasing the fastest when $t = 4$ hours.

Chapter 5 Review Exercises

While solutions for all review exercises are included, expanded explanations are included for every sixth exercise.

1. $P'(x) = -.2P(x)$ implies that $P(x)$ is an exponential function of the form $P(x) = P_0 e^{-.2x}$, where $P_0 = P(0)$ is the atmospheric pressure at sea level. Thus $P_0 = 29.92$, and $P(x) = 29.92 e^{-.2x}$.

2. $P(x) = P_0 e^{kt}$ (t in years, P_0 in herring gulls in 1990)

$$P(13) = 2P_0 = P_0 e^{(k)13} \Rightarrow \ln 2 = \ln e^{13k} \Rightarrow \frac{\ln 2}{13} = k \Rightarrow k \approx 0.0533$$

$$P'(t) = .0533 P(t)$$

3. $10,000 = P_0 e^{(.12)5} = P_0 e^{.6}$

$$P_0 = \frac{10,000}{e^{.6}} \approx \$5488.12$$

4. Solve $3000 = 1000 e^{.1t}$ for t. $\ln 3 = \ln e^{.1t} \Rightarrow \frac{\ln 3}{.1} = t \Rightarrow t \approx 11$ years

5. $\frac{1}{2} = e^{-\lambda(12)}; \frac{\ln .5}{12} = -\lambda \Rightarrow \lambda \approx .058$

6. $.63 = e^{-.00012t} \Rightarrow \frac{\ln .63}{-.00012} = t \Rightarrow t \approx 3850$ years old

7. **(a)** $P(t) = 17e^{kt}$, where $t = 0$ corresponds to January 1, 1990. To find the growth constant, set $P(7) = 19.3$ and solve for k:

$$17e^{k(7)} = 19.3$$

$$e^{7k} = \frac{19.3}{17}$$

$$7k = \ln\left(\frac{19.3}{17}\right)$$

$$k = \frac{1}{7}\ln\left(\frac{19.3}{17}\right) \approx .0181274 \quad \text{or} \quad .0181.$$

Thus $P(t) = 17e^{.0181t}$.

(b) The year 2000 corresponds to $t = 10$. At that time,

$$P(10) = 17e^{.0181(10)}$$

$$\approx 20.4 \text{ million.}$$

(c) The population reaches 25 million when t satisfies

$$17e^{.0181t} = 25$$

$$e^{.0181t} = \frac{25}{17}$$

$$.0181t = \ln\left(\frac{25}{17}\right)$$

$$t = \frac{1}{.0181}\ln\left(\frac{25}{17}\right) \approx 21.3.$$

Since $t = 0$ corresponds to 1990, the population will reach 25 million 21 years later, in 2011.

8. $A(t) = 100,000e^{kt}$

$$A(2) = 117,000 = 100,000e^{2k} \Rightarrow \frac{\ln\frac{11.7}{10}}{2} = k \Rightarrow k \approx .0785 \text{ so it earned } 7.85\%.$$

9. **a.** $A(t) = (10,000e^{.2(5)})e^{.06(5)} = (10,000e)e^{.3}$

$$= 10,000e^{1.3} \approx \$36,693$$

b. $A(t) = 10,000e^{.14(10)} = 10,000e^{1.4}$

$$\approx \$40,552$$

The alternative investment is superior by $\$40,552 - \$36,693 = \$3859$.

10. $P_1(t) = 1000e^{k_1 t}$

$P_1(21) = 2000 = 1000e^{k_1(21)} \Rightarrow \dfrac{\ln 2}{21} = k_1 \Rightarrow k_1 \approx .033$

Thus, $P_1(t) = 1000e^{.033t}$.

$P_2(t) = 710,000e^{k_2 t}$

$P_2(33) = 1,420,000 = 710,000e^{k_2(33)} \Rightarrow \dfrac{\ln 2}{33} = k_2 \Rightarrow k_2 \approx .021$

Thus $P_2(t) = 710,000e^{.021t}$.

Equating P_1 and P_2 and solving for t,

$1000e^{.033t} = 710,000e^{.021t} \Rightarrow 710 = e^{.012t} \Rightarrow \dfrac{\ln 710}{.012} = t \Rightarrow t \approx 547$ minutes

11. The growth constant is .02.

$y' = (.02)(3) = .06$

When the population reaches 3 million people, the population will be growing at the rate of 60,000 people per year.

$100,000 = .02y \Rightarrow y = \dfrac{100,000}{.02} = 5,000,000$

The population level at the growth rate of 100,000 people per year is 5 million people.

12. $y' = .4y \Rightarrow 200,000 = .4y \Rightarrow y = \dfrac{200,000}{.4} = 500,000$

The size of the colony will be 500,000.

$y' = (.4)(1,000,000) = 400,000$

The colony will be growing at the rate of 400,000 bacteria per hour.

13. **(A)** "How fast" concerns $P'(t)$. The time is known, so the answer is (c).

(B) The general form of the function is in (g), $P_0 e^{kt}$.

(C) The question involves the population $P(t)$ and the time is unknown, so the answer is either (a) or (f). The actual size of the population is not given. Rather, it is described as twice the current population. So the answer is (f), which involves $2P(0)$.

(D) The question involves the population $P(t)$, the time is known, and the answer to "what size" requires a value of $P(t)$, not a time. The answer is (b).

(E) The initial size is $P(0)$, which is answer (h).

(F) The exact value of $P(t)$ is specified, and the time is unknown, so the answer is (a).

(G) The question involves the rate of growth, $P'(t)$, and the time is unknown, so the answer is (d).

(H) The differential equation is in (e), $y' = ky$.

14. **(a)** From the graph, $f(5) = 25$ grams.

(b) From the graph, $f(t) = 10$ when $t = 9$ yr.

(c) From the graph, $f(t) = 40$ when $t \approx 3$ yr, so the half-life is about 3 years.

(d) From the graph, $f'(1) = -15$ grams/year.

(e) From the graph, $f'(t) = -5$ when $t = 6$ yr.

15. $A'(t) = rA(t) \Rightarrow 60 = r(1000) \Rightarrow r = \dfrac{60}{1000} = .06 = 6\%$

16. $A'(t) = rA(t) = (.045)(1230) = \$55.35$ per year

17. $\dfrac{f'(t)}{f(t)} = \dfrac{50e^{.2t^2}(.4t)}{50e^{.2t^2}} = .4t$

$\dfrac{f'(10)}{f(10)} = .4(10) = 4 = 400\%$

18. $E(p) = \dfrac{-p(-80p)}{4000 - 40p^2} = \dfrac{80p^2}{4000 - 40p^2} = \dfrac{2}{\frac{100}{p^2} - 1}$

$E(5) = \dfrac{2}{\frac{100}{25} - 1} = \dfrac{2}{4-1} = \dfrac{2}{3} < 1$ so demand is inelastic.

19. This question is based on the elasticity of demand, discussed at the end of Section 5.3. If the elasticity at $p = 8$ is 1.5, then the demand is elastic, because $E(p)$ is greater than 1. In this case, the change in revenue is in the opposite direction of the change in price, for prices close to $p = 8$. So if the price is increased to \$8.16, the revenue will decrease. Furthermore, from the discussion before Example 4 in Section 5.3,

$$\dfrac{[\text{relative rate of change of quantity}]}{[\text{relative rate of change of price}]} = -1.5.$$

The relative increase of price from \$8.00 to \$8.16 is $.16/8.00 = .02$. Therefore the relative change (decrease) of quantity demanded is $(.02)(-1.5) = -.03$. So the demand will fall by about 3%.

20. $f(p) = \dfrac{1}{3p+1}, \ f'(p) = -\dfrac{3}{(3p+1)^2}$

$\dfrac{f'(p)}{f(p)} = \dfrac{-\frac{3}{(3p+1)^2}}{\frac{1}{3p+1}} = -\dfrac{3}{3p+1}$

$\dfrac{f'(1)}{f(1)} = -\dfrac{3}{3+1} = -\dfrac{3}{4} = -75\%$

21. $q = 1000p^2 e^{-.02(p+5)}$

$q' = 1000p^2(e^{-.02(p+5)})(-.02) + (e^{-.02(p+5)})2000p = -1000pe^{-.02(p+5)}(.02p - 2)$

$E(p) = \dfrac{1000p^2 e^{-.02(p+5)}(.02p-2)}{1000p^2 e^{-.02(p+5)}} = .02p - 2$

$E(200) = .02(200) - 2 = 4 - 2 = 2 > 1$

Thus, demand is elastic so a decrease in price will increase revenue.

22. $E(p) = \dfrac{-p(ae^{-bp})(-b)}{ae^{-bp}} = pb$. Thus if $p = \dfrac{1}{b}$, $E(p) = \dfrac{1}{b} \cdot b = 1$.

23. Since for group A, $f'(t) = k(P - f(t))$, it follows that $f(t) = P(1 - e^{-kt})$.

$f(0) = 0 = 100(1 - e^{-k(0)})$ and

$f(13) = 66 = 100(1 - e^{-k(13)}) \Rightarrow .66 = 1 - e^{-13k} \Rightarrow e^{-13k} = .34 \Rightarrow \dfrac{\ln .34}{-13} = k \Rightarrow k \approx .083$

Thus, $f(t) = 100(1 - e^{-.083t})$.

24. $f(t) = \dfrac{M}{1 + Be^{-Mkt}}$

Since 55 is the maximum height for the weed, $M = 55$.

$f(9) = 8 = \dfrac{55}{1 + Be^{-55(9)k}} \Rightarrow 1 + Be^{-55(9)k} = \dfrac{55}{8} \Rightarrow B = \dfrac{47}{8}e^{55(9)k}$

$f(25) = 48 = \dfrac{55}{1 + Be^{-55(25)k}} \Rightarrow 1 + Be^{-55(25)k} = \dfrac{55}{48} \Rightarrow B = \dfrac{7}{48}e^{55(25)k}$

So, $\dfrac{47}{8}e^{55(9)k} = \dfrac{7}{48}e^{55(25)k} \Rightarrow \dfrac{47}{8} = \dfrac{7}{48}e^{880k} \Rightarrow \dfrac{\ln \frac{47 \cdot 48}{8 \cdot 7}}{880} = k \Rightarrow k \approx .0042$ and $-Mk = -.231$

$B = \dfrac{47}{8}e^{55(9)k} \Rightarrow B = \dfrac{47}{8}e^{55(9)(.0042)} \approx 46.98$

Thus, $f(t) = \dfrac{55}{1 + 46.98e^{-.231t}}$.

25. (a) The temperature "after 11 seconds" means the temperature at $t = 11$. This temperature is the y-coordinate of the point on the graph of $f(t)$ at which t is 11, namely, 400°F. (Each horizontal square represents 1 second, and each vertical square represents 100°.)

(b) The "rate of temperature" involves the graph of $f'(t)$. "After 6 seconds" means that $t = 6$. The corresponding y-coordinate is at about −100°per second. Thus, the temperature is decreasing at a rate of 100°/sec.

(c) The temperature of the rod is 200 degrees at the time corresponding to a y-coordinate of 200 on the graph of $f(t)$, namely, at $t = 17$ seconds.

(d) The rod is cooling at the rate of 200°/sec when the y-coordinate of $f'(t)$ is −200, namely, when $t = 2$ sec.

26. Since the culture grows at a rate proportional to its size, $500 = 10,000k \Rightarrow k = .05$. Then, $P' = .05(15,000) = 750$ bacteria per day.

Chapter 6
The Definite Integral

6.1   Antidifferentiation

To find an antiderivative, first make an educated guess and then differentiate the guess. (This may be done mentally if the differentiation is simple.) Your guess should be based on your experience with derivatives. In most cases, functions and their antiderivatives are the same basic type, such as power functions or exponential functions. If your first guess for the antiderivative is correct, you are finished (after you add " + C"). With some practice, your guess usually will be correct except for a constant factor, and you will only need to multiply the guess by a suitable constant. Exercises 25 to 36 show the correct form of the antiderivative. All you have to do is to adjust the constant k correctly.

1. By the constant multiple rule,

$$\frac{d}{dx}\frac{1}{2}x^2 = \frac{1}{2}\cdot\frac{d}{dx}x^2 = \frac{1}{2}\cdot 2x = x.$$

Hence one antiderivative of $f(x) = x$ is $F(x) = \frac{1}{2}x^2$, and all antiderivatives are of the form $F(x) = \frac{1}{2}x^2 + C.$

7. Since $\dfrac{d}{dx}x^4 = 4x^3$,

$$\int 4x^3 dx = x^4 + C.$$

13. Rewriting the integral using Formula 5, we have

$$\int\left(\frac{2}{x}+\frac{x}{2}\right)dx = \int\left(2x^{-1}+\frac{1}{2}x\right)dx = \int 2x^{-1}\,dx + \int\frac{1}{2}x\,dx$$

Since $\dfrac{d}{dx}2\ln|x| = \dfrac{2}{x}$ and $\dfrac{d}{dx}\left(\dfrac{1}{4}x^2\right) = \dfrac{1}{2}x,$

$$\int\left(2x^{-1}+\frac{1}{2}x\right)dx = 2\ln|x|+\frac{1}{4}x^2 + C.$$

19. Since $\dfrac{d}{dx}\left(-\dfrac{3}{2}e^{-2x}\right) = 3e^{-2x},$

$$\int 3e^{-2x}dx = -\frac{3}{2}e^{-2x} + C.$$

25. The equation $\int 5e^{-2t}\,dt = ke^{-2t} + C$ reminds you that e^{-2t} is the correct type of function to use as an antiderivative—you only have to adjust the constant multiple in front of it.

After some practice, you will be able to guess immediately what k should be. At first, however, just compute

$$\frac{d}{dt}ke^{-2t} = -2ke^{-2t}.$$

You want k to make $-2k$ equal to 5 (because there is a factor 5 inside the indefinite integral). To make $-2k = 5$, choose $k = 5/(-2)$, or $k = -5/2$. Check:

$$\frac{d}{dt}\left(\frac{5}{-2}\right)e^{-2t} = \left(\frac{5}{-2}\right) \cdot \frac{d}{dt}e^{-2t}$$
$$= \left(\frac{5}{-2}\right) \cdot (-2)e^{-2t}$$
$$= 5e^{-2t}.$$

Thus $k = -5/2$ is correct, and $(-5/2)e^{-2t}$ is an antiderivative of $5e^{-2t}$. The -2 in the denominator of $\frac{5}{-2}$ anticipates the -2 that will appear when e^{-2t} is differentiated. Once you have the antiderivative $(-5/2)e^{-2t}$, don't forget to add "+ C" to generate all the other antiderivatives.

31. The formula $\int (4-x)^{-1}\,dx = k\ln|4-x| + C$ reminds you of the type of antiderivative you need. If you guess that $k = 1$, then you have forgotten about the -1 that appears when $\ln|4-x|$ is differentiated:

$$\frac{d}{dx}\ln|4-x| = (4-x)^{-1} \cdot (-1).$$

The -1 is the derivative of the function $4-x$ inside the logarithm. To compensate for the -1, take $k = -1$. A check will show why that works:

$$\frac{d}{dx}(-1) \cdot \ln|4-x| = (-1)\frac{d}{dx}\ln|4-x|$$
$$= (-1) \cdot (4-x)^{-1}(-1)$$
$$= (4-x)^{-1}.$$

Thus $-\ln|4-x|$ is an antiderivative of $(4-x)^{-1}$, and the others are found by adding an arbitrary constant.

Helpful Hint: The antiderivatives in Exercises 27–29 and 31–34 are a little harder to find because you must remember how the chain rule works. (Later we will study a general method for handling such problems. For now, however, you may use the technique of simply guessing the general form of an antiderivative.) For instance, to handle Exercise 33, compute

$$\frac{d}{dx}k(3x+2)^5 = 5k(3x+2)^4 \cdot \frac{d}{dx}(3x+2)$$
$$= 5k(3x+2)^4 \cdot 3$$
$$= 15k(3x+2)^4.$$

You want $15k = 1$, so take $k = 1/15$. Perhaps you can do the calculation above mentally. In any case, once you "see" the 5 and the 3 appearing after the differentiation is performed, you can guess how to choose k so that

$$\frac{d}{dx}k(3x+2)^5 = (3x+2)^4.$$

Remember that you can check your final answer by differentiating what you think is the correct antiderivative.

37. Since $\dfrac{d}{dt}\left(\dfrac{2}{5}t^{5/2}\right) = t^{3/2}$, $f(t) = \dfrac{2}{5}t^{5/2} + C$.

Helpful Hint: In Exercises 13–46 and 55–65, each answer is a single function. This function is found by first antidifferentiating and then using additional information to find the proper value of C. On an exam or homework assignment, this value of C should be substituted for C in the antiderivative to obtain the answer.

43. $f(x)$ is an antiderivative of x. Since $\dfrac{d}{dx}\left(\dfrac{1}{2}x^2\right) = x, \dfrac{1}{2}x^2$ is an antiderivative of x and $f(x) = \dfrac{1}{2}x^2 + C$

for some C. To make $f(0) = 3$, we need

$$f(0) = \frac{1}{2}0^2 + C = 3.$$

Therefore, $C = 3$ and $f(x) = \dfrac{1}{2}x^2 + 3$.

49. (a) $\dfrac{d}{dx}\left(\dfrac{1}{x} + C\right) = \dfrac{d}{dx}(x^{-1} + C) = -x^{-2} \neq \ln x$, so (a) is not the answer.

(b) $\dfrac{d}{dx}(x\ln x - x + C) = x\left(\dfrac{1}{x}\right) + \ln x(1) - 1 = \ln x$, (b) is the answer.

(c) $\dfrac{d}{dx}\left(\dfrac{1}{2}(\ln x)^2\right) = \dfrac{1}{2}(2)(\ln x)\left(\dfrac{1}{x}\right) = \dfrac{\ln x}{x} \neq \ln x$, so the answer is not (c).

55. (a) Since velocity is the derivative of position, or in this case, height, to find $s(t)$, the height of the ball at time t, we find the antiderivative of the velocity:

$$s(t) = \int (96 - 32t)dt = 96t - 16t^2 + C.$$

We know the initial height of the ball was 256 feet, so

$$s(0) = 96(0) - 16(0)^2 + C = 256 \Rightarrow C = 256$$

and

$$s(t) = -16t^2 + 96t + 256.$$

(b) To find when the ball hits the ground, we solve $s(t) = 0$:

$$-16t^2 + 96t + 256 = 0$$
$$t^2 - 6t - 16 = 0$$
$$(t+2)(t-8) = 0.$$

Therefore, $t = -2$ or $t = 8$. Since $t = 8$ is the only sensible solution, we conclude the ball hits the ground after 8 seconds.

(c) To find how high the ball goes, we find when $s'(t) = 0$:

$$s'(t) = 96 - 32t = 0$$
$$t = 3.$$

Since $s''(t) = -32 < 0$, $t = 3$ is a relative maximum and the ball reaches a height of

$$s(3) = -16(3)^2 + 96(3) + 256$$
$$= 400 \text{ feet.}$$

61. The marginal profit, *MP*, is the derivative of the profit function. Therefore, to find the profit function, we must antidifferentiate the marginal profit function:

$$P(x) = \int MP\,dx$$
$$= \int (1.30 + .06x - .0018x^2)\,dx$$
$$= 1.30x + .03x^2 - .0006x^3 + C.$$

At a sales level of $x = 0$ per day, the shop will lose \$95. Thus $P(0) = C = -95$ and

$$P(x) = 1.30x + .03x^2 - .0006x^3 - 95.$$

67. We have

$$\int \left(e^{2x} + e^{-x} + \frac{1}{2}x^2 \right) dx = \frac{1}{2}e^{2x} - e^{-x} + \frac{1}{6}x^3 + C.$$

Let $C = 0$, then $f(x) = \frac{1}{2}e^{2x} - e^{-x} + \frac{1}{6}x^3$.

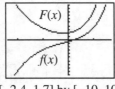

[−2.4, 1.7] by [−10, 10]

6.2 The Definite Integral and Net Change of a Function

We define the **definite integral of *f* from *a* to *b*** as

$$\int_a^b f(x)\,dx = F(b) - F(a),$$

where *f* is a continous function on [*a*, *b*], **F** is an antiderivative of *f* $\left(F'(x) = f(x) \right)$, and *a* and *b* are the limits of integration with *a* as the lower limit and *b* as the upper limit. The number $F(b) - F(a)$ is the **net change** of the function *F*, as *x* varies from *a* to *b*. We use the symbol

$$F(x)\Big|_a^b$$

to represent the net change.

In this section, we will use the definite integral to solve a variety of problems involving net change. In each case, the process is the same:

    (1) Determine an antiderivative for the function.
    (2) Evaluate the definite integral.
    (3) Interpret the result.

1. An antiderivative of $f(x) = 2x - \dfrac{3}{4}$ is $F(x) = x^2 - \dfrac{3}{4}x$. Therefore, we have

$$\int_0^1 \left(2x - \frac{3}{4} \right) dx = \left(x^2 - \frac{3}{4}x \right)\Big|_0^1 = \left(1^2 - \frac{3}{4}(1) \right) - \left(0^2 - \frac{3}{4}(0) \right) = \frac{1}{4}$$

7. First convert the fraction to a sum, then simplify.

$$\int_1^3 \left(\frac{3x - 2x^3 + 4x^5}{4x^7} \right) dx = \int_1^3 \left(\frac{3x}{4x^7} - \frac{2x^3}{4x^7} + \frac{4x^5}{4x^7} \right) dx = \int_1^3 \left(\frac{3}{4}x^{-6} - \frac{1}{2}x^{-4} + x^{-2} \right) dx$$

Now find an antiderivative of the integrand. An antiderivative of $f(x) = \dfrac{3}{4}x^{-6} - \dfrac{1}{2}x^{-4} + x^{-2}$ is

$F(x) = -\dfrac{3}{20x^5} + \dfrac{1}{6x^3} - \dfrac{1}{x}$. Verify this by finding $F'(x)$.

Finally, evaluate $F(x)$ from $a = 1$ to $b = 3$.

$$\left(-\frac{3}{20x^5} + \frac{1}{6x^3} - \frac{1}{x} \right)\Big|_1^3 = \left(-\frac{3}{20(3)^5} + \frac{1}{6(3)^3} - \frac{1}{3} \right) - \left(-\frac{3}{20(1)^5} + \frac{1}{6(1)^3} - \frac{1}{1} \right) = -\frac{59}{180} - \left(-\frac{59}{60} \right) = \frac{59}{90}$$

Thus, $\displaystyle\int_1^3 \left(\frac{3x - 2x^3 + 4x^5}{4x^7} \right) dx = \frac{59}{90}.$

13. First convert the fraction to a sum, then simplify.

$$\int_0^1 \frac{e^x + e^{2x} - 7}{e^{3x}}\,dx = \int_0^1 \left(\frac{1}{e^{2x}} + \frac{1}{e^x} - \frac{7}{e^{3x}} \right) dx = \int_0^1 \left(e^{-2x} + e^{-x} - 7e^{-3x} \right) dx$$

Now find an antiderivative of the integrand. An antiderivative of $f(x) = e^{-2x} + e^{-x} - 7e^{-3x}$ is

$F(x) = -\frac{1}{2}e^{-2x} - e^{-x} + \frac{7}{3}e^{-3x}$. Verify this by finding $F'(x)$.

Finally, evaluate $F(x)$ from $a = 0$ to $b = 1$.

$$\left(-\frac{1}{2}e^{-2x} - e^{-x} + \frac{7}{3}e^{-3x} \right)\Bigg|_0^1 = \left(-\frac{1}{2}e^{-2(1)} - e^{-(1)} + \frac{7}{3}e^{-3(1)} \right) - \left(-\frac{1}{2}e^{-2(0)} - e^{-(0)} + \frac{7}{3}e^{-3(0)} \right)$$

$$= \left(-\frac{1}{2e^2} - \frac{1}{e} + \frac{7}{3e^3} \right) - \left(-\frac{1}{2} - 1 + \frac{7}{3} \right)$$

$$= -\frac{1}{e} - \frac{1}{2e^2} + \frac{7}{3e^3} - \frac{5}{6}$$

Thus, $\int_0^1 \frac{e^x + e^{2x} - 7}{e^{3x}}\,dx = -\frac{1}{e} - \frac{1}{2e^2} + \frac{7}{3e^3} - \frac{5}{6}$.

19. While it's not necessary to do so, the instructions call for us to simplify the expression first, before evaluating the integrals. Whenever possible, try to simplify an expression before evaluating an integral. That usually makes the process easier.

First, apply Property (4), $\int_a^b kf(x)\,dx = k\int_a^b f(x)\,dx$.

$$2\int_1^2 \left(3x + \frac{1}{2}x^2 - x^3 \right) dx + 3\int_1^2 \left(x^2 - 2x + 7 \right) dx = \int_1^2 \left(6x + x^2 - 2x^3 \right) dx + \int_1^2 \left(3x^2 - 6x + 21 \right) dx$$

We can apply Property (2), $\int_a^b \left[f(x) + g(x) \right] dx = \int_a^b f(x)\,dx + \int_a^b g(x)\,dx$, because the limits of integration are the same.

$$\int_1^2 \left(6x + x^2 - 2x^3 \right) dx + \int_1^2 \left(3x^2 - 6x + 21 \right) dx = \int_1^2 \left(6x + x^2 - 2x^3 + 3x^2 - 6x + 21 \right) dx$$

$$= \int_1^2 \left(-2x^3 + 4x^2 + 21 \right) dx$$

Next find an antiderivative of the integrand. An antiderivative of $f(x) = -2x^3 + 4x^2 + 21$ is

$F(x) = -\frac{1}{2}x^4 + \frac{4}{3}x^3 + 21x$. Verify this by finding $F'(x)$.

Finally, evaluate $F(x)$ from $a = 1$ to $b = 2$.

$$\left(-\frac{1}{2}x^4 + \frac{4}{3}x^3 + 21x\right)\Big|_1^2 = \left(-\frac{1}{2}(2)^4 + \frac{4}{3}(2)^3 + 21(2)\right) - \left(-\frac{1}{2}(1)^4 + \frac{4}{3}(1)^3 + 21(1)\right)$$

$$= \frac{134}{3} - \frac{131}{6} = \frac{137}{6}$$

Thus, $2\int_1^2\left(3x + \frac{1}{2}x^2 - x^3\right)dx + 3\int_1^2\left(x^2 - 2x + 7\right)dx = \frac{137}{6}$.

25. We are given $f'(x)$ and are asked to compute $f(1) - f(-1)$. Since $f'(x)$ is an antiderivative of $f(x)$, we will find the value of the definite integral of $f'(x)$ over the limits $a = -1$ to $b = 1$.

$$f(1) - f(-1) = \int_{-1}^1\left(-.5t + e^{-2t}\right)dt = \int_{-1}^1\left(-\frac{1}{2}t + e^{-2t}\right)dt$$

$$= \left(-\frac{1}{4}t^2 - \frac{1}{2}e^{-2t}\right)\Big|_{-1}^1$$

$$= \left(-\frac{1}{4}(1)^2 - \frac{1}{2}e^{-2(1)}\right) - \left(-\frac{1}{4}(-1)^2 - \frac{1}{2}e^{-2(-1)}\right)$$

$$= -\frac{1}{4} - \frac{e^{-2}}{2} - \left(-\frac{1}{4} - \frac{e^2}{2}\right)$$

$$= \frac{e^2 - e^{-2}}{2}$$

31. Let $s(t)$ represent the position function. We know that $s'(t) = v(t) = -32t$, so the change in position is given by

$$s(4) - s(2) = \int_2^4\left(-32t\right)dt = \left(-16t^2\right)\Big|_2^4 = -16(4)^2 - \left(-16(2)^2\right) = -192.$$

The rock fell 192 feet during the time interval $2 \le t \le 4$.

37. Let $T(t)$ represent the value of the investment during the given time interval, 0 to 10 years. Then $T'(t) = R(t)$, and the increase in value is given by

$$T(10) - T(0) = \int_0^{10}T'(t)dt = \int_0^{10}R(t)dt = \int_0^{10}\left(700e^{.07t} + 1000\right)dt$$

$$= \left(\frac{700}{.07}e^{.07t} + 1000t\right)\Big|_0^{10} = \left(10{,}000e^{.07t} + 1000t\right)\Big|_0^{10}$$

$$= \left(10{,}000e^{.07(10)} + 1000(10)\right) - \left(10{,}000e^{.07(0)} + 1000(0)\right)$$

$$\approx 30137.50 - 10{,}000 = 20137.50$$

The investment increased by \$20,137.50.

43. Let $T(t)$ represent the amount of salt in grams during a given time interval. Then $T'(t) = r(t)$, and the amount of salt that was eliminated during the first two minutes is given by

$$T(2) - T(0) = \int_0^2 T'(t)\,dt = \int_0^2 r(t)\,dt = \int_0^2 \left(-\left(t + \frac{1}{2}\right)\right)dt$$

$$= \int_0^2 \left(-t - \frac{1}{2}\right)dt$$

$$= \left(-\frac{t^2}{2} - \frac{1}{2}t\right)\Bigg|_0^2$$

$$= \left(-\frac{2^2}{2} - \frac{1}{2}(2)\right) - \left(-\frac{0^2}{2} - \frac{1}{2}(0)\right)$$

$$= -2 - 1 = -3$$

Three grams of salt were eliminated in the first two minutes.

6.3 The Definite Integral and Area Under a Graph

Sections 6.3 and 6.4 discuss the connection between definite integrals and the areas of regions under a curve.

Riemann (pronounced "Reemahn") sums are introduced here in order to approximate the area under the graph of a nonnegative function, but their importance extends far beyond that. As you will see later, Riemann sums are used to derive some important formulas in applications. Computations with Riemann sums can be time-consuming, but some numerical homework is necessary for a proper understanding. Check with your instructor about how far to carry calculations on an exam. In some cases you may only have to write down the complete Riemann sum (using numbers and not symbols such as Δx), and omit the final arithmetic computation.

Although definite integrals are defined by Riemann sums, they are usually calculated as the net change in an antiderivative. Once you have computed a few Riemann sums, you will probably appreciate the value of the Fundamental Theorem of Calculus.

1. The shaded region is clearly a rectangle. From geometry, we know that the area of a rectangle is given by $A = lw$, so we have $A = 3(2) = 6$ sq units. Using a definite integral, we have

$$A = \int_1^4 2\,dx = (2x)\Big|_1^4 = 2(4) - 2(1) = 6 \text{ sq units.}$$

7. The function is $f(x) = 1/x$. The interval is $1 \le x \le 2$, which is determined by the x-coordinates of the points shown in the figure. The area is the value of

$$\int_1^2 \frac{1}{x}\,dx$$

13. $\int_1^2 \frac{1}{x}\,dx = \ln|x|\Big|_1^2 = \ln 2 - \ln 1 = \ln 2$

19.

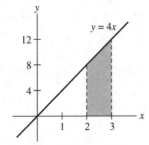

We are seeking the area under the curve $y = 4x$ from $x = 2$ to $x = 3$. This region is shown in the figure above. By Theorem 1, the area is given by the definite integral of the function $y = 4x$ evaluated from $x = 2$ to $x = 3$.

$$\int_2^3 4x \, dx = \left(2x^2\right)\Big|_2^3 = 2(9) - 2(4) = 10 \text{ sq units}$$

25. We are seeking the right endpoint of the region starting at $x = 0$ with area 4 under the curve $y = x^3$. Apply Theorem 1, using b as the right endpoint, then solve for b.

$$\int_0^b x^3 \, dx = 4 \qquad\qquad \text{Apply Theorem 1.}$$

$$\frac{x^4}{4}\Big|_0^b = 4 \qquad\qquad \text{An antiderivative of } x^3 \text{ is } \frac{x^4}{4}.$$

$$\frac{b^4}{4} - 0 = 4 \qquad\qquad \text{Evaluate the antiderivative.}$$

$$b^4 = 16 \Rightarrow b = 2 \quad \text{Solve for } b.$$

31. $a = 0, b = 2, n = 4$, so $\Delta x = \dfrac{b-a}{n} = \dfrac{2-0}{4} = .5$.

The first midpoint is a $a + \dfrac{\Delta x}{2} = 1 + \dfrac{.5}{2} = 1.25$. Subsequent midpoints are found by adding Δx repeatedly. So the other midpoints are:

$$1.25 + .5 = .75$$
$$1.75 + .5 = 2.25$$
$$2.25 + .5 = 2.75.$$

The area is given by

$$\text{Area} = 0.5\left[f(1.25) + f(1.75) + f(2.25) + f(2.75)\right]$$
$$= 0.5\left[(1.25)^2 + (1.75)^2 + (2.25)^2 + (2.75)^2\right]$$
$$= 8.625$$

Helpful Hint: You may use fractions, if you wish, and write 1/4 in place of .25, and so on. However, you can use a calculator more efficiently if you convert to decimals.

37. Since $\Delta x = \dfrac{b-a}{n} = \dfrac{8-0}{4} = 2$, the subintervals have length 2. The first midpoint is at $0 + \dfrac{\Delta x}{2} = 1$. The next midpoint is at $1 + \Delta x = 1 + 2 = 3$, and so on.

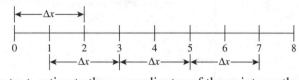

From the figure in the text, estimate the y-coordinates of the points on the curve where $x = 1, 3, 5,$ and 7. Then compute the area.

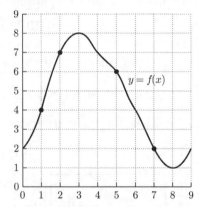

$$A \approx \left[f(1) + f(3) + f(5) + f(7) \right] \Delta x$$
$$= (4 + 8 + 6 + 2) \cdot 2 = 40.$$

43.

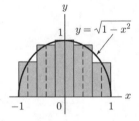

For $-1 \le x \le 1$ and $n = 5$, $\Delta x = \dfrac{b-a}{n} = \dfrac{1-(-1)}{5} = \dfrac{2}{5} = .4$. The first midpoint is at

$-1 + \dfrac{\Delta x}{2} = -1 + \dfrac{.4}{2} = -.8$ The next midpoint is at $-.8 + \Delta x = -.8 + .4 = -.4$, and so on. The midpoints are $-.8, -.4, 0, .4,$ and $.8$. Using a Reimann sum, we compute the area under the curve as follows:

$$\text{Area} = .4\left[f(-.8) + f(-.4) + f(0) + f(.4) + f(.8) \right]$$
$$= .4\left[\left(1-(-.8)^2\right)^{1/2} + \left(1-(-.4)^2\right)^{1/2} + \left(1-(0)^2\right)^{1/2} + \left(1-(.4)^2\right)^{1/2} + \left(1-(.8)^2\right)^{1/2} \right]$$
$$\approx 1.61321$$

The error is

$$1.61321 - 1.57080 = .04241.$$

49. The command fnInt is discussed in the INCORPORATING TECHNOLOGY section at the end of section 6.2.

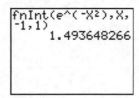

The area under the graph is about 1.494

We include exercise 51 as an example of how to compute a Reimann sum using a graphing calculator.

51. The figure was created on a TI-84 Plus using the program RIEMANN.8xp downloaded from http://www.calcblog.com. Similar programs are available at www.ticalc.org.

Here $\Delta x = (3-1)/20 = .1$. Since we are using midpoints, the left endpoint of the first subinterval is

$$x_1 = 1 + \frac{\Delta x}{2} = 1 + \frac{.1}{2} = 1.05.$$ Refer to the INCORPORATING TECHNOLOGY section on at the end of section 6.3 to see how to compute the Riemann sum in the form

$$\left[f(x_1) + f(x_2) + \cdots + f(x_{20}) \right](.1).$$

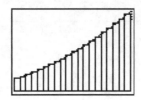

The **seq** command creates the sequence of terms

$$f(x_1),\ f(x_2),\ \ldots,\ f(x_{20})$$

The **sum** command adds the terms and computes

$$f(x_1) + f(x_2) + \cdots + f(x_{20}).$$

After that, multiply the result by $\Delta x = .1$. You should find that

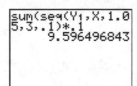

6.4   Areas in the *xy*-Plane

In section 6.3, we used the definite integral to find the area under a continuous, nonnegative function. In this section, we consider cases where the function attains a negative value.

Another goal of this section is to suggest how the change in a function over some interval is related to the area under the graph of the derivative of that function. See Examples 8–10.

1.

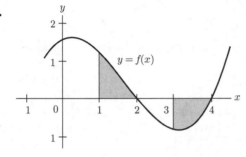

The area is the sum of the area above the *x*-axis and under the curve from $x = 1$ to $x = 2$ and the area below the *x*-axis and above the curve from $x = 3$ and $x = 4$.

$$\int_1^2 f(x)\,dx + \int_3^4 \left(-f(x)\right)dx$$

7.

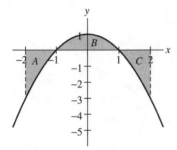

The area is the sum of the areas in regions *A*, *B*, and *C*. Since regions *A* and *C* are below the *x*-axis, we must change the signs of the definite integrals for those regions.

$$A = -\int_{-2}^{-1}\left(1 - x^2\right)dx = -\left(x - \frac{x^3}{3}\right)\Bigg|_{-2}^{-1} = -\left[\left(-1 + \frac{1}{3}\right) - \left(-2 + \frac{8}{3}\right)\right] = -\left(-\frac{4}{3}\right) = \frac{4}{3}$$

$$B = \int_{-1}^{1}\left(1 - x^2\right)dx = \left(x - \frac{x^3}{3}\right)\Bigg|_{-1}^{1} = \left[\left(1 - \frac{1}{3}\right) - \left(-1 + \frac{1}{3}\right)\right] = \frac{4}{3}$$

$$C = -\int_{1}^{2}\left(1 - x^2\right)dx = -\left(x - \frac{x^3}{3}\right)\Bigg|_{1}^{2} = -\left[\left(2 - \frac{8}{3}\right) - \left(1 - \frac{1}{3}\right)\right] = -\left(-\frac{4}{3}\right) = \frac{4}{3}$$

$$A + B + C = \frac{4}{3} + \frac{4}{3} + \frac{4}{3} = 4$$

13. The graph of $y = 2x^2$ is a parabola that opens upward. Its minimum point occurs where $y' = 4x = 0$, that is, at $x = 0$. The minimum point is (0, 0). So the parabola lies entirely below the horizontal line $y = 8$.

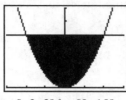

[−3, 3] by [0, 10]

So $f(x) = 2x^2$ and $g(x) = 8$. The area between the two curves from $x = -2$ to $x = 2$ is equal to,

$$\int_{-2}^{2}(8 - 2x^2)\,dx = \left(8x - \frac{2}{3}x^3\right)\Big|_{-2}^{2}$$

$$= 16 - \frac{16}{3} - \left(-16 + \frac{16}{3}\right)$$

$$= 32 - \frac{32}{3} = \frac{96}{3} - \frac{32}{3} = \frac{64}{3}$$

19.

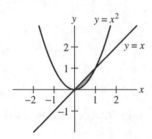

First, we must find the points of intersection of the two curves. Set $x^2 = x$ and solve.

$$x^2 = x$$
$$x^2 - x = 0$$
$$x(x - 1) = 0$$
$$x = 0 \text{ or } x = 1$$

When $x = 0$, $y = 0$, and when $x = 1$, $y = 1$. Thus, the two points of intersection are (0, 0) and (1, 1).

Since $y = x$ is above $y = x^2$ from $x = 0$ to $x = 1$, the area between the curves is given by

$$\int_{0}^{1}\left(x - x^2\right)dx = \left(\frac{1}{2}x^2 - \frac{1}{3}x^3\right)\Big|_{0}^{1} = \frac{1}{2} - \frac{1}{3} = \frac{1}{6}$$

25.

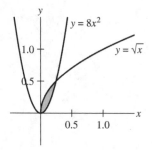

First, we must find the points of intersection of the two curves. Set $8x^2 = \sqrt{x}$ and solve.

$$8x^2 = \sqrt{x}$$

$$64x^4 = x$$

$$64x^4 - x = 0$$

$$x(4x-1)(16x^2 + 4x + 1) = 0$$

$$x = 0 \text{ or } x = \frac{1}{4}$$

(Note that there is no real solution for $16x^2 + 4x + 1 = 0$.)

The two points of intersection are $(0, 0)$ and $\left(\dfrac{1}{4}, \dfrac{1}{2}\right)$.

Since $y = \sqrt{x}$ lies above $y = 8x^2$ from $x = 0$ to $x = \dfrac{1}{4}$, the area between the curves is given by

$$\int_0^{1/4}\left(\sqrt{x} - 8x^2\right)dx = \int_0^{1/4}\left(x^{1/2} - 8x^2\right)dx$$

$$= \left(\frac{2}{3}x^{3/2} - \frac{8}{3}x^3\right)\Bigg|_0^{1/4}$$

$$= \frac{2}{3}\left(\frac{1}{4}\right)^{3/2} - \frac{8}{3}\left(\frac{1}{4}\right)^3$$

$$= \frac{1}{12} - \frac{1}{24} = \frac{1}{24}$$

31. a. Recall from section 6.1 that velocity is the derivative of position, or in this case, height, so we find the antiderivative of the velocity to find $s(t)$, the height of the helicopter at time t. Then we evaluate the definite integral from $t = 0$ to $t = 5$.

$$\int_0^5 (2t + 1)\, dt = \left(t^2 + t\right)\Big|_0^5 = (25 + 5) - 0 = 30 \text{ ft}$$

The helicopter rises 30 ft in the first five seconds.

b.

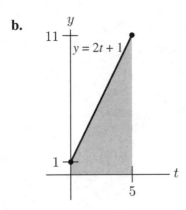

37. a. $\displaystyle\int_0^2 \left(12 + \frac{4}{(t+3)^2}\right) dt = \left(12t - \frac{4}{t+3}\right)\Big|_0^2$

$$= 24 - \frac{4}{5} - \left(-\frac{4}{3}\right)$$

$$= \frac{360}{15} - \frac{12}{15} + \frac{20}{15}$$

$$= \frac{368}{15} \approx 24.5$$

b. The area represents the amount the temperature falls during the first 2 hours.

43. Draw a rough sketch to visualize the area.

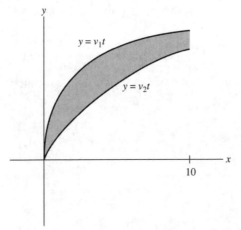

A is the difference between the two heights after 10 seconds.

49. Set $Y_1 = \sqrt{(X+1)}$ and $Y_2 = (X-1)^2$. Deselect all functions except Y_1 and Y_2. You will need a window to graph the functions. Start with one of the basic windows that are predefined on the ZOOM window, and then adjust the range of x-values and y-values as needed. The window shown below is [−.5, 3] by [−1, 2.5]. Notice that between the two intersection points, the graph of Y_1 is above the graph of Y_2. Find the intersection of the two curves, then use **fnInt(** to find the area of the region bounded by the curves.

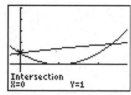 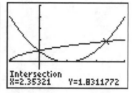

The intersection points are 0 and about 2.35321.

The desired area is the definite integral of $Y_1 - Y_2$ from a to b, where a and b are the x-coordinates of the intersection points. (Refer to the INCORPORATING TECHNOLOGY section at the end of section 6.4 to see how to shade a region under the graph of a function.) Since Y_2 is below Y_1, Y_2 is the first entry in the **Shade(** command.

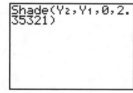 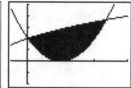

Now use the command **fnInt(** to find the definite integral.

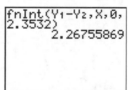

You can also get a fairly good estimate of the desired area without explicitly finding a and b. using the command $\int f(x)dx$ on the CALC menu. Apply the command $\int f(x)dx$ to the function $Y_3 = Y_1 - Y_2$. (Before you integrate Y_3, deselect Y_1 and Y_2.) The area should turn out to be approximately 2.267.

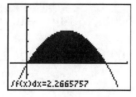

6.5 Applications of the Definite Integral

Although the derivations of the formulas presented in this section rely on Riemann sums, the exercises are worked by substitution into the formulas and do not require analysis in terms of Reimann sums. Example 1 is important.

1. $f(x) = x^2$. By definition, the average value of $f(x)$ between a and b is given by:

$$\text{Average value} = \frac{1}{b-a}\int_a^b f(x)\,dx = \frac{1}{3-0}\int_0^3 x^2\,dx$$

$$= \frac{1}{3}\int_0^3 x^2\,dx = \frac{1}{3}\left[\frac{x^3}{3}\Big|_0^3\right]$$

$$= \frac{1}{3}\left[\frac{3^3}{3}\right] = 3.$$

7. $\text{Average temperature} = \dfrac{1}{12-0}\displaystyle\int_0^{12}\left(47 + 4t - \frac{1}{3}t^2\right)dt$

$$= \frac{1}{12}\left[47t + 2t^2 - \frac{1}{9}t^3\Big|_0^{12}\right]$$

$$= \frac{1}{12}\left[47(12) + 2(12)^2 - \frac{1}{9}(12)^3\right]$$

$$= 47 + 2(12) - \frac{1}{9}(12)^2 = 55°.$$

13. $\text{Consumer's surplus} = \displaystyle\int_A^B [f(x) - B]\,dx, \quad B = f(A).$

$$f(x) = \frac{500}{x+10} - 3$$

$$f(40) = \frac{500}{40+10} - 3 = \frac{500}{50} - 3 = 7$$

$$\text{So the surplus} = \int_0^{40}\left(\frac{500}{x+10} - 3 - 7\right)dx = 500\int_0^{40}\frac{dx}{x+10} - \int_0^{40} 10\,dx$$

$$= 500\ln(x+10)\Big|_0^{40} - 10x\Big|_0^{40} = 500\ln 50 - 500\ln 10 - 400$$

$$= 500\ln 5 - 400 \approx \$404.72.$$

19. To find (A, B) set

$$12 - \frac{x}{50} = \frac{x}{20} + 5$$

$$7 = \frac{x}{20} + \frac{x}{50} = \frac{7x}{100} \Rightarrow x = 100.$$

Thus $A = 100$, and $B = 12 - \frac{100}{50} = 10$.

$$\text{Consumer's surplus} = \int_0^{100} \left(12 - \frac{x}{50} - 10 \right) dx$$

$$= \int_0^{100} \left(2 - \frac{x}{50} \right) dx = 2x - \frac{x^2}{100} \Big|_0^{100}$$

$$= 200 - 100 = \$100.$$

$$\text{Producer's surplus} = \int_0^A (f(A) - f(x)) \, dx$$

$$= \int_0^{100} \left[10 - \left(\frac{x}{20} + 5 \right) \right] dx$$

$$= \int_0^{100} \left(5 - \frac{x}{20} \right) dx = 5x - \frac{x^2}{40} \Big|_0^{100}$$

$$= 500 - \frac{10000}{40} = \$250.$$

25. Future amount $= \int_0^N Pe^{r(N-t)} dt$. We don't know N, and this is what we wish to evaluate. So

$$140,000 = \int_0^N 5000 e^{.1(N-t)} dt$$

$$= 5000 \int_0^N e^{.1N} e^{-.1t} dt$$

hence, $\quad \dfrac{140,000}{5,000} = 28 = \displaystyle\int_0^N e^{.1N} e^{-.1t} dt$

$$= e^{.1N} \int_0^N e^{-.1t} dt = e^{.1N} \left[\frac{e^{-.1t}}{-.1} \Big|_0^N \right]$$

$$= -10 e^{.1N} [e^{-.1N} - 1].$$

So, $\quad 28 = -10 + 10 e^{.1N},$

$\quad\quad\quad 38 = 10 e^{.1N},$

$\quad\quad\quad 3.8 = e^{.1N}.$

Taking the logarithm of both sides, we have $\ln 3.8 = .1N$. Then $10 \ln 3.8 = N$ and $N \approx 13.35$. Thus, it will take about 13.35 years until the value of the investment reaches \$140,000.

31. The volume is given by

$$\int_1^2 \pi (x^2)^2 \, dx = \int_1^2 \pi x^4 \, dx = \frac{\pi}{5} x^5 \bigg|_1^2 = \frac{31\pi}{5} \text{ cubic units.}$$

37. The Riemann sum tells us that $f(x) = x^3$, $n = 4$, and $\Delta x = .5$. Since $a = 8$ and $\Delta x = \dfrac{b-a}{n}$, we have

$$.5 = \frac{b-8}{4},$$

so $2 = b - 8, \quad \text{or} \quad b = 10.$

43. (a) At interest rate r, the average value of the amount in the account over three years is

$$\frac{1}{3} \int_0^3 1000 e^{rt} \, dt = \frac{1}{3} \cdot \frac{1000}{r} e^{rt} \bigg|_0^3 = \frac{1000}{3r} e^{3r} - \frac{1000}{3r}.$$

(b) For the average balance to be $1070.60, the rate r must satisfy the equation

$$\frac{1000}{3r} e^{3r} - \frac{1000}{3r} = 1070.60 \quad \text{or} \quad 1000 e^{3r} - 1000 = 3211.8r$$

To solve for r, set $Y_1 = 1000 * e^{\wedge}(3X) - 1000 - 3211.8X$. Now use the **SOLVER** command (MATH 0). Press △ to display the the edit screen of the **EQUATION SOLVER**. Press CLEAR if necessary, then enter Y_1 (using the VARS menu) to the right of "**eqn: 0 =** ". Next press ENTER. The equation to be solved will be on the first line of the screen and the cursor will be just to the right of "**X =** ". Enter an initial guess of X = .10 (a 10% interest rate). Do not press the ENTER key! You can ignore the next line. Press ALPHA [SOLVE] (above the ENTER key.)The equation solver produces X = .04497… so the desired interest rate is about 4.45%. For more information about using the **EQUATION SOLVER**, refer to the user manual for your calculator.

An alternate way to find the solution is to set $Y_1 = 1000 * e^{\wedge}(3X) - 1000$ and $Y_2 = 3211.8X$, and then find the intersection of the two graphs. The two graphs are close together, so you will need to zoom in.

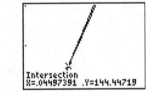

[.0449, .0451] by [144.4, 144.6]

Chapter 6 Review Exercises

While solutions for all review exercises are included, expanded explanations are included for every sixth exercise.

1. The integrand (3^2) is a constant which can be simplified:

$$\int 3^2\, dx = \int 9\, dx.$$

Since
$$\frac{d}{dx}(9x) = 9,$$

$$\int 9\, dx = 9x + C.$$

2. $\int (x^2 - 3x + 2)dx = \frac{1}{3}x^3 - \frac{3}{2}x^2 + 2x + C$

3. $\int \sqrt{x+1}dx = \frac{2}{3}(x+1)^{3/2} + C$

4. $\int \frac{2}{x+4}dx = 2\ln|x+4| + C$

5. $2\int (x^3 + 3x^2 - 1)dx = \frac{1}{2}x^4 + 2x^3 - 2x + C$

6. $\int \sqrt[5]{x+3}dx = \frac{5}{6}(x+3)^{6/5} + C$

7. $\frac{d}{dx}(-2e^{-x/2}) = e^{-x/2}$, so

$$\int e^{-x/2}dx = -2e^{-x/2} + C.$$

8. $\int \frac{5}{\sqrt{x-7}}dx = 10\sqrt{x-7} + C$

9. $\int (3x^4 - 4x^3)\, dx = \frac{3}{5}x^5 - x^4 + C$

10. $\int (2x+3)^7 dx = \frac{1}{16}(2x+3)^8 + C$

11. $\int \sqrt{4-x}\, dx = -\frac{2}{3}(4-x)^{3/2} + C$

12. $\int \left(\frac{5}{x} - \frac{x}{5}\right) dx = 5\ln|x| - \frac{x^2}{10} + C$

13. The integrand $(x+1)^2$ can be simplified:

$$\int_{-1}^{1} (x+1)^2\, dx = \int_{-1}^{1} (x^2 + 2x + 1)\, dx.$$

Since

$$\frac{d}{dx}\left(\frac{x^3}{3} + x^2 + x\right) = x^2 + 2x + 1,$$

$$\int_{-1}^{1} (x^2 + 2x + 1)\, dx = \left(\frac{x^3}{3} + x^2 + x\right)\Bigg|_{-1}^{1} = \left(\frac{1}{3} + 1 + 1\right) - \left(-\frac{1}{3} + 1 - 1\right)$$

$$= \frac{7}{3} - \frac{-1}{3} = \frac{8}{3}.$$

14. $\displaystyle\int_{0}^{1/8} \sqrt[3]{x}\, dx = \frac{3}{4} x^{4/3}\Bigg|_{0}^{1/8} = \frac{3}{64} - 0 = \frac{3}{64}$

15. $\displaystyle\int_{-1}^{2} \sqrt{2x+4}\, dx = \sqrt{2}\int_{-1}^{2} \sqrt{x+2}\, dx$

$$= \frac{2}{3}\sqrt{2}(x+2)^{3/2}\Bigg|_{-1}^{2}$$

$$= \frac{2}{3}\sqrt{2}\left(4^{3/2} - 1\right) = \frac{2}{3}\sqrt{2}(8-1) = \frac{14}{3}\sqrt{2}$$

16. $\displaystyle 2\int_{0}^{1}\left(\frac{2}{x+1} - \frac{1}{x+4}\right) dx = \left[4\ln(x+1) - 2\ln(x+4)\right]\Big|_{0}^{1}$

$$= 4\ln 2 - 2\ln 5 - (4\ln 1 - 2\ln 4)$$

$$= 2\ln\frac{16}{5}$$

17. $\displaystyle\int_{1}^{2}\frac{4}{x^5}\, dx = -\frac{1}{x^4}\Bigg|_{1}^{2} = -\frac{1}{16} - (-1) = \frac{15}{16}$

18. $\displaystyle\frac{2}{3}\int_{0}^{8}\sqrt{x+1}\, dx = \frac{4}{9}(x+1)^{3/2}\Bigg|_{0}^{8} = \frac{4}{9}\left[(8+1)^{3/2} - (0+1)^{3/2}\right] = \frac{4}{9}(27-1) = \frac{104}{9}$

19. $\displaystyle\int_{1}^{4}\frac{1}{x^2}\, dx = \int_{1}^{4} x^{-2}\, dx = -x^{-1}\Big|_{1}^{4} = -\left(\frac{1}{4} - 1\right) = \frac{3}{4}.$

Warning: A common error is to think that $\dfrac{1}{x^3}$ is an antiderivative of $\dfrac{1}{x^2}$. You can avoid this mistake when you switch to the negative exponent notation, $\dfrac{1}{x^2} = x^{-2}$.

Helpful Hint: Many integration problems require finding an antiderivative of a function of the form e^{kx} for some nonzero constant k. For this reason it is helpful to have a method of finding an antiderivative of e^{kx} in your repertoire. Begin by differentiating e^{kx} (use the chain rule):

$$\frac{d}{dx}(e^{kx}) = ke^{kx}.$$

Thus:

$$\frac{d}{dx}\left(\frac{e^{kx}}{k}\right) = e^{kx},$$

$$\text{i.e.,} \quad \int e^{kx}\,dx = \frac{e^{kx}}{k} + C.$$

20. $\displaystyle\int_3^6 e^{2-(x/3)}\,dx = -3e^{2-(x/3)}\Big|_3^6 = -3\left(e^{2-6/3} - e^{2-3/3}\right)$

$$= -3\left(e^{2-2} - e^{2-1}\right) = -3(1-e) = -3+3e = 3(e-1)$$

21. $\displaystyle\int_0^5 (5+3x)^{-1}\,dx = \frac{1}{3}\ln(5+3x)\Big|_0^5$

$$= \frac{1}{3}\ln 20 - \frac{1}{3}\ln 5 = \frac{1}{3}\ln 4$$

22. $\displaystyle\int_{-2}^2 \frac{3}{2e^{3x}}\,dx = -\frac{1}{2e^{3x}}\Big|_{-2}^2 = -\frac{1}{2e^6} - \left(-\frac{1}{2e^{-6}}\right)$

$$= \frac{1}{2}\left(e^6 - e^{-6}\right)$$

23. $\displaystyle\int_0^{\ln 2} \left(e^x - e^{-x}\right)dx = \left(e^x + e^{-x}\right)\Big|_0^{\ln 2} = \left(e^{\ln 2} + e^{-\ln 2}\right) - \left(e^0 + e^0\right)$

$$= \left(2 + \frac{1}{2}\right) - (1+1) = \frac{1}{2}$$

24. $\displaystyle\int_{\ln 2}^{\ln 3} \left(e^x + e^{-x}\right)dx = \left(e^x - e^{-x}\right)\Big|_{\ln 2}^{\ln 3} = \left(e^{\ln 3} - e^{-\ln 3}\right) - \left(e^{\ln 2} - e^{-\ln 2}\right)$

$$= \left(3 - \frac{1}{3}\right) - \left(2 - \frac{1}{2}\right) = \frac{7}{6}$$

25. Begin by simplifying the integrand:

$$\int_0^{\ln 3}\left(\frac{e^x + e^{-x}}{e^{2x}}\right)dx = \int_0^{\ln 3}\left(\frac{e^x}{e^{2x}} + \frac{e^{-x}}{e^{2x}}\right)dx = \int_0^{\ln 3}(e^{-x} + e^{-3x})\,dx.$$

The formula from the helpful hint above tells us that $\dfrac{e^{-x}}{-1}$ is an antiderivative of e^{-x} and $\dfrac{e^{-3x}}{-3}$ is an antiderivative of e^{-3x}. Thus

$$\int_0^{\ln 3}(e^{-x}+e^{-3x})\,dx = \left(-e^{-x}-\frac{e^{-3x}}{3}\right)\Bigg|_0^{\ln 3} = \left(-e^{-\ln 3}-\frac{e^{-3\ln 3}}{3}\right)-\left(-e^0-\frac{e^0}{3}\right)$$

$$= \left(-\frac{1}{3}-\frac{3^{-3}}{3}\right)-\left(-1-\frac{1}{3}\right) = \left(-\frac{1}{3}-\frac{1}{81}\right)+\frac{4}{3}$$

$$= -\frac{28}{81}+\frac{108}{81}=\frac{80}{81}$$

26. $\displaystyle\int_0^1\frac{3+e^{2x}}{e^x}\,dx = \left(-3e^{-x}+e^x\right)\Big|_0^1 = \left(-3e^{-1}+e\right)-\left(-3e^0+e^0\right)$

$$= \left(-3e^{-1}+e\right)-\left(-3+1\right) = 2+e-\frac{3}{e}$$

27. $\displaystyle\int_1^2(3x-2)^{-3}\,dx = \left(-\frac{1}{6}(3x-2)^{-2}\right)\Bigg|_1^2 = -\frac{1}{6}(3\cdot2-2)^{-2}-\left(-\frac{1}{6}(3\cdot1-2)^{-2}\right)$

$$= \left(-\frac{1}{6}\cdot\frac{1}{16}\right)+\frac{1}{6}=\frac{15}{96}=\frac{5}{32}$$

28. $\displaystyle\int_1^9\left(1+\sqrt{x}\right)dx = \left(x+\frac{2}{3}x^{3/2}\right)\Bigg|_1^9 = \left(9+\frac{2}{3}\cdot9^{3/2}\right)-\left(1+\frac{2}{3}\cdot1^{3/2}\right)$

$$= (9+18)-\left(1+\frac{2}{3}\right)=26-\frac{2}{3}=\frac{76}{3}$$

29. $\displaystyle\int_0^1\left(\sqrt{x}-x^2\right)dx = \left(\frac{2}{3}x^{3/2}-\frac{1}{3}x^3\right)\Bigg|_0^1 = \left(\frac{2}{3}\cdot1^{3/2}-\frac{1}{3}\cdot1^3\right)-\left(\frac{2}{3}\cdot0^{3/2}-\frac{1}{3}\cdot0^3\right)$

$$= \left(\frac{2}{3}-\frac{1}{3}\right)-(0-0)=\frac{1}{3}$$

30. $y=x^3$ lies above $y=\dfrac{1}{2}x^3+2x$ on $[-2, 0]$ and below on $[0, 2]$. Thus, we calculate

$$\int_{-2}^0\left[x^3-\left(\frac{1}{2}x^3+2x\right)\right]dx+\int_0^2\left[\frac{1}{2}x^3+2x-x^3\right]dx$$

$$= \int_{-2}^0\left(\frac{1}{2}x^3-2x\right)dx+\int_0^2\left(-\frac{1}{2}x^3+2x\right)dx$$

$$= \left(\frac{1}{8}x^4-x^2\right)\Bigg|_{-2}^0+\left(-\frac{1}{8}x^4+x^2\right)\Bigg|_0^2$$

$$= 0-0-(2-4)+(-2)+4-(0+0)=4$$

31. The shaded region of the graph represents the area between the upper curve $y = e^x$ and the lower curve $y = e^{-x}$ from $x = 0$ to $x = \ln 2$. This area is represented by the integral

$$\int_0^{\ln 2} (e^x - e^{-x})\, dx.$$

Evaluating the integral gives the area of the shaded region of the graph. The formula in the helpful hint preceding the solution to Exercise 25 can be used to find an antiderivative of e^{-x}.

$$\int_0^{\ln 2}(e^x - e^{-x})\, dx = \left[e^x - (-e^{-x}) \right]\Big|_0^{\ln 2}$$

$$= (e^x + e^{-x})\Big|_0^{\ln 2}$$

$$= (e^{\ln 2} + e^{-\ln 2}) - (e^0 + e^0)$$

$$= \left(2 + \frac{1}{2} \right) - (1 + 1)$$

$$= \frac{1}{2}.$$

32. Set $\sqrt{x} = x^2 \Rightarrow x = x^4 \Rightarrow x(x^3 - 1) = 0 \Rightarrow x = 0$ or $x = 1$. Thus, the graphs intersect at $x = 0, 1$. On $(0, 1)$, $y = \sqrt{x}$ lies above $y = x^2$, and below on $(1, 1.21]$. Thus, we calculate

$$\int_0^1 \left[\sqrt{x} - x^2 \right]dx + \int_1^{1.21}\left(x^2 - \sqrt{x} \right)dx = \left(\frac{2}{3}x^{3/2} - \frac{1}{3}x^3 \right)\Big|_0^1 + \left(\frac{1}{3}x^3 - \frac{2}{3}x^{3/2} \right)\Big|_1^{1.21}$$

$$\approx \frac{2}{3} - \frac{1}{3} - (0 - 0) + .5905 - .8873 - \left(\frac{1}{3} - \frac{2}{3} \right) = \frac{1.109561}{3} \approx .370$$

33. $4 - x^2$ and $1 - x^2$ are even, so the area is given by

$$2\int_0^2 (4 - x^2)\, dx - 2\int_0^1 (1 - x^2)\, dx = 2\left[\left(4x - \frac{1}{3}x^3 \right)\Big|_0^2 - \left(x - \frac{1}{3}x^3 \right)\Big|_0^1 \right] = 2\left[\left(8 - \frac{8}{3} \right) - \left(1 - \frac{1}{3} \right) \right] = \frac{28}{3}$$

34. $\int_{1/2}^2 \left[\left(1 - \frac{1}{x} \right) - \left(x^2 - \frac{3}{2}x - \frac{1}{2} \right) \right] dx = \int_{1/2}^2 \left(-x^2 + \frac{3}{2}x + \frac{3}{2} - \frac{1}{x} \right) dx = \left(-\frac{1}{3}x^3 + \frac{3}{4}x^2 + \frac{3}{2}x - \ln x \right)\Big|_{1/2}^2$

$$= -\frac{8}{3} + 3 + 3 - \ln 2 - \left(-\frac{1}{24} + \frac{3}{16} + \frac{3}{4} - \ln \frac{1}{2} \right) = \frac{39}{16} - \ln 4$$

35. $\int_0^1 (e^x - ex)\, dx = \left(e^x - \frac{e}{2}x^2 \right)\Big|_0^1 = e - \frac{e}{2} - (1 - 0) = \frac{e}{2} - 1$

36. $y = 2x^3 - x^2 - 6x$ lies above $y = x^3$ on $[-2, 0]$ and below on $[0, 3]$. Thus, we calculate

$$\int_{-2}^{0}\left[2x^3 - x^2 - 6x - (x^3)\right]dx + \int_{0}^{3}\left[x^3 - (2x^3 - x^2 - 6x)\right]dx$$

$$= \int_{-2}^{0}\left[x^3 - x^2 - 6x\right]dx + \int_{0}^{3}\left[-x^3 + x^2 + 6x\right]dx = \left(\frac{1}{4}x^4 - \frac{1}{3}x^3 - 3x^2\right)\Big|_{-2}^{0} + \left(-\frac{1}{4}x^4 + \frac{1}{3}x^3 + 3x^2\right)\Big|_{0}^{3}$$

$$= 0 - \left(4 + \frac{8}{3} - 12\right) + \left(-\frac{81}{4}\right) + 9 + 27 - 0 = \frac{253}{12}$$

37. A sketch of the graph reveals three points of intersection.

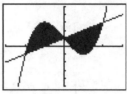

Equate the two expressions in x to obtain

$$x^3 - 3x + 1 = x + 1,$$
$$x^3 - 4x = 0,$$
$$x(x^2 - 4) = 0,$$
$$x(x - 2)(x + 2) = 0.$$

Thus, the graphs intersect at $x = 0, 2, -2$. The total area bounded by the curves is the sum of the areas of the two regions.

$$\text{Area} = \int_{-2}^{0}[(x^3 - 3x + 1) - (x + 1)]dx + \int_{0}^{2}[(x + 1) - (x^3 - 3x + 1)]dx$$

$$= \int_{-2}^{0}(x^3 - 4x)dx + \int_{0}^{2}(-x^3 + 4x)dx$$

$$= \left(\frac{1}{4}x^4 - 2x^2\right)\Big|_{-2}^{0} + \left(-\frac{1}{4}x^4 + 2x^2\right)\Big|_{0}^{2}$$

$$= 0 - (4 - 8) + (-4 + 8) - 0$$

$$= 8.$$

38.
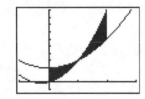

$2x^2 + x = x^2 + 2 \Rightarrow x^2 + x - 2 = 0 \Rightarrow (x + 2)(x - 1) = 0 \Rightarrow x = -2$ or $x = 1$
Thus, on the interval $[0, 2]$, the graphs intersect at $x = 1$. On $(0, 1)$,
$y = x^2 + 2$ lies above $y = 2x^2 + x$ and below on $(1, 2)$.

$$\int_{0}^{1}\left[x^2 + 2 - (2x^2 + x)\right]dx + \int_{1}^{2}\left[2x^2 + x - (x^2 + 2)\right]dx$$

$$= \int_{0}^{1}\left(-x^2 - x + 2\right)dx + \int_{1}^{2}\left(x^2 + x - 2\right)dx = \left(-\frac{1}{3}x^3 - \frac{1}{2}x^2 + 2x\right)\Big|_{0}^{1} + \left(\frac{1}{3}x^3 + \frac{1}{2}x^2 - 2x\right)\Big|_{1}^{2}$$

$$= -\frac{1}{3} - \frac{1}{2} + 2 + \frac{8}{3} + 2 - 4 - \left(\frac{1}{3} + \frac{1}{2} - 2\right) = \frac{6}{3} - \frac{2}{2} + 2 = 3$$

39. $\int (x-5)^2 dx = \frac{1}{3}(x-5)^3 + C$

$f(8) = 2 = \frac{1}{3}(3)^3 + C = 9 + C \Rightarrow C = -7$

$f(x) = \frac{1}{3}(x-5)^3 - 7$

40. $\int e^{-5x} dx = -\frac{1}{5}e^{-5x} + C$

$f(0) = 1 = -\frac{1}{5} + C \Rightarrow C = \frac{6}{5}$

$f(x) = \frac{6}{5} - \frac{1}{5}e^{-5x}$

41. a. $y' = 4t \Rightarrow y = 2t^2 + C$

b. $y' = 4y \Rightarrow y = Ce^{4t}$

c. $y' = e^{4t} \Rightarrow y = \frac{1}{4}e^{4t} + C$

42. Theorem II of section 6.1 states if $F'(x) = 0$ for all x in an interval I, then there is a constant C such that $F(x) = C$ for all x in I.

$$y' = f'(t) = kt(f(t)).$$

Using the hint, we have

$$\frac{d}{dt}\left[f(t)e^{-kt^2/2}\right]$$
$$= f(t)(-kt)e^{-kt^2/2} + f'(x)e^{-kt^2/2}$$
$$= e^{-kt^2/2}\left[-f(t)kt + f'(t)\right]$$
$$= e^{-kt^2/2}\left[-f(t)kt + kt(f(t))\right]$$
$$= e^{-kt^2/2}(0) = 0$$

Since only constant functions have a zero derivative, $f(t)e^{-kt^2/2} = C$ for some C. Thus, $f(t) = Ce^{kt^2/2}$.

43. The cost of producing x tires a day is an antiderivative of the marginal cost function $.04x + 150$. Hence, if the cost function is $C(x)$, then

$$C(x) = \int (.04x + 150)\, dx$$

$$= .02x^2 + 150x + C.$$

If fixed costs are \$500 per day, then the cost of producing no tires is \$500 per day. That is,

$$C(0) = 0 + 0 + C = C = 500.$$

Hence $$C(x) = .02x^2 + 150x + 500 \text{ dollars.}$$

44. $\int_{10}^{20} (400 - 3x^2)\, dx = (400x - x^3)\Big|_{10}^{20} = -3000$

Thus, a loss of \$3000 would result.

45. It represents the total quantity of drug (in cubic centimeters) injected during the first 4 minutes.

46. $v(t) = -9.8t + 20$

a. $\int_{0}^{2} (-9.8t + 20)\, dt = (-4.9t^2 + 20t)\Big|_{0}^{2}$

$$= -19.6 + 40 = 20.4 \text{ m}$$

b.

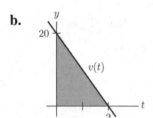

Use the figure below for exercises 47 and 48.

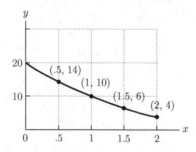

47. $\left[f(0) + f(.5) + f(1) + f(1.5) \right] \Delta x = (20 + 14 + 10 + 6)(.5) = 25$

48. $\left[f(.5) + f(1) + f(1.5) + f(2) \right] \Delta x = (14 + 10 + 6 + 4)(.5) = 17$

49. Here $\Delta x = \dfrac{b-a}{n} = \dfrac{2-0}{2} = 1$, so the midpoints of the two subintervals are given by

$$x_1 = a + \frac{\Delta x}{2} = 0 + \frac{1}{2} = \frac{1}{2},$$

$$x_2 = \frac{1}{2} + \Delta x = \frac{1}{2} + 1 = \frac{3}{2}.$$

By the midpoint rule,

$$\text{Area} \approx \left[f\left(\frac{1}{2}\right) + f\left(\frac{3}{2}\right) \right](1)$$

$$= \left[\frac{1}{\frac{1}{2}+2} + \frac{1}{\frac{3}{2}+2} \right]$$

$$= \frac{2}{5} + \frac{2}{7} = \frac{24}{35} \approx .68571.$$

Computing the integral directly,

$$\int_0^2 \frac{1}{x+2}\,dx = \ln|x+2|\Big|_0^2 = \ln 4 - \ln 2$$

$$= \ln\frac{4}{2} = \ln 2 = .69315 \quad \text{(rounded)}.$$

50. $\Delta x = .2$; the midpoints are .1, .3, .5, .7, .9
Area $= [e^{2(.1)} + e^{2(.3)} + e^{2(.5)} + e^{2(.7)} + e^{2(.9)}](.2) \approx 3.17333$

$$\int_0^1 e^{2x}\,dx = \frac{1}{2}e^{2x}\Big|_0^1 = \frac{1}{2}e^2 - \frac{1}{2} \approx 3.1945$$

51. $p(400) = \sqrt{25 - .04(400)} = 3$

$$\text{C.S.} = \int_0^{400} (\sqrt{25-.04x} - 3)\,dx$$

$$= \left(\frac{2}{-.12}(25-.04x)^{3/2} - 3x \right)\Big|_0^{400}$$

$$= \frac{2}{-.12}(27) - 1200 - \frac{2}{-.12}(125) \approx \$433.33$$

52. $\dfrac{1}{10}\int_0^{10} 3000e^{.04t}\,dt = \int_0^{10} 300e^{.04t}\,dt = 7500e^{.04t}\Big|_0^{10} = 7500(e^{.4} - e^0) \approx \$3688.69$

53. $\dfrac{1}{\frac{1}{2}-\frac{1}{3}}\int_{1/3}^{1/2} \frac{1}{x^3}\,dx = 6\int_{1/3}^{1/2}\frac{1}{x^3}\,dx = 6\left[-\frac{1}{2x^2}\Big|_{1/3}^{1/2} \right] = 6\left(-2+\frac{9}{2}\right) = 15$

54. The sum is approximated by $\displaystyle\int_0^1 3e^{-x}\,dx = -3e^{-x}\Big|_0^1 = -3e^{-1} + 3 = 3(1-e^{-1})$.

55. The region corresponding to

$$\int_a^c f(x)\,dx$$

contains one part above the *x*-axis, with area .68, and one part below the *x*-axis, with area .42. From Section 6.2,

$$\int_a^c f(x)\,dx = .68 - .42 = .26.$$

Similarly,

$$\int_a^d f(x)\,dx = .68 - .42 + 1.70 = 1.96,$$

because the part from *b* to *c* is below the *x*-axis.

56. $\displaystyle\int_0^1 \pi(1-x^2)^2\,dx = \int_0^1 \pi(x^4 - 2x^2 + 1)\ dx = \pi\left(\frac{x^5}{5} - \frac{2}{3}x^3 + x\right)\bigg|_0^1$

$$= \pi\left(\frac{1}{5} - \frac{2}{3} + 1\right) = \frac{8\pi}{15}$$

57. a. Since inventory is decreasing, the slope is $-\dfrac{Q}{A}$. From the graph we can see that $f(t) = Q - \dfrac{Q}{A}t$.

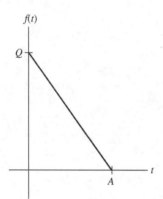

b. $\displaystyle\frac{1}{A}\int_0^A \left(Q - \frac{Q}{A}t\right)dt = \frac{1}{A}\left(Qt - \frac{Q}{2A}t^2\right)\bigg|_0^A = \frac{Q}{A}A - \frac{QA^2}{2A^2} = \frac{Q}{2}$

58. a. $f(t) = Q - \displaystyle\int_0^t rt\ dt = Q - \frac{rt^2}{2}$

b. $0 = Q - \dfrac{rA^2}{2} \Rightarrow r = \dfrac{2Q}{A^2}$

c. $f(t) = Q - \dfrac{\frac{2Q}{A^2}t^2}{2} = Q - \dfrac{Qt^2}{A^2}$

$$\frac{1}{A}\int_0^A \left(Q - \frac{Qt^2}{A^2} \right) dt = \frac{1}{A}\left(Qt - \frac{Qt^3}{3A^2} \right)\Bigg|_0^A$$

$$= Q - \frac{Q}{3} = \frac{2}{3}Q$$

59. a. $g(3)$ is the area under the curve $y = \dfrac{1}{1+t^2}$ from $t = 0$ to $t = 3$.

    b. $g'(x) = \dfrac{1}{1+x^2}$

60. a. $h(0)$ is the area under one-quarter of the unit circle. $h(1)$ is the area under one-half of the unit circle.

    b. $h'(x) = \sqrt{1-x^2}$

61. Observe that the sum is a Riemann sum for the function $f(t) = 5000e^{-.1t}$ from $t = 0$ to $t = 3$. This sum is approximately equal to the value of the definite integral:

$$\int_0^3 5000e^{-.1t}\,dt = -50{,}000e^{-.1t}\Big|_0^3$$

$$= -50{,}000e^{-.3} + 50{,}000$$

$$\approx 13{,}000.$$

62. $\Delta x = \dfrac{1}{n}$, with left endpoints $t_i = i\Delta x$

$$\text{Sum} = \Delta x e^0 + \Delta x e^{\Delta x} + \Delta x e^{2\Delta x} + \cdots + \Delta x e^{(n-1)\Delta x}$$

$$= \Delta x \left[e^{t_0} + e^{t_1} + e^{t_2} + \cdots + e^{t_{n-1}} \right]$$

$$\approx \int_0^1 e^x dx = e - 1$$

63. $\Delta x = \dfrac{1}{n}$, with left endpoints $t_i = i\Delta x$

$$\text{Sum} = \Delta x \left[1 + (1 + \Delta x)^3 + \cdots + (1 + (n-1)\Delta x)^3 \right]$$

$$= \Delta x \left[1 + (1 + t_1)^3 + \cdots + (1 + t_{n-1})^3 \right]$$

$$\approx \int_0^1 (1+x)^3 dx = \frac{(1+x)^4}{4}\Bigg|_0^1 = \frac{15}{4}$$

64. By figure 3(a), the average value of $f(x) = 4$ on $2 \le x \le 6$.

65. True; $3 \le f(x) \le 4$

$$\int_0^5 3\, dx \le \int_0^5 f(x)\, dx \le \int_0^5 4\, dx$$

$$15 \le \int_0^5 f(x)\, dx \le 20 \text{, so } 3 \le \frac{1}{5}\int_0^5 f(x)\, dx \le 4$$

66. a. $\int_{t_1}^{t_2} (20-4t_1)\, dt = (20-4t_1)t\Big|_{t_1}^{t_2} = (20-4t_1)t_2 - (20-4t_1)t_1 = (20-4t_1)\Delta t$

b. Let $R(t) =$ the amount of water added up to time t. Then $R'(t) = r(t)$ and

$\int_0^5 r(t)\, dt = R(5) - R(0) =$ the total amount of water added to the tank from $t = 0$ to $t = 5$.

67. Amount of water used between 1960 and 1995:

$$= \int_0^{35} 860e^{.04t}\, dt$$

$$= 860\frac{e^{.04t}}{.04}\Big|_0^{35} = 21{,}500(e^{.04(35)} - e^0)$$

$$= 21{,}500(e^{1.4} - 1) = 65{,}687 \text{ km}^3.$$

68. $\int_0^1 4500e^{.09(1-t)}\, dt = -50{,}000e^{.09(1-t)}\Big|_0^1 = -50{,}000\left(e^{.09(1-1)} - e^{.09(1-0)}\right) \approx \$4708.71$

69. $\int (3x^2 - 2x + 1)dx = x^3 - x^2 + x + C$

Now $f(1) = 1$, so

$(x^3 - x^2 + x + C)\Big|_{x=1} = 1 \Rightarrow 1 - 1 + 1 + C = 1 \Rightarrow C = 0$

So, $f(x) = x^3 - x^2 + x$.

70. The slope of the line connecting $(0, 0)$ and (a, a^2) is $m = \frac{a^2 - 0}{a - 0} = a$. The equation of the line is

$$y - 0 = a(x - 0) \Rightarrow y = ax.$$

The shaded area is equal to 1, so we have

$$\int_0^a \left(ax - x^2\right)dx = 1$$

$$\left(\frac{a}{2}x^2 - \frac{1}{3}x^3\right)\Big|_0^a = 1$$

$$\frac{a^3}{2} - \frac{a^3}{3} - 0 = 1 \Rightarrow \frac{a^3}{6} = 1 \Rightarrow a = \sqrt[3]{6}$$

71. $\displaystyle\int_0^{b^2}\sqrt{x}\,dx+\int_0^b x^2\,dx=\frac{2}{3}x^{3/2}\Big|_0^{b^2}+\frac{1}{3}x^3\Big|_0^b$

$$=\frac{2}{3}(b^2)^{3/2}-0+\frac{1}{3}b^3-0$$

$$=\frac{2}{3}|b|^3+\frac{1}{3}b^3\qquad\left(\text{since }(b^2)^{1/2}=|b|\right)$$

$$=\frac{2}{3}b^3+\frac{1}{3}b^3\qquad\left(|b|=b\text{ since }b\text{ is positive.}\right)$$

$$=b^3$$

72. $\displaystyle\int_0^{b^n}\sqrt[n]{x}\,dx+\int_0^b x^n\,dx=\left(\frac{n}{n+1}x^{\frac{n+1}{n}}\right)\Big|_0^{b^n}+\left(\frac{1}{n+1}x^{n+1}\right)\Big|_0^b$

$$=\left(\frac{n}{n+1}\left(b^n\right)^{\frac{n+1}{n}}\right)-0+\left(\frac{1}{n+1}b^{n+1}-0\right)$$

$$=\frac{n}{n+1}b^{n+1}+\frac{1}{n+1}b^{n+1}=b^{n+1}$$

$$\left[\text{Note: }\left(b^n\right)^{1/n}=\begin{cases}b,\ n\text{ is odd}\\|b|,\ n\text{ is even}\end{cases},\text{ but }|b|=b\text{ since }b\text{ is positive. So, }\left(b^n\right)^{1/n}=b.\right]$$

73. $\displaystyle\int_0^1(\sqrt{x}-x^2)\,dx=\int_0^1(x^{1/2}-x^2)\,dx$

$$=\int_0^1 x^{1/2}\,dx-\int_0^1 x^2\,dx$$

$$=\left(\frac{x^{3/2}}{\frac{3}{2}}-\frac{x^3}{3}\right)\Big|_0^1$$

$$=\left(\frac{2}{3}x^{3/2}-\frac{x^3}{3}\right)\Big|_0^1$$

$$=\left(\frac{2}{3}1^{3/2}-\frac{1^3}{3}\right)-\left(\frac{2}{3}0^{3/2}-\frac{0^3}{3}\right)$$

$$=\left(\frac{2}{3}-\frac{1}{3}\right)-0=\frac{1}{3}.$$

74. $\displaystyle\int_0^1\left(\sqrt[n]{x}-x^n\right)dx=\left(\frac{n}{n+1}x^{\frac{n+1}{n}}-\frac{1}{n+1}x^{n+1}\right)\Big|_0^1=\left(\frac{n}{n+1}-\frac{1}{n+1}\right)-0=\frac{n-1}{n+1}$

Chapter 7
Functions of Several Variables

7.1   Examples of Functions of Several Variables

You should be sure that you completely understand the meanings of the variables in Examples 2 and 3. These examples will be used to illustrate the concepts introduced in later sections.

1. $f(x, y) = x^2 - 3xy - y^2$. So

$$f(5, 0) = (5)^2 - 3(5)(0) - (0)^2 = 25,$$
$$f(5, -2) = (5)^2 - 3(5)(-2) - (-2)^2$$
$$= 25 + 30 - 4 = 51,$$
$$f(a, b) = a^2 - 3ab - b^2.$$

7. $C(x, y, z)$ is the cost of materials for the rectangular box with dimensions x, y, z in feet.

[Cost] = [Cost of top and bottom] + [Cost of four sides]

   = [Total area of top and bottom]·[Cost per sq. ft.] + [Total area of four sides]·[Cost per sq. ft.]

   = $[xy + xy] \cdot 3 + [xz + yz + xz + yz] \cdot 5$

   = $6xy + 10xz + 10yz$

   = $C(x, y, z)$.

13. $T = f(r, v, x) = \dfrac{r}{100}(.40v - x)$

(a) $v = 200,000$, $x = 5000$, $r = 2.5$,

$$T = \frac{r}{100}[.4v - x],$$
$$T = \frac{2.5}{100}[.4(200,000) - 5000]$$
$$= \$1875.00.$$

(b) $v = 200,000$, $x = 5000$, $r = 3$,

$$T = \frac{3.00}{100}[.4(200,000) - 5000]$$
$$= \$2250$$

The tax due also increases by 20% (or 1/5), since

$$1875.00 + \frac{1}{5}(1875.00) = \$2250.00.$$

19. A level curve has the form $f(x, y) = C$ (C is arbitrary). Rearrange the equation $y = 3x - 4$ so that all terms involving x and y are on the left-hand side:

$$y - 3x = -4.$$

So,

$$y - 3x = C, \text{ some constant.}$$

Thus $f(x, y) = y - 3x$.

25. The graph in Exercise 25 is matched with the level curves in (c). Imagine slicing "near the top" of the four humps. You should visualize a cross section of four circular-like figures and as you move further down the z-axis these circular-like figures become larger and larger. Similarly, Exercise 23 is matched with (d), Exercise 24 is matched with (b), and Exercise 26 is matched with (a).

7.2 Partial Derivatives

Any letter can be used as a variable of differentiation, even y. As a warm-up to this section, look over the following ordinary derivatives.

$$\text{If } f(y) = ky^2, \quad \text{then } \frac{df}{dy} = 2ky.$$

$$\text{If } g(y) = 3y^2x, \quad \text{then } \frac{dg}{dy} = 6yx.$$

Notice that x is a constant in the formula for $g(y)$ because the notation $g(y)$ means that y is the only variable in the function.

1. $\dfrac{\partial}{\partial x}5xy = \dfrac{\partial}{\partial x}[5y]x = 5y,$ [treat $5y$ as a constant]

$\dfrac{\partial}{\partial y}5xy = \dfrac{\partial}{\partial y}[5x]y = 5x.$ [treat $5x$ as a constant]

7. $\dfrac{\partial}{\partial x}(2x - y + 5)^2 = 2(2x - y + 5) \cdot \dfrac{\partial}{\partial x}(2x - y + 5)$

$\qquad\qquad = 2(2x - y + 5) \cdot (2 - 0 + 0)$ [treat y as a constant]

$\qquad\qquad = 4(2x - y + 5).$

$$\frac{\partial}{\partial y}(2x - y + 5)^2 = 2(2x - y + 5) \cdot \frac{\partial}{\partial y}(2x - y + 5)$$

$$= 2(2x - y + 5) \cdot (0 - 1 + 0) \quad \text{[treat } 2x \text{ as a constant]}$$

$$= -2(2x - y + 5).$$

13. $f(L, K) = 3\sqrt{LK} = 3(LK)^{1/2} = 3L^{1/2}K^{1/2}$. Treat 3 and $K^{1/2}$ as constants:

$$\frac{\partial}{\partial L}3L^{1/2}K^{1/2} = \frac{\partial}{\partial L}[3K^{1/2}]L^{1/2} = [3K^{1/2}]\frac{1}{2}L^{-1/2}$$

$$= \frac{3}{2}K^{1/2}L^{-1/2} = \frac{3\sqrt{K}}{2\sqrt{L}}.$$

19. $\dfrac{\partial}{\partial x}(x^2 + 2xy + y^2 + 3x + 5y) = 2x + 2y + 0 + 3 + 0 = 2x + 2y + 3.$

Thus $\qquad \dfrac{\partial f}{\partial x}(2, -3) = 2(2) + 2(-3) + 3 = 4 - 6 + 3 = 1.$

$$\frac{\partial}{\partial y}(x^2 + 2xy + y^2 + 3x + 5y) = 0 + 2x + 2y + 0 + 5 = 2x + 2y + 5.$$

Thus $\qquad \dfrac{\partial f}{\partial y}(2, -3) = 2(2) + 2(-3) + 5 = 4 - 6 + 5 = 3.$

25. $f(x, y) = 200\sqrt{6x^2 + y^2}$

    (a) Marginal productivity of labor is given by $\dfrac{\partial f}{\partial x}$.

$$f(x, y) = 200\sqrt{6x^2 + y^2} = 200(6x^2 + y^2)^{1/2}.$$

$$\frac{\partial f}{\partial x} = 200\left(\frac{1}{2}\right)(6x^2 + y^2)^{-1/2} \cdot \frac{\partial}{\partial x}(6x^2 + y^2)$$

$$= 100(6x^2 + y^2)^{-1/2}(12x) = \frac{1200x}{\sqrt{6x^2 + y^2}}.$$

$$\frac{\partial f}{\partial x}(10, 5) = \frac{1200(10)}{\sqrt{6(10)^2 + (5)^2}} = \frac{12,000}{\sqrt{625}} = \frac{12,000}{25} = 480.$$

Here is what the value 480 of the marginal productivity of labor really means:

At $(x, y) = (10, 5)$, if the amount x of labor changes slightly from 10 units and y stays fixed at 5 units, production increases by 480 times the amount of change in the amount of labor.

See the discussion in Example 5.

Marginal productivity of capital is given by $\dfrac{\partial f}{\partial y}$.

$$\frac{\partial f}{\partial y} = 200\left(\frac{1}{2}\right)(6x^2 + y^2)^{-1/2} \cdot \frac{\partial f}{\partial y}(6x^2 + y^2)$$

$$= 100(6x^2 + y^2)^{-1/2} \cdot (2y) = \frac{200y}{\sqrt{6x^2 + y^2}}.$$

$$\frac{\partial f}{\partial y}(10, 5) = \frac{200(5)}{\sqrt{6(10)^2 + (5)^2}} = \frac{1000}{\sqrt{625}} = \frac{1000}{25} = 40.$$

The value 40 for the marginal productivity of capital means that:

For $(x, y) = (10, 5)$, production changes at a rate 40 times the change in capital, when x stays fixed at 10 units and the amount y of capital changes slightly from 5 units.

This interpretation and the one above illustrate what is written in the box preceding Example 6. The next two parts of the exercise make this discussion more concrete.

(b) Let h represent a small number (positive or negative). If labor is changed by h units from 10 to $10 + h$ units and if capital is fixed at 5 units, then the quantity of goods produced will change by approximately $480 \cdot h$ units of goods.

(c) Suppose labor decreases from 10 to 9.8 units (while capital stays fixed at 5 units). Then the h from part (b) is $-.5$, and production will change by approximately $480(-.5)$ units. That is, production will *decrease* by about 240 units.

31.
$$V = \frac{.08T}{P} = .08TP^{-1},$$

$$\frac{\partial V}{\partial P} = -(.08)TP^{-2},$$

$$\frac{\partial V}{\partial P}(20, 300) = -(.08)(300)(20)^{-2} = \frac{-(.08)(300)}{400} = -.06,$$

$$\frac{\partial V}{\partial T} = .08P^{-1},$$

$$\frac{\partial V}{\partial T}(20, 300) = .08(20)^{-1} = \frac{.08}{20} = .004.$$

$\dfrac{\partial V}{\partial P}(20, 300)$ represents the rate of change of volume, when temperature is fixed at 300 and pressure is allowed to vary near 20. That is, if the pressure is increased by 1 small unit, volume will decrease by about .06 unit.

$\dfrac{\partial V}{\partial T}(20, 300)$ represents the rate of change of volume when pressure is fixed at 20 and temperature is allowed to vary near 300. That is, if temperature is increased by 1 small unit, volume will increase by approximately .004 unit.

37. $f(x, y) = 3x^2 + 2xy + 5y$

$$f(1+h, 4) - f(1, 4) = 3(1+h)^2 + 2(1+h)(4) + 5(4) - [3(1)^2 + 2(1)(4) + 5(4)]$$
$$= 3(1+h^2+2h) + 8 + 8h + 20 - 3 - 8 - 20$$
$$= 3 + 3h^2 + 6h + 8h - 3$$
$$= 3h^2 + 14h.$$

7.3   Maxima and Minima of Functions of Several Variables

Finding relative extreme points for functions of several variables is accomplished in much the same way as for functions of one variable. In Chapter 2, you found relative maximum and minimum points by differentiating and then solving one equation for *x*. In this section, you will differentiate (with partials) and then solve a system of two equations in two unknowns.

Three types of systems of equations arise in this section.

First type: Each equation involves just one variable. An example is:

$$\begin{cases} x^2 + 1 = 5, \\ 3y - 4 = 11. \end{cases}$$

To solve this system, solve each equation separately for its variable. The solutions to the system consist of all combinations of these individual solutions. In this example above, the first equation has the two solutions $x = 2$ and $x = -2$. The second equation has the single solution $y = 5$. Therefore, the system has two solutions: $x = 2$, $y = 5$ and $x = -2$, $y = 5$. [*Note:* These two solutions can also be written as (2, 5) and (−2, 5).]

Second type: One equation contains both variables, the other equation contains just one variable. An example is

$$\begin{cases} 2x + 4y = 26, \\ 3y - 4 = 11. \end{cases}$$

To solve this system, solve the equation having just one variable, substitute this value into the other equation, and then solve that equation for the remaining variable. In the example above, the second equation has the solution $y = 5$. Substitute 5 for *y* into the first equation and obtain $2x + 4(5) = 26$ or $x = 3$. Therefore, the system has the solution $x = 3$, $y = 5$.

Third type: Both equations contain both variables. An example is

$$\begin{cases} -9x + y = 3, \\ \quad y + 2 = 4x. \end{cases}$$

Two methods for solving such systems are the EQUATE METHOD and the SOLVE-SUBSTITUTE METHOD.

With the EQUATE METHOD, both equations are solved for y, the two expressions for y are equated and solved for x, and the value of x is substituted into one of the equations to obtain the value of y. For the system above, begin the writing

$$\begin{cases} y = 3 + 9x, \\ y = 4x - 2. \end{cases}$$

Next, equate the two expressions for y and solve:

$$3 + 9x = 4x - 2,$$
$$5x = -5,$$
$$x = -1.$$

Finally, substitute $x = -1$ into the first equation: $y = 3 + 9(-1) = -6$. Therefore, the solution is $x = -1$, $y = -6$.

With the SOLVE-SUBSTITUTE METHOD, one equation is solved for y as an expression in x, this expression is substituted into the other equation, that equation is solved for x, and the value(s) of x is substituted into the expression for y to obtain the value(s) of y. For the system above, proceed as follows:

Solve first equation for y: $\qquad\qquad\qquad y = 3 + 9x.$

Substitute into second equation: $\quad (3 + 9x) + 2 = 4x.$

Solve for x: $\qquad\qquad\qquad\qquad\qquad x = -1.$

Substitute the x value into the equation for y:

$$y = 3 + 9(-1) = -6.$$

1. $\dfrac{\partial}{\partial x}(x^2 - 3y^2 + 4x + 6y + 8) = 2x - 0 + 4 + 0 + 0 = 2x + 4,$

$\dfrac{\partial}{\partial y}(x^2 - 3y^2 + 4x + 6y + 8) = 0 - 6y + 0 + 6 + 0 = -6y + 6.$

To find possible relative maxima or minima, set the first partial derivatives equal to zero.

$$\frac{\partial f}{\partial x} = 2x + 4 = 0, \qquad \frac{\partial f}{\partial y} = -6y + 6 = 0,$$
$$2x = -4, \qquad\qquad -6y = -6,$$
$$x = \frac{-4}{2} = -2, \qquad\qquad y = 1.$$

Thus $f(x, y)$ has a possible relative maximum or minimum at $(-2, 1)$.

7. $\dfrac{\partial}{\partial x}\left(\dfrac{1}{3}x^3 - 2y^3 - 5x + 6y - 5\right) = x^2 - 0 - 5 + 0 - 0 = x^2 - 5,$

$\dfrac{\partial}{\partial y}\left(\dfrac{1}{3}x^3 - 2y^3 - 5x + 6y - 5\right) = 0 - 6y^2 - 0 + 6 - 0 = -6y^2 + 6.$

Again, set the first partial derivatives equal to zero.

$$\frac{\partial f}{\partial x} = x^2 - 5 = 0, \qquad \frac{\partial f}{\partial y} = -6y^2 + 6 = 0,$$

$$x^2 = 5, \qquad\qquad 6y^2 = 6,$$

$$x = \sqrt{5}, \quad x = -\sqrt{5}, \qquad y = 1, \quad y = -1.$$

There are four points (x, y) where there is a possible relative maximum or minimum:

$$(\sqrt{5}, 1), \quad (-\sqrt{5}, 1) \quad (\sqrt{5}, -1), \quad (-\sqrt{5}, -1).$$

13. $f(x, y) = 2x^2 - x^4 - y^2,$

$$\frac{\partial f}{\partial x} = 4x - 4x^3, \quad \frac{\partial^2 f}{\partial x^2} = 4 - 12x^2,$$

$$\frac{\partial f}{\partial y} = -2y, \quad \frac{\partial^2 f}{\partial y^2} = -2,$$

$$\frac{\partial^2 f}{\partial x \partial y} = 0.$$

$$D(x, y) = \frac{\partial^2 f}{\partial x^2} \cdot \frac{\partial^2 f}{\partial y^2} - \left(\frac{\partial^2 f}{\partial x \partial y}\right)^2$$

$$= (4 - 12x^2) \cdot (-2) - 0$$

$$= -2(4 - 12x^2).$$

Use the second-derivative test for functions of two variables. Look first at $(-1, 0)$:

$$D(-1, 0) = -2(4 - 12), \text{ which is positive; and}$$

$$\frac{\partial^2 f}{\partial x^2}(-1, 0) = 4 - 12(-1)^2, \text{ which is negative.}$$

Hence $f(x, y)$ has a relative maximum at $(-1, 0)$.

For $(0, 0)$: $D(0, 0) = -2(4 - 0) = -8$, which is negative, so $f(x, y)$ has neither a relative maximum nor a relative minimum at $(0, 0)$.

Finally at $(1, 0)$: $D(1, 0) = -2(4 - 12)$, which is positive, and

$$\frac{\partial^2 f}{\partial x^2}(1, 0) = 4 - 12 = -8, \text{ which is negative.}$$

Therefore again by the test, $f(x, y)$ has a relative maximum at $(1, 0)$.

19. $\dfrac{\partial}{\partial x}(-2x^2 + 2xy - y^2 + 4x - 6y + 5) = -4x + 2y + 4,$

$\dfrac{\partial}{\partial y}(-2x^2 + 2xy - y^2 + 4x - 6y + 5) = 2x - 2y - 6.$

Set these partials equal to zero and solve:

$$\frac{\partial f}{\partial x} = -4x + 2y + 4 = 0, \quad \frac{\partial f}{\partial y} = 2x - 2y - 6 = 0,$$

$$2y = 4x - 4, \qquad\qquad 2y = 2x - 6,$$

$$y = 2x - 2, \qquad\qquad y = x - 3.$$

Equate these two expressions for y:

$$2x - 2 = x - 3,$$

$$x = -1.$$

Substitute $x = -1$ into $y = x - 3$ and obtain $y = x - 3 = -1 - 3 = -4$. So there is a possible relative maximum or minimum at $(-1, -4)$. Use the second derivative test to determine which, if either, is the case.

$$\frac{\partial^2 f}{\partial x^2} = \frac{\partial}{\partial x}(-4x + 2y + 4) = -4,$$

$$\frac{\partial^2 f}{\partial y^2} = \frac{\partial}{\partial y}(2x - 2y - 6) = -2,$$

$$\frac{\partial^2 f}{\partial x \partial y} = \frac{\partial}{\partial x}(2x - 2y - 6) = 2.$$

Therefore, $D(x, y) = (-4)(-2) - (2)^2 = 8 - 4 = 4,$ and hence $D(-1, -4) = 4$. Since $4 > 0, f(x, y)$ indeed has a relative maximum or minimum at $(-1, -4)$. Since $\frac{\partial^2 f}{\partial x^2}(-1, -4) = -4 < 0,$ $f(x, y)$ has a relative maximum at $(-1, -4)$.

25. $f(x, y) = 2x^2 + y^3 - x - 12y + 7$

$\dfrac{\partial f}{\partial x} = 4x - 1, \quad \dfrac{\partial f}{\partial y} = 3y^2 - 12.$

Set the two partials equal to 0 and solve:

$$4x - 1 = 0, \quad 3y^2 - 12 = 0,$$

$$4x = 1, \qquad 3y^2 = 12,$$

$$x = \frac{1}{4}, \qquad y^2 = 4,$$

$$y = 2, -2.$$

So there are two possible relative maxima and/or minima: $\left(\frac{1}{4}, 2\right), \left(\frac{1}{4}, -2\right).$

For the second derivative test, compute

$$\frac{\partial^2 f}{\partial x^2} = \frac{\partial}{\partial x}(4x-1) = 4,$$

$$\frac{\partial^2 f}{\partial y^2} = \frac{\partial}{\partial y}(3y^2 - 12) = 6y,$$

$$\frac{\partial^2 f}{\partial x \partial y} = \frac{\partial}{\partial x}(3y^2 - 12) = 0.$$

$$D(x, y) = 4(6y) - 0^2$$

$$= 24y.$$

Now $D\left(\frac{1}{4}, 2\right) = 24(2) = 48 > 0,$ so $f(x, y)$ does indeed have a relative maximum or minimum at $\left(\frac{1}{4}, 2\right)$. Since

$$\frac{\partial^2 f}{\partial x^2}\left(\frac{1}{4}, 2\right) = 4 > 0,$$

there is a relative minimum at $\left(\frac{1}{4}, 2\right)$.

Finally, $D\left(\frac{1}{4}, -2\right) = 24(-2) = -48 < 0,$ indicating that $\left(\frac{1}{4}, -2\right)$ is a saddle point; this is neither a relative maximum nor a relative minimum.

31. Note that the revenue can be expressed by $10x + 9y$, since the company sells x units of Product I for 10 dollars each, and y units of Product II for 9 dollars each. Since profit = (revenue) − (cost), the profit function is

$$P(x, y) = 10x + 9y - [400 + 2x + 3y + .01(3x^2 + xy + 3y^2)]$$

$$= 8x + 6y - .03x^2 - .01xy - .03y^2 - 400.$$

Now proceed as in Exercises 19 and 25 above.

$$\frac{\partial P}{\partial x} = 8 - .06x - .01y,$$

$$\frac{\partial P}{\partial y} = 6 - .01x - .06y.$$

$$8 - .06x - .01y = 0, \qquad 6 - .01x - .06y = 0,$$

$$.01y = 8 - .06x, \qquad .06y = 6 - .01x,$$

$$y = 800 - 6x, \qquad y = 100 - \frac{1}{6}x.$$

So,

$$800 - 6x = 100 - \frac{1}{6}x, \qquad 700 = \frac{35}{6}x, \qquad x = 120,$$

and

$$y = 800 - 6x, \qquad y(120) = 800 - 6(120) = 800 - 720 = 80.$$

So (120, 80) is the only possible maximum. For the second derivative test, compute

$$\frac{\partial^2 P}{\partial x^2} = -.06, \qquad \frac{\partial^2 P}{\partial y^2} = -.06, \qquad \frac{\partial^2 P}{\partial x \partial y} = -.01.$$

Therefore, $D(x, y) = (-.06)(-.06) - (-.01)^2 = .0035$. Thus $D(120, 80) = .0035 > 0$ and (120, 80) is indeed a maximum or minimum. Since $\frac{\partial^2 P}{\partial x^2}(120, 80) = -.06 < 0$, $(120, 80)$ is a maximum. Thus profit is maximized by manufacturing and selling 120 units of Product I and 80 units of Product II.

7.4   Lagrange Multipliers and Constrained Optimization

Constrained optimization problems are frequently encountered in economics, operations research, and science; the Greek letter λ commonly appears whenever Lagrange multipliers are discussed. Think of λ as just another (new) symbol for a variable. The following results illustrate derivatives using λ.

$$\text{If } f(\lambda) = 5\lambda, \qquad \text{then} \qquad \frac{\partial f(\lambda)}{\partial \lambda} = 5.$$

$$\text{If } f(\lambda) = \lambda k, \qquad \text{then} \qquad \frac{\partial f(\lambda)}{\partial \lambda} = k.$$

Partial derivatives of functions involving λ are not difficult to compute once the unfamiliarity of using a Greek letter is overcome.

The problems in this section require that systems of three and four variables be solved. The method following Example 1 applies to all equations in x, y, and λ. Systems with the additional variable z can usually be solved by mimicking the solution to Example 4.

1. Construct the function

$$F(x, y, \lambda) = x^2 + 3y^2 + 10 + \lambda(8 - x - y).$$

Set the first partial derivatives equal to zero and solve for λ to obtain

$$\frac{\partial F}{\partial x} = 2x - \lambda = 0, \quad \text{so} \quad \lambda = 2x,$$

$$\frac{\partial F}{\partial y} = 6y - \lambda = 0, \quad \text{so} \quad \lambda = 6y,$$

$$\frac{\partial F}{\partial \lambda} = 8 - x - y = 0.$$

Equate $\lambda = 2x$ and $\lambda = 6y$ to obtain $2x = 6y$, or $x = 3y$. Next, substitute $3y$ for x into the third equation and solve for y:

$$8 - (3y) - y = 0,$$
$$8 - 4y = 0,$$
$$4y = 8,$$
$$y = 2.$$
$$\lambda = 6y \Rightarrow \lambda = 6(2) = 12$$

Finally, $x = 3y = 3(2) = 6$. So the minimum is at $(6, 2)$, and the minimum value is

$$F(6, 2) = 6^2 + 3(2^2) + 10$$
$$= 36 + 12 + 10 = 58.$$

7. Minimize the function, $f(x, y) = x + y$ subject to the constraint

$$xy = 25, \quad \text{or} \quad xy - 25 = 0.$$

$F(x, y, \lambda) = x + y + \lambda(xy - 25)$ is the Lagrange function. For a relative extremum, set the first partial derivatives equal to zero:

$$\left.\begin{array}{l} \dfrac{\partial F}{\partial x} = 1 + \lambda y = 0 \\[2mm] \dfrac{\partial F}{\partial y} = 1 + \lambda x = 0 \end{array}\right\} \Rightarrow \lambda y = \lambda x, \quad \text{or} \quad y = x, \qquad (1)$$

$$\frac{\partial F}{\partial \lambda} = xy - 25 = 0 \qquad \text{(substitute in equation (1))},$$

$$x^2 - 25 = 0, \quad \text{or} \quad x = \pm 5.$$

If $x = 5$ (since x must be positive): $y = 5$ by equation (1). Therefore, $x = 5$, $y = 5$.

13. Maximize $P(x, y) = 3x + 4y$ subject to the constraint

$$9x^2 + 4y^2 - 18,000 = 0, \ x \geq 0, \ y \geq 0.$$

$F(x, y, \lambda) = 3x + 4y + \lambda(9x^2 + 4y^2 - 18,000)$. Set the first partial derivatives equal to zero:

$$\frac{\partial F}{\partial x} = 3 + 18\lambda x = 0,$$

$$\frac{\partial F}{\partial y} = 4 + 8\lambda y = 0,$$

$$\frac{\partial F}{\partial \lambda} = 9x^2 + 4y^2 - 18,000 = 0. \quad (1)$$

Thus, $3 + 18\lambda x = 0$ and $4 + 8\lambda y = 0$, which give

$$1 + 6\lambda x = 0, \quad \text{and} \quad 1 + 2\lambda y = 0,$$

respectively. Therefore,

$$1 + 6\lambda x = 1 + 2\lambda y,$$
$$6\lambda x = 2\lambda y,$$
$$3x = y.$$

Substitute $3x$ for y in equation (1):

$$9x^2 + 4(3x)^2 - 18,000 = 0,$$
$$45x^2 = 18,000,$$
$$x^2 = 400, \quad \text{or} \quad x = \pm 20.$$

Since $y = 3x$, $y = \pm 60$. But x, $y \geq 0$ so,

$$x = 20, \quad \text{and} \quad y = 60.$$

19. Let $F(x, y, z, \lambda) = 3x + 5y + z - x^2 - y^2 - z^2 + \lambda(6 - x - y - z)$, and set the first partial derivatives equal to zero.

$$\frac{\partial F}{\partial x} = 3 - 2x - \lambda = 0,$$

$$\frac{\partial F}{\partial y} = 5 - 2y - \lambda = 0,$$

$$\frac{\partial F}{\partial z} = 1 - 2z - \lambda = 0.$$

This provides three expressions for λ:

$$\lambda = 3 - 2x,$$

$$\lambda = 5 - 2y,$$

$$\lambda = 1 - 2z.$$

Equate the first two equations and solve for x in terms of y:

$$3 - 2x = 5 - 2y,$$

$$-2x = 2 - 2y,$$

$$x = -1 + y.$$

Next, equate the second two equations and solve for z in terms of y:

$$5 - 2y = 1 - 2z,$$

$$2z = 2y - 4,$$

$$z = y - 2.$$

Finally, substitute these expressions for x and for z into the equation

$$\frac{\partial F}{\partial \lambda} = 6 - x - y - z = 0.$$

$$6 - (-1 + y) - y - (y - 2) = 0$$

$$6 + 1 - y - y - y + 2 = 0$$

$$9 - 3y = 0$$

$$9 = 3y, \quad \text{or} \quad y = 3.$$

Thus, for $y = 3$, $x = -1 + y = -1 + 3 = 2$ and $z = y - 2 = 3 - 2 = 1$. The function is maximized when $x = 2$, $y = 3$, and $z = 1$.

25. Throughout the problem use the general expression $f(x, y)$ for the production function, which we wish to maximize. The constraint is the fact that the cost of labor plus the cost of capital is c dollars, that is:

$$ax + by = c, \quad \text{or} \quad ax + by - c = 0.$$

Construct the function

$$F(x, y, \lambda) = f(x, y) + \lambda(ax + by - c).$$

Take partial derivatives and set them equal to zero.

$$\frac{\partial F}{\partial x} = \frac{\partial f}{\partial x} + \lambda a = 0,$$

$$\frac{\partial F}{\partial y} = \frac{\partial f}{\partial y} + \lambda b = 0.$$

Solve each equation for λ and get

$$\lambda = -\frac{1}{a}\frac{\partial f}{\partial x} \quad \text{and} \quad \lambda = -\frac{1}{b}\frac{\partial f}{\partial y}.$$

And equating these expressions for λ, we conclude that

$$-\frac{1}{a}\frac{\partial f}{\partial x} = -\frac{1}{b}\frac{\partial f}{\partial y}$$

$$b\frac{\partial f}{\partial x} = a\frac{\partial f}{\partial y}$$

$$\frac{\frac{\partial f}{\partial x}}{\frac{\partial f}{\partial y}} = \frac{a}{b}.$$

7.5   The Method of Least Squares

Each problem in this section requires the solution of a system of two equations in two unknowns. These systems differ from those previously studied in three ways.

1. The variables (that is, the unknowns) will always be A and B, rather then x and y.
2. The equations are linear. That is, they have the form $aA + bB = c$, where a, b, and c are constants.
3. The systems have exactly one solution. This solution is best found by multiplying one equation by a constant and subtracting it from the other equation to eliminate one of the variables. The value of the other variable is then easily found. By substituting this value into one of the equations, the value of the first variable is easily found.

1. The given points are (1, 3), (2, 6), (3, 8), and (4, 6) with straight line $y = 1.1x + 3$. When $x = 1, 2, 3, 4$ the corresponding y-coordinates are $1.1 + 3$, $2(1.1) + 3$, $3(1.1) + 3$, $4(1.1) + 3$ or 4.1, 5.2, 6.3, and 7.4, respectively. Then

$$E = E_1^2 + \cdots + E_4^2 = (A+B-3)^2 + (2A+B-6)^2 + (3A+B-8)^2 + (4A+B-6)^2$$

$$= (1.1+3-3)^2 + (2(1.1)+3-6)^2 + (3(1.1)+3-8)^2 + (4(1.1)+3-6)^2 = 6.7.$$

7. Let the straight line be $y = Ax + B$. Then

$$E = (A+B-9)^2 + (2A+B-8)^2 + (3A+B-6)^2 + (4A+B-3)^2$$

To minimize the error E, take partial derivatives with respect to A and B, and set them equal to zero.

$$\frac{\partial E}{\partial A} = 2(A+B-9)+2(2A+B-8)\cdot 2+2(3A+B-6)\cdot 3+2(4A+B-3)\cdot 4$$

$$= (2+8+18+32)A+(2+4+6+8)B-(18+32+36+24)$$

$$= 60A+20B-110=0.$$

$$\frac{\partial E}{\partial B} = 2(A+B-9)+2(2A+B-8)+2(3A+B-6)+2(4A+B-3)$$

$$= (2+4+6+8)A+(2+2+2+2)B-(18+16+12+6)$$

$$= 20A+8B-52=0.$$

To solve the system of simultaneous linear equations

$$60A+20B=110$$
$$20A+8B=52,$$

multiply the second equation by -3, and add it to the first equation:

$$60A+20B=110$$
$$-60A-24B=-156$$
$$-4B=-46, \quad B=11.5.$$

Then find A by substituting this value of B into the first equation:

$$60A+20(11.5)=110, \qquad 60A=110-230, \qquad A=-2.$$

The least-squares line is $y=-2x+11.5$.

13. (a) The data points are $(5, 3.35)$, $(10, 3.80)$, $(15, 4.25)$, $(20, 5.15)$, $(25, 5.15)$, and $(30, 7.25)$. The table for these data is

x	y	xy	x^2
5	3.35	16.75	25
10	3.80	38	100
15	4.25	63.75	225
20	5.15	103	400
25	5.15	128.75	625
30	7.25	217.5	900
$\sum x = 105$	$\sum y = 28.95$	$\sum xy = 567.75$	$\sum x^2 = 2275$

The formulas for A and B are:

$$A = \frac{N\cdot\Sigma xy - \Sigma x\cdot\Sigma y}{N\cdot\Sigma x^2 - (\Sigma x)^2} = \frac{6\cdot 567.75 - 105\cdot 28.95}{6\cdot 2275 - (105)^2} = \frac{366.75}{2625} \approx .139714$$

$$B = \frac{\Sigma y - A\cdot\Sigma x}{N} = \frac{28.95 - .139714\cdot 105}{6} \approx 2.38.$$

The least-squares line is $y = .139714x + 2.38$. The number of decimal places to use in the equation of the line should be determined by the accuracy of the original data. In this problem, the data have three significant figures, so a reasonable answer for the least-squares line might be $y = .140x + 2.38$, or even $y = .14x + 2.4$.

Alternatively, we can use a graphing calculator to solve the problem. Enter the data, then use the linear regression function to determine the equation of the line of best fit.

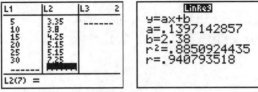

Thus, we obtain the equation $y = .140x + 2.38$.

(b) Since 1998 corresponds to $x = 18$, the best estimate for the minimum wage in 1998 wass $.140(18) + 2.38 = \$4.90$ per hour.

(c) Solve $.140x + 2.38 = 10$ and obtain $x = 54.4$, that is, 55 years after 1980. The minimum wage will reach $10 per hour in 2035.

7.6 Double Integrals

The problems in this section amount to performing two integrations, once with respect to y as the variable and once with respect to x. Antidifferentiation with respect to y takes a little getting used to. It is just the reverse operation to taking partial derivatives with respect to y.

1. First evaluate the inner integral.

$$\int_0^1 e^{x+y}dy = e^{x+y}\Big|_0^1 = e^{x+1} - e^x.$$

[The dy in the integral says that y is the variable of integration when evaluating e^{x+y} at 0 and 1.] Now evaluate the outer integral.

$$\int_0^1 (e^{x+1} - e^x)dx = (e^{x+1} - e^x)\Big|_0^1 = (e^2 - e) - (e - 1) = e^2 - 2e + 1.$$

7. First evaluate the inner integral.

$$\int_x^{2x} (x+y)dy = \left(xy + \frac{1}{2}y^2\right)\Big|_x^{2x} = \left[x(2x) + \frac{1}{2}(2x)^2\right] - \left[x \cdot x + \frac{1}{2}x^2\right]$$

$$= \left[2x^2 + 2x^2\right] - x^2 - \frac{1}{2}x^2 = \frac{5}{2}x^2.$$

Now evaluate the outer integral.

$$\int_{-1}^1 \frac{5}{2}x^2 dx = \frac{5}{6}x^3\Big|_{-1}^1 = \frac{5}{6} - \left(-\frac{5}{6}\right) = \frac{5}{3}.$$

13. The desired volume is given by the double integral:

$$\iint_R (x^2 + y^2)dx\,dy,$$

which is equivalent to the iterated integral:

$$\int_1^3 \left(\int_0^1 (x^2 + y^2)\,dy\right)dx.$$

First evaluate the inner integral:

$$\int_0^1 (x^2 + y^2)dy = \left(x^2 y + \frac{1}{3} y^3 \right)\Big|_0^1 = \left(x^2 + \frac{1}{3} \right) - (0+0) = x^2 + \frac{1}{3}.$$

Then evaluate the outer integral:

$$\int_1^3 \left(x^2 + \frac{1}{3} \right)dx = \left(\frac{1}{3} x^3 + \frac{1}{3} x \right)\Big|_1^3 = (9+1) - \left(\frac{1}{3} + \frac{1}{3} \right) = \frac{28}{3} = 9\frac{1}{3}.$$

Chapter 7 Review Exercises

While solutions for all review exercises are included, expanded explanations are included for every sixth exercise.

1. $f(2,9) = \dfrac{2\sqrt{9}}{1+2} = \dfrac{6}{3} = 2, \quad f(5,1) = \dfrac{5\sqrt{1}}{1+5} = \dfrac{5}{6}, \quad f(0,0) = \dfrac{0\sqrt{0}}{1+0} = \dfrac{0}{1} = 0.$

2. $f(x,\, y,\, z) = x^2 e^{y/z}$

   $f(-1,\, 0,\, 1) = (-1)^2 e^{0/1} = 1e^0 = 1$

   $f(1,\, 3,\, 3) = 1^2 e^{3/3} = e$

   $f(5,\, -2,\, 2) = 5^2 e^{-2/2} = \dfrac{25}{e}$

3. $f(A,\, t) = Ae^{.06t}$

   $f(10,\, 11.5) = 10e^{(.06)(11.5)} \approx 19.94$

   Ten dollars increases to approximately 20 dollars in 11.5 years.

4. $f(x, y, \lambda) = xy + \lambda(5 - x - y)$
   $f(1, 2, 3) = (1)(2) + 3(5 - 1 - 2) = 2 + 3(2) = 8$

5. $f(x,\, y) = 3x^2 + xy + 5y^2$

   $\dfrac{\partial f}{\partial x} = 6x + y; \quad \dfrac{\partial f}{\partial y} = x + 10y$

6. $f(x,\, y) = 3x - \dfrac{1}{2} y^4 + 1$

   $\dfrac{\partial f}{\partial x} = 3; \quad \dfrac{\partial f}{\partial y} = -2y^3$

7. $f(x,\, y) = e^{x/y} = e^{xy^{-1}},$

   $\dfrac{\partial f}{\partial x} = y^{-1} e^{xy^{-1}} = \dfrac{1}{y} e^{x/y}, \qquad$ (treat y as a constant)

   $\dfrac{\partial f}{\partial y} = e^{xy^{-1}} \left[\dfrac{\partial}{\partial y} xy^{-1} \right] = -xy^{-2} e^{xy^{-1}} = -\dfrac{x}{y^2} e^{x/y}, \quad$ (treat x as a constant).

8. $f(x, y) = \dfrac{x}{x - 2y}$

$$\frac{\partial f}{\partial x} = \frac{(1)(x - 2y) - (x)(1)}{(x - 2y)^2} = \frac{-2y}{(x - 2y)^2}$$

$$\frac{\partial f}{\partial y} = \frac{(0)(x - 2y) - (-2)(x)}{(x - 2y)^2} = \frac{2x}{(x - 2y)^2}$$

9. $f(x, y, z) = x^3 - yz^2$

$$\frac{\partial f}{\partial x} = 3x^2; \quad \frac{\partial f}{\partial y} = -z^2; \quad \frac{\partial f}{\partial z} = -2yz$$

10. $f(x, y, \lambda) = xy + \lambda(5 - x - y) = xy + 5\lambda - x\lambda - y\lambda$

$$\frac{\partial f}{\partial x} = y - \lambda; \quad \frac{\partial f}{\partial y} = x - \lambda; \quad \frac{\partial f}{\partial \lambda} = 5 - x - y$$

11. $f(x, y) = x^3 y + 8$

$$\frac{\partial f}{\partial x} = 3x^2 y; \quad \frac{\partial f}{\partial x}(1, 2) = 3(1)^2(2) = 6$$

$$\frac{\partial f}{\partial y} = x^3; \quad \frac{\partial f}{\partial y}(1, 2) = (1)^3 = 1$$

12. $f(x, y, z) = (x + y)z = xz + yz$

$$\frac{\partial f}{\partial y} = z; \quad \frac{\partial f}{\partial y}(2, 3, 4) = 4$$

13. $f(x, y) = x^5 - 2x^3 y + \dfrac{1}{2} y^4$

In order to find the second partial derivatives, first determine the first partial derivatives with respect to x and y.

$$\frac{\partial f}{\partial x} = 5x^4 - 6x^2 y, \quad \frac{\partial f}{\partial y} = -2x^3 + 2y^3.$$

Differentiating each of these with respect to x and y, we have

$$\frac{\partial^2 f}{\partial x^2} = \frac{\partial}{\partial x}\left(\frac{\partial f}{\partial x}\right) = \frac{\partial}{\partial x}(5x^4 - 6x^2 y) = 20x^3 - 12xy,$$

$$\frac{\partial^2 f}{\partial y^2} = \frac{\partial}{\partial y}\left(\frac{\partial f}{\partial y}\right) = \frac{\partial}{\partial y}(-2x^3 + 2y^3) = 6y^2,$$

$$\frac{\partial^2 f}{\partial x \partial y} = \frac{\partial}{\partial x}\left(\frac{\partial f}{\partial y}\right) = \frac{\partial}{\partial x}(-2x^3 + 2y^3) = -6x^2,$$

$$\frac{\partial^2 f}{\partial y \partial x} = \frac{\partial}{\partial y}\left(\frac{\partial f}{\partial x}\right) = \frac{\partial}{\partial y}(5x^4 - 6x^2 y) = -6x^2.$$

14. $f(x, y) = 2x^3 + x^2y - y^2$

$$\frac{\partial f}{\partial x} = 6x^2 + 2xy \Rightarrow \frac{\partial^2 f}{\partial x^2} = 12x + 2y$$

$$\frac{\partial f}{\partial y} = x^2 - 2y \Rightarrow \frac{\partial^2 f}{\partial y^2} = -2$$

$$\frac{\partial^2 f}{\partial x^2}(1, 2) = 16; \quad \frac{\partial^2 f}{\partial y^2}(1, 2) = -2$$

$$\frac{\partial^2 f}{\partial x \partial y} = 2x; \quad \frac{\partial^2 f}{\partial x \partial y}(1, 2) = 2$$

15. $f(p, t) = -p + 6t - .02pt$

$$\frac{\partial f}{\partial p} = -1 - .02t \; ; \; \frac{\partial f}{\partial p}(25, 10,000) = -201$$

$$\frac{\partial f}{\partial t} = 6 - .02p; \frac{\partial f}{\partial t}(25, 10,000) = 5.5$$

At the level $p = 25$, $t = 10,000$, an increase in price of \$1 will result in a loss in sales of approximately 201 calculators, and an increase in advertising of \$1 will result in the sale of approximately 5.5 additional calculators.

16. The crime rate increases with increased unemployment and decreases with increased social services and police force size.

17. $f(x, y) = -x^2 + 2y^2 + 6x - 8y + 5$

$$\frac{\partial f}{\partial x} = -2x + 6; \quad \frac{\partial f}{\partial y} = 4y - 8$$

$-2x + 6 = 0 \Rightarrow x = 3; 4y - 8 = 0 \Rightarrow y = 2$
The only possibility is $(x, y) = (3, 2)$.

18. $f(x, y) = x^2 + 3xy - y^2 - x - 8y + 4$

$$\frac{\partial f}{\partial x} = 2x + 3y - 1; \frac{\partial f}{\partial y} = 3x - 2y - 8$$

$\left.\begin{array}{l} 2x + 3y = 1 \\ 3x - 2y = 8 \end{array}\right\} \begin{array}{l} x = 2 \\ y = -1 \end{array}$

The only possibility is $(x, y) = (2, -1)$.

19. $f(x, y) = x^3 + 3x^2 + 3y^2 - 6y + 7$

$$\frac{\partial}{\partial x}(x^3 + 3x^2 + 3y^2 - 6y + 7) = 3x^2 + 6x,$$

$$\frac{\partial}{\partial y}(x^3 + 3x^2 + 3y^2 - 6y + 7) = 6y - 6.$$

To find possible relative maxima or minima, set the first derivatives equal to zero.

$$\frac{\partial f}{\partial x} = 3x^2 + 6x = 0, \qquad \frac{\partial f}{\partial y} = 6y - 6 = 0,$$

$$3x(x+2) = 0, \qquad\qquad 6y = 6,$$

$$x = 0, \quad x = -2, \qquad\qquad y = 1.$$

Thus $f(x, y)$ has a possible relative maximum or minimum at $(0, 1)$ and at $(-2, 1)$.

20. $f(x,\ y) = \dfrac{1}{2}x^2 + 4xy + y^3 + 8y^2 + 3x + 2$

$\dfrac{\partial f}{\partial x} = x + 4y + 3$

$\dfrac{\partial f}{\partial y} = 4x + 3y^2 + 16y$

$x + 4y + 3 = 0 \Rightarrow x = -4y - 3$

$4x + 3y^2 + 16y = 0 \Rightarrow 4(-4y-3) + 3y^2 + 16y = 0 \Rightarrow -16y - 12 + 3y^2 + 16y = 0 \Rightarrow y = \pm 2$

$x = -4(-2) - 3 = 5;\ x = -4(2) - 3 = -11;$

$(x, y) = (-11, 2),\ (5, -2)$

21. $f(x,\ y) = x^2 + 3xy + 4y^2 - 13x - 30y + 12$

$\dfrac{\partial f}{\partial x} = 2x + 3y - 13 \Rightarrow \dfrac{\partial^2 f}{\partial x^2} = 2$

$\dfrac{\partial f}{\partial y} = 3x + 8y - 30 \Rightarrow \dfrac{\partial^2 f}{\partial y^2} = 8;\ \dfrac{\partial^2 f}{\partial x \partial y} = 3$

$\left.\begin{matrix} 2x + 3y = 13 \\ 3x + 8y = 30 \end{matrix}\right\} \begin{matrix} x = 2 \\ y = 3 \end{matrix}$

$D(x, y) = \dfrac{\partial^2 f}{\partial x^2} \cdot \dfrac{\partial^2 f}{\partial y^2} - \left(\dfrac{\partial^2 f}{\partial x \partial y}\right)^2 \Rightarrow$

$D(2, 3) = 2 \cdot 8 - 3^2 > 0$ and $\dfrac{\partial^2 f}{\partial x^2} > 0,$

so $f(x, y)$ has a relative minimum at $(2, 3)$.

22. $f(x,\ y) = 7x^2 - 5xy + y^2 + x - y + 6$

$\dfrac{\partial f}{\partial x} = 14x - 5y + 1;\ \dfrac{\partial^2 f}{\partial x^2} = 14$

$\dfrac{\partial f}{\partial y} = -5x + 2y - 1;\ \dfrac{\partial^2 f}{\partial y^2} = 2;\ \dfrac{\partial^2 f}{\partial x \partial y} = -5$

$\left.\begin{matrix} 14x - 5y = -1 \\ -5x + 2y = 1 \end{matrix}\right\} \begin{matrix} x = 1 \\ y = 3 \end{matrix}$

$D(x, y) = \dfrac{\partial^2 f}{\partial x^2} \cdot \dfrac{\partial^2 f}{\partial y^2} - \left(\dfrac{\partial^2 f}{\partial x \partial y}\right)^2 \Rightarrow D(1, 3) = 14 \cdot 2 - (-5)^2 > 0$ and $\dfrac{\partial^2 f}{\partial x^2} > 0,$ so $f(x, y)$ has a relative

minimum at $(1, 3)$.

23. $f(x, y) = x^3 + y^2 - 3x - 8y + 12$

$\dfrac{\partial f}{\partial x} = 3x^2 - 3; \dfrac{\partial^2 f}{\partial x^2} = 6x;$

$\dfrac{\partial f}{\partial y} = 2y - 8; \dfrac{\partial^2 f}{\partial y^2} = 2; \dfrac{\partial^2 f}{\partial x \partial y} = 0$

$3x^2 = 3 \Rightarrow x = \pm 1$

$2y = 8 \Rightarrow y = 4$

$D(x, y) = \dfrac{\partial^2 f}{\partial x^2} \cdot \dfrac{\partial^2 f}{\partial y^2} - \left(\dfrac{\partial^2 f}{\partial x \partial y}\right)^2 \Rightarrow$

$D(1, 4) = 6(1)(2) - 0^2 > 0$ and $\dfrac{\partial^2 f}{\partial x^2}(1, 4) > 0,$ so $f(x, y)$ has a relative minimum at $(1, 4)$.

$D(-1, 4) = 6(-1)(2) - 0^2 < 0$, so $f(x, y)$ has neither a maximum nor a minimum at $(-1, 4)$.

24. $f(x, y, z) = x^2 + 4y^2 + 5z^2 - 6x + 8y + 3$

$\dfrac{\partial f}{\partial x} = 2x - 6 = 0 \Rightarrow x = 3$

$\dfrac{\partial f}{\partial y} = 8y + 8 = 0 \Rightarrow y = -1$

$\dfrac{\partial f}{\partial z} = 10z = 0 \Rightarrow z = 0$

$f(x, y, z)$ must assume its minimum value at $(3, -1, 0)$.

25. Construct the function

$$F(x, y, \lambda) = 3x^2 + 2xy - y^2 + \lambda(5 - 2x - y).$$

Set the first partial derivatives equal to zero and solve for λ:

$$\dfrac{\partial F}{\partial x} = 6x + 2y - 2\lambda = 0 \Rightarrow 2\lambda = 6x + 2y \Rightarrow \lambda = 3x + y.$$

$$\dfrac{\partial F}{\partial y} = 2x - 2y - \lambda = 0 \Rightarrow \lambda = 2x - 2y.$$

Next, equate the two expressions for λ:

$$3x + y = 2x - 2y \quad \text{or} \quad x = -3y.$$

The equation $\dfrac{\partial F}{\partial \lambda} = 5 - 2x - y = 0$ is the constraint. Substitute $-3y$ for x and obtain:

$$5 - 2(-3y) - y = 0,$$
$$5 + 5y = 0,$$
$$5y = -5, \quad \text{or} \quad y = -1.$$

Hence, $x = -3y = -3(-1) = 3$, so the maximum is at $(3, -1)$, and the maximum value is $f(3, -1) = 3(9) + 2(3)(-1) - 1 = 27 - 6 - 1 = 20.$

26. $F(x, y, \lambda) = -x^2 - 3xy - \dfrac{1}{2}y^2 + y + 10 + \lambda(10 - x - y)$

$$\left.\begin{array}{l}\dfrac{\partial F}{\partial x} = -2x - 3y - \lambda = 0 \\[2mm] \dfrac{\partial F}{\partial y} = -3x - y + 1 - \lambda = 0 \\[2mm] \dfrac{\partial F}{\partial \lambda} = 10 - x - y = 0\end{array}\right\}\left.\begin{array}{l}\lambda = -2x - 3y \\[2mm] \lambda = -3x - y + 1 \\[2mm] x + y = 10\end{array}\right\}\left.\begin{array}{l}x - 2y = 1 \\[2mm] x + y = 10\end{array}\right\}\begin{array}{l}x = 7 \\[2mm] y = 3\end{array}$$

27. $F(x, y, z, \lambda) = 3x^2 + 2y^2 + z^2 + 4x + y + 3z + \lambda(4 - x - y - z)$

$$\left.\begin{array}{l}\dfrac{\partial F}{\partial x} = 6x + 4 - \lambda = 0 \\[2mm] \dfrac{\partial F}{\partial y} = 4y + 1 - \lambda = 0 \\[2mm] \dfrac{\partial F}{\partial z} = 2z + 3 - \lambda = 0 \\[2mm] \dfrac{\partial F}{\partial \lambda} = 4 - x - y - z = 0\end{array}\right\}\left.\begin{array}{l}6x + 4 = 4y + 1 \\[2mm] 4y + 1 = 2z + 3 \\[2mm] x + y + z = 4\end{array}\right\}\left.\begin{array}{l}x = \dfrac{2}{3}y - \dfrac{1}{2} \\[2mm] z = 2y - 1 \\[2mm] \dfrac{2}{3}y - \dfrac{3}{2} + y + 2y = 4\end{array}\right\}\begin{array}{l}x = \dfrac{1}{2} \\[2mm] y = \dfrac{3}{2} \\[2mm] z = 2\end{array}$$

28. The problem is to minimize $x + y + z$ subject to $xyz = 1000$. $(x > 0, y > 0, z > 0)$
$F(x, y, \lambda) = x + y + z + \lambda(1000 - xyz)$ (Assuming $x \neq 0, y \neq 0, z \neq 0$)

$$\left.\begin{array}{l}\dfrac{\partial F}{\partial x} = 1 - \lambda yz \\[2mm] \dfrac{\partial F}{\partial y} = 1 - \lambda xz \\[2mm] \dfrac{\partial F}{\partial z} = 1 - \lambda xy \\[2mm] \dfrac{\partial F}{\partial \lambda} = 1000 - xyz\end{array}\right\}\left.\begin{array}{l}\lambda = \dfrac{1}{yz} \\[2mm] \lambda = \dfrac{1}{xz} \\[2mm] \lambda = \dfrac{1}{xy} \\[2mm] xyz = 1000\end{array}\right\}\left.\begin{array}{l}yz = xz \\[2mm] xz = xy \\[2mm] xyz = 1000\end{array}\right\}\left.\begin{array}{l}x = y = z \\[2mm] xyz = 1000\end{array}\right\}\begin{array}{l}x = 10 \\[2mm] y = 10 \\[2mm] z = 10\end{array}$$

The optimal dimensions are 10 in. $\times$ 10 in. $\times$ 10 in.

29.

The problem is to maximize xy subject to $2x + y = 40$.
$F(x, y, \lambda) = xy + \lambda(40 - 2x - y)$

$$\left.\begin{array}{l}\dfrac{\partial F}{\partial x} = y - 2\lambda = 0 \\[2mm] \dfrac{\partial F}{\partial y} = x - \lambda = 0 \\[2mm] \dfrac{\partial F}{\partial \lambda} = 40 - 2x - y = 0\end{array}\right\}\left.\begin{array}{l}\lambda = \dfrac{1}{2}y \\[2mm] \lambda = x \\[2mm] 2x + y = 40\end{array}\right\}\left.\begin{array}{l}x = \dfrac{1}{2}y \\[2mm] 2y = 40\end{array}\right\}\begin{array}{l}x = 10 \text{ ft} \\[2mm] y = 20 \text{ ft}\end{array}$$

The dimensions of the garden should be 10 ft $\times$ 20 ft.

30. Maximize xy subject to $2x + y = 41$.

$$F(x, y, \lambda) = xy + \lambda(41 - 2x - y)$$

$$\left.\begin{array}{l} \dfrac{\partial F}{\partial x} = y - 2\lambda = 0 \\[2ex] \dfrac{\partial F}{\partial y} = x - \lambda = 0 \\[2ex] \dfrac{\partial F}{\partial \lambda} = 41 - 2x - y = 0 \end{array}\right\} \left.\begin{array}{l} \lambda = \dfrac{1}{2}y \\[2ex] \lambda = x \\[2ex] 2x + y = 41 \end{array}\right\} \left.\begin{array}{l} x = \dfrac{1}{2}y \\[2ex] 2y = 41 \end{array}\right\} \left.\begin{array}{l} x = 10.25 \text{ ft} \\[2ex] y = 20.5 \text{ ft} \end{array}\right\}$$

The new area is $xy = (10.25)(20.5) = 210.125$ sq ft.

The increase in area (compared with the area in Exercise 29) is $210.125 - (10)(20) = 10.125$, which is approximately the value of λ.

31. Let the straight line by $y = Ax + B$. When $x = 1, 2, 3$, the corresponding y-coordinates of the points of the line are $A + B$, $2A + B$, $3A + B$, respectively. Therefore, the squares of the vertical distances from the line to the points $(1, 1)$, $(2, 3)$, and $(3, 6)$ are

$$E_1^2 = (A + B - 1)^2,$$
$$E_2^2 = (2A + B - 3)^2,$$
$$E_3^2 = (3A + B - 6)^2.$$

Thus the least-squares error is

$$f(A, B) = E_1^2 + E_2^2 + E_3^2 = (A + B - 1)^2 + (2A + B - 3)^2 + (3A + B - 6)^2$$

To minimize $f(A, B)$, take partial derivatives with respect to A and B, and set them equal to zero.

$$\frac{\partial f}{\partial A} = 2(A + B - 1) + 2(2A + B - 3)(2) + 2(3A + B - 6)(3)$$
$$= (2A + 2B - 2) + (8A + 4B - 12) + (18A + 6B - 36)$$
$$= 28A + 12B - 50 = 0$$

$$\frac{\partial f}{\partial B} = 2(A + B - 1) + 2(2A + B - 3) + 2(3A + B - 6)$$
$$= (2A + 2B - 2) + (4A + 2B - 6) + (6A + 2B - 12)$$
$$= 12A + 6B - 20 = 0$$

To find A and B, you must solve the system of simultaneous linear equations

$$28A + 12B = 50,$$
$$12A + 6B = 20.$$

Subtract 2 times the second equation from the first equation and obtain

$$4A + 10, \quad \text{and} \quad A = \frac{5}{2}.$$

Hence

$$6B = 20 - 12\left(\frac{5}{2}\right) = -10, \quad \text{and} \quad B = -\frac{5}{3}.$$

Therefore, the straight line that minimizes the least-squares error is $y = \dfrac{5}{2}x - \dfrac{5}{3}$.

32. Let the straight line be $y = Ax + B$.

$E_1^2 = (A + B - 1)^2$; $E_2^2 = (3A + B - 4)^2$; $E_3^2 = (5A + B - 7)^2$

Let $f(A, B) = E_1^2 + E_2^2 + E_3^2 = (A + B - 1)^2 + (3A + B - 4)^2 + (5A + B - 7)^2$.

$\dfrac{\partial f}{\partial A} = 2(A + B - 1) + 2(3A + B - 4)(3) + 2(5A + B - 7)(5) = 70A + 18B - 96$

$\dfrac{\partial f}{\partial B} = 2(A + B - 1) + 2(3A + B - 4) + 2(5A + B - 7) = 18A + 6B - 24$

Setting $\dfrac{\partial f}{\partial A}$ and $\dfrac{\partial f}{\partial B}$ equal to zero we obtain the system $\begin{cases} 70A + 18B = 96 \\ 18A + 6B = 24 \end{cases}$.

Then $A = \dfrac{3}{2}$ and $B = -\dfrac{1}{2}$ so the line with the best least-squares fit to the data points is $y = \dfrac{3}{2}x - \dfrac{1}{2}$.

33. Let the straight line be $y = Ax + B$.

$E_1^2 = (0A + B - 1)^2$; $E_2^2 = (A + B + 1)^2$; $E_3^2 = (2A + B + 3)^2$; $E_4^2 = (3A + B + 5)^2$

Let $f(A, B) = E_1^2 + E_2^2 + E_3^2 + E_4^2 = (0A + B - 1)^2 + (A + B + 1)^2 + (2A + B + 3)^2 + (3A + B + 5)^2$.

$\dfrac{\partial f}{\partial A} = 2(A + B + 1) + 2(2A + B + 3)(2) + 2(3A + B + 5)(3) = 28A + 12B + 44$

$\dfrac{\partial f}{\partial B} = 2(B - 1) + 2(A + B + 1) + 2(2A + B + 3) + 2(3A + B + 5) = 12A + 8B + 16$

Setting $\dfrac{\partial f}{\partial A}$ and $\dfrac{\partial f}{\partial B}$ equal to zero we obtain the system: $\begin{cases} 28A + 12B = -44 \\ 12A + 8B = -16 \end{cases}$.

Then $A = -2$ and $B = 1$ so the line with the best least-squares fit to the data points is $y = -2x + 1$.

34. $\displaystyle\int_0^1 \left(\int_0^4 (x\sqrt{y} + y) \, dy \right) dx = \int_0^1 \left(\frac{2}{3}xy^{3/2} + \frac{1}{2}y^2 \Big|_{y=0}^{4} \right) dx = \int_0^1 \left(\frac{16}{3}x + 8 \right) dx = \frac{8}{3}x^2 + 8x \Big|_0^1 = \frac{8}{3} + 8 = \frac{32}{3}$

35. $\displaystyle\int_0^5 \left(\int_1^4 (2xy^4 + 3) \, dy \right) dx = \int_0^5 \left(\frac{2}{5}xy^5 + 3y \Big|_{y=1}^{4} \right) dx = \int_0^5 \left(\frac{2046}{5}x + 9 \right) dx = \frac{1023}{5}x^2 + 9x \Big|_0^5$

$$= 5115 + 45 = 5160$$

36. $\displaystyle\int_1^3 \left(\int_0^4 (2x + 3y)dx \right) dy = \int_1^3 \left(x^2 + 3xy \Big|_{x=0}^{4} \right) dy = \int_1^3 (16 + 12y) \, dy = 16y + 6y^2 \Big|_1^3 = 80$

37. $\displaystyle\iint_R 5dxdy$ represents the volume of a box with dimensions $(4 - 0) \times (3 - 1) \times 5$.

So, $\displaystyle\iint_R 5dxdy = 4 \cdot 2 \cdot 5 = 40$.

38.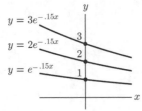

Chapter 8
The Trigonometric Functions

Trigonometric functions are useful for describing physical processes that are cyclical or periodic. The goal of this short chapter is to study basic properties of these functions. We realize that many of our readers have never studied trigonometry, and so we mention a few simple connections with right triangles. However, our main interest is in calculus, not in trigonometry.

8.1 Radian Measure of Angles

All of the exercises in this text use radian measure in connection with trigonometric functions.

1. $30° = 30 \times \dfrac{\pi}{180}$ radians $= \dfrac{\pi}{6}$ radian

$120° = 120 \times \dfrac{\pi}{180}$ radians $= \dfrac{2\pi}{3}$ radians

$315° = 315 \times \dfrac{\pi}{180}$ radians $= \dfrac{7\pi}{4}$ radians.

7. The angle described by this figure consists of one full revolution plus three quarters of a revolution. That is

$$t = 2\pi + \frac{3}{4}(2\pi) = 2\pi + \frac{3}{2}\pi = \frac{7\pi}{2}.$$

13. See the figures below. Here is the reasoning that leads to those figures. First, since $\pi/2$ is one quarter-revolution of the circle, $3\pi/2$ is three quarter-revolutions. Next, since $\pi/4$ is one eighth-revolution of the circle, $3\pi/4$ is three eighth-revolutions. Notice also that $3\pi/4$ is exactly one half the angle described in the first case. Finally, since π is a one half-revolution of the circle, 5π is five half-revolutions or two-and-one-half revolutions.

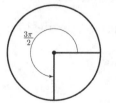  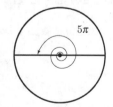

8.2   The Sine and the Cosine

The most important pictures in this section are on page 399. Years from now if someone asks you what the sine function is, you should think of the graphs in Figures 12 and 13, not the triangles in Figure 3, or even the circle in Figure 7. It is the regular, fluctuating shape of the graphs of cos t and sin t that make them useful in mathematical models. Nevertheless, your instructor will want you to have a basic understanding of how cos t and sin t are defined, and that is the purpose of the other figures in this section.

Exercises 1–20 help you learn the basic definitions:

$$\cos t = \frac{x}{r}, \quad \text{and} \quad \sin t = \frac{y}{r},$$

where x and y are the coordinates of a point that determines an angle of t radians, and r is the distance of the point from the origin. Exercises 21–38 help you learn the alternative definitions of cos t and sin t as the x- and y-coordinates of an appropriate point on the unit circle. The alternative definitions enable you to understand some important properties of the sine and cosine. Be sure to find out which of these properties you must memorize. [Formula (7), for example, is only used later for proofs of derivative formulas, so you may not need to learn it.]

1. $\sin t = \dfrac{\text{opposite}}{\text{hypotenuse}} = \dfrac{1}{2}, \quad \cos t = \dfrac{\text{adjacent}}{\text{hypotenuse}} = \dfrac{\sqrt{3}}{2}.$

7. First compute r.

$$r = \sqrt{x^2 + y^2} = \sqrt{(-2)^2 + 1^2} = \sqrt{4+1} = \sqrt{5}.$$

Then

$$\sin t = \frac{y}{r} = \frac{1}{\sqrt{5}}, \quad \cos t = \frac{x}{r} = \frac{-2}{\sqrt{5}}.$$

13. First compute

$$\sin t = \frac{b}{c} = \frac{5}{13} = .385.$$

One way to find t is to look in the "sin t" column of a table of trigonometric functions. A number in this column close to .385 is .38942, in the row corresponding to $t = .4$. From this you can conclude that $t \approx .4$. For more accuracy, use a scientific calculator. Check to make sure that angle measurement is set for radians, and use the "inverse sine" function to compute $\sin^{-1}.385 = .39521$.

19. Since "a" is given and it is the side *adjacent* to the given angle t, use the formula

$$\cos t = \frac{\text{adjacent}}{\text{hypotenuse}}.$$

In this exercise

$$\cos .5 = \frac{2.4}{c}.$$

From a calculator, $\cos .5 = .87758$, so

$$.87758 = \frac{2.4}{c},$$

$$.87758c = 2.4,$$

$$c = \frac{2.4}{.87758} = 2.7348 \approx 2.7$$

One way to find b, now that you know c, is to use the formula

$$\sin t = \frac{b}{c}.$$

That is,

$$\sin .5 = \frac{b}{2.7348},$$

$$.47943 = \frac{b}{2.7348},$$

$$2.73(.47943) = b,$$

$$b = 1.311 \approx 1.3$$

Alternatively, using the Pythagorean theorem, $(2.4)^2 + b^2 = (2.7348)^2$,

$$5.76 + b^2 = 7.4791,$$

$$b^2 = 1.7191,$$

$$b = 1.311 \approx 1.3$$

25. On the unit circle, locate the point P that is determined by an angle of $-\frac{5\pi}{8}$ radians. The

x-coordinate of P is $\cos\left(-\frac{5\pi}{8}\right)$. There is another point Q on the unit circle with the same

x-coordinate. (See the figure.) Let t be the radian measure of the angle determined by Q. Then

$\cos t = \cos\left(-\frac{5\pi}{8}\right)$ because Q and P have the same x-coordinate. Also, $0 \le t \le \pi$. From the

symmetry of the diagram, it is clear that $t = \frac{5\pi}{8}$. Also, since $\cos(-t) = \cos t$ and $0 \le \frac{5\pi}{8} \le \pi$, we

have $t = \frac{5\pi}{8}$.

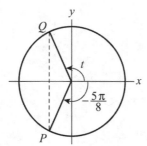

31. The equation $\sin t = -\sin(\pi/6)$ relates the y-coordinates of two points on the unit circle. One point P corresponds to an angle of $\pi/6$ radian. (See the figure.) The other point Q must be on the right half of the unit circle, since $-\dfrac{\pi}{2} \le t \le \dfrac{\pi}{2}$. But its y-coordinate must be the negative of the y-coordinate of P. From the symmetry of the figure, $t = -\dfrac{\pi}{6}$. Another method is to recall that $-\sin t = \sin(-t)$. In particular,

$$-\sin\left(\frac{\pi}{6}\right) = \sin\left(\frac{-\pi}{6}\right).$$

Thus $-\pi/6$ works for t. Since $-\pi/6$ is between $-\pi/2$ and $\pi/2$, $t = -\pi/6$.

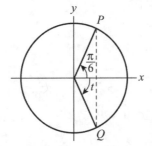

37. Think of $\sin t$ as the y-coordinate of the point P on the unit circle that is determined by an angle of t radians. From the figures below conclude that $\sin 5\pi = 0$ and $\sin(-2\pi) = 0$.

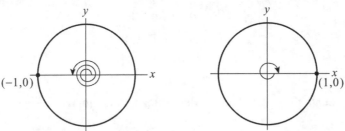

Since $\dfrac{17\pi}{2} = 8\pi + \dfrac{\pi}{2} = 4(2\pi) + \dfrac{\pi}{2}$, the number $\dfrac{17\pi}{2}$ corresponds to four complete revolutions of the circle plus one quarter-revolution. So P is the point $(0, 1)$ and $\sin\dfrac{17\pi}{2} = 1$.

Since $-\dfrac{13\pi}{2} = -6\pi - \dfrac{\pi}{2} = 3(-2\pi) - \dfrac{\pi}{2}$, the number $-\dfrac{13\pi}{2}$ corresponds to three complete revolutions of the circle in the negative direction plus one quarter-revolution in the negative direction. So P is the point $(0, -1)$ and $\sin\left(-\dfrac{13\pi}{2}\right) = -1$.

8.3 Differentiation and Integration of sin *t* and cos *t*

The significance of the derivative formula

$$\frac{d}{dt}\sin t = \cos t$$

is revealed by Figures 1 and 2 in the text. That is, for each number t, the value of cos t gives the slope of the sine curve at the specified value of t. This fundamental property of the derivative as a slope function is worth contemplating because it will be crucial for your work in Chapter 10.

Most of the exercises in this section are routine drill problems. Exercises 35–46 are just a "warm-up" to remind you of antiderivatives. We will concentrate on problems such as these in Section 9.1.

Helpful Hint: *Don't* memorize antiderivative formulas for cos t and sin t, because you might confuse them with the derivative formulas. Just memorize where the minus sign goes in the derivative formulas, and then use this knowledge to check your work whenever you need to "guess" an antiderivative involving a sine or cosine. (See our solution to Exercise 37, below.)

1. By the chain rule,

$$\frac{d}{dt}(\sin 4t) = (\cos 4t) \cdot \frac{d}{dt}4t = 4\cos 4t.$$

7. $\dfrac{d}{dt}(t + \cos \pi t) = \dfrac{d}{dt}t + \dfrac{d}{dt}\cos \pi t$

$$= 1 + (-\sin \pi t)\frac{d}{dt}\pi t$$

$$= 1 - \pi \sin \pi t.$$

13. By repeated application of the chain rule,

$$\frac{d}{dx}\sin \sqrt{x-1} = \cos \sqrt{x-1} \cdot \frac{d}{dx}\sqrt{x-1}$$

$$= \cos \sqrt{x-1} \cdot \frac{d}{dx}(x-1)^{1/2}$$

$$= \cos \sqrt{x-1} \cdot \left[\frac{1}{2}(x-1)^{-1/2}\right]\frac{d}{dx}(x-1)$$

$$= \cos \sqrt{x-1} \cdot \left[\frac{1}{2}(x-1)^{-1/2}\right]$$

$$= \frac{\cos \sqrt{x-1}}{2\sqrt{x-1}}.$$

Helpful Hint: Instructors like problems such as Exercise 19 because they force students to understand the notation. To avoid mistakes, be sure to include the first step shown in the solution above.

19. By repeated application of the chain rule,

$$\frac{d}{dx}\cos^2 x^3 = \frac{d}{dx}(\cos x^3)^2$$

$$= 2\cos x^3 \cdot \frac{d}{dx}\cos x^3$$

$$= 2\cos x^3(-\sin x^3) \cdot \frac{d}{dx}x^3$$

$$= 2\cos x^3(-\sin x^3)3x^2$$

$$= -6x^2 \cos x^3 \sin x^3.$$

25. By the quotient rule,

$$\frac{d}{dt}\left(\frac{\sin t}{\cos t}\right) = \frac{\cos t(\cos t) - \sin t(-\sin t)}{\cos^2 t}$$

$$= \frac{\cos^2 t + \sin^2 t}{\cos^2 t} = \frac{1}{\cos^2 t}$$

31. In order to find the slope of the line tangent to the graph of $y = \cos 3x$ at $x = 13\pi/6$, first find the derivative of the equation.

$$\frac{d}{dx}(\cos 3x) = -(\sin 3x)(3) = -3\sin 3x$$

Now evaluate the derivative for $x = 13\pi/6$ to determine the slope.

$$\text{slope} = -3\sin 3\left(\frac{13\pi}{6}\right) = -3(1) = -3.$$

37. Our first guess is that an antiderivative of $\cos\frac{x}{7}$ would involve the sine function, namely $\sin\frac{x}{7}$. We may differentiate our guess (not the original function) to check our reasoning:

$$\frac{d}{dx}\sin\frac{x}{7} = \cos\frac{x}{7}\cdot\left(\frac{1}{7}\right) = \frac{1}{7}\cos\frac{x}{7}.$$

We need to adjust our guess so the constant in front of our derivative becomes $-\frac{1}{2}$. We adjust our guess by multiplying it by $-\frac{7}{2}$. Now

$$\frac{d}{dx}\left(-\frac{7}{2}\sin\frac{x}{7}\right) = \left(-\frac{7}{2}\cos\frac{x}{7}\right)\frac{1}{7} = -\frac{1}{2}\cos\frac{x}{7}.$$

Thus $-\frac{7}{2}\sin\frac{x}{7}$ is an antiderivative of $-\frac{1}{2}\cos\frac{x}{7}$. Therefore,

$$\int -\frac{1}{2}\cos\frac{x}{7} = -\frac{7}{2}\sin\frac{x}{7} + C.$$

43. Our first guess is that an antiderivative of $\sin(4x + 1)$ should involve the cosine function, namely, $\cos(4x + 1)$. We may differentiate our guess (*not* the original function) to check our reasoning:

$$\frac{d}{dx}\cos(4x+1) = -\sin(4x+1)\cdot 4 = -4\sin(4x+1).$$

This derivative is −4 times what it should be. So, we adjust our guess by dividing it by −4. Now

$$\frac{d}{dx}\left[-\frac{1}{4}\cos(4x+1)\right] = -\frac{1}{4}\cdot\frac{d}{dx}\cos(4x+1)$$

$$= -\frac{1}{4}[-\sin(4x+1)\cdot 4]$$

$$= \sin(4x+1).$$

Thus $-\frac{1}{4}\cos(4x+1)$ *is* an antiderivative of $\sin(4x + 1)$. Hence,

$$\int \sin(4x+1)dx = -\frac{1}{4}\cos(4x+1)+C.$$

49. We are looking for

$$\lim_{h\to 0}\frac{\sin\left(\frac{\pi}{2}+h\right)-1}{h},$$

or equivalently

$$\lim_{h\to 0}\frac{\sin\left(\frac{\pi}{2}+h\right)-\sin\left(\frac{\pi}{2}\right)}{h}.$$

Since

$$f'(x) = \lim_{h\to 0}\frac{f(a+h)-f(a)}{h},$$

we see that $f(x) = \sin x$ and $a = \frac{\pi}{2}$. Therefore,

$$\frac{d}{dx}\sin x\Big|_{x=\pi/2} = \cos x\Big|_{x=\pi/2} = \cos\frac{\pi}{2} = 0$$

$$\Rightarrow \lim_{h\to 0}\frac{\sin\left(\frac{\pi}{2}+h\right)-1}{h} = 0.$$

8.4   The Tangent and Other Trigonometric Functions

You may need to memorize the discussion of how to get a derivative formula for tan t. (See the *Manual* solution to Exercise 25 in Section 8.3.) This is a nice application of the quotient rule. Of course you should also memorize the formula:

$$\frac{d}{dt}\tan t = \sec^2 t.$$

1. Since

$$\cos t = \frac{\text{adjacent}}{\text{hypotenuse}} \quad \text{and} \quad \sec t = \frac{1}{\cos t},$$

we have

$$\sec t = \frac{\text{hypotenuse}}{\text{adjacent}}.$$

7. $\tan t = \frac{y}{x} = \frac{2}{-2} = -1$. To find $\sec t$ you need r.

$$r = \sqrt{x^2 + y^2} = \sqrt{(-2)^2 + 2^2} = \sqrt{4+4}$$

$$= \sqrt{8} = 2\sqrt{2},$$

$$\sec t = \frac{r}{x} = \frac{2\sqrt{2}}{-2} = -\sqrt{2}.$$

13. Note that $\sec t = 1/\cos t = (\cos t)^{-1}$. Hence,

$$\frac{d}{dt}\sec t = \frac{d}{dt}(\cos t)^{-1} = (-1)(\cos t)^{-2} \cdot \frac{d}{dt}\cos t$$

$$= (-1)(\cos t)^{-2}(-\sin t) = \frac{\sin t}{\cos^2 t}.$$

This answer is acceptable. It may be written in other equivalent forms, such as $\sin t \sec^2 t$, or

$$\frac{\sin t}{\cos t} \cdot \frac{1}{\cos t} = \tan t \sec t.$$

19. By the chain rule,

$$f'(x) = \frac{d}{dx} 3\tan(\pi - x)$$

$$= 3\sec^2(\pi - x) \cdot \frac{d}{dx}(\pi - x)$$

$$= -3\sec^2(\pi - x).$$

25. By the product rule,

$$y' = \frac{d}{dx}(x\tan x)$$

$$= x\cdot\frac{d}{dx}(\tan x)+(\tan x)\cdot\frac{d}{dx}x$$

$$= x\sec^2 x+\tan x.$$

31. From the chain rule and the solution to Exercise 13,

$$y' = \frac{d}{dt}\ln(\tan t+\sec t) = \frac{1}{\tan t+\sec t}\cdot\frac{d}{dt}(\tan t+\sec t)$$

$$= \frac{1}{\tan t+\sec t}\left(\frac{d}{dt}\tan t+\frac{d}{dt}\sec t\right) = \frac{1}{\tan t+\sec t}(\sec^2 t+\tan t\sec t)$$

$$= \frac{(\sec t+\tan t)\sec t}{\tan t+\sec t} = \sec t.$$

37. In this section of the text it is shown that the derivative of *tan(x)* is *sec²(x)*. Hence *tan(x)* is an antiderivative of *sec²(x)*, so the integration can proceed as follows:

$$\int_{-\pi/4}^{\pi/4}\sec^2 x\,dx = (\tan x)\Big|_{-\pi/4}^{\pi/4} = \tan\left(\frac{\pi}{4}\right)-\tan\left(\frac{-\pi}{4}\right)$$

$$= 1-(-1) = 2.$$

Chapter 8 Review Exercises

Use the chapter checklist to review the main definitions and facts. You may also need to memorize some of the following facts: (1) the graphs of sin *t* and cos *t* (knowing the values of sin *t* and cos *t* when $t = 0$, $\pi/2, \pi, 3\pi/2,$ and $2\pi,$ for example); (2) selected trigonometric identities from Section 8.2; and (3) the definitions of sec *t*, cot *t*, and csc *t*. Check with your instructor.

While solutions for all review exercises are included, expanded explanations are included for every sixth exercise.

1. The angle described is three quarter-revolutions of the circle or $3\left(\frac{\pi}{2}\text{ radians}\right) = \frac{3\pi}{2}$ radians.

2. $t = -\frac{7\pi}{2}$

3. $t = -\frac{3\pi}{4}$

4.

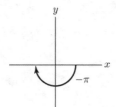

5.

6.

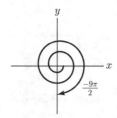

7. First find r.

$$r = \sqrt{x^2 + y^2} = \sqrt{3^2 + 4^2} = \sqrt{25} = 5.$$

Therefore,

$$\sin t = \frac{y}{r} = \frac{4}{5} = .8, \quad \cos t = \frac{x}{r} = \frac{3}{5} = .6, \quad \tan t = \frac{y}{x} = \frac{4}{3}.$$

8. $(-.6)^2 + (.8)^2 = (\text{hyp.})^2 \Rightarrow \text{hyp.} = 1$

$\sin t = \dfrac{\text{opp.}}{\text{hyp.}} = .8 \,;\; \cos t = \dfrac{\text{adj.}}{\text{hyp.}} = -.6; \quad \tan t = \dfrac{\text{opp.}}{\text{adj.}} = \dfrac{.8}{-.6} = -\dfrac{4}{3}$

9. $(-.6)^2 + (-.8)^2 = (\text{hyp.})^2 \Rightarrow \text{hyp.} = 1$

$\sin t = \dfrac{\text{opp.}}{\text{hyp.}} = -.8; \; \cos t = \dfrac{\text{adj.}}{\text{hyp.}} = -.6; \quad \tan t = \dfrac{\text{opp.}}{\text{adj.}} = \dfrac{-.8}{-.6} = \dfrac{4}{3}$

10. $3^2 + (-4)^2 = (\text{hyp.})^2; \text{hyp.} = 5$

$\sin t = \dfrac{\text{opp.}}{\text{hyp.}} = -\dfrac{4}{5}; \; \cos t = \dfrac{\text{adj.}}{\text{hyp.}} = \dfrac{3}{5}; \quad \tan t = \dfrac{\text{opp.}}{\text{adj.}} = -\dfrac{4}{3}$

11. $\sin t = \dfrac{1}{5}; (\text{opp.})^2 + (\text{adj.})^2 = (\text{hyp.})^2$

$1 + (\text{adj.})^2 = 25 \Rightarrow \text{adj.} = \pm\sqrt{24} = \pm 2\sqrt{6}$

$\cos t = \dfrac{\text{adj.}}{\text{hyp.}} = \pm\dfrac{2\sqrt{6}}{5}$

12. $\cos t = -\dfrac{2}{3}; (\text{opp.})^2 + (\text{adj.})^2 = (\text{hyp.})^2$

$(\text{opp.})^2 + 4 = 9 \Rightarrow \text{opp.} = \pm\sqrt{5}; \sin t = \dfrac{\pm\sqrt{5}}{3}$

13. $\sin t = \cos t$ whenever the point on the unit circle defined by the angle t has identical x and y coordinates (i.e., the point lies on the line $x = y$). The points on the unit circle with this property are $(1/\sqrt{2}, 1/\sqrt{2})$ and $(-1/\sqrt{2}, -1/\sqrt{2})$. The four angles between -2π and 2π that correspond to these points are $-\dfrac{3\pi}{4}, -\dfrac{7\pi}{4}, \dfrac{\pi}{4}, \dfrac{5\pi}{4}$.

14. $\dfrac{3\pi}{4}, \dfrac{7\pi}{4}, -\dfrac{\pi}{4}, -\dfrac{5\pi}{4}$

15. negative

16. positive

17. Let r be the length of the rafter needed to support the roof.

$r^2 = (15)^2 + \left[15(\tan 23°)\right]^2 \Rightarrow r \approx 16.3$ ft

18. Let t be the height of the tree.
$t = 60(\tan 53°) \approx 79.62$ feet

19. $f'(t) = \dfrac{d}{dt} 3\sin t = 3\cos t.$

20. $f(t) = \sin 3t; \dfrac{d}{dt}\sin 3t = (\cos 3t)(3) = 3\cos 3t$

21. $f(t) = \sin\sqrt{t} = \sin t^{1/2}$

$\dfrac{d}{dt}\sin t^{1/2} = \left(\cos\left(t^{1/2}\right)\right)\left(\dfrac{1}{2}\right)t^{-1/2} = \dfrac{\cos\sqrt{t}}{2\sqrt{t}}$

22. $f(t) = \cos t^3$

$\dfrac{d}{dt}\cos t^3 = (-\sin t^3)(3t^2) = -3t^2\sin t^3$

23. $g(x) = x^3\sin x$

$\dfrac{d}{dx}(x^3\sin x) = x^3\cos x + 3x^2\sin x$

24. $g(x) = \sin(-2x) \cos 5x$

$$\frac{d}{dx}\left[\sin(-2x)\cos(5x)\right] = \sin(-2x)(-\sin(5x))(5) + \cos(-2x)(-2)\cos(5x)$$
$$= -5\sin(5x)\sin(-2x) - 2\cos(-2x)\cos(5x)$$
$$= 5\sin(5x)\sin(2x) - 2\cos(5x)\cos(2x)$$

25. By the quotient rule,

$$f'(x) = \frac{d}{dx}\left(\frac{\cos 2x}{\sin 3x}\right)$$
$$= \frac{\sin(3x) \cdot \frac{d}{dx}\cos 2x - \cos(2x) \cdot \frac{d}{dx}\sin 3x}{(\sin 3x)^2}$$
$$= \frac{\sin(3x)(-2\sin 2x) - 3\cos(2x)\cos(3x)}{\sin^2(3x)}$$
$$= -\frac{2\sin(3x)\sin(2x) + 3\cos(2x)\cos(3x)}{\sin^2(3x)}$$

26. $f(x) = \dfrac{\cos x - 1}{x^3}$

$$\frac{d}{dx}\left(\frac{\cos x - 1}{x^3}\right) = \frac{(-\sin x)(x^3) - 3x^2(\cos x - 1)}{x^6} = \frac{-x^3 \sin x - 3x^2(\cos x - 1)}{x^6}$$

27. $f(x) = \cos^3 4x$

$$\frac{d}{dx}\cos^3 4x = (3\cos^2 4x)(-\sin 4x)(4) = -12\cos^2 4x \sin 4x$$

28. $f(x) = \tan^3 2x$

$$\frac{d}{dx}\tan^3 2x = 3(\tan^2 2x)(\sec^2 2x)(2) = 6\tan^2 2x \sec^2 2x$$

29. $y = \tan(x^4 + x^2)$

$$\frac{d}{dx}\tan(x^4 + x^2) = \left(\sec^2\left(x^4 + x^2\right)\right)\left(4x^3 + 2x\right) = (4x^3 + 2x)\sec^2(x^4 + x^2)$$

30. $y = \tan e^{-2x}$

$$\frac{d}{dx}\tan e^{-2x} = (\sec^2 e^{-2x})e^{-2x}(-2) = -2e^{-2x}\sec^2\left(e^{-2x}\right)$$

31. By the chain rule,

$$y' = \frac{d}{dx}\sin(\tan x) = \cos(\tan x)\frac{d}{dx}\tan x$$
$$= \cos(\tan x)\sec^2 x.$$

32. $y = \tan(\sin x)$

$$\frac{d}{dx}\tan(\sin x) = \sec^2(\sin x)\cos x$$

33. $y = \sin x \tan x$

$$\frac{d}{dx}[\sin x \tan x] = \sin x \sec^2 x + \cos x \tan x = \sin x \sec^2 x + \sin x$$

34. $y = (\ln x)\cos x$

$$\frac{d}{dx}[(\ln x)\cos x] = \ln x(-\sin x) + \frac{1}{x}\cos x = \frac{\cos x}{x} - (\ln x)\sin x$$

35. $y = \ln(\sin x)$

$$\frac{d}{dx}\ln(\sin x) = \frac{1}{\sin x}(\cos x) = \cot x$$

36. $y = \ln(\cos x)$

$$\frac{d}{dx}\ln(\cos x) = \frac{1}{\cos x}(-\sin x) = -\tan x$$

37. By the product and chain rules,

$$y' = \frac{d}{dx}e^{3x}\sin^4 x$$

$$= e^{3x}(4\sin^3 x)\cdot\frac{d}{dx}\sin x + (\sin^4 x)\cdot 3e^{3x}$$

$$= 4e^{3x}\sin^3 x\cos x + 3e^{3x}\sin^4 x.$$

38. $y = \sin^4 e^{3x}$

$$\frac{d}{dx}\sin^4 e^{3x} = \left(4\sin^3 e^{3x}\right)\left(\cos e^{3x}\right)\left(e^{3x}\right)(3) = 12e^{3x}\left(\cos e^{3x}\right)\left(\sin^3 e^{3x}\right)$$

39. $f(t) = \dfrac{\sin t}{\tan 3t}$

$$\frac{d}{dt}\left(\frac{\sin t}{\tan 3t}\right) = \frac{\cos t \tan 3t - (\sec^2 3t)(3)\sin t}{\tan^2 3t} = \frac{\cos t \tan 3t - 3\sin t \sec^2 3t}{\tan^2 3t}$$

40. $f(t) = \dfrac{\tan 2t}{\cos t}$

$$\frac{d}{dt}\left(\frac{\tan 2t}{\cos t}\right) = \frac{(\sec^2 2t)(2)\cos t - (-\sin t)\tan 2t}{\cos^2 t} = \frac{2\cos t \sec^2 2t + \sin t \tan 2t}{\cos^2 t}$$

41. $f(t) = e^{\tan t}$

$$\frac{d}{dt}e^{\tan t} = e^{\tan t}(\sec^2 t)$$

42. $f(t) = e^t \tan t$

$$\frac{d}{dt}[e^t \tan t] = e^t (\sec^2 t) + e^t \tan t = e^t (\sec^2 t + \tan t)$$

43. $f(t) = \sin^2 t.$

$$f'(t) = \frac{d}{dt}(\sin t)^2$$

$$= (2 \sin t) \cdot \frac{d}{dt} \sin t$$

$$= 2 \sin t \cos t.$$

$$f''(t) = \frac{d}{dt}(2 \sin t \cdot \cos t)$$

$$= 2 \left[(\sin t) \cdot \frac{d}{dt} \cos t + (\cos t) \cdot \frac{d}{dt} \sin t \right]$$

$$= 2 [(\sin t)(-\sin t) + \cos t \cdot \cos t]$$

$$= 2(\cos^2 t - \sin^2 t).$$

44. $y = 3 \sin 2t + \cos 2t$

$y' = 3(\cos 2t)(2) + (-\sin 2t)(2) = 6 \cos 2t - 2 \sin 2t$

$y'' = 6[(-\sin 2t)(2)] - 2(\cos 2t)(2) = -12 \sin 2t - 4 \cos 2t = -4(3 \sin 2t + \cos 2t)$

$-4y = -12 \sin 2t - 4 \cos 2t$

Therefore y'' and $-4y$ are equal.

45. $f(s, t) = \sin s \cos 2t; \quad \dfrac{\partial f}{\partial s} = \cos s \cos 2t$

$$\frac{\partial f}{\partial t} = \sin s \, (-\sin 2t)(2) = -2 \sin s \sin 2t$$

46. $z = \sin wt; \quad \dfrac{\partial z}{\partial w} = t \cos wt; \quad \dfrac{\partial z}{\partial t} = w \cos wt$

47. $f(s, t) = t \sin st; \quad \dfrac{\partial f}{\partial s} = t(\cos st)(t) = t^2 \cos st$

$$\frac{\partial f}{\partial t} = t(\cos st)(s) + (1) \sin st = st \cos st + \sin st$$

48. $\sin(s + t) = \sin s \cos t + \cos s \sin t$

$$\frac{\partial}{\partial t} \sin(s + t) = (\cos(s + t))$$

$$\frac{\partial}{\partial t}[\sin s \cos t + \cos s \sin t] = \sin s(-\sin t) + \cos s \cos t = \cos s \cos t - \sin s \sin t$$

Thus, $\cos(s + t) = \cos s \cos t - \sin s \sin t.$

49. $y' = \dfrac{d}{dt} \tan t = \sec^2 t.$

The slope of the tangent line when $t = \dfrac{\pi}{4}$ is

$$\sec^2\left(\frac{\pi}{4}\right) = \left(\frac{1}{\cos\frac{\pi}{4}}\right)^2 = \left(\frac{1}{\frac{\sqrt{2}}{2}}\right)^2 = 2.$$

At $t = \dfrac{\pi}{4}$, $y = \tan\dfrac{\pi}{4} = 1$, so $\left(\dfrac{\pi}{4}, 1\right)$ is a point on the line. Hence, the equation of the line is

$$y - 1 = 2\left(t - \frac{\pi}{4}\right).$$

50. a. $f(t) = \sin t + \cos t$

$f'(t) = \cos t - \sin t$

$\cos t - \sin t = 0 \Rightarrow \cos t = \sin t \Rightarrow t = \dfrac{\pi}{4}, \dfrac{5\pi}{4}$

b. $f''(t) = -\cos t - \sin t$

$f''\left(\dfrac{\pi}{4}\right) = -\sqrt{2}$, so the curve is concave down at $t = \dfrac{\pi}{4}$. $f''\left(\dfrac{5\pi}{4}\right) = \sqrt{2}$, so the curve is concave

up at $t = \dfrac{5\pi}{4}$.

c. $f''(t) = -\cos t - \sin t = 0 \Rightarrow -\cos t = \sin t \Rightarrow t = \dfrac{3\pi}{4}, \dfrac{7\pi}{4}$

The inflection points are at $\left(\dfrac{3\pi}{4}, 0\right)$ and $\left(\dfrac{7\pi}{4}, 0\right)$.

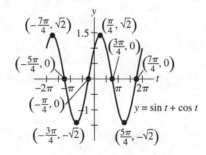

51.

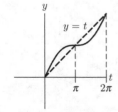

52. $y = 2 + \sin 3t$

Area under the curve is $\int_0^{\pi/2} (2 + \sin 3t) \, dt$.

$$\int_0^{\pi/2} (2 + \sin 3t) \, dt = \left[2t - (\cos 3t)\left(\frac{1}{3}\right) \right]_0^{\pi/2} = \left[2t - \frac{1}{3}\cos 3t \right]_0^{\pi/2} = \pi - \frac{1}{3}(0) - \left(0 - \frac{1}{3} \right) = \frac{1}{3} + \pi$$

53. The desired area is

$$\int_0^\pi \sin t \, dt + \int_\pi^{2\pi} (-\sin t) \, dt = (-\cos t)\Big|_0^\pi + \cos t \Big|_\pi^{2\pi} = 2 + 2 = 4$$

54. The desired area is

$$\int_0^{\pi/2} \cos t \, dt + \int_{\pi/2}^{3\pi/2} (-\cos t) \, dt = \sin t \Big|_0^{\pi/2} + (-\sin t)\Big|_{\pi/2}^{3\pi/2} = 1 + 2 = 3$$

55. It is easy to check that the line $y = x$ is tangent to the graph of $y = \sin x$ at $x = 0$. From the graph of $y = \sin x$, it is clear that $y = \sin x$ lies below $y = x$ for $x \geq 0$. So, the area between these two curves from $x = 0$ to $x = \pi$ is given by

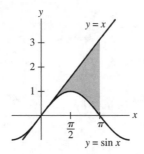

$$[\text{Area}] = \int_0^\pi (x - \sin x) \, dx$$

$$= \left[\frac{x^2}{2} + \cos x \right]_0^\pi = \left[\frac{\pi^2}{2} + \cos \pi \right] - [0 + \cos 0]$$

$$= \frac{\pi^2}{2} + (-1) - [0 + 1] = \frac{\pi^2}{2} - 2.$$

Exercises 56–58 refer to the function $V(t) = 3 + .05 \sin\left(160\pi t - \frac{\pi}{2} \right)$, where $V(t)$ is the lung volume in liters and t is measured in minutes.

56. a. $V(0) = 2.95,\ V\left(\frac{1}{320} \right) = 3,\ V\left(\frac{1}{160} \right) = 3.05, V\left(\frac{1}{80} \right) = 2.95$

b. $V'(t) = .05\cos\left(160\pi t - \dfrac{\pi}{2}\right)(160\pi)$

Setting $V'(t) = 0$ gives $\cos\left(160\pi t - \dfrac{\pi}{2}\right) = 0 \Rightarrow$

$160\pi t - \dfrac{\pi}{2} = \dfrac{\pi}{2}, \dfrac{3\pi}{2}, \dfrac{5\pi}{2}, \ldots \Rightarrow 160\pi t = \pi, 2\pi, 3\pi, \ldots \Rightarrow t = \dfrac{1}{160}, \dfrac{2}{160}, \dfrac{3}{160}, \ldots$

$V''(t) = -8\pi\sin\left(160\pi t - \dfrac{\pi}{2}\right)(160\pi) = -1280\pi^2 \sin\left(160\pi t - \dfrac{\pi}{2}\right)$

At $t = \dfrac{1}{160}$, $V''(t) < 0$, so this value of t gives a relative maximum, and the maximum lung

volume is $V\left(\dfrac{1}{160}\right) = 3.05$ liters.

57. a. $V'(t) = 8\pi\cos\left(160\pi t - \dfrac{\pi}{2}\right)$

b. Inspiration (in the first cycle) occurs from $t = 0$ to $t = \dfrac{1}{160}$. To find the maximum rate of

inspiration, we need to find the maximum of $V'(t)$ on $\left[0, \dfrac{1}{160}\right]$.

$V''(t) = -1280\pi^2 \sin\left(160\pi t - \dfrac{\pi}{2}\right)$

Setting $V''(t) = 0$ gives $160\pi t - \dfrac{\pi}{2} = 0, \pi, 2\pi, \ldots \Rightarrow 160\pi t = \dfrac{\pi}{2}, \dfrac{3\pi}{2}, \dfrac{5\pi}{2}, \ldots \Rightarrow$

$t = \dfrac{1}{320}, \dfrac{3}{320}, \dfrac{5}{320}, \ldots$ Among these values, only $\dfrac{1}{320}$ is within $\left[0, \dfrac{1}{160}\right]$.

Since $V'(t) = 0$ at the end points of the interval, $t = \dfrac{1}{320}$ must give the maximum value of

$V'(t)$. Thus, the maximum rate of air flow is

$V'\left(\dfrac{1}{320}\right) = 8\pi$ liters/min.

c. The average value of $V'(t)$ on $\left[0, \dfrac{1}{160}\right]$ is

$\dfrac{1}{\frac{1}{160}}\displaystyle\int_0^{1/160} V'(t)\,dt = 160\int_0^{1/160} 8\pi\cos\left(160\pi t - \dfrac{\pi}{2}\right) dt = -8\sin\left(160\pi t - \dfrac{\pi}{2}\right)\Big|_0^{1/160}$

$= 8(1+1) = 16$ liters/min

58. During one minute, there will be 80 inspirations. Each inspiration represents
$$V\left(\frac{1}{160}\right) - V(0) = 3 + .05 - (3 - .05) = .1 \text{ liter.}$$ Therefore, the minute volume is $.1(80) = 8$ liters. In Exercise 57(b), we found that the peak respiratory flow was 8π liters/min. Thus the first statement is verified. In Exercise 57(c), we found the mean inspiratory flow to be $16 = 8 \cdot 2$ liters/min, verifying the second statement.

59. $\int \sin(\pi - x)dx = \cos(\pi - x) + C$

60. $\int (3\cos 3x - 2\sin 2x)dx = \sin 3x + \cos 2x + C$

61. It may be helpful to review Example 5(b) of Section 8.3 before attempting this problem. To integrate, we will need to find an antiderivative of $cos(6x)$. As a first guess, we could try $sin(6x)$. Differentiating shows
$$\frac{d}{dx}(\sin 6x) = 6\cos 6x,$$
which is 6 times too much. So we divide our original guess by 6 to find that $\frac{1}{6}\sin(6x)$ is a correct antiderivative of $cos(6x)$. (Check!) The integration proceeds as follows:
$$\int_0^{\pi/2} \cos(6x)dx = \left(\frac{1}{6}\sin(6x)\right)\Big|_0^{\pi/2} = \left(\frac{1}{6}\sin(3\pi)\right) - \left(\frac{1}{6}\sin(0)\right)$$
$$= 0 - 0 = 0.$$

62. $\int \cos(6 - 2x)dx = -\frac{1}{2}\sin(6 - 2x) + C$

63. $\int_0^{\pi}[x - 2\cos(\pi - 2x)]dx = \left[\frac{1}{2}x^2 + \sin(\pi - 2x)\right]\Big|_0^{\pi} = \frac{\pi^2}{2} + 0 - (0) = \frac{\pi^2}{2}$

64. $\int_{-\pi}^{\pi}(\cos 3x + 2\sin 7x)dx = \left(\frac{1}{3}\sin 3x - \frac{2}{7}\cos 7x\right)\Big|_{-\pi}^{\pi} = 0 - \left(-\frac{2}{7}\right) - \left[0 - \left(-\frac{2}{7}\right)\right] = 0$

65. $\int \sec^2 \frac{x}{2}dx = 2\tan\frac{x}{2} + C$

66. $\int 2\sec^2 2x\,dx = \tan 2x + C$

67. The shaded region A_1 represents the region between the upper curve ($y = \cos x$) and the lower curve ($y = \sin x$) from $x = 0$ to $x = \pi/4$. Thus the area of A_1 can be computed by evaluating the integral
$$\int_0^{\pi/4}(\cos x - \sin x)dx.$$
Use the facts that $sin(x)$ is an antiderivative of $cos(x)$ and $cos(x)$ is an antiderivative of $-sin(x)$ to integrate:

$$\int_0^{\pi/4}(\cos x-\sin x)dx=(\sin x+\cos x)\Big|_0^{\pi/4}$$

$$=\left(\sin\frac{\pi}{4}+\cos\frac{\pi}{4}\right)-(\sin 0+\cos 0)$$

$$=\left(\frac{\sqrt2}{2}+\frac{\sqrt2}{2}\right)-(0+1)$$

$$=\sqrt2-1.$$

68. $\int_0^{\pi/4}\sin x\,dx+\int_{\pi/4}^{\pi/2}\cos x\,dx=(-\cos x)\Big|_0^{\pi/4}+(\sin x)\Big|_{\pi/4}^{\pi/2}=-\frac{\sqrt2}{2}-(-1)+1-\left(\frac{\sqrt2}{2}\right)=2-\sqrt2$

69. $\int_{\pi/4}^{\pi/2}(\sin x-\cos x)dx+\int_{\pi/2}^{\pi}\sin x\,dx=(-\cos x-\sin x)\Big|_{\pi/4}^{\pi/2}+(-\cos x)\Big|_{\pi/2}^{\pi}$

$$=0-1-\left(-\frac{\sqrt2}{2}-\frac{\sqrt2}{2}\right)+[-(-1)]-(0)=\sqrt2$$

70. $\int_{\pi/2}^{\pi}(0-\cos x)dx=-\int_{\pi/2}^{\pi}\cos x\,dx=-(\sin x)\Big|_{\pi/2}^{\pi}=-[0-(1)]=1$

71. Average $=\dfrac{1}{b-a}\int_a^b f(t)dt=\dfrac{1}{2\pi-0}\int_0^{2\pi}\left(1+\sin 2t-\frac13\cos 2t\right)dt=\dfrac{1}{2\pi}\left[t-\frac12\cos 2t-\frac16\sin 2t\right]\Big|_0^{2\pi}$

$$=\frac{1}{2\pi}\left[2\pi-\frac12-0-\left(0-\frac12-0\right)\right]=1$$

72. $\dfrac{1}{\pi-0}\int_0^{\pi}(t-\cos 2t)dt=\dfrac{1}{\pi}\left[\frac12 t^2-\frac12\sin 2t\right]\Big|_0^{\pi}=\dfrac{1}{\pi}\left[\frac{\pi^2}{2}-0-(0-0)\right]=\dfrac{\pi}{2}$

73. Using the formula for the average value of a function on an interval given in Section 6.5, we see that the answer to this problem can be computed by evaluating

$$\frac{1}{\frac{3\pi}{4}-0}\int_0^{3\pi/4}\left[1000+200\sin\left(2\left(t-\frac{\pi}{4}\right)\right)\right]dt.$$

This can be partially evaluated as follows:

$$\frac{1}{3\pi/4-0}\int_0^{3\pi/4}\left[1000+200\sin\left(2\left(t-\frac{\pi}{4}\right)\right)\right]dt$$

$$= \frac{1}{3\pi/4}\left[\int_0^{3\pi/4} 1000\,dt + 200\left(\int_0^{3\pi/4} \sin\left(2\left(t - \frac{\pi}{4}\right)\right)dt\right)\right]$$

$$= \frac{4}{3\pi}\left[\left(1000t\big|_0^{3\pi/4}\right) + 200\left(\int_0^{3\pi/4} \sin\left(2\left(t - \frac{\pi}{4}\right)\right)dt\right)\right]$$

$$= \frac{4}{3\pi}\left[750\pi + 200\left(\int_0^{3\pi/4} \sin\left(2\left(t - \frac{\pi}{4}\right)\right)dt\right)\right]$$

$$= 1000 + \frac{800}{3\pi}\left(\int_0^{3\pi/4} \sin\left(2\left(t - \frac{\pi}{4}\right)\right)dt\right).$$

All that remains is to evaluate the integral

$$\int_0^{3\pi/4} \sin\left(2\left(t - \frac{\pi}{4}\right)\right)dt$$

and simplify. To find an antiderivative of

$$\sin\left(2\left(t - \frac{\pi}{4}\right)\right),$$

it is helpful to refer to the method used in the solution to Exercise 61 and Example 5(b) of Section 8.3 to observe that

$$-\frac{1}{2}\cos(2t)$$

is an antiderivative of $\sin(2t)$.

Use the chain rule to verify that

$$\cos\left(t - \frac{\pi}{4}\right)$$

is an antiderivative of $\sin\left(t - \frac{\pi}{4}\right)$. Thus

$$-\frac{1}{2}\cos\left(2\left(t - \frac{\pi}{4}\right)\right)$$

is a reasonable (and correct) guess for an antiderivative of $\sin\left(2\left(t - \frac{\pi}{4}\right)\right)$. Thus the remaining integral can be evaluated as follows:

$$\int_0^{3\pi/4} \sin\left(2\left(t - \frac{\pi}{4}\right)\right)dt = \left(-\frac{1}{2}\cos\left(2\left(t - \frac{\pi}{4}\right)\right)\right)\Bigg|_0^{3\pi/4}$$

$$= \left(-\frac{1}{2}\cos\left(2\left(\frac{\pi}{2}\right)\right)\right) - \left(-\frac{1}{2}\cos\left(\frac{-2\pi}{4}\right)\right)$$

$$= \left(-\frac{1}{2}(-1)\right) - 0 = \frac{1}{2}.$$

Finally, the average value can now be computed:

$$1000+\frac{800}{3\pi}\left(\int_0^{3\pi/4}\sin\left(2\left(t-\frac{\pi}{4}\right)\right)dt\right)=1000+\frac{800}{3\pi}\left(\frac{1}{2}\right)$$

$$=1000+\frac{400}{3\pi}\approx1042.44$$

74. $\dfrac{1}{0-(-\pi)}\displaystyle\int_{-\pi}^{0}(\cos t+\sin t)dt=\dfrac{1}{\pi}\Big[\sin t-\cos t\Big]\Big|_{-\pi}^{0}=\dfrac{1}{\pi}\big[0-1-(0-(-1))\big]=-\dfrac{2}{\pi}$

75. Substitute $\tan^2 x=\sec^2 x-1$.

$\displaystyle\int\tan^2 x\,dx=\int(\sec^2 x-1)dx=\tan x-x+C$

76. Substitute $\tan^2 3x=\sec^2 3x-1$.

$\displaystyle\int\tan^2 3x\,dx=\int(\sec^2 3x-1)dx=\frac{1}{3}\tan 3x-x+C$

77. Substitute $1+\tan^2 x=\sec^2 x$.

$\displaystyle\int(1+\tan^2 x)dx=\int\sec^2 x\,dx=\tan x+C$

78. Substitute $1+\tan^2 x=\sec^2 x$.

$\displaystyle\int(2+\tan^2 x)dx=\int(1+1+\tan^2 x)dx=\int(1+\sec^2 x)dx=x+\tan x+C$

79. First, use identity (1) of Section 8.4, $\tan^2 x+1=\sec^2 x,$ to transform the integral before evaluating it, as suggested in the hint:

$$\int_0^{\pi/4}\tan^2 x\,dx=\int_0^{\pi/4}(\sec^2 x-1)\,dx.$$

Next, use the fact that $\tan x$ is an antiderivative of $\sec^2 x$ to evaluate the new integral:

$$\int_0^{\pi/4}\tan^2 x\,dx=\int_0^{\pi/4}(\sec^2 x-1)dx$$
$$=\int_0^{\pi/4}(\sec^2 x)dx-\int_0^{\pi/4}1\,dx$$
$$=(\tan x-x)\Big|_0^{\pi/4}$$
$$=1-\frac{\pi}{4}-(0-0)=1-\frac{\pi}{4}$$

80. Substitute $\tan^2 x=\sec^2 x-1$.

$\displaystyle\int_0^{\pi/4}(2+2\tan^2 x)dx=\int_0^{\pi/4}\big[2+2(\sec^2 x-1)\big]dx=\int_0^{\pi/4}2\sec^2 x\,dx=2\tan x\Big|_0^{\pi/4}=2-0=2$

Chapter 9
Techniques of Integration

9.1 Integration by Substitution

The technique of integration by substitution is best learned by trial and error. The more problems you work, the fewer errors you will make in your first guess at a substitution. The basic strategy is to arrange the integrand in the form $f(g(x))g'(x)$. (This is possible for every problem in this section.) Look for the most complicated part of the integrand that may be expressed as a composite function, and let u be the "inside" function $g(x)$. Then proceed as discussed in the text.

Remember that you can check your answers to an antidifferentiation problem by differentiating your answer. (You ought to do this at least mentally for every indefinite integral you find.) We urge you to work *all* the problems (odd and even) in Section 9.1. Use the solutions in this *Guide* to check your work, *not* to show you how to start a problem.

1. Let $u = x^2 + 4$. Then $du = \dfrac{d}{dx}(x^2 + 4)dx = 2x\,dx$. So

$$\int 2x(x^2 + 4)^5\,dx = \int u^5 du$$

$$= \frac{1}{6}u^6 + C$$

$$= \frac{1}{6}(x^2 + 4)^6 + C.$$

7. Let $u = 4 - x^2$. Then $du = \frac{d}{dx}(4 - x^2)dx = -2x\,dx$. So,

$$\int x\sqrt{4 - x^2}\,dx = \int \left(-\frac{1}{2}\right)(-2)x\sqrt{4 - x^2}\,dx$$

$$= -\frac{1}{2}\int \sqrt{4 - x^2}\,\underbrace{(-2x)dx}_{du} = -\frac{1}{2}\int \sqrt{u}\,du$$

$$= -\frac{1}{2}\int u^{1/2}du = -\frac{1}{2}\cdot\frac{2}{3}u^{3/2} + C$$

$$= -\frac{1}{3}u^{3/2} + C = -\frac{1}{3}(4 - x^2)^{3/2} + C.$$

13. Let $u = \ln(2x)$. Then $du = \dfrac{d}{dx}[\ln(2x)]dx = \dfrac{1}{2x} \cdot 2 dx = \dfrac{1}{x} dx$. Thus

$$\int \frac{\ln 2x}{x} dx = \int (\ln 2x) \cdot \frac{1}{x} dx = \int u \, du$$

$$= \frac{u^2}{2} + C = \frac{(\ln 2x)^2}{2} + C.$$

19. $\ln \sqrt{x} = \ln x^{1/2} = \dfrac{1}{2} \ln x$. This gives us the new form:

$$\int \frac{\ln \sqrt{x}}{x} dx = \int \frac{\frac{1}{2} \ln x}{x} dx = \frac{1}{2} \int \frac{\ln x}{x} dx.$$

So we first need to find $\displaystyle\int \frac{\ln x}{x} dx$.

Let $u = \ln x$. Then

$$du = \frac{d}{dx} \ln x \, dx = \frac{1}{x} dx.$$

Hence,

$$\int \frac{\ln x}{x} dx = \int u \, du = \frac{u^2}{2} + C = \frac{(\ln x)^2}{2} + C.$$

Finally,

$$\int \frac{\ln \sqrt{x}}{x} dx = \frac{1}{2} \int \frac{\ln x}{x} dx = \frac{1}{2} \left[\frac{(\ln x)^2}{2} + C \right] = \frac{(\ln x)^2}{4} + C.$$

25. A good strategy to use when faced with an unfamiliar integral is to ask yourself, "Can I rewrite the integrand in some way to make the problem simpler?" You should recall that $\ln x^2 = 2 \ln x$. That fact is helpful here:

$$\int \frac{1}{x \ln x^2} dx = \int \frac{1}{2x \ln x} dx = \frac{1}{2} \int \frac{1}{x \ln x} dx.$$

Let $u = \ln x$. Then $du = \dfrac{1}{x} dx$, and

$$\int \frac{1}{x \ln x^2} dx = \frac{1}{2} \int \frac{1}{x} \cdot \frac{1}{\ln x} dx = \frac{1}{2} \int \frac{1}{u} du = \frac{1}{2} \ln |u| + C = \frac{1}{2} \ln |\ln x| + C.$$

If you do not happen to notice the initial simplification of the integral shown, you can make the substitution $u = \ln x^2$. In this case,

$$du = \frac{d}{dx} (\ln x^2) \, dx = \frac{1}{x^2} \cdot 2x \, dx = \frac{2}{x} dx.$$

Then

$$\int \frac{1}{x \ln x^2} dx = \frac{1}{2} \int \frac{2}{x} \cdot \frac{1}{\ln x^2} dx = \frac{1}{2} \int \frac{1}{u} du = \frac{1}{2} \ln |\ln x^2| + C.$$

You may find it strange that this antiderivative is similar to the first one but is not quite the same. (Both antiderivatives are correct.) How can that be possible? How must two antiderivatives of the same function be related?

31. Let $u = 1 + 2e^x$. Then $du = \dfrac{d}{dx}(1 + 2e^x)dx = 2e^x dx$. Then

$$\int \frac{e^x}{1+2e^x}dx = \frac{1}{2}\int \frac{1}{1+2e^x}\cdot 2e^x dx = \frac{1}{2}\int \frac{1}{u}du$$

$$= \frac{1}{2}\ln|u| + C$$

$$= \frac{1}{2}\ln|1 + 2e^x| + C$$

$$= \frac{1}{2}\ln\left(1 + 2e^x\right) + C$$

37. If $f'(x) = \dfrac{x}{\sqrt{x^2+9}}$, then $f(x) = \displaystyle\int \frac{x}{\sqrt{x^2+9}}dx$. Let $u = x^2 + 9$. Then $du = 2x\,dx$, so

$$\int \frac{x}{\sqrt{x^2+9}}dx = \int \frac{\left(\frac{1}{2}\right)2x}{\sqrt{x^2+9}}dx = \frac{1}{2}\int \frac{du}{\sqrt{u}} = \frac{1}{2}\int u^{-1/2}du$$

$$= \frac{1}{2}2u^{1/2} + C = u^{1/2} + C = \sqrt{x^2+9} + C.$$

But $f(4) = 8$, so $8 = \sqrt{4^2+9} + C$ which implies that $C = 3$. Therefore, $f(x) = \sqrt{x^2+9} + 3$.

43. Let $u = \sin x$. Thus $du = \dfrac{d}{dx}(\sin x)dx = \cos x\,dx$. So,

$$\int \sin x \cos x\,dx = \int u\,du$$

$$= \frac{u^2}{2} + C = \frac{(\sin x)^2}{2} + C.$$

Note: Another substitution is $u = \cos x$. Then $du = -\sin x\,dx$, and

$$\int \sin x \cos x\,dx = \int -\cos x(-\sin x)dx = \int (-u)\,du$$

$$= -\frac{u^2}{2} + C = -\frac{(\cos x)^2}{2} + C.$$

How can the two functions $\dfrac{(\sin x)^2}{2}$ and $-\dfrac{(\cos x)^2}{2}$ both be antiderivatives of $\sin x \cos x$?

49. Let $u = 2 - \sin 3x$. Then

$$du = \frac{d}{dx}(2 - \sin 3x)dx = -3\cos 3x\, dx.$$

So,

$$\int \frac{\cos 3x}{\sqrt{2 - \sin 3x}}dx = \int (2 - \sin 3x)^{-1/2}\left(-\frac{1}{3}\right)(-3\cos 3x)\, dx$$

$$= -\frac{1}{3}\int u^{-1/2}du = -\frac{1}{3}(2u^{1/2}) + C$$

$$= -\frac{2}{3}(2 - \sin 3x)^{1/2} + C.$$

9.2   Integration by Parts

Exercises 1–24 are all solved using integration by parts. This technique, like the method of substitution, is learned by working many problems. There is another method for integration by parts that some students may have seen elsewhere. It uses u in place of $f(x)$ and dv in place of $g(x)\, dx$. Years ago we tried teaching both methods and we found that the one we now present in the text is easier to learn and use. Once you have mastered integration by parts and by substitution, you are ready to try Exercises 25–36. These problems are more difficult because you have to decide which method to use!

(a) Always consider the method of substitution first. This will work if the integrand is the product of two functions and one of the functions is the derivative of the "inside" of the other function (except maybe for a constant multiple).

(b) If the integrand is the product of two unrelated functions, try integration by parts.

(c) Some antiderivatives cannot be found by either of the methods we have discussed. We will never give you such a problem to solve, but you should be careful if you look in another text for more problems to work.

1. Let $f(x) = x,$ $\quad g(x) = e^{5x},$

$\quad f'(x) = 1,$ $\quad G(x) = \frac{1}{5}e^{5x}.$ [Stop and check that $G'(x) = g(x)$.]

Then

$$\int xe^{5x}dx = \frac{1}{5}xe^{5x} - \int \frac{1}{5}e^{5x}dx$$

$$= \frac{1}{5}xe^{5x} - \frac{1}{5}\int e^{5x}dx$$

$$= \frac{1}{5}xe^{5x} - \frac{1}{5}\cdot\frac{1}{5}e^{5x} + C$$

$$= \frac{1}{5}xe^{5x} - \frac{1}{25}e^{5x} + C.$$

Helpful Hint: The various steps in your calculation of $\int xe^{5x}dx$ should be connected by equals signs, showing how you move from one step to the next. It requires a little more effort to rewrite the first term $\frac{1}{5}xe^{5x}$ on each line, but the solution is not correct otherwise. A careless student might write

$$\int xe^{5x}dx = \frac{1}{5}xe^{5x} - \int \frac{1}{5}e^{5x}dx$$

$$\boxed{=} \qquad -\frac{1}{5}\int e^{5x}dx$$

$$\boxed{\text{errors!}} \quad \boxed{=} \qquad -\frac{1}{5}\cdot\frac{1}{5}e^{5x} + C$$

$$\boxed{=} \qquad -\frac{1}{25}e^{5x} + C.$$

Avoid such "shortcuts" in your *homework* as well as on tests. Develop the habit of writing proper mathematical solutions on your homework. This will help to generate correct patterns of thought when you take an exam.

7. Let $f(x) = x$, $\quad g(x) = \dfrac{1}{\sqrt{x+1}} = (x+1)^{-1/2}$,

$\qquad f'(x) = 1$, $\quad G(x) = 2(x+1)^{1/2}$. [Stop and check that $G'(x) = g(x)$.]

Then

$$\int \frac{x}{\sqrt{x+1}}dx = 2x(x+1)^{1/2} - \int 2(x+1)^{1/2}dx$$

$$= 2x(x+1)^{1/2} - 2\int (x+1)^{1/2}dx$$

$$= 2x(x+1)^{1/2} - 2\cdot\frac{2}{3}(x+1)^{3/2} + C$$

$$= 2x(x+1)^{1/2} - \frac{4}{3}(x+1)^{3/2} + C.$$

13. Let $f(x) = x$, $\quad g(x) = \sqrt{x+1} = (x+1)^{1/2}$,

$\qquad f'(x) = 1$, $\quad G(x) = \dfrac{2}{3}(x+1)^{3/2}$.

Then

$$\int x\sqrt{x+1}\,dx = \frac{2}{3}x(x+1)^{3/2} - \int \frac{2}{3}(x+1)^{3/2}dx$$

$$= \frac{2}{3}x(x+1)^{3/2} - \frac{2}{3}\int (x+1)^{3/2}dx$$

$$= \frac{2}{3}x(x+1)^{3/2} - \frac{2}{3}\cdot\frac{2}{5}(x+1)^{5/2} + C$$

$$= \frac{2}{3}x(x+1)^{3/2} - \frac{4}{15}(x+1)^{5/2} + C.$$

19. Let $f(x) = \ln 5x$, $\qquad g(x) = x$,

$\qquad f'(x) = \dfrac{5}{5x} = \dfrac{1}{x}$, $\qquad G(x) = \dfrac{x^2}{2}$. $\quad$ See Example 7 for a similar problem.

Then

$$\int x \ln 5x \, dx = \frac{x^2}{2} \ln 5x - \int \frac{1}{x} \cdot \frac{x^2}{2} \, dx$$

$$= \frac{x^2}{2} \ln 5x - \frac{1}{2} \int x \, dx$$

$$= \frac{x^2}{2} \ln 5x - \frac{1}{2} \cdot \frac{1}{2} x^2 + C$$

$$= \frac{x^2}{2} \ln 5x - \frac{1}{4} x^2 + C.$$

25. Since $x(x+5)^4$ is the product of two functions, neither of which is a multiple of the derivative of the other, try integration by parts.

Set $\qquad\qquad\qquad\qquad f(x) = x, \qquad g(x) = (x+5)^4,$

$\qquad\qquad\qquad\qquad f'(x) = 1, \qquad G(x) = \dfrac{1}{5}(x+5)^5.$

Then

$$\int x(x+5)^4 \, dx = \frac{1}{5} x(x+5)^5 - \int \frac{1}{5}(x+5)^5 \, dx$$

$$= \frac{1}{5} x(x+5)^5 - \frac{1}{5} \cdot \frac{1}{6}(x+5)^6 + C$$

$$= \frac{1}{5} x(x+5)^5 - \frac{1}{30}(x+5)^6 + C.$$

31. Since $x^2 + 1$ has $2x$ as a derivative, and $2x$ is a multiple of x, use substitution to evaluate the integral. Let $u = x^2 + 1$. Then $du = 2x \, dx$. So,

$$\int x \sec^2(x^2 + 1) \, dx = \frac{1}{2} \int 2x \sec^2(x^2 + 1) \, dx = \frac{1}{2} \int \sec^2 u \, du$$

$$= \frac{1}{2} \tan u + C = \frac{1}{2} \tan(x^2 + 1) + C.$$

37. The slope is $\dfrac{x}{\sqrt{x+9}}$, so

$$f'(x) = \frac{x}{\sqrt{x+9}} \quad \text{and} \quad f(x) = \int \frac{x}{\sqrt{x+9}} \, dx.$$

We can integrate by parts. Since $f(x)$ already denotes a function, write $h(x)$ in place of the usual $f(x)$ in the formula for integration by parts. That is, let

$$h(x) = x, \quad g(x) = (x+9)^{-1/2},$$
$$h'(x) = 1, \quad G(x) = 2(x+9)^{1/2}.$$

Then

$$\int x(x+9)^{-1/2}\,dx = x \cdot 2(x+9)^{1/2} - \int 1 \cdot 2(x+9)^{1/2}\,dx$$
$$= 2x(x+9)^{1/2} - \int 2(x+9)^{1/2}\,dx$$
$$= 2x(x+9)^{1/2} - 2 \cdot \frac{2}{3}(x+9)^{3/2} + C.$$

Now, $(0, 2)$ is on the graph of $f(x)$, so $2 = f(0)$. That is,

$$2 = 2(0)(0+9)^{1/2} - \frac{4}{3}(0+9)^{3/2} + C$$
$$2 = -\frac{4}{3}(9)^{3/2} + C = -\frac{4}{3}(27) + C = -36 + C.$$

Thus $C = 38$, and $f(x) = 2x(x+9)^{1/2} - \frac{4}{3}(x+9)^{3/2} + 38$.

9.3 Evaluation of Definite Integrals

One purpose of this section is to give you lots of practice choosing between integration by parts and by substitution. Another purpose is to show how these procedures may be simplified somewhat when working with definite integrals.

A common mistake in this section is to make a change of variable in a definite integral and not to change the limits of integration. For example, here is an *incorrect* evaluation of the integral

$$\int_0^1 2x(x^2+1)^5\,dx.$$

Let $u = x^2 + 1$ and $du = 2x\,dx.$

Then

$$\int_0^1 (x^2+1)^5 \cdot 2x\,dx \boxed{=} \int_0^1 u^5\,du = \frac{1}{6}u^6 \Big|_0^1$$

$$\boxed{\text{error!}} \qquad = \frac{1}{6}(1)^6 - \frac{1}{6}(0)^6 = \frac{1}{6}.$$

The error in this calculation was made on the first line. The integral on the left equals $21/2$, not $1/6$. See Example 1.

Here is another *incorrect* evaluation whose final answer is correct. However, the work contains two errors and therefore does not justify the final answer. As before, let $u = x^2 + 1$, $du = 2x\,dx$.

Then

$$\int_0^1 (x^2+1)^5 \cdot 2x\, dx \boxed{=} \int_0^1 u^5 du = \frac{1}{6}u^6 \Big|_0^1$$

$$\boxed{\text{errors!}} \qquad \boxed{=} \frac{1}{6}(x^2+1)^6 \Big|_0^1 = \frac{1}{6}(2)^6 - \frac{1}{6}(1)^6$$

$$= \frac{63}{6} = \frac{21}{2}.$$

The symbol $\frac{1}{6}u^6\Big|_0^1$ represents the net change in $\frac{1}{6}u^6$ over the interval $0 \le u \le 1$. This number is not the same as the net change in $\frac{1}{6}(x^2+1)^6$ over the interval $0 \le x \le 1$.

The second incorrect evaluation above may be corrected by omitting the limits of integration at first, that is, by considering an antiderivative instead of a definite integral. Thus,

$$\int (x^2+1)^5 \cdot 2x\, dx = \int u^5 du = \frac{1}{6}u^6 + C = \frac{1}{6}(x^2+1)^6 + C,$$

and hence,

$$\int_0^1 (x^2+1)^5 2x\, dx = \frac{1}{6}(x^2+1)^6 \Big|_0^1 = \frac{1}{6}(2)^6 - \frac{1}{6}(1)^6 = \frac{21}{2}.$$

This is the first solution method discussed in Example 1 in the text.

1. Let $u = 2x - 5$. Then $du = \dfrac{d}{dx}(2x-5)\,dx = 2\,dx$.

$$\text{If } x = \frac{5}{2}, \text{ then } u = 2\left(\frac{5}{2}\right) - 5 = 0.$$

$$\text{If } x = 3, \text{ then } u = 2(3) - 5 = 1.$$

Therefore

$$\int_{5/2}^3 2(2x-5)^{14}\, dx = \int_0^1 u^{14} du = \frac{1}{15}u^{15}\Big|_0^1$$

$$= \frac{1}{15}(1)^{15} - 0 = \frac{1}{15}.$$

7. Since (x^2-9) has $2x$ as a derivative, and since this is a multiple of x, use the substitution $u = x^2 - 9$. Then $du = 2x\,dx$. If $x = 3$, then $u = 3^2 - 9 = 0$; if $x = 5$, then $u = 5^2 - 9 = 16$.

$$\int_3^5 x\sqrt{x^2-9}\, dx = \frac{1}{2}\int_3^5 2x\sqrt{x^2-9}\, dx = \frac{1}{2}\int_0^{16} \sqrt{u}\, du$$

$$= \frac{1}{2}\cdot\frac{2}{3}u^{3/2}\Big|_0^{16} = \frac{1}{3}\cdot 16^{3/2} - 0 = \frac{64}{3}.$$

13. Since x^3 has $3x^2$ as a derivative, and since this is a multiple of x^2, use the substitution $u = x^3$. Then $du = 3x^2 dx$. If $x = 1$, then $u = 1^3 = 1$, and if $x = 3$, then $u = 3^3 = 27$.

$$\int_1^3 x^2 e^{x^3} dx = \frac{1}{3} \int_1^3 3x^2 e^{x^3} dx = \frac{1}{3} \int_1^{27} e^u du$$

$$= \frac{1}{3} e^u \Big|_1^{27} = \frac{1}{3} e^{27} - \frac{1}{3} e = \frac{1}{3}(e^{27} - e).$$

19. Since the integrand is the product of two unrelated functions, use the technique of integration by parts. Let $f(x) = x$, and $g(x) = \sin \pi x$. Then

$$f'(x) = 1, \quad G(x) = -\frac{1}{\pi} \cos \pi x. \quad [\text{Check that } G'(x) = g(x).]$$

Then,

$$\int x \sin \pi x \, dx = -\frac{1}{\pi} x \cos \pi x - \int \left(-\frac{1}{\pi} \cos \pi x \right) dx$$

$$= -\frac{1}{\pi} x \cos \pi x + \frac{1}{\pi} \int \cos \pi x \, dx$$

$$= -\frac{1}{\pi} x \cos \pi x + \frac{1}{\pi} \left(\frac{1}{\pi} \sin \pi x \right) + C$$

$$= -\frac{1}{\pi} x \cos \pi x + \frac{1}{\pi^2} \sin \pi x + C.$$

So,

$$\int_0^1 x \sin \pi x \, dx = \left(-\frac{1}{\pi} x \cos \pi x + \frac{1}{\pi^2} \sin \pi x \right) \Big|_0^1$$

$$= \left[-\frac{1}{\pi} \cos \pi + \frac{1}{\pi^2} \sin \pi \right] - \left[-\frac{1}{\pi}(0) + \frac{1}{\pi^2} \sin 0 \right]$$

$$= \left[-\frac{1}{\pi}(-1) + 0 \right]$$

$$= \frac{1}{\pi}.$$

25. First, find where $y = x\sqrt{4 - x^2}$ cuts the x-axis ($y = 0$):

$$0 = x\sqrt{4 - x^2},$$

$$x = 0, \quad \text{or} \quad \sqrt{4 - x^2} = 0 \Rightarrow 4 - x^2 = 0,$$

$$x^2 = 4, \quad \text{or} \quad x = \pm 2.$$

The area of the portion from $x = 0$ to $x = 2$ is given by

$$\int_0^2 x\sqrt{4 - x^2} \, dx.$$

To find $\int x\sqrt{4-x^2}\,dx$, let

$$u = 4 - x^2,$$

$$du = \frac{d}{dx}(4-x^2)dx = -2x\,dx.$$

So,

$$\int x\sqrt{4-x^2}\,dx = \int \left(-\frac{1}{2}\right)(-2x)\sqrt{4-x^2}\,dx$$

$$= -\frac{1}{2}\int \sqrt{u}\,du = -\frac{1}{2}\int u^{1/2}\,du$$

$$= -\frac{1}{2}\frac{u^{3/2}}{\frac{3}{2}} = -\frac{1}{3}u^{3/2} + C$$

$$= -\frac{1}{3}(4-x^2)^{3/2} + C.$$

Therefore, the area from $x = 0$ to $x = 2$ is

$$\int_0^2 x\sqrt{4-x^2}\,dx = -\frac{1}{3}(4-x^2)^{3/2}\Big|_0^2$$

$$= -\frac{1}{3}(4-4)^{3/2} + \frac{1}{3}(4)^{3/2} = \frac{8}{3}.$$

By symmetry, the area from $x = -2$ to $x = 0$ is also $\dfrac{8}{3}$. The total area is $\dfrac{16}{3}$.

9.4 Approximation of Definite Integrals

It will be helpful for you to review Section 6.2 before you study Section 9.4. When you work the exercises in Section 9.4, do not be concerned if your answers differ slightly (in, say, the fourth or fifth significant figure) from the answers in the text. When numerical calculations are made that involve several steps, any decision (ours or yours) to "round off" intermediate answers can affect the final answer. Check with your instructor about how many significant figures you should retain in your calculations.

1. The interval has length $5 - 3 = 2$. When this interval is divided into 5 subintervals of equal length, each subinterval will have length $\Delta x = \frac{2}{5} = .4$. Therefore, the end points of the subintervals are:

$$a_0 = 3,$$
$$a_1 = 3 + .4 = 3.4,$$
$$a_2 = 3.4 + .4 = 3.8,$$
$$a_3 = 3.8 + .4 = 4.2,$$
$$a_4 = 4.2 + .4 = 4.6,$$
$$a_5 = 4.6 + .4 = 5.$$

7. If $n = 2$, then $\Delta x = \dfrac{(4-0)}{2} = 2$. The first midpoint is $x_1 = a + \dfrac{\Delta x}{2} = 0 + \dfrac{2}{2} = 1$. The second midpoint is

$x_2 = 1 + \Delta x = 1 + 2 = 3$. If $f(x) = x^2 + 5$, then $f(1) = 6$, $f(3) = 14$. By the midpoint rule,

$$\int_0^4 (x^2 + 5)dx \approx [f(1) + f(3)] \cdot 2$$

$$= (6 + 14) \cdot 2$$

$$= 40.$$

If $n = 4$, then $\Delta x = \dfrac{(4-0)}{4} = 1$. The first midpoint is $x_1 = a + \dfrac{\Delta x}{2} = 0 + \dfrac{1}{2} = .5$. Since the other

midpoints are spaced 1 unit apart, $x_2 = 1.5, x_3 = 2.5$, and $x_4 = 3.5$. Hence,

$$\int_0^4 (x^2 + 5)dx \approx [f(x_1) + f(x_2) + f(x_3) + f(x_4)]\Delta x$$

$$= \{[(.5)^2 + 5] + [(1.5)^2 + 5] + [(2.5)^2 + 5] + [(3.5)^2 + 5]\}(1).$$

On an exam, some instructors may permit you to stop at this point, because the expression above shows that you know how to use the midpoint rule. (Check with your instructor.) Only arithmetic remains. For homework, of course, you should complete the calculation, and obtain

$$\int_0^4 (x^2 + 5)dx \approx \{5.25 + 7.25 + 11.25 + 17.25\}(1)$$

$$= 41.$$

Evaluating the integral directly,

$$\int_0^4 (x^2 + 5)\, dx = \left(\frac{1}{3}x^3 + 5x \right)\Bigg|_0^4$$

$$= \frac{64}{3} + 20 = 41\frac{1}{3}.$$

13. For $n = 3, \Delta x = \dfrac{5-1}{3} = \dfrac{4}{3}$. The endpoints of the three subintervals begin at $a + 0 = 1$ and

are spaced $\dfrac{4}{3}$ units apart. Thus $a_1 = 1 + \dfrac{4}{3} = \dfrac{7}{3}, a_2 = \dfrac{7}{3} + \dfrac{4}{3} = \dfrac{11}{3}$, and $a_3 = \dfrac{11}{3} + \dfrac{4}{3} = \dfrac{15}{3} = 5$. By the

trapezoidal rule,

$$\int_1^5 \frac{1}{x^2}dx \approx \left[f(1) + 2f\left(\frac{7}{3}\right) + 2f\left(\frac{11}{3}\right) + f(5) \right]\left(\frac{4}{3}\right) \cdot \frac{1}{2}$$

$$= \left[1 + 2 \cdot \frac{1}{\left(\frac{7}{3}\right)^2} + 2 \cdot \frac{1}{\left(\frac{11}{3}\right)^2} + \frac{1}{5^2} \right]\left(\frac{4}{3}\right) \cdot \frac{1}{2}.$$

At this point, all substitutions into the trapezoidal rule are finished; only arithmetic remains. (Check to see if you can stop here on an exam.) The fractions above become quite messy, and it is best to use decimal approximations. We'll use five decimal places. (Check to see how many you should use on an exam.)

$$\int_1^5 \frac{1}{x^2}dx \approx [1 + .36735 + .14876 + .04](.66667)$$

$$= 1.03741.$$

Evaluating the integral directly produces

$$\int_1^5 \frac{1}{x^2}\,dx = \int_1^5 x^{-2}\,dx = -x^{-1}\Big|_1^5$$

$$= -\frac{1}{5} - (-1) = \frac{4}{5} = .8.$$

19. When $n = 5,$ we recommend that you use decimals. In this exercise,

$$\Delta x = \frac{(5-2)}{5} = \frac{3}{5} = .6.$$

It is often helpful to draw a picture. This is particularly true here because you need both the endpoints and midpoints of the subintervals, and the numbers involved are not as simple as in some exercises. Draw a line segment and, beginning at one end, mark off five equal subintervals. (This is easier than trying to divide one large interval into five equal parts.) Label the left endpoint $a_0 = 2,$ and repeatedly add $\Delta x = .6$ to get the other endpoints.

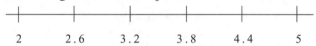

The first midpoint is $2 + \dfrac{\Delta x}{2} = 2 + .3 = 2.3.$ Repeatedly add $\Delta x = .6$ to get the other midpoints.

Use the midpoint rule for $f(x) = xe^x$ and obtain

$$\int_2^5 xe^x\,dx \approx [2.3e^{2.3} + 2.9e^{2.9} + 3.5e^{3.5} + 4.1e^{4.1} + 4.7e^{4.7}](.6)$$

$$\approx (955.69661)(.6) = 573.41797 = M.$$

Using the trapezoidal rule,

$$\int_2^5 xe^x\,dx \approx [2e^2 + (2)2.6e^{2.6} + (2)3.2e^{3.2} + (2)3.8e^{3.8}$$

$$+ (2)4.4e^{4.4} + 5e^5](.6) \cdot \frac{1}{2}$$

$$\approx (2040.36019)(.6)(.5)$$

$$= 612.10806 = T.$$

Use the values of M and T in Simpson's rule:

$$\int_2^5 xe^x\,dx \approx \frac{2M + T}{3}$$

$$= \frac{2(573.41797) + 612.10806}{3}$$

$$= 586.31467 = S.$$

To evaluate the integral directly, use the technique of integration by parts. Set

$$f(x) = x, \qquad g(x) = e^x,$$
$$f'(x) = 1, \qquad G(x) = e^x.$$

Then,

$$\int_2^5 xe^x dx = xe^x \Big|_2^5 - \int_2^5 e^x dx$$

$$= (5e^5 - 2e^2) - e^x \Big|_2^5$$

$$= 5e^5 - 2e^2 - (e^5 - e^2)$$

$$= 4e^5 - e^2$$

$$\approx 586.26358.$$

25. Let $f(x)$ represent the distance (in feet) from the shore to the property line as x runs from the top to the bottom of the diagram, and is measured in feet.

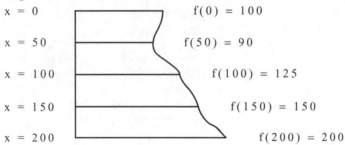

So by the trapezoidal rule,

$$[\text{Area of Property}] = \int_0^{200} f(x)dx$$

$$\approx [f(0) + 2f(50) + 2f(100) + 2f(150) + f(200)]\left(\frac{50}{2}\right)$$

$$= [100 + 180 + 250 + 300 + 200]25$$

$$= (1030)25$$

$$= 25,750 \text{ square feet.}$$

31. (a) Note from Figure 13(a) in the text that the height of the triangle is $k - h$. The area of the triangle is $\frac{1}{2}(k - h)\ell$, and the area of the rectangle is $h\ell$. Therefore, the area of the trapezoid is given by

$$A = \frac{1}{2}(k - h)\ell + h\ell = \left[\frac{1}{2}k - \frac{1}{2}h + h\right]\ell$$

$$= \left[\frac{1}{2}h + \frac{1}{2}k\right]\ell = \frac{1}{2}(h + k)\ell.$$

(b) Note that $h = f(a_0), k = f(a_1)$, and $\ell = \Delta x$. Therefore,

$$A = \frac{1}{2}[(a_0) + f(a_1)]\Delta x.$$

(c) The area of the first trapezoid is:

$$A = \frac{1}{2}[f(a_0) + f(a_1)]\Delta x = [f(a_0) + f(a_1)]\frac{\Delta x}{2}.$$

Similarly, the respective areas of the second, third, and fourth rectangles are

$$A_2 = [f(a_1) + f(a_2)]\frac{\Delta x}{2},$$

$$A_3 = [f(a_2) + f(a_3)]\frac{\Delta x}{2},$$

$$A_4 = [f(a_3) + f(a_4)]\frac{\Delta x}{2}.$$

So, the sum of the areas is given by

$$A_1 + A_2 + A_3 + A_4 = \left([f(a_0) + f(a_1)] + [(f(a_1) + f(a_2)] + [f(a_2) + f(a_3)] + [f(a_3) + f(a_4)]\right)\frac{\Delta x}{2}$$

$$= \left(f(a_0) + 2f(a_1) + 2f(a_2) + 2f(a_3) + f(a_4)\right)\frac{\Delta x}{2}.$$

This quantity is the value of the trapezoidal rule with $n = 4$ for

$$\int_a^b f(x)\,dx.$$

The combined area of the trapezoids is approximately equal to the area under the graph of $f(x)$ from $x = a$ to $x = b$.

37. In this problem, $f(x) = 1/x$, $a = 1$, $b = 11$, $n = 10$, and $\Delta x = 1$. The formula for $f(x)$ is so simple, there is no need to use it to define Y_1. Just use it directly in the **sum (seq(...))** command. When you compute the midpoint and trapezoidal estimates, store the results to variables, which you might as well call M and T.

Midpoint rule: $M = \left[\dfrac{1}{1.5} + \dfrac{1}{2.5} + \cdots + \dfrac{1}{10.5}\right] \cdot 1.$

```
sum(seq(1/X,X,1.
5,10.5,1))*1
          2.361749156
Ans→M
          2.361749156
```

Trapezoidal rule: $T = \left[\dfrac{1}{1} + 2\left(\dfrac{1}{2}\right) + \cdots + 2\left(\dfrac{1}{10}\right) + \dfrac{1}{11}\right] \cdot \dfrac{1}{2}.$

```
(1+sum(seq(2/X,X       (2M+T)/3
,2,10,1))+1/11)/              2.399307037
2
          2.474422799
Ans→T
          2.474422799
```

The exact answer is

$$\int_1^{11} \frac{1}{x}\,dx = \ln x\Big|_1^{11}$$

$$= \ln 11 - \ln 1$$

$$= 2.397895273.$$

If you store the exact answer ($\ln 11$) in a variable, say E, then the following error calculations are easily made:

Error using the midpoint rule: $|M - E| = 0.036146117,$

Error using the trapezoidal rule: $|T - E| = 0.076527526,$

Error using Simpson's rule: $|(2M + T)/3 - E| = 0.001411764,$

Notice that the exact answer lies between M and T, and the error for T is about twice the error for M, as usually happens.

9.5 Some Applications of the Integral

The purpose of this section is to show how fairly difficult integration problems can arise easily in applications. The Riemann sum argument in Example 3 is well worth studying, but your instructor may not expect you to reproduce it every time you work an exercise. (Check with your instructor.)

1. Use the text's formula for present value, with $K(t) = 35,000, T_1 = 0, T_2 = 5,$ and $r = .07.$

$$[\text{present value}] = \int_0^5 35000e^{-.07t}\,dt = \frac{35000e^{-.07t}}{-.07}\Big|_0^5$$

$$= -500,000[e^{-.07(5)} - e^0]$$

$$= -500,000[e^{-.35} - 1]$$

$$\approx \$147,656.$$

7. (a) $[\text{present value}] = \int_0^2 (30 + 5t)e^{-.1t}\,dt.$

(b) First, find the appropriate antiderivative, using integration by parts:

$$f(t) = 30 + 5t \qquad g(t) = e^{-.1t}$$

$$f'(t) = 5 \qquad\qquad G(t) = -10e^{-.1t}$$

$$\int (30 + 5t)e^{-.1t}\,dt = (30 + 5t)(-10e^{-.1t}) - \int 5(-10e^{-.1t})\,dt$$

$$= (-300 - 50t)e^{-.1t} + 50\int e^{-.1t}\,dt$$

$$= (-300 - 50t)e^{-.1t} + 50(-10e^{-.1t}) + C$$

$$= (-800 - 50t)e^{-.1t} + C.$$

Then use this to evaluate the definite integral.

$$\int_0^2 (30 + 5t)e^{-.1t}\,dt = ((-800 - 50t)e^{-.1t})\Big|_0^2$$

$$= -900e^{-.2} - (-800e^0)$$

$$\approx 63.14.$$

The present value of the stream of earnings for the next two years is about 63.1 million dollars.

13. **(a)** Area of ring is $2\pi t(\Delta t)$. Population density is $40e^{-.5t}$ thousand per square mile. Thus the population living in the ring is:

$$2\pi t(\Delta t)40e^{-.5t} = 80\pi t(\Delta t)e^{-.5t} \text{ thousand people.}$$

(b) $\dfrac{dP}{dt}$, or $P'(t)$.

(c) It represents the number of people who live between 5 miles from the city center and $(5+\Delta t)$ miles from the city center.

(d) $P(t+\Delta t) - P(t) = 80\pi t(\Delta t)e^{-.5t}$ from (a), so

$$\frac{P(t+\Delta t) - P(t)}{\Delta t} \approx P'(t)$$

$$= 80\pi te^{-.5t}.$$

(e) $P(b) - P(a) = \displaystyle\int_a^b P'(t)\,dt$, (by the Fundamental Theorem of Calculus)

$$= \int_a^b 80\pi te^{-.5t}\,dt.$$

9.6 Improper Integrals

Although some instructors may disagree, we feel that the evaluation of an improper integral such as

$$\int_0^\infty e^{-x}dx$$

should *not* use notation such as $-e^{-x}\big|_0^\infty$ or $e^{-\infty} + e^{-0}$. In the context of Section 9.6, the value of a function "at infinity" is not well-defined. The proper notation should involve a limit, such as $\lim\limits_{b\to\infty}\left(1-e^{-b}\right)$. You should use two steps to evaluate an improper integral:

  i.   Compute the net change in the antiderviative over a finite interval, such as from $x=1$ to $x=b$.
  ii. Find the limit of the result in (i) as $b \to \infty$.

1. As b gets large, $\frac{5}{b}$ approaches zero. That is, $\lim\limits_{b\to\infty}\dfrac{5}{b}=0$.

7. As b gets large, $\sqrt{b+1}$ gets large, without bound. Hence $\dfrac{1}{\sqrt{b+1}} = (b+1)^{-1/2}$ approaches zero, and thus $2-(b+1)^{-1/2}$ approaches 2.

13. The area is $\displaystyle\int_2^\infty x^{-2}dx$. First, compute

$$\int_2^b x^{-2}dx = -x^{-1}\big|_2^b$$

$$= -b^{-1} - (-2^{-1}) = \frac{1}{2} - \frac{1}{b}.$$

Then, take the limit as $b \to \infty$, namely,

$$\int_2^\infty x^{-2} dx = \lim_{b\to\infty}\left(\frac{1}{2}-\frac{1}{b}\right)=\frac{1}{2}.$$

19. $\displaystyle\int_1^b (14x+18)^{-4/5}\,dx = \frac{1}{14}(5)(14x+18)^{1/5}\Big|_1^b$ (Check the antiderivative.)

$$=\frac{5}{14}(14b+18)^{1/5}-\frac{5}{14}(32)^{1/5}.$$

As $b\to\infty$, the quantity $\frac{5}{14}(14b+18)^{1/5}$ grows without bound, so the region under the graph of $(14x+18)^{-4/5}$ cannot be assigned any finite number as its area.

25. Take $b>0$ and compute

$$\int_0^b e^{2x}dx = \frac{1}{2}e^{2x}\Big|_0^b = \frac{1}{2}e^{2b}-\frac{1}{2}.$$

Now consider the limit as $b\to\infty$. As $b\to\infty$, the number $\frac{1}{2}e^{2b}-\frac{1}{2}$ can be made larger than any specified number. Therefore,

$$\int_0^b e^{2x}dx$$

has no limit as $b\to\infty$, so $\displaystyle\int_0^\infty e^{2x}dx$ is divergent.

31. $\displaystyle\int_0^b 6e^{1-3x}dx = \frac{6}{-3}e^{1-3x}\Big|_0^b$

$$=-2e^{1-3b}-(-2e^1)=2e-2e^{1-3b}.$$

As $b\to\infty$, $1-3b$ becomes extremely negative, so $2e^{1-3b}\to 0$.
Thus,

$$\int_0^\infty 6e^{1-3x}dx = \lim_{b\to\infty}(2e-2e^{1-3b})=2e.$$

37. First consider the indefinite integral $\displaystyle\int 2x(x^2+1)^{-3/2}\,dx$ and let $u=x^2+1$, $du=2x\,dx$.

Thus

$$\int 2x(x^2+1)^{-3/2}dx = \int u^{-3/2}du = -2u^{-1/2}+C$$

$$=-2(x^2+1)^{-1/2}+C.$$

Using this antiderivative of $2x(x^2+1)^{-3/2}$, compute

$$\int_0^b 2x(x^2+1)^{-3/2}dx = -2(x^2+1)^{-1/2}\Big|_0^b$$

$$=-2(b^2+1)^{-1/2}-(-2\cdot 1).$$

As $b \to \infty$, $(b^2 + 1)^{-1/2} \to 0$, so

$$\int_0^\infty 2x(x^2 + 1)^{-3/2}\,dx = \lim_{b\to\infty}(2 - 2(b^2+1)^{-1/2}) = 2.$$

43. Let $u = e^{-x} + 2$ and $du = -e^x dx$, so that

$$\int \frac{e^{-x}}{(e^{-x}+2)^2}\,dx = \int \frac{-1}{u^2}\,du = u^{-1} + C$$
$$= (e^{-x}+2)^{-1} + C.$$

As $b \to \infty$, $e^{-b} \to 0$, and $\frac{1}{e^{-b}+2} \to \frac{1}{2}$. Thus,

$$\int_0^\infty \frac{e^{-x}}{(e^{-x}+2)^2}\,dx = \lim_{b\to\infty}(e^{-x}+2)^{-1}\Big|_0^b$$
$$= \lim_{b\to\infty}[(e^{-b}+2)^{-1} - 3^{-1}]$$
$$= \frac{1}{2} - \frac{1}{3} = \frac{1}{6}.$$

49. When the rate of income is a constant K dollars per year, the antiderivative of Ke^{-rt} is easy to find:

$$[\text{capital value}] = \lim_{b\to\infty}\int_0^b Ke^{-rt}\,dt = \lim_{b\to\infty}\left(-\frac{1}{r}Ke^{-rt}\right)\Big|_0^b$$
$$= \lim_{b\to\infty}\left(-\frac{K}{r}e^{-rb} + \frac{K}{r}e^0\right)$$
$$= -\frac{K}{r}\left(\lim_{b\to\infty}e^{-rb}\right) + \frac{K}{r} = \frac{K}{r}.$$

Chapter 9 Review Exercises

Sections 9.1–9.3 and 9.6 provide the tools for solving many of the problems in the next three chapters. It is essential that you master the techniques in Sections 9.1 and 9.2. Review Exercises 19–36 will help you learn to recognize which technique to use.

While solutions for all review exercises are included, expanded explanations are included for every sixth exercise.

1. Since $3x^2$ has $6x$ as a derivative, and this is a multiple of x, use the technique of substitution.
Let $u = 3x^2$, then $du = 6x\,dx$. Then

$$\int x\sin 3x^2\,dx = \frac{1}{6}\int 6x\sin 3x^2\,dx$$
$$= \frac{1}{6}\int \sin u\,du$$
$$= -\frac{1}{6}\cos u + C$$
$$= -\frac{1}{6}\cos 3x^2 + C.$$

2. $\int \sqrt{2x+1}\,dx = \dfrac{1}{3}(2x+1)^{3/2} + C$

3. $\int x(1-3x^2)^5\,dx$

Let $u = 1-3x^2$, $du = -6x\,dx$. Then

$$\int x(1-3x^2)^5\,dx = -\frac{1}{6}\int u^5\,du = -\frac{1}{36}u^6 + C$$

$$= -\frac{1}{36}(1-3x^2)^6 + C$$

4. $\int \dfrac{(\ln x)^5}{x}\,dx$

Let $u = \ln x$, $du = \dfrac{1}{x}\,dx$. Then

$$\int \frac{(\ln x)^5}{x}\,dx = \int u^5\,du = \frac{1}{6}u^6 + C$$

$$= \frac{1}{6}(\ln x)^6 + C$$

5. $\int \dfrac{(\ln x)^2}{x}\,dx$

Let $u = \ln x$, $du = \dfrac{1}{x}\,dx$. Then

$$\int \frac{(\ln x)^2}{x}\,dx = \int u^2\,du = \frac{1}{3}u^3 + C = \frac{1}{3}(\ln x)^3 + C$$

6. $\int \dfrac{1}{\sqrt{4x+3}}\,dx = \dfrac{1}{2}(4x+3)^{1/2} + C$

7. The derivative of $4-x^2$ is a multiple of x (the factor next to the square root). So make the substitution

$$u = 4-x^2,$$
$$du = -2x\,dx.$$

Then

$$\int x\sqrt{4-x^2}\,dx = -\frac{1}{2}\int (-2x)\sqrt{4-x^2}\,dx$$

$$= -\frac{1}{2}\int \sqrt{u}\,du$$

$$= -\frac{1}{2}\cdot\frac{2}{3}u^{3/2} + C$$

$$= -\frac{1}{3}(4-x^2)^{3/2} + C.$$

8. $\int x \sin 3x \, dx$

Use integration by parts with $f(x) = x$,

$g(x) = \sin 3x$. Then $f'(x) = 1$, $G(x) = -\frac{1}{3}\cos 3x$ and

$$\int x \sin 3x \, dx = -\frac{1}{3}x \cos 3x + \frac{1}{3}\int \cos 3x \, dx$$

$$= -\frac{1}{3}x \cos 3x + \frac{1}{9}\sin 3x + C$$

9. $\int x^2 e^{-x^3} \, dx$

Let $u = -x^3$, $du = -3x^2 dx$. Then

$$\int x^2 e^{-x^3} \, dx = -\frac{1}{3}\int e^u \, du = -\frac{1}{3}e^u + C$$

$$= -\frac{1}{3}e^{-x^3} + C$$

10. $\int \frac{x \ln(x^2+1)}{x^2+1} dx$

Let $u = \ln(x^2 + 1)$, $du = \frac{2x}{x^2+1}dx$.

Then $$\int \frac{x \ln(x^2+1)}{x^2+1} dx = \frac{1}{2}\int u \, du = \frac{1}{4}u^2 + C$$

$$= \frac{1}{4}(\ln(x^2+1))^2 + C$$

11. $\int x^2 \cos 3x \, dx$

Use integration by parts with $f(x) = x^2$,

$g(x) = \cos 3x$. Then $f'(x) = 2x$, $G(x) = \frac{1}{3}\sin 3x$ and $\int x^2 \cos 3x \, dx = \frac{1}{3}x^2 \sin 3x - \frac{2}{3}\int x \sin 3x \, dx$.

To evaluate $\int x \sin 3x \, dx$ integrate by parts again:

$$\int x \sin 3x \, dx = -\frac{1}{3}x \cos 3x + \frac{1}{3}\int \cos 3x \, dx$$

$$= -\frac{1}{3}x \cos 3x + \frac{1}{9}\sin 3x + C_1.$$

Thus, $$\int x^2 \cos 3x \, dx$$

$$= \frac{1}{3}x^2 \sin 3x + \frac{2}{9}x \cos 3x - \frac{2}{27}\sin 3x + C$$

12. $\int \dfrac{\ln(\ln x)}{x \ln x} dx$

Let $u = \ln(\ln x)$, $du = \dfrac{1}{x \ln x} dx$.

$$\int \frac{\ln(\ln x)}{x \ln x} dx = \int u\, du = \frac{u^2}{2} + C$$

$$= \frac{1}{2}(\ln(\ln x))^2 + C$$

13. First simplify the integrand, using the algebraic property $\ln x^2 = 2\ln x$:

$$\int \ln x^2 dx = 2\int \ln x dx.$$

Then recall that $\int \ln x\, dx$ is easily found by integration by parts. Use $f(x) = \ln x$ and $g(x) = 1$. The details are in Example 7 of Section 9.2. Using the results of that example,

$$\int \ln x^2 dx = 2\int \ln x dx = 2(x \ln x - x) + C.$$

Helpful Hint: Integrals of the form $\int x^k \ln x dx$ are found by integration by parts, with $f(x) = \ln x$ and $g(x) = x^k$.

14. $\int x\sqrt{x+1}dx$

Use integration by parts with $f(x) = x$, $g(x) = \sqrt{x+1}$. Then $f'(x) = 1$, $G(x) = \dfrac{2}{3}(x+1)^{3/2}$ and

$$\int x\sqrt{x+1}dx = \frac{2}{3}x(x+1)^{3/2} - \frac{2}{3}\int (x+1)^{3/2} dx$$

$$= \frac{2}{3}x(x+1)^{3/2} - \frac{4}{15}(x+1)^{5/2} + C.$$

15. $\int \dfrac{x}{\sqrt{3x-1}} dx$

Use integration by parts with $f(x) = x$, $g(x) = (3x-1)^{-1/2}$. Then $f'(x) = 1$, $G(x) = \dfrac{2}{3}(3x-1)^{1/2}$ and

$$\int \frac{x}{\sqrt{3x-1}} dx$$

$$= \frac{2}{3}x(3x-1)^{1/2} - \frac{2}{3}\int (3x-1)^{1/2} dx$$

$$= \frac{2}{3}x(3x-1)^{1/2} - \frac{4}{27}(3x-1)^{3/2} + C$$

16. $\int x^2 \ln x^2 dx$

Use integration by parts with $f(x) = \ln x^2$, $g(x) = x^2$. Then $f'(x) = \dfrac{2x}{x^2} = \dfrac{2}{x}$, and $G(x) = \dfrac{x^3}{3}$.

$$\int x^2 \ln x^2 dx = \frac{x^3}{3} \ln x^2 - \int \frac{2x^2}{3} dx$$

$$= \frac{x^3}{3} \ln x^2 - \frac{2}{9} x^3 + C.$$

17. $\int \dfrac{x}{(1-x)^5} dx$

Use integration by parts with $f(x) = x$, $g(x) = (1-x)^{-5}$. Then $f'(x) = 1$, $G(x) = \dfrac{1}{4}(1-x)^{-4}$ and

$$\int \frac{x}{(1-x)^5} dx = \frac{1}{4} x(1-x)^{-4} - \frac{1}{4} \int (1-x)^{-4} dx$$

$$= \frac{1}{4} x(1-x)^{-4} - \frac{1}{12}(1-x)^{-3} + C.$$

18. $\int x(\ln x)^2 dx$

Use integration by parts with $f(x) = (\ln x)^2$,

$g(x) = x$. Then $f'(x) = \dfrac{2\ln x}{x}$, $G(x) = \dfrac{x^2}{2}$ and

$$\int x(\ln x)^2 dx = \frac{x^2(\ln x)^2}{2} - \int x \ln x \, dx$$

$$= \frac{x^2(\ln x)^2}{2} - \left[\frac{x^2}{2} \ln x - \int \frac{x}{2} dx \right] \text{ (using parts again)}$$

$$= \frac{x^2(\ln x)^2}{2} - \frac{x^2}{2} \ln x + \frac{x^2}{4} + C$$

$$= \frac{x^2}{2} \left[(\ln x)^2 - \ln x + \frac{1}{2} \right] + C.$$

19. Since the integrand is the product of two functions, x and e^{2x}, neither of which is a multiple of the derivative of the other, use integration by parts. Set

$$f(x) = x, \quad g(x) = e^{2x}.$$

20. Integration by parts: $f(x) = x - 3$, $g(x) = e^{-x}$

21. Substitution: $u = \sqrt{x+1}$

22. Substitution: $u = x^3 - 1$

23. Substitution: $u = x^4 - x^2 + 4$

24. Integration by parts: $f(x) = \ln \sqrt{5-x} = \dfrac{1}{2} \ln(5-x)$, $g(x) = 1$

25. Since the integrand is the product of two functions, e^{-x} and $(3x-1)^2$, and neither is a multiple of the derivative of the other, use integration by parts. Set

$$f(x) = (3x-1)^2, \qquad g(x) = e^{-x}.$$

In order to complete the problem, you must integrate by parts a second time. Note that if you interchange $f(x)$ and $g(x)$ above, integration by parts will not yield a solution.

26. Substitution: $u = 3 - x^2$

27. Integration by parts: $f(x) = 500 - 4x$,
$g(x) = e^{-x/2}$

28. Integration by parts: $f(x) = \ln x$,
$g(x) = x^{5/2}$

29. Integration by parts: $f(x) = \ln(x+2)$,
$g(x) = \sqrt{x+2}$

30. Repeated integration by parts, starting with
$f(x) = (x+1)^2, \ g(x) = e^{3x}$

31. Observe that the derivative of $x^2 + 6x$ is $2x + 6$, which is a multiple of $x + 3$. Use the technique of substitution, setting
$u = x^2 + 6x$.

32. Substitution: $u = \sin x$

33. Substitution: $u = x^2 - 9$

34. Integration by parts: $f(x) = 3 - x$,
$g(x) = \sin 3x$

35. Substitution: $u = x^3 - 6x$

36. Substitution: $u = \ln x$

37. Since $x^2 + 1$ has $2x$ as a derivative, use the technique of substitution. Let $u = x^2 + 1$. Then $du = 2x\,dx$. If $x = 0$, then $u = 1$, and if $x = 1$, then $u = 2$. So,

$$\int_0^1 \frac{2x}{(x^2+1)^3}\,dx = \int_1^2 \frac{1}{u^3}\,du = \int_1^2 u^{-3}\,du$$

$$= -\frac{1}{2}u^{-2}\bigg|_1^2 = -\frac{1}{2}\left(\frac{1}{4}\right) + \frac{1}{2} = \frac{3}{8}.$$

38. $\int_0^{\pi/2} x\sin 8x\,dx$

Using integration by parts with $f(x) = x$,

$g(x) = \sin 8x$, we have $f'(x) = 1$, $G(x) = -\frac{1}{8}\cos 8x$ and

$$\int_0^{\pi/2} x\sin 8x\,dx = -\frac{1}{8}x\cos 8x\bigg|_0^{\pi/2} + \frac{1}{8}\int_0^{\pi/2}\cos 8x\,dx$$

$$= \frac{1}{8}\left[-x\cos 8x + \frac{1}{8}\sin 8x\right]_0^{\pi/2}$$

$$= \frac{1}{8}\left[-\frac{\pi}{2} - 0\right]$$

$$= -\frac{\pi}{16}$$

39. $\displaystyle\int_0^2 xe^{-(1/2)x^2}\,dx$

Let $u=-\dfrac{1}{2}x^2$, $du=-x\,dx$. When $x=0$, $u=0$; when $x=2$, $u=-2$.

$$\int_0^2 xe^{-(1/2)x^2}\,dx = -\int_0^{-2} e^u\,du = -e^u\Big|_0^{-2} = 1-e^{-2}$$

40. $\displaystyle\int_{1/2}^1 \frac{\ln(2x+3)}{2x+3}\,dx$

Let $u=\ln(2x+3)$, $du=\dfrac{2}{2x+3}\,dx$. When $x=\dfrac{1}{2}$, $u=\ln(4)$, when $x=1$, $u=\ln 5$.

$$\int_{1/2}^1 \frac{\ln(2x+3)}{2x+3}\,dx = \frac{1}{2}\int_{\ln 4}^{\ln 5} u\,du = \frac{u^2}{4}\Big|_{\ln 4}^{\ln 5}$$

$$= \frac{1}{4}\Big[(\ln 5)^2 - (\ln 4)^2\Big]$$

41. $\displaystyle\int_1^2 xe^{-2x}\,dx$

Use integration by parts with $f(x)=x$, $g(x)=e^{-2x}$. Then $f'(x)=1$, $G(x)=-\dfrac{1}{2}e^{-2x}$

$$\int_1^2 xe^{-2x}\,dx = -\frac{1}{2}xe^{-2x}\Big|_1^2 + \frac{1}{2}\int_1^2 e^{-2x}\,dx$$

$$= \left(-\frac{1}{2}xe^{-2x} - \frac{1}{4}e^{-2x}\right)\Big|_1^2$$

$$= -\frac{1}{2}\left[e^{-2x}\left(x+\frac{1}{2}\right)\Big|_1^2\right]$$

$$= -\frac{1}{2}\left[\frac{5}{2}e^{-4} - \frac{3}{2}e^{-2}\right] = \frac{3}{4}e^{-2} - \frac{5}{4}e^{-4}$$

42. $\displaystyle\int_1^2 x^{-3/2}\ln x\,dx$

Use integration by parts with $f(x)=\ln x$, $g(x)=x^{-3/2}$. Then $f'(x)=\dfrac{1}{x}$, $G(x)=-2x^{-1/2}$ and

$$\int_1^2 x^{-3/2}\ln x\,dx = -2x^{-1/2}\ln x\Big|_1^2 + \int_1^2 2x^{-3/2}\,dx$$

$$= \left(-2x^{-1/2}\ln x - 4x^{-1/2}\right)\Big|_1^2$$

$$= -2\left[x^{-1/2}(\ln x+2)\Big|_1^2\right]$$

$$= -2\left[\frac{\ln 2+2}{\sqrt{2}} - 2\right]$$

$$= -\sqrt{2}\ln 2 - 2\sqrt{2} + 4$$

43. The interval has length $9 - 1 = 8$ and $n = 4$, so $\Delta x = \dfrac{8}{4} = 2$. The subintervals have endpoints $a_0 = 1$, $a_1 = 3, a_2 = 5, a_3 = 7$, and $a_4 = 9$, and midpoints $x_1 = 2, x_2 = 4, x_3 = 6$, and $x_4 = 8$.

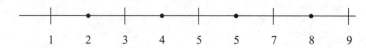

Using the midpoint rule, with $f(x) = \dfrac{1}{\sqrt{x}} = x^{-1/2}$,

$$\int_1^9 f(x)\,dx \approx M = [f(2) + f(4) + f(6) + f(8)] \cdot 2$$

$$= \left[\frac{1}{\sqrt{2}} + \frac{1}{\sqrt{4}} + \frac{1}{\sqrt{6}} + \frac{1}{\sqrt{8}} \right](2)$$

$$= [.70711 + .50000 + .40825 + .35355] \cdot 2$$

$$= 3.93782.$$

Using the trapezoidal rule,

$$\int_1^9 f(x)\,dx \approx T = [f(1) + 2f(3) + 2f(5) + 2f(7) + f(9)] \cdot \frac{2}{2}$$

$$= \left[\frac{1}{\sqrt{1}} + \frac{2}{\sqrt{3}} + \frac{2}{\sqrt{5}} + \frac{2}{\sqrt{7}} + \frac{1}{\sqrt{9}} \right] \frac{2}{2}$$

$$= [1 + 1.15470 + .89443 + .75593 + .33333]$$

$$= 4.13839.$$

Using Simpson's rule,

$$\int_1^9 \frac{1}{\sqrt{x}}\,dx \approx \frac{2M + T}{3} = 4.00468.$$

44. $\displaystyle\int_0^{10} e^{\sqrt{x}}\,dx;\ n = 5,\ \Delta x = \dfrac{(10 - 0)}{5} = 2$

Midpoint rule: $x_1 = 1,\ x_2 = 3,\ x_3 = 5,\ x_4 = 7,\ x_5 = 9$

$$\int_0^{10} e^{\sqrt{x}}\,dx \approx \left[e^{\sqrt{1}} + e^{\sqrt{3}} + e^{\sqrt{7}} + e^{\sqrt{9}} \right] 2 \approx 103.81310$$

Trapezoidal rule: $a_0 = 0,\ a_1 = 2,\ a_2 = 4,\ a_3 = 6,\ a_4 = 8,\ a_5 = 10$

$$\int_0^{10} e^{\sqrt{x}}\,dx \approx \left[e^{\sqrt{0}} + 2e^{\sqrt{2}} + 2e^{\sqrt{4}} + 2e^{\sqrt{6}} + 2e^{\sqrt{8}} + e^{\sqrt{10}} \right] \frac{2}{2} \approx 104.63148$$

Simpson's rule: $\displaystyle\int_0^{10} e^{\sqrt{x}}\,dx \approx \dfrac{2(103.8130) + 104.63148}{3} \approx 104.08589$

45. $\int_1^4 \dfrac{e^x}{x+1} dx$; $n = 5$, $\Delta x = \dfrac{3}{5} = .6$

Midpoint rule: $x_1 = 1.3$, $x_2 = 1.9$, $x_3 = 2.5$, $x_4 = 3.1$, $x_5 = 3.7$

$$\int_1^4 \dfrac{e^x}{x+1} dx \approx \left[\dfrac{e^{1.3}}{2.3} + \dfrac{e^{1.9}}{2.9} + \dfrac{e^{2.5}}{3.5} + \dfrac{e^{3.1}}{4.1} + \dfrac{e^{3.7}}{4.7} \right](.6) \approx 12.84089$$

Trapezoidal rule: $a_0 = 1$, $a_1 = 1.6$, $a_2 = 2.2$, $a_3 = 2.8$, $a_4 = 3.4$, $a_5 = 4$

$$\int_1^4 \dfrac{e^x}{x+1} dx \approx \left[\dfrac{e}{2} + \dfrac{2e^{1.6}}{2.6} + \dfrac{2e^{2.2}}{3.2} + \dfrac{2e^{2.8}}{3.8} + \dfrac{2e^{3.4}}{4.4} + \dfrac{e^4}{5} \right](.3) \approx 13.20137$$

Simpson's rule: $\quad \int_1^4 \dfrac{e^x}{x+1} dx \approx \dfrac{2(12.84089) + 13.20137}{3} \approx 12.96105$

46. $\int_{-1}^1 \dfrac{1}{1+x^2} dx$; $n = 5$, $\Delta x = \dfrac{2}{5} = .4$

Midpoint rule: $x_1 = -.8$, $x_2 = -.4$, $x_3 = 0$, $x_4 = .4$, $x_5 = .8$

$$\int_{-1}^1 \dfrac{1}{1+x^2} dx \approx \left[\dfrac{1}{1+(-.8)^2} + \dfrac{1}{1+(-.4)^2} + \dfrac{1}{1+0^2} + \dfrac{1}{1+(.4)^2} + \dfrac{1}{1+(.8)^2} \right](.4) \approx 1.57746$$

Trapezoidal rule: $a_0 = -1$, $a_1 = -.6$, $a_2 = -.2$, $a_3 = .2$, $a_4 = .6$, $a_5 = 1$

$$\int_{-1}^1 \dfrac{1}{1+x^2} dx \approx \left[\dfrac{1}{1+(-1)^2} + \dfrac{2}{1+(-.6)^2} + \dfrac{2}{1+(-.2)^2} + \dfrac{2}{1+(.2)^2} + \dfrac{2}{1+(.6)^2} + \dfrac{1}{1+1^2} \right](.2) \approx 1.55747$$

Simpson's rule: $\quad \int_{-1}^1 \dfrac{1}{1+x^2} dx \approx \dfrac{2(1.57746) + 1.55747}{3} \approx 1.57080$

47. $\int_0^\infty e^{6-3x} dx = \lim\limits_{b \to \infty} \int_0^b e^{6-3x} dx$

Let $u = 6 - 3x$, $du = -3 dx$. When $x = 0$, $u = 6$; when $x = b$, $u = 6 - 3b$.

$$\int_0^b e^{6-3x} dx = -\dfrac{1}{3} \int_6^{6-3b} e^u du = -\dfrac{1}{3} e^u \Big|_6^{6-3b} = -\dfrac{1}{3} e^{6-3b} + \dfrac{1}{3} e^6$$

Thus $\quad \int_0^\infty e^{6-3x} dx = \lim\limits_{b \to \infty} \left[-\dfrac{1}{3} e^{6-3b} + \dfrac{e^6}{3} \right] = \dfrac{e^6}{3}.$

48. $\int_1^\infty x^{-2/3} dx = \lim\limits_{b \to \infty} \int_1^b x^{-2/3} dx$

$$= \lim\limits_{b \to \infty} \left[3x^{1/3} \Big|_1^b \right] = \lim\limits_{b \to \infty} [3b^{1/3} - 3]$$

As $b \to \infty$, $3b^{1/3}$ increases without bound. Thus, $\int_1^\infty x^{-2/3} dx$ diverges.

49. The derivative of $x^2 + 4x - 2$ is $2x + 4$, which is a multiple of $x + 2$. Use the technique of substitution. Let $u = x^2 + 4x - 2$ and $du = (2x + 4)dx$.

$$\int \frac{x+2}{x^2+4x-2}\,dx = \frac{1}{2}\int \frac{2x+4}{x^2+4x-2}\,dx = \frac{1}{2}\int \frac{1}{u}\,du$$

$$= \frac{1}{2}\ln|u| = \frac{1}{2}\ln|x^2+4x-2| + C.$$

Hence,

$$\int_1^b \frac{x+2}{x^2+4x-2}\,dx = \frac{1}{2}\ln|x^2+4x-2|\,\bigg|_1^b$$

$$= \frac{1}{2}\ln|b^2+4b-2| - \frac{1}{2}\ln 3.$$

As $b \to \infty$, the number $|b^2 + 4b - 2|$ gets large, without bound, and so $\ln|b^2 + 4b - 2|$ gets large, without bound. Therefore,

$$\frac{1}{2}\ln|b^2+4b-2| - \frac{1}{2}\ln 3$$

can be made larger than any specific number.
Therefore,

$$\int_1^b \frac{x+2}{x^2+4x-2}\,dx$$

has no limit as $b \to \infty$ and

$$\int_1^\infty \frac{x+2}{x^2+4x-2}\,dx \text{ is divergent.}$$

50. $\displaystyle\int_0^\infty x^2 e^{-x^3}\,dx = \lim_{b\to\infty}\int_0^b x^2 e^{-x^3}\,dx$

Let $u = -x^3$, then $du = -3x^2\,dx$.

When $x = 0$, $u = 0$; when $x = b$, $u = -b^3$.

$$\int_0^b x^2 e^{-x^3}\,dx = -\frac{1}{3}\int_0^{-b^3} e^u\,du = -\frac{1}{3}\left[e^{-b^3} - 1\right]$$

$$= \frac{1}{3} - \frac{e^{-b^3}}{3}$$

Thus $\displaystyle\int_0^\infty x^2 e^{-x^3}\,dx = \lim_{b\to\infty}\left[\frac{1}{3} - \frac{e^{-b^3}}{3}\right] = \frac{1}{3}$.

51. $\displaystyle\int_{-1}^\infty (x+3)^{-5/4}\,dx = \lim_{b\to\infty}\int_{-1}^b (x+3)^{-5/4}\,dx$

$$= \lim_{b\to\infty}\left[-4(x+3)^{-1/4}\,\bigg|_{-1}^b\right]$$

$$= \lim_{b\to\infty}\left[2^{7/4} - 4(b+3)^{-1/4}\right]$$

$$= 2^{7/4}$$

52. $\displaystyle\int_{-\infty}^{0}\frac{8}{(5-2x)^3}\,dx = \lim_{b\to-\infty}\int_{b}^{0}\frac{8}{(5-2x)^3}\,dx$

$$= \lim_{b\to-\infty}\left[2(5-2x)^{-2}\Big|_{b}^{0}\right]$$

$$= \lim_{b\to-\infty}\left[\frac{2}{25}-2(5-2b)^{-2}\right]$$

$$= \frac{2}{25}$$

53. $\displaystyle\int_{1}^{\infty}xe^{-x}dx = \lim_{b\to\infty}\int_{1}^{b}xe^{-x}dx$

Using integration by parts with $f(x)=x$, $g(x)=e^{-x}$; $f'(x)=1$, $G(x)=-e^{-x}$ gives

$$\int_{1}^{b}xe^{-x}dx = -xe^{-x}\Big|_{1}^{b}+\int_{1}^{b}e^{-x}dx$$

$$= (-xe^{-x}-e^{-x})\Big|_{1}^{b}$$

$$= 2e^{-1}-be^{-b}-e^{-b}$$

Thus $\displaystyle\int_{1}^{\infty}xe^{-x}dx = \lim_{b\to\infty}\left[\frac{2}{e}-be^{-b}-e^{-b}\right]=\frac{2}{e}\left(\text{since }\lim_{b\to\infty}be^{-b}=0\right).$

54. $\displaystyle\int_{0}^{\infty}xe^{-kx}dx = \lim_{b\to\infty}\int_{0}^{b}xe^{-kx}dx$

Using integration by parts with $f(x)=x$, $g(x)=e^{-kx}$; $f'(x)=1$, $G(x)=-\dfrac{1}{k}e^{-kx}$, we have

$$\int_{0}^{b}xe^{-kx}dx = \frac{-x}{k}e^{-kx}\Big|_{0}^{b}+\frac{1}{k}\int_{0}^{b}e^{-kx}dx$$

$$= \left(-\frac{x}{k}e^{-kx}-\frac{1}{k^2}e^{-kx}\right)\Big|_{0}^{b}$$

$$= \frac{1}{k^2}-\frac{b}{k}e^{-kb}-\frac{1}{k^2}e^{-kb}$$

Thus $\displaystyle\int_{0}^{\infty}xe^{-kx}dx = \lim_{b\to\infty}\left[\frac{1}{k}\left(\frac{1}{k}-be^{-kb}-\frac{1}{k}e^{-kb}\right)\right]$

$$= \frac{1}{k^2}\left(\text{since }\lim_{b\to\infty}be^{-kb}=0\right).$$

55. Use the text's formula for present value given in Section 9.5, with $K(t) = 50e^{-.08t}$, $r = .12$, $T_1 = 0$, and $T_2 = 4$.

$$[\text{present value}] = \int_0^4 50e^{-.08t} \cdot e^{-.12t} dt.$$

Integration by parts is not necessary this time because the exponential functions can be combined:

$$[\text{present value}] = \int_0^4 50e^{-.20t} dt = -250e^{-.2t}\Big|_0^4$$
$$= -112.332 + 250 = 137.668.$$

The present value of the continuous stream of income over the next four years is $137,668 (to the nearest dollar).

56. Using the method of Example 3, section 9.5, the total tax revenue is

$$\int_0^{10} (2\pi t)50e^{-t/20} dt = 100\pi \int_0^{10} te^{-t/20} dt$$
$$= 100\pi \left[-20te^{-t/20}\Big|_0^{10} + \int_0^{10} 20e^{-t/20} dt \right]$$
$$= 100\pi \left[-20te^{-t/20} - 400e^{-t/20}\Big|_0^{10} \right]$$
$$= 100\pi \left[400 - e^{-.5}(600) \right]$$
$$\approx \$11,335 \text{ thousand}$$

57. **(a)** $M(t_1)\Delta t + \cdots + M(t_n)\Delta t \approx \int_0^2 M(t)dt$

(b) $M(t_1)e^{-.1t_1}\Delta t + \cdots + M(t_n)e^{-.1t_n}\Delta t \approx \int_0^2 M(t)e^{-.1t} dt$

58. $80,000 + \int_0^{\infty} 50,000e^{-rt} dt$

Chapter 10
Differential Equations

10.1 Solutions of Differential Equations

This section lays the foundation for the chapter. You should read the text and the examples several times. The high point of the chapter is Section 10.6, which presents a wide variety of applications. Our students find this material extremely interesting, and we know that they can master it if they are given adequate preparation. So each section includes a few problems that teach you how to read and understand a verbal problem about a differential equation. Be sure to try Exercise 19 or 20.

1. Replace y by $f(t) = \frac{3}{2}e^{t^2} - \frac{1}{2}$, and y' by $f'(t) = 3te^{t^2}$.

   Then

   $$y' - 2ty = 3te^{t^2} - 2t\left[\frac{3}{2}e^{t^2} - \frac{1}{2}\right]$$
   $$= 3te^{t^2} - [3te^{t^2} - t] = t.$$

   This is true for all t, so $f(t) = \frac{3}{2}e^{t^2} - \frac{1}{2}$ is a solution.

7. The derivative of the constant function $f(t) = 3$ is the constant function $f'(t) = 0$. If you replace y by $f(t) = 3$ and y' by $f'(t) = 0$, then

   $$y' = 6 - 2y,$$
   $$0 = 6 - 2(3).$$

   Both sides are zero for all t, so $f(t) = 3$ is a solution.

13. The differential equation $y' = .2(160 - y)$ describes a relationship between the acceleration (the derivative of the downward velocity) and the downward velocity, at each moment during the free fall. At a time when the velocity is 60 ft/sec, the acceleration is
    $$y' = .2(160 - 60) = .2(100) = 20 \text{ ft/sec per second.}$$

19. Let $f(t)$ be the amount of capital invested. Then $f'(t)$ is the rate of net investment. We are told that

    [rate of net investment]   is proportional to $(C - \text{[capital investment]})$.

    So there is a constant of proportionality, k, such that
    $$f'(t) = k\{C - f(t)\}.$$

To determine the sign of k, suppose that at some time $f(t)$ is larger than C. Then $C - f(t)$ is negative. Should $f'(t)$ be positive or negative? That is, should $f(t)$ be increasing or decreasing? Well, if the amount of $f(t)$ is larger than the optimum level C, then it is reasonable to want $f(t)$ to decrease down towards C. So when $C - f(t)$ is negative, we want $f'(t)$ to be negative, too. This means that k must be positive.

$$f'(t) = k\{C - f(t)\},$$

$$[\text{neg}] = [\text{pos}] \cdot [\text{neg}].$$

You would arrive at the same conclusion that k is positive if you supposed that $f(t)$ is less than C. The differential equation is simplified by writing y for $f(t)$ and y' for $f'(t)$, namely,

$$y' = k(C - y), \qquad k > 0.$$

25. The initial condition $f(0) = 1500$ tells us that the point (0, 1500) is on the graph of the solution curve $y = f(t)$. Starting at this point in the ty-plane of Figure 5(a) of the text and tracing a curve tangent to the line segments in the slope field, we obtain a solution curve through the point (0, 1500) that appears to be similar to the solution through (0, 1000), shifted vertically up 500 units.

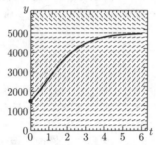

The key point to notice about the slope field in Figure 5(a) is that above the line $y = 5000$, the tangent lines pictured are negative. This can also be verified in general by observing that y' is always negative whenever a value of y greater than 5000 is plugged into the equation $y' = .0002y(5000 - y)$. For this reason, any solution curve to the differential equation $y' = .0002y(5000 - y)$ that contains a point (t, y) with $y < 5000$ will never go above the line $y = 5000$. Thus, in this scenario, $f(t)$ (the number of infected people) will never exceed 5000.

31. (a) Set $Y_1 = 10 + 500*e\wedge(-.2X)$ and deselect all other functions. Set Xmin = 0, Xmax = 30, Ymin = −75, and Ymax = 550, then graph the function. The graph shows a decreasing function in the first quadrant with a vertical asymptote at the y-axis, and a horizontal asymptote at the x-axis.

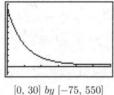

[0, 30] *by* [−75, 550]

(b) We are given that $y = 10 + 500e^{-.2t}$ is a solution to the differential equation $y' = .2(10 - y)$. This exercise is asking us to use the calculator to check that this is indeed the case when $t = 5$. Note that, based on the solution to part (a), we can expect the value of $f'(5)$ to be a negative number, since the curve $y = f(t)$ has a negative derivative (i.e., is heading downhill) when $t = 5$.

When $t = 5$, $y = f(t) = 10 + 500e^{-.2(5)}$. Thus, when $t = 5$, we have

$$.2(10 - y) = .2(10 - (10 + 500e^{-.2(5)})).$$

Using a graphing calculator to evaluate this, we find that
$$.2(10 - y) \approx -36.78794412$$

when $t = 5$. To evaluate $f'(5)$ we first find the formula for $f'(t)$. A straightforward application of the chain rule gives

$$f'(t) = -100e^{-.2t}.$$

Thus $f'(5) = -100e^{-.2(5)}$. Evaluating this with a graphing calculator, we find that

$$f'(5) \approx -36.78794412$$

Thus we have confirmed (to eight decimal places, for $t = 5$) that $y = 10 + 500e^{-.2t}$ is a solution to the differential equation $y' = .2(10 - y)$.

Alternatively, set $Y_1 = f(t)$, and then use nDeriv in the MATH menu as shown below.

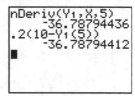

10.2 Separation of Variables

This is probably the most difficult section in Chapter 10 because it requires substantial integration and algebraic skills. Each example illustrates one or two situations you will encounter in the exercises. After you study Examples 1, 2, and 3, you may work on Exercises 1–6, 13–15, and 17, respectively.

Remember to use the examples properly. Study them carefully and then try to work the exercises *without referring to the examples*. If you really get stuck, take a peek at the appropriate example, but do not study the entire solution. As a last resort, peek at a solution in this *Manual* if a solution is available. You simply must train yourself to work independently. If you yield to the temptation to "copy" our solutions, you will not survive an examination on this material.

Exercises 33–38 are here to help you prepare for Section 10.6. You will be rewarded later if you spend some time on these exercises now.

1. Separate the variables and integrate both sides with respect to t.

$$y^2 \frac{dy}{dt} = 5 - t,$$

$$\int y^2 \frac{dy}{dt} dt = \int (5 - t) dt,$$

$$\int y^2 dy = \int (5 - t) dt.$$

One antiderivative of y^2 is $\frac{1}{3}y^3$. One antiderivative of $5-t$ is $5t-\frac{1}{2}t^2$. Since antiderivatives of the same function differ by a constant,

$$\frac{1}{3}y^3 = 5t - \frac{1}{2}t^2 + C_1,$$

for some constant C_1. Then

$$y^3 = 3\left(5t - \frac{1}{2}t^2 + C_1\right).$$

Since C_1 is arbitrary, so is $3C_1$, and it is easier to write C in place of $3C_1$. Therefore,

$$y^3 = 15t - \frac{3}{2}t^2 + C, \quad \text{and} \quad y = \left(15t - \frac{3}{2}t^2 + C\right)^{1/3}.$$

7. $y' = \left(\dfrac{t}{y}\right)^2 e^{t^3} \Rightarrow \dfrac{dy}{dt} = \dfrac{t^2}{y^2}e^{t^3}$ By separating the variables,

$$y^2\frac{dy}{dt} = t^2 e^{t^3}$$

$$\int y^2\frac{dy}{dt}\,dt = \int t^2 e^{t^3}\,dt$$

$$\int y^2\,dy = \int t^2 e^{t^3}\,dt.$$

One antiderivative of y^2 is $\frac{1}{3}y^3$, and similarly an antiderivative of $t^2 e^{t^3}$ is $\frac{1}{3}e^{t^3}$. (This is found by a substitution, $u = t^3$.) So

$$\frac{1}{3}y^3 = \frac{1}{3}e^{t^3} + C.$$

Hence, $y^3 = e^{t^3} + C$ (another constant C), and finally, $y = (e^{t^3} + C)^{1/3}$.

13. Write y' as $\dfrac{dy}{dt}$, integrate both sides of the equation with respect to t, and "cancel" the dt's on the left side of the equation.

$$y'e^y = te^{t^2} \Rightarrow e^y\frac{dy}{dt} = te^{t^2}$$

$$\int e^y\,dy = \int te^{t^2}\,dt.$$

One antiderivative of e^y is e^y. To find an antiderivative of te^{t^2}, let $u = t^2$. Then $\dfrac{du}{dt} = 2t$ and $du = 2t\,dt$.

$$te^{t^2}\,dt = \frac{1}{2}e^{t^2}(2t\,dt) = \frac{1}{2}e^u\,du \Rightarrow \int te^{t^2}\,dt = \frac{1}{2}\int e^u\,du = \frac{1}{2}e^u + C = \frac{1}{2}e^{t^2} + C$$

Thus,

$$e^y + C_1 = \frac{1}{2}e^{t^2} + C_2 \Rightarrow e^y = \frac{1}{2}e^{t^2} + C$$

To solve for y, take the natural logarithm of both sides:

$$\ln(e^y) = \ln\left(\frac{1}{2}e^{t^2} + C\right).$$

Simplify to arrive at the solution:

$$y = \ln\left(\frac{1}{2}e^{t^2} + C\right).$$

Note: To check that this is indeed the correct solution, differentiate to find y' and plug the formulas for y and y' into the equation $y'e^y = te^t$.

19. $y' = 2te^{-2y} - e^{-2y}$, $y(0) = 3$.

The equation can be rewritten as

$$\frac{dy}{dt} = 2te^{-2y} - e^{-2y} = e^{-2y}(2t - 1).$$

Applying the method of separation of variables,

$$\frac{1}{e^{-2y}}\frac{dy}{dt} = 2t - 1, \qquad \left(\frac{1}{e^{-2y}} = e^{2y}\right),$$

$$\int e^{2y}\frac{dy}{dt}\,dt = \int (2t - 1)\,dt,$$

$$\int e^{2y}\,dy = \int (2t - 1)\,dt,$$

$$\frac{1}{2}e^{2y} = t^2 - t + C.$$

Hence, $e^{2y} = 2t^2 - 2t + C$. (This C is actually twice the old C.) Take the logarithm of both sides.

$$\ln e^{2y} = \ln(2t^2 - 2t + C)$$

$$2y = \ln(2t^2 - 2t + C).$$

Therefore,

$$y = \frac{1}{2}\ln(2t^2 - 2t + C).$$

Next, choose C to make $y(0) = 3$. Set $t = 0$ and $y = 3$:

$$3 = \frac{1}{2}\ln(2\cdot 0^2 - 2(0) + C) = \frac{1}{2}\ln C,$$

$$6 = \ln C,$$

$$e^6 = C.$$

So finally,

$$y = \frac{1}{2}\ln(2t^2 - 2t + e^6).$$

25. $\dfrac{dy}{dt} = \dfrac{t+1}{ty}$, $t > 0$, $y(1) = -3$. Separating the variables:

$$y\frac{dy}{dt} = \frac{t+1}{t} = 1 + \frac{1}{t},$$

$$\int y\frac{dy}{dt}\,dt = \int \left(1 + \frac{1}{t}\right)dt,$$

$$\int y\,dy = \int \left(1 + \frac{1}{t}\right)dt.$$

Therefore, $\dfrac{y^2}{2} = t + \ln|t| + C.$

Since $t > 0$, rewrite $\ln|t|$ as $\ln t$, so

$$\frac{y^2}{2} = t + \ln t + C,$$

$$y^2 = 2t + 2\ln t + C.$$

(C is now two times the value of the old C.)

$$y = \pm\sqrt{2t + 2\ln t + C}.$$

Since $y(1) = -3$, and since a square root cannot be negative, use the "minus" form:

$$y = -\sqrt{2t + 2\ln t + C},$$

$$-3 = -\sqrt{2(1) + 2\ln(1) + C},$$

$$-3 = -\sqrt{2 + 0 + C}, \quad \text{(square both sides)}$$

$$9 = 2 + C, \quad \text{or} \quad C = 7.$$

Therefore, $\qquad\qquad y = -\sqrt{2t + 2\ln t + 7}.$

31. The constant function $y = 0$ is a solution since this makes both sides of the equation equal zero. If $y \neq 0$ you may divide by y to obtain

$$\frac{1}{y}\frac{dy}{dp} = -\frac{1}{2}\cdot\frac{1}{p+3}$$

$$\int \frac{1}{y}\frac{dy}{dp}\,dp = \int\left(-\frac{1}{2}\cdot\frac{1}{p+3}\right)dp$$

$$\int \frac{1}{y}\,dy = -\frac{1}{2}\int \frac{1}{p+3}\,dp$$

$$\ln|y| = -\frac{1}{2}\ln|p+3| + C \quad (C \text{ a constant})$$

$$\ln|y| = \ln|p+3|^{-1/2} + C$$

$$|y| = e^{\ln|p+3|^{-1/2} + C} = |p+3|^{-1/2}\cdot e^C$$

$$y = \pm e^C|p+3|^{-1/2}$$

The general solution (including the constant solution) has the form

$$y = A\,|\,p+3\,|^{-1/2}, \qquad A \text{ any constant.}$$

Note, however, that both the price p and the sales volume y should be positive quantities. So the only solutions that make some sense economically have the form

$$y = A(p+3)^{-1/2}, \quad A > 0.$$

37. $\dfrac{dy}{dt} = -ay \ln \dfrac{y}{b}$. Note that y cannot be zero, because the logarithm would not be defined. However,

$\ln \dfrac{y}{b}$ is zero when $y = b$. So, the only constant solution is $y = b$. If $y \neq b$, you may separate the variables:

$$\frac{1}{y \ln \frac{y}{b}} \cdot \frac{dy}{dt} = -a,$$

$$\int \frac{1}{y \ln \frac{y}{b}} \cdot \frac{dy}{dt}\,dt = \int -a\,dt,$$

$$\int \frac{1}{y \ln \frac{y}{b}}\,dy = \int -a\,dt = -at + C.$$

To evaluate the left-hand side, let

$$u = \ln \frac{y}{b} \Rightarrow du = \frac{1}{\left(\frac{y}{b}\right)} \cdot \frac{1}{b}\,dy = \frac{1}{y}\,dy.$$

Then

$$\int \frac{1}{\ln \frac{y}{b}} \cdot \frac{1}{y}\,dy = \int \frac{1}{u}\,du = \ln|u| + C,$$

$$= \ln\left|\ln \frac{y}{b}\right| + C.$$

Therefore, setting the two sides equal,

$$\ln\left|\ln \frac{y}{b}\right| = -at + C \qquad \text{(different } C\text{)}.$$

Next, take the exponential of both sides to produce:

$$\left|\ln \frac{y}{b}\right| = e^{-at+C} = e^C \cdot e^{-at} = Ce^{-at}\,(c = e^C) \qquad \text{(different } C \text{ again)}$$

$$\left|\ln \frac{y}{b}\right| = Ce^{-at} \qquad \text{(where } C \text{ is positive)}$$

$$\ln \frac{y}{b} = \pm Ce^{-at}$$

Write $\pm C$ as simply C, where now C can be positive or negative. The constant solution $y = b$ corresponds to $C = 0$. So the general solution, with an arbitrary C, is

$$\frac{y}{b} = e^{Ce^{-at}}, \quad \text{and} \quad y = be^{Ce^{-at}}.$$

10.3 First-Order Linear Differential Equations

Before you start this section, review the product rule for differentiation and the chain rule. For instance, make sure you understand the following identity: If y and $A(t)$ are functions of t, then

$$\frac{d}{dt}\left[e^{A(t)}y\right] = e^{A(t)}y' + ye^{A(t)}A'(t).$$

In this section, $A(t)$ stands for an antiderivative of a function $a(t)$. So that $A'(t) = a(t)$ and the preceding identity can be written as

$$\frac{d}{dt}\left[e^{A(t)}y\right] = e^{A(t)}y' + ye^{A(t)}a(t).$$

Exercises 1–6 are warm-up exercises. They do not require the full power of the method of this section. You should do as many of them as you can, in order to prepare for the remaining exercises in this section.

1. In this exercise, we are not asked to solve the equation $y' - 2y = t$. All we need is the integrating factor. To get this integrating factor you must first put the equation in standard form $y' + a(t)y = b(t)$. The given equation is already in standard form, with $a(t) = -2$. The integrating factor is obtained from $a(t)$ in two steps:

    Step 1: Find an antiderivative $A(t)$ of $a(t)$. In this case,

    $$A(t) = \int a(t)\,dt = \int (-2)\,dt = -2t.$$

    You have to keep in mind that all we need is one antiderivative, so do not include an arbitrary constant when evaluating $A(t)$. That is, do not write $A(t) = -2t + C$. Just set $C = 0$ and take $A(t) = -2t$.

    Step 2: Form the integrating factor $e^{A(t)}$. This is a straightforward step, since you have $A(t) = -2t$, then the integrating factor is e^{-2t}, and you are done.

7. We will solve the equation by following the step-by-step method described in the text.

    Step 1: Put the equation in standard form $(y' + a(t)y = b(t))$. The equation is already in standard form $y' + y = 1$ where $a(t) = 1$.

    Step 2: Find an integrating factor. This step is like Exercise 1. First find an antiderivative of $a(t)$.

    $$A(t) = \int 1\,dt = t.$$

    Note how we did not include an arbitrary constant. Now form the integrating factor $e^{A(t)} = e^t$.

Step 3: Multiply the equation through by the integrating factor and then simplify. (This is a crucial step in working the product rule backward, as you will now see.)

$$e^t(y' + y) = e^t \cdot 1$$

$$\overbrace{e^t y' + e^t y}^{\frac{d}{dt}[e^t y]} = e^t$$

$$\frac{d}{dt}[e^t y] = e^t.$$

We used the product rule in expressing $e^t y' + e^t y = \frac{d}{dt}[e^t y]$.

Step 4: Solve the differential equation. Integrate both sides of the last equation and remember that the integral cancels the derivative:

$$\frac{d}{dt}[e^t y] = e^t \quad \Rightarrow \quad e^t y = \int e^t dt = e^t + C.$$

(Here it is important to include the arbitrary constant when evaluating the integral, since we want all possible answers and not just one.) Solve for y by multiplying by e^{-t}:

$$e^{-t} e^t y = e^{-t}(e^t + C) \quad \Rightarrow \quad y = 1 + Ce^{-t}.$$

13. Follow the steps of Exercise 7.

    Step 1: Put the equation in standard form $(y' + a(t)y = b(t))$. The equation is already in standard form

    $$y' + \frac{1}{10+t} y = 0$$

    where $a(t) = \dfrac{1}{10+t}$.

    Step 2: Find an integrating factor.

    $$A(t) = \int \frac{1}{10+t} dt = \ln|10+t| = \ln(10+t)$$

    because $t > 0$ so $10+t > 0$. The integrating factor is

    $$e^{A(t)} = e^{\ln(10+t)} = 10 + t.$$

    Step 3: Multiply the equation through by the integrating factor and then simplify.

    $$(10+t)\left(y' + \frac{1}{10+t} y \right) = (10+t) \cdot 0$$

    $$\overbrace{(10+t)y' + y}^{\frac{d}{dt}[(10+t)y]} = 0$$

    $$\frac{d}{dt}[(10+t)y] = 0.$$

Step 4: Integrate both sides:

$$\frac{d}{dt}[(10+t)y] = 0 \quad \Rightarrow \quad (10+t)y = C,$$

because 0 is the derivative of constant functions. Finally, solve for y by multiplying by $\frac{1}{10+t}$:

$$(10+t)y = C \quad \Rightarrow \quad y = \frac{C}{10+t}.$$

Note that the choice $C = 0$ yields the function $y = 0$, which is clearly a solution of the differential

equation $y' + \frac{1}{10+t}y = 0$.

19. Follow the step-by-step method of integrating factors.

Step 1: Put the equation in standard form $(y' + a(t)y = b(t))$. The equation is already in standard form

$$y' + y = 2 - e^t, \quad \text{where } a(t) = 1.$$

Step 2: Find an integrating factor.

$$A(t) = \int 1\, dt = t.$$

The integrating factor is $e^{A(t)} = e^t$.

Step 3: Multiply the equation through by the integrating factor and then simplify.

$$e^t(y' + y) = e^t \cdot (2 - e^t)$$

$$\overbrace{e^t y' + e^t y}^{\frac{d}{dt}[e^t y]} = 2e^t - e^{2t}$$

$$\frac{d}{dt}[e^t y] = 2e^t - e^{2t}.$$

Step 4: Integrate both sides:

$$e^t y = \int (2e^t - e^{2t})dt = 2e^t - \frac{1}{2}e^{2t} + C.$$

Finally, solve for y:

$$y = e^{-t}\left(2e^t - \frac{1}{2}e^{2t} + C\right) = 2 - \frac{1}{2}e^t + Ce^{-t}.$$

25. Solving an initial value problem is really like solving two separate problems. First, you must solve the equation

$$y' + y = e^{2t}.$$

Your solution will contain an arbitrary constant C. To each value of C corresponds one solution of the differential equation. Next, use the initial condition $y(0) = 1$ to determine the unique solution of the initial value problem.

To solve the equation, use an integrating factor. The equation is in standard form with $a(t) = 1$. So, $A(t) = t$ and the integrating factor is e^t. Multiply the equation by the integrating factor and simplify to get

$$\frac{d}{dt}[e^t y] = e^t \cdot e^{2t} \quad \Rightarrow \quad \frac{d}{dt}[e^t y] = e^{3t}.$$

Integrate both sides and then solve for y:

$$e^t y = \int e^{3t} dt = \frac{1}{3}e^{3t} + C \quad \Rightarrow \quad y = \frac{1}{3}e^{2t} + Ce^{-t}.$$

This is the general solution of the differential equation. It consists of infinitely many functions; one for each value of C. Next, we choose the constant C in order to satisfy the initial condition $y(0) = -1$:

$$-1 = y(0) \quad \Rightarrow \quad -1 = \frac{1}{3}e^{2\cdot 0} + Ce^{-0} = \frac{1}{3} + C \quad \Rightarrow \quad C = -\frac{4}{3}.$$

Thus the (unique) solution of the initial value problem is $y = \frac{1}{3}e^{2t} - \frac{4}{3}e^{-t}$.

10.4 Applications of First-Order Linear Differential Equations

There is no doubt that word problems are the most challenging problems in calculus since there is no step-by-step method to do these problems. You have to read each question very carefully and then try to identify the kind of problem. Once you have identified the problem, you can look back at some of the examples that we have solved in the text and model your answer after its solution. The goal is to reach a point where you can do the exercises without looking at the solved examples in your text.

1. **(a)** Since this question is about rate of change, we answer it by looking at the differential equation. Let $P(t)$ denote the amount of money in the account at time t. We know from Example 1 that $P(t)$ satisfies the differential equation

    $$y' - .06y = 2400$$

    The rate of change of $P(t)$ is $P'(t)$. From the differential equation

    $$P'(t) = .06P(t) + 2400. \tag{1}$$

    When the amount in the bank, $P(t)$, is \$30,000, the rate of change is

    $$P' = .06(30,000) + 2400 = 4200 \text{ dollars per year.}$$

    Note that we did not need to know t in order to answer this question.

    (b) Here we are told that $P(t)$ is growing at a rate twice as fast as the rate of the annual contributions. That is the rate of growth is $2 \times 2400 = 4800$ dollars per year. Plugging this value of P' into the equation (1) and solving for P, we find

    $$4800 = .06P(t) + 2400 \quad \Rightarrow \quad P(t) = \frac{4800 - 2400}{.06} = 40,000 \text{ dollars.}$$

    Thus \$40,000 was in the account when the account was growing at the rate of \$4800 per year. Here again note that we answered the question without knowing the value of t.

(c) In this part, we are given that $P(t) = \$40,000$ and we are asked to find t. From Example 1, we have

$$P(t) = -40,000 + 41,000e^{.06t}.$$

Plugging $P(t) = 40,000$ and solving for t, we find

$$
\begin{aligned}
40,000 &= -40,000 + 41,000e^{.06t} \\
80,000 &= 41,000e^{.06t} \\
\frac{80,000}{41,000} &= e^{.06t} \\
\ln\left(\frac{80}{41}\right) &= .06t \\
t &= \frac{1}{.06}\ln\left(\frac{80}{41}\right) \approx 11.1 \text{ years.}
\end{aligned}
$$

Thus it takes approximately 11.1 years or 11 years and 2 months for the account to reach $40,000 dollars.

7. This is a problem about paying back a loan. The closest example that we have to direct us through the solution is Example 2 in the text. You should read it very carefully and understand it thoroughly before you attempt to solve this exercise.

Let $P(t)$ denote the amount that the person owes at time t (in years from the time the loan was taken). We are asked to determine the value of k, the rate of annual payments, if the loan is to be paid in full in exactly 10 years; that is, if $P(10) = 0$. To answer this question, we must find a formula for $P(t)$ first.

Two influences act on $P(t)$: The interest that is added, and the payments that are subtracted (remember that a payment subtracts from the amount owed). Thus, the differential equation satisfied by $P(t)$ is

$$y' = .075y - k, \quad \text{or} \quad y' - .075y = -k,$$

where the last equation is in standard form. We are also given important information about $P(t)$, namely, that the initial amount owed was 100,000. Hence $y(0) = 100,000$. So $P(t)$ is the solution of the initial value problem

$$y' - .075y = -k \quad y(0) = 100,000.$$

We solve this problem using the integrating factor method from the previous section:

$$a(t) = -.075$$
$$A(t) = \int a(t)\,dt = -.075t$$

Integrating factor $= e^{A(t)} = e^{-.075t}$

Multiplying the equation by the integrating factor and simplifying, we obtain

$$e^{-.075t}(y' - .075y) = -ke^{-.075t}$$
$$\frac{d}{dt}[e^{-.075t}y] = -ke^{-.075t}.$$

Integrating with respect to t and solving for y, we find

$$e^{-.075t}y = \int -ke^{-.075t}dt = \frac{k}{.075}e^{-.075t} + C$$

$$y = e^{.075t}\left(\frac{k}{.075}e^{-.075t} + C\right)$$

$$y = \frac{k}{.075} + Ce^{.075t}.$$

Note that the solution contains an arbitrary constant C (as it should) and it also depends on k the rate of annual payments. The value of C will be determined from the initial condition: $P(0) = 100,000$. Plugging $t = 0$ into the formula for y, we find

$$y(0) = 100,000 = \frac{k}{.075} + Ce^{.075 \cdot 0} = \frac{k}{.075} + C$$

$$C = 100,000 - \frac{k}{.075}.$$

Plugging this value of C into the formula for y, we obtain

$$P(t) = \frac{k}{.075} + \overbrace{(100,000) - \frac{k}{.075}}^{=C}e^{.075t}.$$

This is the formula of $P(t)$ in terms of the annual payments k.

Now if the loan is to be paid in full in 10 years, this implies that $P(10) = 0$. Hence,

$$0 = P(10) = \frac{k}{.075} + \left(100,000 - \frac{k}{.075}\right)e^{.075 \cdot 10}$$

$$-\frac{k}{.075} = \left(100,000 - \frac{k}{.075}\right)e^{.75}$$

$$\frac{k}{.075}e^{.75} - \frac{k}{.075} = 100,000e^{.75}$$

$$k\left(\frac{e^{.75}}{.075} - \frac{1}{.075}\right) = 100,000e^{.75}$$

$$k = \frac{100,000e^{.75}}{\frac{e^{.75}}{.075} - \frac{1}{.075}} \approx 14,214.4 \text{ dollars per year.}$$

Thus in order for the $100,000 loan to be paid in full in 10 years, the rate of payments should be about $14,214.40 per year (or $1184.50 per month).

13. We have to solve the initial value problem

$$y' = .1(10 - y), \quad y(0) = 350.$$

Put the equation in standard form and solve it using the step by step method of integrating factor:

$$y' + .1y = 1 \qquad \text{(Equation in standard form)}$$

$$a(t) = .1, \quad A(t) = .1t$$

$$\text{Integrating factor} = e^{A(t)} = e^{.1t}$$

$$e^{.1t}(y' + .1y) = e^{.1t} \qquad \text{(Multiply by the integrating factor.)}$$

$$\frac{d}{dt}[e^{.1t}y] = e^{.1t} \qquad \text{(Simplify the equation.)}$$

$$e^{.1t}y = \int e^{.1t}dt = \frac{1}{.1}e^{.1t} + C \quad \text{(Integrate.)}$$

$$y = 10 + Ce^{-.1t} \qquad \text{(Multiply by } e^{-.1t}.)$$

To satisfy the initial condition $y(0) = 350$, we must have

$$350 = y(0) = 10 + Ce^{-.1 \cdot 0} = 10 + C \quad \text{or} \quad C = 340.$$

Thus the temperature of the rod at any time t is given by $f(t) = 10 + 340e^{-.1t}$.

19. **(a)** In this problem, the concentration of creatinine in the dialysate solution at any time t (in hours) is denoted by $f(t)$ and measured by grams per liter. The concentration $f(t)$ satisfies the differential equation.

$$y' = k(110 - y).$$

We are told that when the concentration was $f(t) = 75$ grams per liter, it was rising at the rate of 10 grams per liter per hour. So, when $y = 75$, $y' = 10$. Plugging these values into the differential equation and solving for k, we find

$$10 = k(110 - 75) \quad \Rightarrow \quad k = \frac{10}{35} = \frac{2}{7} \approx .286.$$

(b) Using the value of k that we just found, we obtain the differential equation satisfied by $f(t)$

$$y' = \frac{2}{7}(110 - y).$$

This equation expresses the rate of change of $f(t)$ as a function of $f(t)$. We also know that initially the concentration in the dialysate is 0. Hence $y(0) = 0$. The rate of change of the concentration after 4 hours of dialysis is given in part (a). It is 10 grams per liter per hour. To obtain the rate of change of the concentration at the beginning of the session, we set $t = 0$ in the differential equation and use $y(0) = 0$:

$$y'(0) = \frac{2}{7}(110 - y(0)) = \frac{2}{7}110 \approx 31.43 \text{ grams per liter per hour.}$$

Thus at the beginning of the session, the creatinine substance was filtering into the dialysate solution at the rate of 31.43 grams per liter per hour, which is over three times the rate after four hours of dialysis. So if we replace the solution after four hours by a fresh solution, we would triple the rate of creatinine clearance from the body.

25. (a) Let $f(t)$ (in milligrams) denote the amount of morphine in the body at time t in hours. The rate of change of $f(t)$ is affected by two influences: the rate at which the body removes the substance, and the rate at which morphine is injected in the body. The first rate of change is given by $-.35f(t)$, since the body removes the substance at a rate proportional to the amount of the substance present at time t, with constant of proportionality $k = .35$. The second rate of change is given to us: It is t milligrams per hour. Thus $f(t)$ satisfies the differential equation

$$y' = -.35y + t.$$

(b) Since the body was free of morphine at the beginning of the infusion, we have the initial condition $y(0) = 0$. Putting the equation in standard form, we obtain the initial value problem:

$$y' + .35y = t \quad y(0) = 0.$$

To determine $f(t)$, we must solve this initial value problem. We use an integrating factor:

$$a(t) = .35, \quad A(t) = .35t$$

$$\text{Integrating factor} = e^{A(t)} = e^{.35t}$$

$$e^{.35t}(y' + .35y) = e^{.35t}t \qquad \text{(Multiply by the integrating factor.)}$$

$$\frac{d}{dt}[e^{.35t}y] = e^{.35t}t \qquad \text{(Simplify the equation).}$$

$$e^{.35t}y = \int e^{.35t}t\,dt = \frac{1}{(.35)^2}e^{.35t}(.35t - 1) + C \qquad \text{(Integrate.)}$$

In evaluating the last integral, you can use integration by parts or appeal to (1) in Section 10.4, with $a = 1, b = 0, c = .35$. Multiplying both sides by $e^{-.35t}$, we obtain

$$y = \frac{1}{(.35)^2}(.35t - 1) + Ce^{-.35t}$$

In order to satisfy the initial condition, we must have

$$0 = y(0) = \frac{1}{(.35)^2}(-1) + C$$

$$C = \frac{1}{(.35)^2} = \frac{1}{.1225}$$

Hence the amount of morphine in the body at time t is

$$f(t) = \frac{1}{(.35)^2}(.35t - 1) + \frac{1}{(.35)^2}e^{-.35t}$$

$$= \frac{1}{(.35)^2}(.35t - 1 + e^{-.35t}).$$

After 8 hours, there were $f(8) = \dfrac{1}{(.35)^2}\left(.35(8) - 1 + e^{-.35(8)}\right) \approx 15.2$ mg of morphine in the body.

10.5 Graphing Solutions of Differential Equations

We hope you get to study this section! Some instructors omit it because they think it is too hard for first-year calculus students. (No other book at this level attempts to teach the material.) But we have found that our students do as well on the qualitative theory as on almost any material in the second half of the text. The secret is to spend about one week on this section.

One of the difficulties here is learning to sketch *yz*-graphs where the *y*-axis is horizontal. The difficulty is real though it is mainly psychological. To help you get used to *yz*-graphs, we have included them for Exercises 7–20. Study them there so you will be able to produce similar graphs in Exercises 21–37.

1. In graphing the autonomous equations described in Exercises 1–6, you may find it helpful to fill in the blanks in the following sentences with either *increases* or *decreases*:

    As *t* increases, *y* _____.

    As *y* increases, $z = y'$ _____.

    In sketching the *yz*-graph for Exercise 1, a key point to keep in mind is that $z = y'$, i.e., *z* represents the derivative (or slope) of *y*. Thus the statement

    "the slope of y is always positive,"

    a statement about the *ty*-graph, can be translated to

    "z is always positive," (1)

    a statement about the *yz*-graph. Similarly, the statement

    "as t increases, the slope of y becomes less positive"

    can be translated to

    "as t increases, z decreases."

    Since *y* increases as *t* increases (be careful here), we can thus conclude:

    "as y increases, z decreases." (2)

    Fortified with the information in (1) and (2), we are now ready to sketch the *yz*-graph. Begin with the point $(y, z) = (1, z(1))$ corresponding to $t = 0$. From (1), we know that the value of $z(1)$ is positive. (See Figure 1(a)). As *t* moves from 0 to 4, the value of *y* will increase, and thus our graph will move to the right from the starting point $(1, z(1))$. From (2), we know that the graph will move downward as *y* increases, so we can indicate the initial direction of the graph with an arrow as in Figure 1(b).

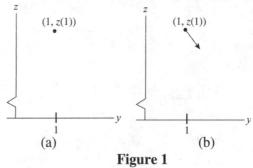

Figure 1

From (1) we know that the graph will not descend below $y = 0$ in the domain $(1, y(3))$. A possible *yz*-graph is sketched in Figure 2. Note that we do not necessarily know that the graph is linear, however.

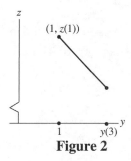

Figure 2

This is a graph that shows the behavior of y' (remember, $z = y'$). We can use this graph to sketch the particular solution of the equation. Start with the fact that $(0, 1)$ is on the graph of the particular solution (Figure 3).

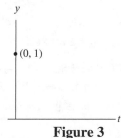

Figure 3

The domain is $0 \le t \le 3$. What happens as t moves from 0 towards 3? Since y' is positive, we know that y increases as t increases. Denote this with an upward arrow, as in Figure 4(a). Next, we need to determine whether the curve will be concave up or concave down. We can answer this by answering the question: "what happens to y' as y increases?" Referring to Figure 2, we see that y' decreases as y increases. Thus the slope of y is decreasing, so the curve is concave down, as in Figure 4(b).

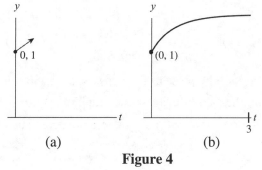

Figure 4

Since y' is always positive, the autonomous equation $y' = g(y)$ does not have any constant solutions.

7. Given $y' = 3 - \frac{1}{2}y$, let $g(y) = 3 - \frac{1}{2}$. The graph of $z = g(y)$ is shown in Figure 5(a). Set $3 - \frac{1}{2}y = 0$ and find $y = 6$. Thus $g(y)$ has a zero when $y = 6$. Therefore the constant function $y = 6$ is a solution of the differential equation. See Figure 5(b).

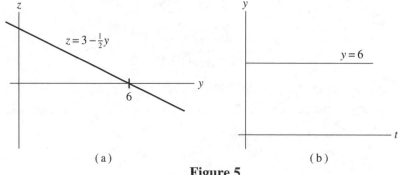

(a)   (b)

Figure 5

To sketch the solution corresponding to $y(0) = 4$, we locate this initial value on the y-axes in Figure 6(a) and Figure 6(b), and note that the z-coordinate in Figure 6(a) is positive when $y = 4$. That is, the derivative is positive when $y = 4$. So, we place an upward arrow at the initial point in Figure 6(b).

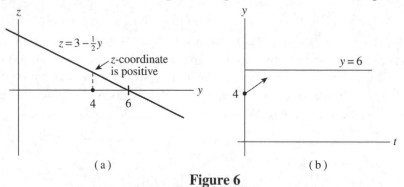

(a)   (b)

Figure 6

From Figure 2b, the y-values will increase as time passes, so y will move to the right on the yz-graph. See Figure 3(a). As a result, the z-coordinate on the graph of $z = 3 - \frac{1}{2}$ will become less positive. That is, the slope of the solution curve will become less positive. Thus the solution curve is concave down, as in Figure 3(b).

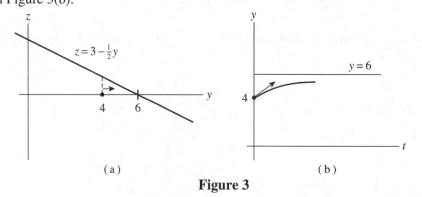

(a)   (b)

Figure 3

To obtain the graph of the solution satisfying $y(0) = 8$, begin by making the sketch in Figure 4. First, note that the z-coordinate in Figure 4(a) is negative when $y = 8$. (That is, the derivative is negative when $y = 8$.) So place a downward arrow at $y = 8$ in Figure 4(b).

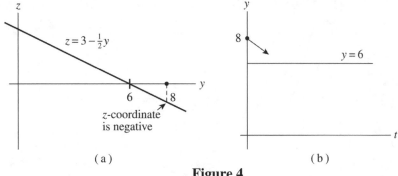

Figure 4

The downward arrow in Figure 4(b) shows that the y-values will decrease as time passes. This means that y will move to the *left* on the yz-graph. See Figure 5(a). As a result, the z-coordinates on the yz-graph will become less negative and the *slopes* on the ty-graph will become less negative. (Reread the preceding explanation because this is a key idea.)

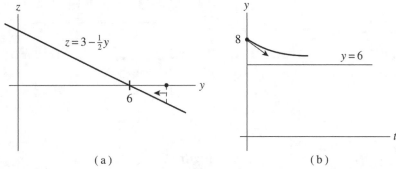

Figure 5

The solution curve in Figure 5(b) is concave up because as time passes the negative slopes on the curve become less negative.

13. First, sketch the graph of $z = y^3 - 9y$. Find the zeros of $g(y) = y^3 - 9y$ by setting $g(y) = 0$ and solving for y.

$$y^3 - 9y = 0,$$
$$y(y^2 - 9) = 0,$$
$$y(y + 3)(y - 3) = 0.$$

The graph of $z = y^3 - 9y$ crosses the y-axis at $y = 0$, $y = \pm 3$. Next, to find where the graph has a relative maximum and relative minimum, set $\dfrac{dz}{dy} = 0$ and solve for y.

$$\frac{d}{dy}(y^3 - 9y) = 3y^2 - 9 = 0.$$

Thus $3y^2 = 9$, $y^2 = 3$, and $y = \pm\sqrt{3}$. See Figure 6(a). The constant solutions are shown in Figure 6(b).

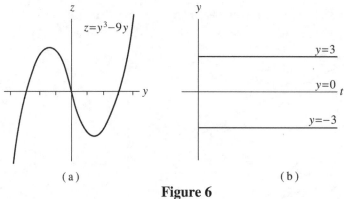

(a)   (b)

Figure 6

Figure 7 shows the sketches needed to obtain the graphs of the solutions such that $y(0) = -4$, $y(0) = -1$, and $y(0) = 4$. Note that the solution where $y(0) = -1$ cannot cross the constant solution $y = 0$.

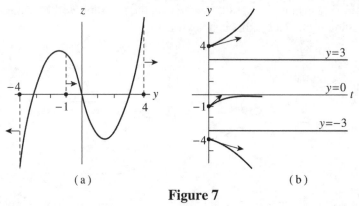

(a)   (b)

Figure 7

Figure 8 shows the work for the solution with $y(0) = 2$. Since the initial z-coordinate in Figure 8(a) is negative, the y-values of the solution curve will decrease and y will move to the *left* on the yz-graph. At first, the z-coordinate will become more negative (i.e., the slope of the solution curve will become more negative). Then, as y moves to the left past $y = \sqrt{3}$, the z-coordinates on the yz-graph will become less negative (i.e., the slope of the solution curve will become less negative). Thus the solution curve will have an inflection point at $y = \sqrt{3}$, as in Figure 8(b).

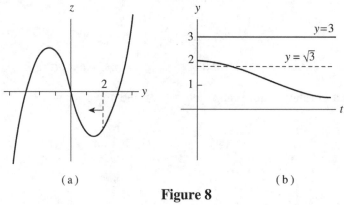

(a)   (b)

Figure 8

19. The graph of $z = g(y)$ shown in Figure 9(a) has zeros at $y = 1$ and $y = 6$. So, the constant solutions of $y' = g(y)$ are $y = 1$ and $y = 6$. The relative maximum of $g(y)$ occurs when $y = 4$, so a solution curve will have an inflection point whenever it crosses the dashed line shown in Figure 9(b). The inflection point at $y = 2$ on the graph of $g(y)$ has no influence on the general shape of solution curves, so you should ignore it.

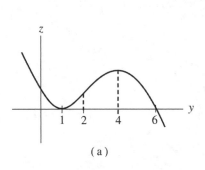

(a)

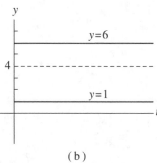

(b)

Figure 9

Figure 10 shows the work for the solution of $y' = g(y)$ such that $y(0) = 0$, $y(0) = 1.2$, $y(0) = 5$, and $y(0) = 7$.

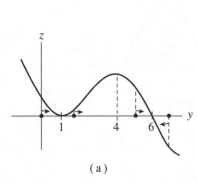

(a)

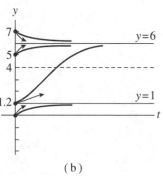

(b)

Figure 10

Warning: Although the function $g(y)$ in Exercise 19 has a minimum at $y = 1$, the minimum value is zero. Notice that Rule 7 after Example 3 in Section 10.5 applies only to *nonzero* relative maximum or minimum points.

25. The graph of $z = y^2 - 3y - 4$ is a parabola. It opens upward because the coefficient of y^2 is positive. To find where the graph crosses the y-axis, solve

$$y^2 - 3y - 4 = 0,$$
$$(y + 1)(y - 4) = 0,$$
$$y = -1, \quad \text{or} \quad y = 4.$$

Set $\dfrac{dz}{dy} = 0$ and find that $2y - 3 = 0$ and $y = \frac{3}{2} = 1.5$. Thus the parabola has a minimum at $y = 1.5$.

This is enough information to produce the initial sketch in Figure 11(b).

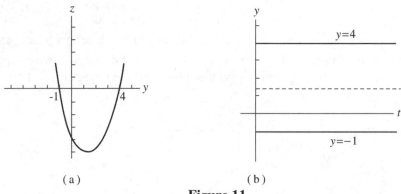

(a)   (b)

Figure 11

Figure 12 shows the work for the solutions of the equation $y' = y^2 - 3y - 4$ such that $y(0) = 0$ and $y(0) = 3$.

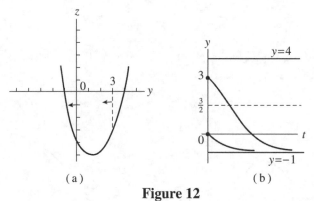

(a)   (b)

Figure 12

31. The graph of $z = \dfrac{1}{y}$ is shown in Figure 13(a). The graph does not cross the y-axis, so there are no constant solutions of $y' = \dfrac{1}{y}$. The solution satisfying $y(0) = 1$ is an increasing curve because the z-coordinate of the initial point on the yz-graph is positive. However, as y increases on the yz-graph, the z-coordinates become less positive; i.e., the slopes of the solution curve become less positive. A similar solution holds for the solution satisfying $y(0) = -1$, where the slopes are negative and become less negative. See Figure 13(b).

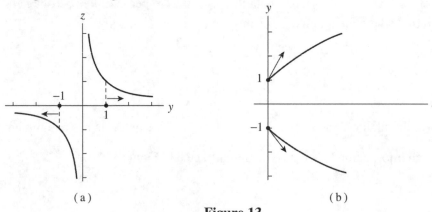

(a)   (b)

Figure 13

37. Let $f(t)$ be the height of the sunflower at time t. Then $f'(t)$ is the rate at which the sunflower is growing at time t. You are told that this rate is proportional to the product of its height and the difference between its height at maturity and its current height. If you let H be the sunflower's height at maturity, then

$$f'(t) = kf(t)[H - f(t)] \qquad (1)$$

or equivalently,

$$y' = ky(H - y). \qquad (2)$$

Before you can sketch a solution, you must determine whether k is positive or negative. Clearly, $f(t)$ is increasing, so $f'(t)$ is positive. Also, $f(t)$ is less than H, so the factor $H - f(t)$ is positive. From this it follows that k must be positive.

From (2) you must sketch

$$z = ky(H - y) = kHy - ky^2.$$

The graph of this equation is a parabola which opens down, with zeros at 0 and H, and maximum at $H/2$. See Figure 14(a).

The initial height of the sunflower is $y(0)$, which is slightly greater than 0. Since z is positive here, this solution is increasing as it leaves the initial point. As y moves to the right, the z-coordinate becomes more positive until y reaches $H/2$. After that, the z-coordinate becomes less positive, and hence the curve has an inflection point at $H/2$, and is concave down after that point. Also, it is clear from equation (1) that the constant functions $y = 0$ and $y = H$ are solutions. See Figure 14(b).

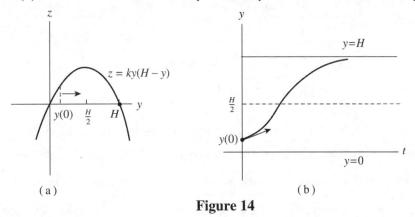

(a)   (b)

Figure 14

Helpful Hint: Exercise 37 is a key exercise. Work on it now! You will see more exercises like this in Section 10.6.

10.6 Applications of Differential Equations

Here we are! If you have been faithful in your work on the "word problems" in Sections 10.1–10.5, you should be ready for Section 10.6. The exercises fall into two categories:

1. Problems involving one constant of proportionality, and
2. "One-compartment" problems.

The first category is analyzed carefully in the first two pages of the section. Most of Section 5.4 concerned applications of the first category of problems. One of the most important applications there involved logistic growth, and this subject is discussed again in Section 10.6.

One-compartment problems are very common in applications. Multi-compartment problems also arise. You can read about them in *Mathematical Techniques for Physiology and Medicine*, by William Simon (New York: Academic Press, 1972). This book was at one time the text for a course at the Rochester School of Medicine and Dentistry.

The material on population genetics was suggested to us by one of our students. While in our class, she was taking a course on population genetics where she saw a qualitative analysis of certain differential equations. Only the yz-graphs were shown in her textbook, *Population Genetics*, by C. C. Li (Chicago: University of Chicago Press, 1995). Our student used her knowledge from our calculus course to provide the ty-graphs of the solution curves. Now you, too, can have a glimpse into this fascinating topic.

1. **(a)** $\dfrac{dN}{dt} = N(1-N), N(0) = .75$

   The carrying capacity is $K = 1$ and the intrinsic rate is $r = 1$.

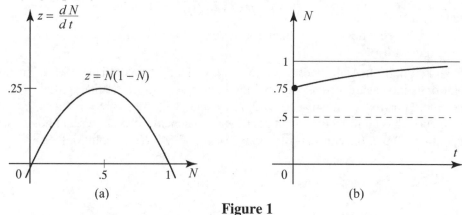

   (a)   (b)

   Figure 1

   (b) The graph of $z = \dfrac{dN}{dt}$ is shown in Figure 1(a).

   (c) The zeros of $z = N(1-N) = N - N^2$ are 0 and 1. Hence the constant solutions to the differential equation are $N = 0$ and $N = 1$, and the concavity of certain solutions may change at $N = \dfrac{0+1}{2} = .5$. The constant solutions $N = 0$ and $N = 1$ are shown in Figure 1(b), along with the dashed line $N = .5$ indicating where the concavity of certain solutions may change.

   (d) When $N = .75, \dfrac{dN}{dt} = (.75)(1 - .75)$ is positive, the solution curve corresponding to the initial condition $N(0) = .75$ is always increasing. The curve will not pass through the line $N = 1$, and the curve's concavity will not change since it will not pass through $N = .5$. Thus we sketch a curve starting at the point $(0, .75)$ that is increasing, concave down, and approaches the horizontal asymptote $N = 1$. See Figure 1(b).

7. Let $y = f(t)$ be the percentage of the population having the information at time t. Then $0 \le f(t) \le 100$, and $100 - f(t)$ is the percentage that does not have the information. The problem says that

$$
\begin{bmatrix} \text{the rate of} \\ \text{spread of} \\ \text{information} \end{bmatrix}
\begin{bmatrix} \text{is} \\ \text{proportional} \\ \text{to} \end{bmatrix}
\begin{bmatrix} \text{precentage} \\ \text{not having} \\ \text{information} \end{bmatrix}.
$$

Thus,

$$f'(t) = k \cdot [100 - f(t)] \tag{1}$$

for some constant k. Equivalently

$$y' = k(100 - y). \tag{2}$$

Before you can sketch a solution, you must determine whether k is positive or negative. Clearly, $f(t)$ is increasing, so $f'(t)$ is positive. Also, $f(t)$ is not more than 100%, so $100 - f(t)$ is also positive. From this it follows that k is positive.

From (2) you must sketch

$$z = k(100 - y)$$
$$= 100k - ky,$$

where k is positive. The graph of this equation is a straight line with negative slope. Also, when $y = 100$ we have $z = 0$. See Figure 2(a) below. For the solution where $y(0) = 1$, use the method in Section 10.4 and obtain the curve in Figure 2(b).

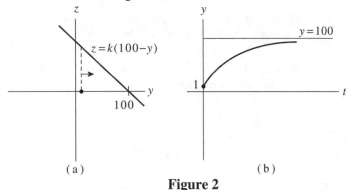

(a)   (b)

Figure 2

13. Let $y = f(t)$ be the amount of substance A at time t. Then $f'(t)$ is the rate at which substance A is converted into substance B. You are told that this rate is proportional to the square of the amount of A. Thus,

$$f'(t) = k[f(t)]^2 \tag{1}$$

where k is a constant of proportionality. Equivalently,

$$y' = ky^2. \tag{2}$$

Before you can sketch a solution, you must determine whether k is positive or negative. Clearly, $f(t)$ is decreasing, so $f'(t)$ is negative. Since $f^2(t)$ is never negative, k must be negative.

From (2) you must sketch,

$$z = ky^2,$$

where k is negative, which determines a parabola that opens down. See Figure 3(a) below. To sketch a solution where $y(0)$ is some positive number (representing the initial amount of substance A), use the method in Section 10.4 and obtain the curve in Figure 3(b).

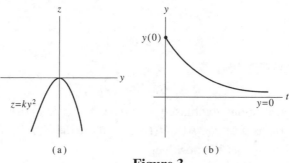

Figure 3

19. **(a)** At first, ignore the deposits to the account. In Section 5.2, we showed that if no deposits or withdrawals are made, then $f(t)$ satisfies the equation

$$y' = .05y.$$

That is, the account grows at a rate proportional to the amount in the account. We conclude that interest is being added to the account at a rate proportional to the amount in the account.

Now suppose that continuous deposits are being made to the same account at the rate of $10,000 per year. There are two influences on the way the amount of money in the account changes—the rate at which interest is added and the rate at which money is deposited. Let $f(t)$ be the amount of money in the account at time t. Then the rate of change of $f(t)$ is the net effect of these two influences. That is, $f(t)$ satisfies the equation

$$f'(t) = .05f(t) + 10,000,$$

or equivalently,

$$y' = .05y + 10,000. \tag{1}$$

At time $t = 0$, assume there is no money in the account. Hence $y(0) = 0$.

(b) We note that $y = \dfrac{-10,000}{.05} = -200,000$ is a constant solution, since it makes both sides of equation (1) equal zero. Assuming that $y \neq -200,000$, use separation of variables:

$$y' = .05(y + 200,000),$$

$$\int \frac{1}{(y + 200,000)} \frac{dy}{dt} dt = \int .05 \, dt,$$

$$\int \frac{1}{(y + 200,000)} dy = \int .05 \, dt,$$

$$\ln|y + 200,000| = .05t + C, \qquad C \text{ a constant,}$$

$$|y + 200,000| = e^{.05t + C} = e^C \cdot e^{.05t}.$$

If $y \neq -200,000$, then $y + 200,000$ is either positive or negative for all t. Thus,

$$y = \pm e^C \cdot e^{.05t} - 200,000.$$

The general solution (including the constant solution) has the form

$$y = Ae^{.05t} - 200,000, \qquad A \text{ a constant.}$$

Use the fact that $y(0) = 0$ to obtain $A = 200,000$. Hence,

$$y = 200,000e^{.05t} - 200,000 = 200,000(e^{.05t} - 1).$$

The amount in the account after 5 years is given by

$$y(t) = 200,000(e^{.25} - 1) \approx \$56,805$$

25. First sketch

$$z = -.0001q^2(1-q).$$

Clearly $z = 0$ when $q = 0$ or $q = 1$. Note that $z = .0001q^3 - .0001q^2$, which shows that the graph is a cubic curve. To find relative extreme points set $\dfrac{dz}{dq} = 0$.

$$.0003q^2 - .0002q = 0,$$
$$.0001q(3q - 2) = 0.$$

So $q = 0$, or $3q - 2 = 0$ and $q = 2/3$. Since $\dfrac{d^2z}{dq^2} = .0006q - .0002 = .0002(3q - 1)$, you can see

that $\dfrac{d^2z}{dq^2}$ is negative at $q = 0$ and positive at $q = \dfrac{2}{3}$.

Here is a rough sketch of the graph.

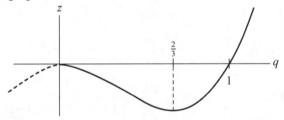

Figure 4

You are asked to find a solution when $q(0)$ is slightly less than 1, so you should assume $q(0) > 2/3$. Using the method of Section 10.4, we obtain

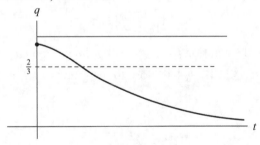

Figure 5

10.7 Numerical Solution of Differential Equations

1. If $f(t)$ is a solution of $y' = ty - 5$, then

$$f'(t) = tf(t) - 5,$$

for all t in the domain of $f(t)$. The graph of $f(t)$ passing through (2, 4) means that $f(2) = 4$. Set $t = 2$ in the equation above, and find that

$$f'(2) = 2f(2) - 5,$$
$$= 2 \cdot 4 - 5 = 3.$$

So, the slope of the graph is 3 at $t = 2$.

Helpful Hint: Make sure you understand the solutions to Exercises 1–4. They make good test questions and they prepare you for both Euler's method and the theory in Section 10.7.

7. Here $g(t, y) = 2t - y + 1, a = 0, b = 2, y(0) = 5$, and $h = \dfrac{2-0}{4} = \dfrac{1}{2}$. Starting with $(t_0, y_0) = (0, 5)$, compute $g(0, 5) = 2(0) - 5 + 1 = -4$. Thus,

$$t_1 = \frac{1}{2}, \quad y_1 = 5 + (-4) \cdot \frac{1}{2} = 3.$$

Next, $g\left(\frac{1}{2}, 3\right) = 2\left(\frac{1}{2}\right) - 3 + 1 = -1$, so

$$t_2 = 1, \quad y_2 = 3 + (-1)\frac{1}{2} = \frac{5}{2}.$$

Next $g\left(1, \frac{5}{2}\right) = 2(1) - \frac{5}{2} + 1 = \frac{1}{2}$, so

$$t_3 = \frac{3}{2}, \quad y_3 = \frac{5}{2} + \left(\frac{1}{2}\right)\frac{1}{2} = \frac{11}{4}.$$

And finally, $g\left(\frac{3}{2}, \frac{11}{4}\right) = 2\left(\frac{3}{2}\right) - \frac{11}{4} + 1 = \frac{5}{4}$, so

$$t_4 = 2,$$
$$y_4 = \frac{11}{4} + \left(\frac{5}{4}\right)\frac{1}{2} = \frac{27}{8}.$$

Thus the approximation to the solution $f(t)$ is given by the polygonal path shown in the answer section of the text. The last point $\left(2, \frac{27}{8}\right)$ is close to the graph of $f(t)$ at $t = 2$, so $f(2) \approx \frac{27}{8}$.

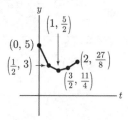

Helpful Hint: The fractions in Exercise 7 were rather simple. In many cases decimals are easier to use, particularly when h is smaller than $\frac{1}{4}$. Of course, a calculator becomes indispensable in such cases.

13. $y' = .5(1 - y)(4 - y)$

To generate the following graphs, set the TI83/TI84 in sequence mode and then use the following in the sequence **Y=** editor.

Set **vMin** equal to the value of $y(0)$.

a. $y(0) = -1$; solution is type C: increasing, concave down, and asymptotic to the line $y = 1$.

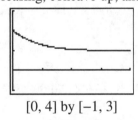

[0, 4] by [−2, 2]

b. $y(0) = 1$; solution is type A: constant solution.

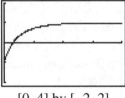

[0, 4] by [−2, 2]

c. $y(0) = 2$; solution is type E: decreasing, concave up, and asymptotic to the line $y = 1$.

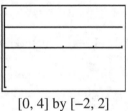

[0, 4] by [−1, 3]

d. $y(0) = 3.9$; solution is type B: decreasing, has an inflection point, and asymptotic to the line $y = 1$.

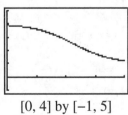

[0, 4] by [−1, 5]

e. $y(0) = 4.1$; solution is type D: concave up and increasing indefinitely.

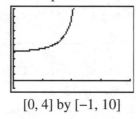

[0, 4] by [−1, 10]

Review of Chapter 10

Except for one or two differential equations in Section 10.1, all of the equations have been *first-order* equations that involve only y', y, and perhaps some functions of t. We have three methods for analyzing these equations: (1) separation of variables to find explicit solutions; (2) Euler's method to find an approximate solution; and (3) qualitative analysis of the equation. The chart below lists the methods that may be applied to first-order differential equations.

Form of the Equation	Methods of Solution
$y' = g(t, y)$	Euler's method
$y' = p(t)q(y)$	Euler's method, separation of variables
$y' = g(y)$	Euler's method, separation of variables, qualitative analysis

On an exam, your instructor might ask you to use more than one method on the same differential equation. This chart will help you anticipate what kinds of questions might be asked.

Chapter 10 Review Exercises

While solutions for all review exercises are included, expanded explanations are included for every sixth exercise.

1. Use separation of variables to obtain

$$\int y^2 \frac{dy}{dt} dt = \int (4t^3 - 3t^2 + 2)dt,$$

$$\int y^2 dy = \int (4t^3 - 3t^2 + 2)dt,$$

$$\frac{1}{3}y^3 = t^4 - t^3 + 2t + C_1,$$

$$y^3 = 3(t^4 - t^3 + 2t + C_1),$$

$$y^3 = 3(t^4 - t^3 + 2t) + C, \quad C \text{ a constant.}$$

Hence,

$$y = \sqrt[3]{3t^4 - 3t^3 + 6t + C}.$$

2.
$$\frac{y'}{t+1} = y+1$$

$$\int (y+1)^{-1} dy = \int (t+1)\, dt \quad (y \neq -1)$$

$$\ln|y+1| = \frac{t^2}{2} + t + C$$

$$y = -1 + Ae^{t^2/2+t}$$

3.
$$y' = \frac{y}{t} - 3y, t > 0$$

$$\int y^{-1} dy = \int \left(\frac{1}{t} - 3\right) dt \quad (y \neq 0)$$

$$\ln|y| = \ln|t| - 3t + C \Rightarrow y = Ate^{-3t}$$

4. $(y')^2 = t \Rightarrow y' = \pm\sqrt{t}$

$$\int dy = \pm\int t^{1/2} dt \Rightarrow y = \pm\frac{2}{3} t^{3/2} + C$$

5. $y = 7y' + ty'$, $y(0) = 3$

$$\int y^{-1} dy = \int (7+t)^{-1} dt \quad (y \neq 0)$$

$$\ln|y| = \ln|7+t|^{-1} + C$$

$$y = A(7+t)$$

$y(0) = 3 = 7A$ so $A = \frac{3}{7}$ and the solution is $y = \frac{3}{7}(7+t) = 3 + \frac{3}{7}t$.

6. $y' = te^{t+y}$, $y(0) = 0$

$$\int e^{-y} dy = \int te^t dt$$

$$-e^{-y} = te^t - e^t + C_1$$

$$y = -\ln(-te^t + e^t + C)$$

$y(0) = 0 = -\ln(1 + C)$, so $C = 0$ and the solution is $y = -\ln(-te^t + e^t)$.

7. Use separation of variables to obtain

$$yy' = 6t^2 - t,$$

$$\int y\frac{dy}{dt} dt = \int (6t^2 - t)\, dt,$$

$$\int y\, dy = \int (6t^2 - t)\, dt,$$

$$\frac{1}{2} y^2 = 2t^3 - \frac{1}{2}t^2 + C_1,$$

$$y^2 = 2\left(2t^3 - \frac{1}{2}t^2\right) + C, \quad C \text{ a constant.}$$

Hence,

$$y = \pm\sqrt{4t^3 - t^2 + C}.$$

Since $y(0) = 7$, select the positive square root. (If $y(0)$ were negative here, you would use the negative square root.) Since $y(0) = \sqrt{C} = 7$, we have $C = 49$. Therefore,

$$y = \sqrt{4t^3 - t^2 + 49}.$$

8. $y' = 5 - 8y$, $y(0) = 1$

$$y' = -8\left(-\frac{5}{8} + y\right)$$

$$\int\left(y - \frac{5}{8}\right)^{-1} dy = \int -8 \, dt$$

$$\ln\left|y - \frac{5}{8}\right| = -8t + C \Rightarrow y = \frac{5}{8} + Ae^{-8t}$$

$y(0) = 1 = \frac{5}{8} + A$, so $A = \frac{3}{8}$ and the solution is $y = \frac{5}{8} + \frac{3}{8}e^{-8t}$.

9. $y' - \dfrac{2}{1-t}y = (1-t)^4 \Rightarrow$

$$a(t) = -\frac{2}{1-t}, \ b(t) = (1-t)^4$$

$$A(t) = -\int \frac{2}{1-t} \, dt = 2\ln|1-t| = \ln(1-t)^2$$

$$e^{A(t)} = e^{\ln(1-t)^2} = (1-t)^2$$

$$e^{-A(t)} = e^{-\ln(1-t)^2} = \frac{1}{(1-t)^2}$$

$$y = e^{-A(t)}\left[\int e^{A(t)}b(t)dt + C\right] = \frac{1}{(1-t)^2}\left[\int (1-t)^2(1-t)^4 \, dt + C\right]$$

$$= \frac{1}{(1-t)^2}\left[\int (1-t)^6 \, dt + C\right] = \frac{1}{(1-t)^2}\left[-\frac{1}{7}(1-t)^7 + C\right]$$

$$= -\frac{1}{7}(1-t)^5 + \frac{C}{(1-t)^2} \ \text{ or } y = \frac{1}{7}(t-1)^5 + \frac{C}{(t-1)^2}$$

10. $y' - \dfrac{1}{2(1+t)}y = 1 + t \Rightarrow$

$$a(t) = -\frac{1}{2(1+t)}, \ b(t) = 1 + t$$

$$A(t) = -\frac{1}{2}\int \frac{1}{1+t} \, dt = -\frac{1}{2}\ln|1+t| = -\frac{1}{2}\ln(1+t), \ (t \geq 0)$$

$$= \ln\left[(1+t)^{-1/2}\right]$$

$$e^{A(t)} = e^{\ln\left[(1+t)^{-1/2}\right]} = (1+t)^{-1/2}$$

$$e^{-A(t)} = e^{-\ln\left[(1+t)^{-1/2}\right]} = (1+t)^{1/2} \quad y = e^{-A(t)}\left[\int e^{A(t)}b(t)dt + C\right]$$

$$= (1+t)^{1/2}\left[\int (1+t)^{-1/2}(1+t)dt + C\right] = (1+t)^{1/2}\left[\int (1+t)^{1/2}dt + C\right]$$

$$= (1+t)^{1/2}\left[\frac{2}{3}(1+t)^{3/2} + C\right] = \frac{2}{3}(1+t)^2 + C\sqrt{1+t}$$

11. slope $= \dfrac{dy}{dx} = x+y; \; y' = x+y; \; y'-y = x$

$a(x) = -1; \; b(x) = x$

$A(x) = \int (-1)dx = -x; \; e^{A(x)} = e^{-x}; \; e^{-A(x)} = e^x$

$y = e^{-A(x)}\left[\int e^{A(x)}b(x)dx + C\right] = e^x\left[\int e^{-x}(x)dx + C\right] = e^x\left[(-x-1)e^{-x} + C\right] = -x-1+Ce^x$

$y(0) = 0 = 0-1+Ce^0; \; C = 1$

The solution is then $y = -x-1+e^x$.

12. a. $\dfrac{dP}{dt} = k(D-S) = k(12-3.3P)$

$$\left.\frac{dP}{dt}\right|_{D=10,\,S=20} = -1 = k(10-20); \; k = .1$$

$$\frac{dP}{dt} = .1(12-3.3P) = 1.2-.33P$$

b. $\dfrac{1}{1.2-.33P}\dfrac{dP}{dt} = 1$

$$\int \frac{1}{1.2-.33P}dP = \int dt$$

$$-\frac{1}{.33}\ln|1.2-.33P| = t+C$$

$$1.2-.33P = A_1 e^{-.33t} \quad (A_1 \neq 0)$$

$$P = \frac{1.2}{.33} - Ae^{-.33t}$$

$$P(0) = 1 = \frac{1.2}{.33} - Ae^0 \Rightarrow A = \frac{.87}{.33}$$

The solution is $P = \dfrac{1}{.33}(1.2-.87e^{-.33t})$, or $P = \dfrac{1}{11}(40-29e^{-.33t})$.

13. Suppose $f(t)$ is a solution of $y' = (2 - y)e^{-y}$ and t_0 is a number such that $f(t_0) = 3$. If $y = f(t_0) = 3$, then

$$y' = (2 - y)e^{-y}$$
$$= (2 - 3)e^{-3}$$
$$= -e^{-3} = \frac{-1}{e^3}.$$

This is a negative number, so we conclude that y' is negative when $y = 3$. Thus $f(t)$ is decreasing when $t = t_0$, i.e., $f(t)$ is decreasing when $f(t) = 3$.

14. Note that the constant solution $y = 1$ is a solution to the given equation. Since this solution satisfies the initial condition $y(0) = 1$, it must be the desired particular solution.

15. $z = 2 \cos y$

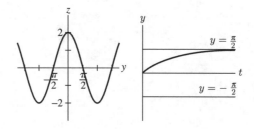

16. $z = 5 + 4y - y^2$

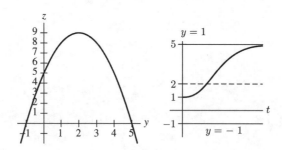

17. $z = y^2 + y$

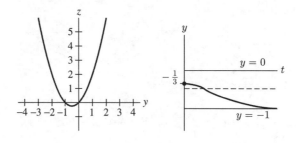

18. $z = y^2 - 2y + 1$

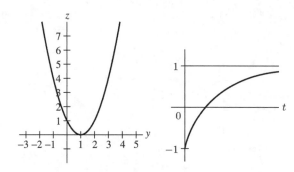

19. From $y' = \ln y$ we obtain the function $z = \ln y,$ whose graph is shown in Figure 1. Thus $g(y)$ has a zero at $y = 1.$ Therefore, the constant function $y = 1$ is a solution of the differential equation.

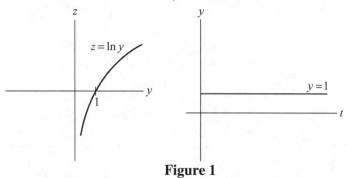

Figure 1

To sketch the solution corresponding to $y(0) = 2,$ observe that when $y = 2,$ $g(y)$ is positive. Hence the solution is increasing as it leaves the initial point. As y moves to the right, the z-coordinate gets more positive, and hence the solution is concave up, and will continue to increase without bound. See Figure 2.

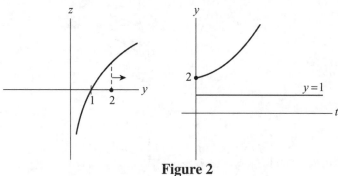

Figure 2

20. $z = \cos y + 1$

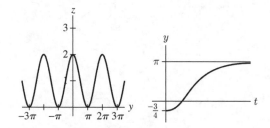

21. $z = \dfrac{1}{y^2 + 1}$

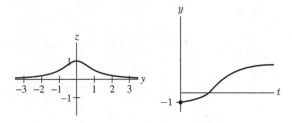

22. $z = \dfrac{3}{y + 3}$

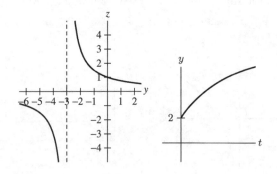

23. $z = .4y^2(1 - y)$

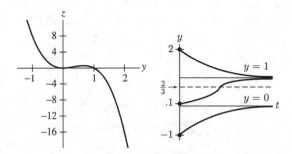

24. $z = y^3 - 6y^2 + 9y$

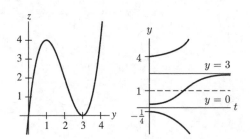

25. (a) Let $N = f(t)$ be the city's population at time t. Then $f'(t)$ is the rate of change of the population of the city at time t. Since the birth rate is 3.5%, and the death rate is 2%, and 3000 people leave the city each year,

$$f'(t) = .035 f(t) - .02 f(t) - 3000$$
$$= .015 f(t) - 3000,$$

or equivalently,

$$N' = .015N - 3000.$$

(b) We seek a constant function N which satisfies the differential equation

$$N' = .015N - 3000$$
$$= .015(N - 200,000).$$

Clearly, the constant function $N = 200,000$ satisfies the equation since it makes both sides equal to zero. However, it is unlikely that a city could maintain this constant population in practice, for the graph indicates that the constant solution is unstable.

Note that $z = .015N - 3000$ is the graph of a line with positive slope, and with a zero at $N = 200,000$. Using the method of Section 10.4, we obtain

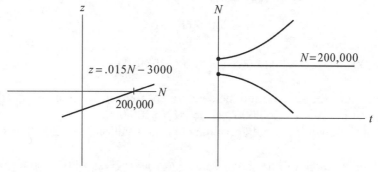

Figure 3

26. (a) $\left(10 - \dfrac{1}{4} y\right)$ represents the amount of unreacted substance A present and $\left(15 - \dfrac{3}{4} y\right)$ that of B.

(b) $k > 0$ since the amount of C is increasing.

(c)

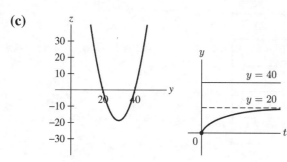

27. Let $f(t)$ be the balance in the account after t years. Then $f(t)$ satisfies the differential equation
$y' = .05y - 2000$, $y(0) = 20,000$. Thus $y' = .05(y - 40,000)$.

$$\int (y - 40,000)^{-1}\, dy = \int .05\, dt$$

$$\ln|y - 40,000| = .05t + C$$

$$y = 40,000 + Ae^{.05t}$$

$y(0) = 20,000 = 40,000 + A$, so $A = -20,000$ and $f(t) = 40,000 - 20,000e^{.05t}$.

We want to find t so that $f(t) = 0$, i.e. $40,000 = 20,000e^{.05t}$,

$2 = e^{.05t}$, $\ln 2 = .05t$

$t = 20 \ln 2 \approx 13.86294$ years.

28. Let $f(t)$ be the balance in the savings account after t years and let $M = f(0)$ be the initial amount. Then $f(t)$ satisfies the differential equation
$y' = .06y - 12,000 = .06(y - 200,000)$,

$y(0) = M$. Thus, $\int (y - 200,000)^{-1}dy = \int .06\, dt$

$$\ln|y - 200,000| = .06t + C$$

$$y = 200,000 + Ae^{.06t}$$

Now $y(0) = M = 200,000 + A$, so
$A = M - 200,000$ and $f(t) = 200,000 + (M - 200,000)e^{.06t}$.

a. If the initial amount M is to fund the endowment forever, we must have $f(t) > 0$ for all t. This happens first in case $M - 200,000 \geq 0 \Rightarrow M \geq \$200,000$.
(A \$200,000 endowment would give the constant solution $f(t) = 200,000$.)

b. Using the expression for $f(t)$ above, setting $f(20) = 0$ gives $-200,000 = (M - 200,000)e^{1.2} \Rightarrow$

$-\dfrac{200,000}{e^{1.2}} + 200,000 = M \Rightarrow M \approx \$139,761.16$.

29. *Euler's Method*: $y' = 2e^{2t-y}$, $y(0) = 0$, $n = 4$; $h = .5$. The iterates are

t_1	y_1
0	0
.5	1
1	2
1.5	3
2	4

So the estimate is $f(2) \approx 4$.

Exact Solution: $\int e^y dy = \int 2e^{2t} dt \Rightarrow e^y = e^{2t} + C \Rightarrow$

$y = \ln(e^{2t} + C)$

$y(0) = 0 = \ln(1 + C)$, so $C = 0$ and the exact solution is $y = \ln(e^{2t}) = 2t$. Then $f(2) = 4$ and the estimate above is exact.

30. $y' = \dfrac{t+1}{y}$, $y(0) = 1$

Euler's Method: $n = 3$, $h = \dfrac{1}{3}$. The iterates are

t_1	y_1
0	1
$\frac{1}{3}$	$\frac{4}{3}$
$\frac{2}{3}$	$\frac{5}{3}$
1	2

So $y(1) \approx 2$.
Exact Solution:

$$\int y\,dy = \int (t+1)\,dt$$

$$\frac{y^2}{2} = \frac{t^2}{2} + t + C_1$$

$$y = \pm\sqrt{t^2 + 2t + C}$$

$y(0) = 1 = \sqrt{C}$, so $C = 1$ and the exact solution is $y = \sqrt{t^2 + 2t + 1} = \sqrt{(t+1)^2} = |t+1|$.
Thus $y(1) = 2$ and the above estimate is exact.

31. Here $g(t, y) = .1y(20 - y)$, $a = 0$, $b = 3$, $y_0 = 2$, and $h = \dfrac{3-0}{6} = .5$. Compute:

$t_0 = 0$, $y_0 = 2$, $g(0, 2) = .2(20 - 2) = 3.6$,

$t_1 = .5$, $y_1 = 2 + (3.6)(.5) = 3.8$, $g(1/2, 3.8) = .38(20 - 3.8) = 6.16$,

$t_2 = 1$, $y_2 = 3.8 + (6.16)(.5) = 6.9$, $g(1, 6.9) = .69(20 - 6.9) = 9.04$,

$t_3 = 1.5$, $y_3 = 6.9 + (9.04)(.5) = 11.4$, $g(3/2, 11.4) = 1.14(20 - 11.4) = 9.8$,

$t_4 = 2$, $y_4 = 11.4 + (9.8)(.5) = 16.3$, $g(2, 16.3) = 1.63(20 - 16.3) = 6.03$,

$t_5 = 2.5$, $y_5 = 16.3 + (6.03)(.5) = 19.3$, $g(5/2, 19.3) = 1.93(20 - 19.3) = 1.35$,

$t_6 = 3$, $y_6 = 19.3 + (1.35)(.5) = 19.98$.

The polygonal path connecting the points $(t_0, y_0), \ldots, (t_6, y_6)$ is shown below.

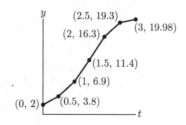

32. $y' = \dfrac{1}{2} y(y-10)$; $y(0) = 9$; $n = 5, h = .2$

t_1	y_1
0	9
.2	8.1
.4	6.561
.6	4.30467
.8	1.85302
1	.34337

$y(1) \approx .34337$

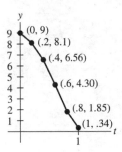

Chapter 11
Taylor Polynomials and Infinite Series

11.1 Taylor Polynomials

Taylor polynomials are often used in applications to approximate more complex functions, and they have an important connection with the Taylor series, to be discussed in Section 11.5.

The phrase "at $x = 0$" in "Taylor polynomial of $f(x)$ at $x = 0$" indicates that the coefficients of the polynomial are computed by evaluating $f(x)$ and its derivatives at $x = 0$. The phrase "at $x = 0$" does *not* mean that the x in the polynomial must be zero. However, in an application, the values of x are usually taken to be close to zero. In such cases, the remainder theorem is useful for estimating how close the values of a Taylor polynomial are to the values of the function $f(x)$.

1. $f(x) = \sin x, \qquad f(0) = 0,$

    $f'(x) = \cos x, \qquad f'(0) = 1,$

    $f''(x) = -\sin x, \qquad f''(0) = 0,$

    $f^{(3)}(x) = -\cos x, \qquad f^{(3)}(0) = -1.$

    Therefore, the third Taylor polynomial at $x = 0$ is

    $$p_3(x) = 0 + \frac{1}{1!}x + \frac{0}{2!}x^2 + \frac{-1}{3!}x^3$$

    $$= x - \frac{1}{6}x^3.$$

7. $f(x) = xe^{3x}, \qquad\qquad\qquad f(0) = 0,$

    $f'(x) = \underbrace{3xe^{3x} + e^{3x}}_{\text{from product rule}}, \qquad\qquad f'(0) = 1,$

    $f''(x) = 9xe^{3x} + 3e^{3x} + 3e^{3x}, \qquad f''(0) = 6,$

    $f^{(3)}(x) = \underbrace{27xe^{3x} + 9e^{3x}}_{\text{from product rule}} + 9e^{3x} + 9e^{3x} \quad f^{(3)}(0) = 27.$

    Therefore,

    $$p_3(x) = 0 + \frac{1}{1!}x + \frac{6}{2!}x^2 + \frac{27}{3!}x^3$$

    $$= x + 3x^2 + \frac{9}{2}x^3.$$

13. Since

$$f(x) = f'(x) = f''(x) = \cdots = f^{(n)}(x) = e^x,$$

$$f(0) = f'(0) = f''(0) = \cdots = f^{(n)}(0) = e^0 = 1.$$

Therefore, the *n*th Taylor polynomial for $f(x) = e^x$ at $x = 0$ is:

$$p_n(x) = 1 + \frac{1}{1!}x + \frac{1}{2!}x^2 + \frac{1}{3!}x^3 + \cdots + \frac{1}{n!}x^n$$

$$= 1 + x + \frac{1}{2}x^2 + \frac{1}{3!}x^3 + \cdots + \frac{1}{n!}x^n.$$

Helpful Hint: If your class plans to cover Section 11.5, then you should memorize the formulas for the Taylor polynomials of e^x and $\dfrac{1}{1-x}$, as described in Examples 2 and 3.

19.
$$f(x) = \cos x, \qquad f(\pi) = -1,$$
$$f'(x) = -\sin x, \qquad f'(\pi) = 0,$$
$$f''(x) = -\cos x, \qquad f''(\pi) = 1,$$
$$f^{(3)}(x) = \sin x, \qquad f^{(3)}(\pi) = 0,$$
$$f^{(4)}(x) = \cos x, \qquad f^{(4)}(\pi) = -1.$$

Thus the third Taylor polynomial at $x = \pi$ is

$$p_3(x) = -1 + \frac{0}{1!}(x-\pi) + \frac{1}{2!}(x-\pi)^2 + \frac{0}{3!}(x-\pi)^3$$

$$= -1 + \frac{1}{2}(x-\pi)^2,$$

and the fourth Taylor polynomial is

$$p_4(x) = -1 + \frac{0}{1!}(x-\pi) + \frac{1}{2!}(x-\pi)^2 + \frac{0}{3!}(x-\pi)^3 + \frac{-1}{4!}(x-\pi)^4$$

$$= -1 + \frac{1}{2}(x-\pi)^2 - \frac{1}{24}(x-\pi)^4.$$

Notice that $p_3(x)$ is a polynomial of degree 2, not degree 3, because $f^{(3)}(\pi) = 0$. See Practice Problem 1.

25. Be sure to try the Practice Problems before you attempt this problem. You should realize from Practice Problem 2 that if $f(x)$ is a polynomial of degree 3, then the third Taylor polynomial of $f(x)$ at $x = 0$ is $f(x)$ itself. This is true because $f(x)$ is a polynomial that agrees with $f(x)$ and all of its derivatives at $x = 3$.

Now write the abstract *formula* for the third Taylor polynomial of $f(x)$, set that equal to $f(x)$, and equate corresponding coefficients. From this you should be able to discover what the values of $f''(0)$ and $f'''(0)$ must be. Try to do this before reading the rest of this solution.

The third Taylor polynomial of $f(x)$ is

$$p_3(x) = f(0) + \frac{f'(0)}{1}x + \frac{f''(0)}{2!}x^2 + \frac{f'''(0)}{3!}x^3$$

and

$$f(x) = 3 + 4x + \left(-\frac{5}{2!}\right)x^2 + \left(\frac{7}{3!}\right)x^3.$$

Since $p_3(x)$ equals $f(x)$ for all x, the polynomials must have the same coefficients. That is, $f(0) = 3$, $f'(0) = 4$, $\frac{f''(0)}{2!} = -\frac{5}{2!}$ and $\frac{f'''(0)}{3!} = \frac{7}{3!}$. Accordingly, $f''(0) = -5$ and $f'''(0) = 7$.

31. From Example 3, you should know that the fourth Taylor polynomial of $Y_1 = 1/(1-X)$ is the polynomial given by

$$\mathbf{Y_2 = 1 + X + X^2 + X^3 + X^4.}$$

When you graph Y_1 and Y_2 in the window $[-1, 1]$ by $[-1, 5]$, you will probably find that their graphs appear identical for x between about $-.68$ and $.53$, or perhaps $.55$. So b is about $.55$.

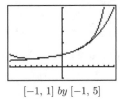

$[-1, 1]$ by $[-1, 5]$

You can use **TRACE** to display the y-values of points on the two curves with the same x-coordinates. At $x = .5106383$, for instance, $Y_1 - Y_2 \approx 2.04348 - 1.97253 = .07095$. Another method for comparing values of Y_1 and Y_2 is shown in Figure 5 on page 518 in the text.

11.2 The Newton-Raphson Algorithm

A calculator is indispensable for this section. In order to memorize the Newton-Raphson algorithm, it is important to work several problems by hand using a calculator. If you have access to an IBM-compatible computer, you should experiment with Visual Calculus, which will show you geometrically how the algorithm works.

1. $\sqrt{5}$ is a zero of the function $x^2 - 5$. Note that $\sqrt{5}$ lies between 2 and 3 because $2^2 < 5 < 3^2$. For an initial approximation, take $x_0 = 2$. (The value $x_0 = 3$ would work just as well.) Since $f'(x) = 2x$, we have

$$x_1 = x_0 - \frac{x_0^2 - 5}{2x_0} = 2 - \frac{(2)^2 - 5}{2(2)} = 2 - \frac{-1}{4} = \frac{9}{4} = 2.25$$

$$x_2 = 2.25 - \frac{(2.25)^2 - 5}{2(2.25)} = 2.25 - \frac{.0625}{4.5} = 2.2361$$

$$x_3 = 2.2361 - \frac{(2.2361)^2 - 5}{2(2.2361)} = 2.2361 - \frac{.00014}{4.4722} = 2.23607.$$

7. Here $f(x) = \sin x + x^2 - 1$ and $x_0 = 0$. First, compute $f'(x) = \cos x + 2x$, and then use the Newton-Raphson algorithm:

$$x_1 = x_0 - \frac{\sin x_0 + x_0^2 - 1}{\cos x_0 + 2x_0} = 0 - \frac{\sin 0 + 0^2 - 1}{\cos 0 + 2(0)} = 0 - \frac{-1}{1} = 1$$

$$x_2 = 1 - \frac{\sin 1 + 1^2 - 1}{\cos(1) + 2(1)} = 1 - \frac{.84147}{2.54030} = .66875$$

$$x_3 = .66875 - \frac{.62001 + .44723 - 1}{.78460 + 1.3375} = .66875 - .03168 = .63707.$$

13. Let i be the monthly rate of interest. The present value of an amount A to be received in k months is $A(1 + i)^{-k}$. Therefore, you must solve the following equation for i:

$$\begin{bmatrix} \text{amount of initial} \\ \text{investment} \end{bmatrix} = \begin{bmatrix} \text{sum of present} \\ \text{values of returns} \end{bmatrix}$$

$$500 = 100(1+i)^{-1} + 200(1+i)^{-2} + 300(1+i)^{-3}.$$

Multiply both sides by $(1 + i)^3$, take all terms to the left, and obtain

$$500(1+i)^3 - 100(1+i)^2 - 200(1+i) - 300 = 0.$$

Let $x = 1 + i$, and solve the resulting equation by the Newton-Raphson algorithm with $x_0 = 1.1$.

$$f(x) = 500x^3 - 100x^2 - 200x - 300 = 0,$$

$$f'(x) = 1500x^2 - 200x - 200.$$

$$x_1 = 1.1 - \frac{500(1.1)^3 - 100(1.1)^2 - 200(1.1) - 300}{1500(1.1)^2 - 200(1.1) - 200}$$

$$\approx 1.1 - .018 = 1.082$$

$$x_2 = 1.082 - \frac{500(1.082)^3 - 100(1.082)^2 - 200(1.082) - 300}{1500(1.082)^2 - 200(1.082) - 200}$$

$$\approx 1.082 - (-.00008) \approx 1.082.$$

Therefore, the solution is $x = 1.082$. Hence, $i = .082$ and the investment had an internal rate of return of 8.2% per month.

19. The slope of the tangent line is 4, so $f'(3) = 4$. Also, $f(3) = 4(3) + 5 = 17$. If the initial guess is 3, then the Newton-Raphson algorithm produces:

$$x_1 = 3 - \frac{f(3)}{f'(3)} = 3 - \frac{17}{4} = \frac{-5}{4}.$$

25. When $f(x) = x^{1/3}$, and $f'(x) = \frac{1}{3}x^{-2/3}$, the Newton-Raphson formula is

$$x_1 = x_0 - \frac{f(x_0)}{f'(x_0)} = x_0 - \frac{x_0^{1/3}}{\left(\frac{1}{3}\right)x_0^{-2/3}}$$

$$= x_0 - 3x_0 = -2x_0.$$

Similarly, $x_2 = -2x_1, x_3 = -2x_2, \ldots$, and $x_{n+1} = -2x_n$. If $x_0 = 1$, then the Newton-Raphson algorithm produces the sequence 1, -2, 4, -8, and so on. This sequence does *not* converge to the actual root, $r = 0$. Can you draw a picture, based on Figure 10(a) in the text, to show what is happening?

11.3 Infinite Series

Most students who take this calculus course will see an application of infinite series in some other course. More often than not, the application will involve a geometric series.

Make sure that you completely master the material in this section. It will help you to understand the concepts in the sections to follow.

1. $1 + \frac{1}{6} + \frac{1}{6^2} + \frac{1}{6^3} + \frac{1}{6^4} + \cdots = \left(\frac{1}{6}\right)^0 + \left(\frac{1}{6}\right)^1 + \left(\frac{1}{6}\right)^2 + \left(\frac{1}{6}\right)^3 + \left(\frac{1}{6}\right)^4 + \cdots.$

Thus $a = 1$, $r = \frac{1}{6}$, and the series converges to

$$\frac{a}{1-r} = \frac{1}{1-\frac{1}{6}} = \frac{1}{\frac{5}{6}} = \frac{6}{5}.$$

7. Find r by dividing any term by the preceding term. So,

$$r = \frac{\frac{1}{5^4}}{\frac{1}{5}} = \frac{5}{5^4} = \frac{1}{5^3} = \frac{1}{125}.$$

The first term of the series is $a = \frac{1}{5}$, so the sum of the series is

$$\frac{\frac{1}{5}}{1 - \frac{1}{125}} = \frac{1}{5} \cdot \frac{125}{124} = \frac{25}{124}.$$

13. Find r by dividing any term by the preceding term. So,

$$r = \frac{4}{5}.$$

The first term of the series is $a = 5$, so the sum of the series is

$$\frac{5}{1 - \frac{4}{5}} = 5 \cdot 5 = 25.$$

19. The technique here is to find a rational number that represents $.01\overline{1011}$, and then to add 4 to this number. First, write

$$.01\overline{1011} = .011 + .000011 + .000000011 + \ldots$$

$$= \frac{11}{1000} + \frac{11}{1000^2} + \frac{11}{1000^3} + \cdots,$$

which is a geometric series with $a = \frac{11}{1000}$ and $r = \frac{1}{1000}$. The sum of the series is:

$$\frac{\frac{11}{1000}}{1 - \frac{1}{1000}} = \frac{11}{1000} \cdot \frac{1000}{999} = \frac{11}{999}.$$

Then, add 4 to the sum.

$$4.01\overline{1011} = 4 + \frac{11}{999} = \frac{3996}{999} + \frac{11}{999} = \frac{4007}{999}.$$

25. (a) Construct a sum of the present value of all future payments. That is, express capital value as

$$100 + 100(1.01)^{-1} + 100(1.01)^{-2} + \cdots = \sum_{k=0}^{\infty} 100(1.01)^{-k}.$$

(b) This is a geometric series with $a = 100$ and $r = (1.01)^{-1} = \frac{1}{1.01}$, (so $|r| < 1$). So the sum of the series is

$$\frac{100}{1 - \frac{1}{1.01}} = 100 \frac{1.01}{.01} = 10,100.$$

Thus the capital value of the perpetuity is $10,100.

31. At the end of the first "day" (a 24-hour period), 25% of the original dose of M mg has been eliminated, and only .75M remains. Immediately after the next dose, the body contains $M + .75M$ mg of the drug. At the end of the second day, only 75% of that $M + .75M$ remains, which may be written as .75($M + .75M$), or $.75M + (.75)^2 M$. When a dose of M mg is given again, the new amount in the body is

$$M + .75M + (.75)^2 M \qquad \text{(after the third dose)}.$$

After n full days, just after the next dose is given, the amount in the body is

$$M + .75M + (.75)^2 M + \cdots + (.75)^n M \quad \text{milligrams}.$$

This is a partial sum of a geometric series whose initial term is M and whose ratio is $r = .75$. When n is large, the value of this partial sum will be very close to the sum of the infinite series

$$\sum_{k=0}^{\infty} M(.75)^k = \frac{M}{1 - .75} = \frac{M}{.25} = 4M.$$

To make this amount equal to 20 mg, M must be 5 mg. This is the "maintenance dose."

37. $\sum_{j=1}^{\infty} 5^{-2j} = \frac{1}{5^2} + \frac{1}{5^4} + \frac{1}{5^6} + \frac{1}{5^8} + \cdots$.

This is a geometric series with $a = 1/5^2$ and $r = 1/5^2$. So the sum of the series is

$$\frac{\frac{1}{25}}{1 - \frac{1}{25}} = \frac{1}{25} \cdot \frac{25}{24} = \frac{1}{24}.$$

43. The first entry in the **sum(seq(. . .))** command displays the general term, (2/3)^X, from which you can see that $a = 2/3$ and the ratio is $r = 2/3$. The remaining entries in the **sum(seq(. . . X, 0, 10, 1))** command show that the "counter" X, begins at 0 and ends at 10, with increment (step size) 1. (On the TI-83 and TI-86, the value for the increment may be omitted.) Thus the calculator is computing the partial sum

$$\left(\frac{2}{3}\right)^0 + \left(\frac{2}{3}\right)^1 + \left(\frac{2}{3}\right)^2 + \left(\frac{2}{3}\right)^3 + \cdots + \left(\frac{2}{3}\right)^{10} = 1 + \left(\frac{2}{3}\right)^1 + \left(\frac{2}{3}\right)^2 + \left(\frac{2}{3}\right)^3 + \cdots + \left(\frac{2}{3}\right)^{10}.$$

The exact sum of the corresponding geometric series is $1 + \dfrac{a}{1 - r} = 1 + \dfrac{\frac{2}{3}}{1 - \frac{2}{3}} = 1 + \dfrac{\frac{2}{3}}{\frac{1}{3}} = 3$.

49. Unfortunately, the T1-82 **sum(seq(. . .))** command is limited to 99 terms, so let's begin there. You should find that

$$\text{sum(seq(X\^{}-2, X, 1, 99, 1))} \approx 1.634884$$

(to six decimal places). Also, $\pi^2/6 = 1.644934,$ which is a difference of about .01. With 99 replaced by 999, the computation is much slower, but the result 1.6439934 differs from $\pi^2/6$ only by about .001.

The result on a TI-84 is shown below.

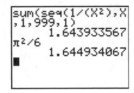

11.4 Series with Positive Terms

The integral test provides a nice link between infinite series and improper integrals. Your understanding of both topics should be strengthened by studying them together. It may be helpful to review Section 9.6 before you begin the exercises in Section 11.4.

Problems involving the comparison test can be very difficult. Knowing what series to use for comparison is a skill that requires a lot of experience—more than we expect you to obtain in this course. We have made Exercises 21–26 more reasonable by suggesting one or two series for comparison.

1. To apply the integral test, compute

$$\int_1^\infty \frac{3}{\sqrt{x}}\, dx = \int_1^\infty 3x^{-1/2}\, dx = \lim_{b \to \infty} \int_1^b 3x^{-1/2}\, dx$$

$$= \lim_{b \to \infty} \left(6x^{1/2}\Big|_1^b \right) = \lim_{b \to \infty} (6b^{1/2} - 6) = \infty.$$

The series

$$\sum_{k=1}^\infty \frac{3}{\sqrt{k}}$$

is divergent, because the corresponding improper integral is divergent.

7. Study the improper integral

$$\int_2^\infty \frac{x}{(x^2 + 1)^{3/2}}\, dx.$$

Begin by finding an antiderivative of $x(x^2 + 1)^{-3/2}$ using the method of substitution. Set $u = x^2 + 1$ and $du = 2x\, dx$. Then

$$\int \frac{x}{(x^2 + 1)^{3/2}}\, dx = \frac{1}{2} \int (x^2 + 1)^{-3/2} \cdot 2x\, dx = \frac{1}{2} \int u^{-3/2} du$$

$$= \frac{1}{2}(-2u^{-1/2}) = -(x^2 + 1)^{-1/2} + C.$$

Next,

$$\int_2^b \frac{x}{(x^2+1)^{3/2}}\,dx = -(x^2+1)^{-1/2}\Big|_2^b = -\frac{1}{\sqrt{b^2+1}} - \left(-\frac{1}{\sqrt{2^2+1}}\right).$$

Finally,

$$\int_2^\infty \frac{x}{(x^2+1)^{3/2}}\,dx = \lim_{b\to\infty}\left(\frac{1}{\sqrt{5}} - \frac{1}{\sqrt{b^2+1}}\right) = \frac{1}{\sqrt{5}}.$$

Since the improper integral is convergent, the corresponding infinite series is also convergent.

Warning: The sum of the series in Exercise 7 is *not* $\dfrac{1}{\sqrt{5}}$. The integral test only shows that the series converges.

13. Evaluate

$$\int_1^\infty xe^{-x^2}\,dx.$$

To find an antiderivative for xe^{-x^2}, either guess e^{-x^2} and adjust the guess to obtain $-\frac{1}{2}e^{-x^2}$ or use substitution with $u=-x^2$, $du=-2x\,dx$, and compute

$$\int xe^{-x^2}\,dx = -\frac{1}{2}\int\left(-2xe^{-x^2}\right)dx = -\frac{1}{2}\int e^u\,du = -\frac{1}{2}e^u + C = -\frac{1}{2}e^{-x^2} + C.$$

Next, compute

$$\int_1^\infty xe^{-x^2}\,dx = \lim_{b\to\infty}\int_1^b xe^{-x^2}\,dx = \lim_{b\to\infty}\left(-\frac{1}{2}e^{-x^2}\Big|_1^b\right)$$

$$= \lim_{b\to\infty}\left(-\frac{1}{2}e^{-b^2} - \left(-\frac{1}{2}e^{-1}\right)\right)$$

$$= \lim_{b\to\infty}\left(-\frac{1}{2}e^{-b^2} + \frac{1}{2}e^{-1}\right) = \frac{1}{2}e^{-1}.$$

Since the improper integral is convergent, so is the infinite series $\displaystyle\sum_{k=1}^\infty ke^{-k^2}$.

19. Evaluate the integral

$$\int_1^\infty \frac{x}{e^x}\,dx = \int_1^\infty xe^{-x}\,dx$$

using integration by parts. Set

$$f(x) = x, \quad g(x) = e^{-x},$$
$$f'(x) = 1, \quad G(x) = -e^{-x}.$$

Then

$$\int \frac{x}{e^x}\,dx = -xe^{-x} - \int \left(-e^{-x}\right)dx = -xe^{-x} + \int e^{-x}dx$$

$$= -xe^{-x} - e^{-x} + C$$

and

$$\int_1^b \frac{x}{e^x}\,dx = (-xe^{-x} - e^{-x})\Big|_1^b$$

$$= (-be^{-b} - e^{-b}) - (-e^{-1} - e^{-1})$$

$$= 2e^{-1} - be^{-b} - e^{-b}.$$

For the final step, use the fact (given in the text) that $\lim_{b \to \infty} be^{-b} = 0$.

$$\int_1^\infty \frac{x}{e^x}\,dx = \lim_{b \to \infty} (2e^{-1} - be^{-b} - e^{-b}) = 2e^{-1} - 0 - 0.$$

Since the improper integral is convergent, so is the infinite series $\displaystyle\sum_{k=1}^\infty \frac{k}{e^k}$.

25. First compare

$$\sum_{k=1}^\infty \frac{1}{5^k}\cos^2\left(\frac{k\pi}{4}\right) = \frac{1}{5}\cos^2\left(\frac{\pi}{4}\right) + \frac{1}{25}\cos^2\left(\frac{\pi}{2}\right) + \frac{1}{125}\cos^2\left(\frac{3\pi}{4}\right) + \cdots$$

with

$$\sum_{k=1}^\infty \cos^2\left(\frac{k\pi}{4}\right) = \cos^2\left(\frac{\pi}{4}\right) + \cos^2\left(\frac{\pi}{2}\right) + \cos^2\left(\frac{3\pi}{4}\right) + \cdots.$$

Note that each term of the first series is less than the corresponding term of the second series. Since $\left|\cos\dfrac{k\pi}{4}\right| = \dfrac{1}{\sqrt{2}}$ when k is odd, and $\left|\cos\dfrac{k\pi}{4}\right| = 0$ or 1 (alternately) when k is even, the second series has the form

$$\frac{1}{2} + 0 + \frac{1}{2} + 1 + \frac{1}{2} + 0 + \frac{1}{2} + 1 + \cdots.$$

Since these terms keep repeating, the series diverges. We have learned nothing about the convergence of the first series. Next, compare the first series with

$$\sum_{k=1}^\infty \frac{1}{5^k} = \frac{1}{5} + \frac{1}{5^2} + \frac{1}{5^3} + \frac{1}{5^4} + \cdots,$$

which is a convergent geometric series because $|r| = \frac{1}{5} < 1$. Now, since

$$\frac{1}{5^k}\cos^2\left(\frac{k\pi}{4}\right) \le \frac{1}{5^k} \quad \text{for all values of } k,$$

each term of $\sum 5^{-k}$ is greater than the corresponding term of the first series shown above. Since $\sum 5^{-k}$ converges, so does the first series.

31. $\displaystyle\sum_{k=0}^{\infty}\frac{8^k+9^k}{10^k}=\frac{8^0+9^0}{10^0}+\frac{8^1+9^1}{10^1}+\frac{8^2+9^2}{10^2}+\cdots$

$$=\left(\frac{8^0}{10^0}+\frac{9^0}{10^0}\right)+\left(\frac{8^1}{10^1}+\frac{9^1}{10^1}\right)+\left(\frac{8^2}{10^2}+\frac{9^2}{10^2}\right)+\cdots$$

$$=\sum_{k=0}^{\infty}\left(\frac{8^k}{10^k}+\frac{9^k}{10^k}\right)$$

$$=\sum_{k=0}^{\infty}\frac{8^k}{10^k}+\sum_{k=0}^{\infty}\frac{9^k}{10^k}\quad\text{by Exercise 29,}$$

$$=\sum_{k=0}^{\infty}\left(\frac{8}{10}\right)^k+\sum_{k=0}^{\infty}\left(\frac{9}{10}\right)^k.$$

Each term in this sum is a convergent geometric series. The first has $a=1$ and $r=.8$, and its sum is $\dfrac{a}{1-r}=\dfrac{1}{1-.8}=5$. The second has $a=1$ and $r=.9$, and its sum is $\dfrac{1}{1-.9}=10$. By Exercise 29, the original series is convergent, and its sum is $5+10=15$.

11.5 Taylor Series

One of the main ideas in this section is that a Taylor series is a function. Be sure to read the text on pages 540–541 and 544. You need to know the general formula for a Taylor series at $x=0$ and the specific formulas for e^x and $1/(1-x)$. Check with your instructor to see if you also need to memorize the Taylor series at $x=0$ for $\cos x$ and $\sin x$. Study Examples 3, 5, and 6 carefully to get ideas for the solutions of Exercises 5–27.

1. Compute:

$$f(x)=\frac{1}{2x+3}=(2x+3)^{-1},\qquad f(0)=\frac{1}{3},$$

$$f'(x)=-2(2x-3)^{-2},\qquad f'(0)=\frac{-2}{9}=-\frac{2}{9},$$

$$f''(x)=8(2x+3)^{-3},\qquad f''(0)=\frac{8}{27},$$

$$f^{(3)}(x)=-48(2x+3)^{-4},\qquad f^{(3)}(0)=\frac{-48}{81}=-\frac{16}{27},$$

$$f^{(4)}(x)=384(2x+3)^{-5},\qquad f^{(4)}(0)=\frac{384}{243}=\frac{128}{81}.$$

Then write

$$f(x) = \frac{1}{3} + \frac{-\frac{2}{9}}{1!}x + \frac{\frac{8}{27}}{2!}x^2 + \frac{-\frac{16}{27}}{3!}x^3 + \frac{\frac{128}{81}}{4!}x^4 + \cdots$$

$$= \frac{1}{3} - \frac{2}{9}x + \frac{4}{27}x^2 - \frac{8}{81}x^3 + \frac{16}{243}x^4 + \cdots$$

$$= \frac{1}{3} - \frac{2}{9}x + \frac{2^2}{3^3}x^2 - \frac{2^3}{3^4}x^3 + \frac{2^4}{3^5}x^4 - \cdots$$

7. In the Taylor series at $x = 0$ for $\dfrac{1}{1-x}$, replace x by $-x^2$ to obtain

$$\frac{1}{1-(-x^2)} = \frac{1}{1+x^2}$$

$$= 1 - x^2 + x^4 - x^6 + \cdots.$$

13. In the Taylor series at $x = 0$ for e^x, replace x by $-x$ to obtain

$$e^{-x} = 1 + (-x) + \frac{1}{2!}(-x)^2 + \frac{1}{3!}(-x)^3 + \frac{1}{4!}(-x)^4 + \cdots.$$

Hence,

$$1 - e^{-x} = 1 - \left(1 - x + \frac{1}{2!}x^2 - \frac{1}{3!}x^3 + \frac{1}{4!}x^4 - \cdots \right)$$

$$= x - \frac{1}{2!}x^2 + \frac{1}{3!}x^3 - \frac{1}{4!}x^4 + \cdots.$$

19. In the Taylor series at $x = 0$ for $\cos x$, replace x by $3x$ to obtain

$$\cos 3x = 1 - \frac{9}{2!}x^2 + \frac{81}{4!}x^4 - \frac{729}{6!}x^6 + \cdots.$$

Now differentiate both sides and divide by -3 to obtain

$$-3\sin 3x = -\frac{18}{2!}x + \frac{4(81)}{4!}x^3 - \frac{6(729)}{6!}x^5 + \cdots.$$

$$\sin 3x = \frac{6}{2!}x - \frac{4(27)}{4!}x^3 + \frac{6(243)}{6!}x^5 - \cdots$$

$$= 3x - \frac{27}{3!}x^3 + \frac{243}{5!}x^5 - \cdots$$

$$= 3x - \frac{3^3}{3!}x^3 + \frac{3^5}{5!}x^5 - \cdots.$$

25. In the Taylor series expansion at $x = 0$ for $\dfrac{1}{\sqrt{1+x}}$, replace x by $-x$ to obtain

$$\frac{1}{\sqrt{1-x}} = 1 + \frac{1}{2}x + \frac{1 \cdot 3}{2 \cdot 4}x^2 + \frac{1 \cdot 3 \cdot 5}{2 \cdot 4 \cdot 6}x^3.$$

31. If you are reading this before you work the exercise, stop and study Practice Problem 4. Then try the exercise again before reading on with the solution. We observe that the term in the series containing x^5 is

$$\frac{f^{(5)}(0)}{5!}x^5 = \frac{2}{5}x^5.$$

So

$$\frac{f^{(5)}(0)}{5!} = \frac{2}{5}, \quad \text{and} \quad f^{(5)}(0) = (5!)\frac{2}{5} = (4!)2 = 24 \cdot 2 = 48.$$

37. In the Taylor expansion at $x = 0$ for $\frac{1}{1-x}$ replace x by $-x^3$ to obtain

$$\frac{1}{1+x^3} = 1 - x^3 + x^6 - x^9 + \cdots.$$

Then integrate both sides, term by term:

$$\int \frac{1}{1+x^3}\,dx = \int (1 - x^3 + x^6 - x^9 + \cdots)\,dx$$

$$= \left[x - \frac{1}{4}x^4 + \frac{1}{7}x^7 - \frac{1}{10}x^{10} + \cdots \right] + C.$$

43. Since $e^x = 1 + x + \frac{1}{2}x^2 + \frac{1}{6}x^3 + \cdots$, for all x, we have $e^x > \frac{1}{6}x^3$ when $x > 0$. The idea now is to find a function which approaches 0 as $x \to \infty$, with the property that for every value of x greater than 0, this function is greater than the function $x^2 e^{-x}$. From the inequality above, we have

$$\frac{1}{e^x} < \frac{6}{x^3}$$

and so

$$x^2 e^{-x} < x^2 \cdot \frac{6}{x^3} = \frac{6}{x}.$$

As $x \to \infty$, $\dfrac{6}{x}$ approaches 0. Since $\dfrac{6}{x} > x^2 e^{-x}$ for all $x > 0$, it follows that $x^2 e^{-x}$ must approach 0 as $x \to \infty$.

Chapter 11 Review Exercises

While solutions for all review exercises are included, expanded explanations are included for every sixth exercise.

1. $f(x) = x(x+1)^{3/2},$ $f(0) = 0,$

$f'(x) = \dfrac{3}{2}x(x+1)^{1/2} + (x+1)^{3/2},$ $f'(0) = 1,$

$f''(x) = \dfrac{3}{4}x(x+1)^{-1/2} + \dfrac{3}{2}(x+1)^{1/2} + \dfrac{3}{2}(x+1)^{1/2}$

$\qquad = \dfrac{3}{4}x(x+1)^{-1/2} + 3(x+1)^{1/2}.$ $f''(0) = 3.$

Hence,

$$p_2(x) = 0 + \frac{1}{1!}x + \frac{3}{2!}x^2 = x + \frac{3}{2}x^2.$$

2. $f(x) = (2x+1)^{3/2};\ f'(x) = 3(2x+1)^{1/2}$

$f''(x) = 3(2x+1)^{-1/2};\ f'''(x) = -3(2x+1)^{-3/2};$

$f^{(4)}(x) = 9(2x+1)^{-5/2}$

$$p_4(x) = 1 + 3x + \frac{3x^2}{2!} - \frac{3x^3}{3!} + \frac{9x^4}{4!} = 1 + 3x + \frac{3}{2}x^2 - \frac{1}{2}x^3 + \frac{3}{8}x^4$$

3. For all $n \ge 3$, $p_n(x) = x^3 - 7x^2 + 8.$

4. $f(x) = \dfrac{2}{2-x} = \dfrac{1}{1 - \frac{x}{2}} = 1 + \dfrac{x}{2} + \left(\dfrac{x}{2}\right)^2 + \left(\dfrac{x}{2}\right)^3 + \cdots + \left(\dfrac{x}{2}\right)^n + \cdots$

So $p_n(x) = 1 + \dfrac{x}{2} + \dfrac{1}{2^2}x^2 + \dfrac{1}{2^3}x^3 + \cdots + \dfrac{1}{2^n}x^n.$

5. $f(x) = x^2,\ f'(x) = 2x,\ f''(x) = 2,$

$f^{(n)}(x) = 0$ for $n \ge 3.$

$p_3(x) = 3^2 + 2(3)(x-3) + \dfrac{2}{2!}(x-3)^2 + 0 = 9 + 6(x-3) + (x-3)^2$

6. $f(x) = e^x = f^{(n)}(x)$ for all $n \ge 1.$

$p_3(x) = e^2 + e^2(x-2) + \dfrac{e^2}{2!}(x-2)^2 + \dfrac{e^2}{3!}(x-2)^3$

7. $f(t) = -\ln(\cos 2t),$ $\qquad\qquad\qquad\qquad$ $f(0) = -\ln(1) = 0,$

$\qquad$ $f'(t) = \dfrac{2\sin 2t}{\cos 2t},$ $\qquad\qquad\qquad\qquad$ $f'(0) = 0,$

$\qquad$ $f''(t) = \dfrac{(\cos 2t)(4\cos 2t) + (2\sin 2t)(2\sin 2t)}{\cos^2 2t}$

$\qquad\qquad$ $= \dfrac{4\cos^2 2t + 4\sin^2 2t}{\cos^2 2t} = 4 + \dfrac{4\sin^2 2t}{\cos^2 2t},$ $\quad$ $f''(0) = 4.$

Hence,

$$p_2(t) = 0 + \frac{0}{1!}t + \frac{4}{2!}t^2 = 2t^2.$$

The area under the graph of $y = f(t)$ is approximately equal to the area under the graph of $y = p_2(t)$. Thus,

$$[\text{Area}] \approx \int_0^{1/2} 2t^2\,dt = \frac{2}{3}t^3\bigg|_0^{1/2} = \frac{2}{3}\cdot\frac{1}{8} = \frac{1}{12}.$$

8. $f(x) = \tan x,\ f'(x) = \sec^2 x,$

$\qquad$ $f''(t) = (2\sec x)\sec x\tan x = 2\sec^2 x\tan x$

$\qquad$ $p_2(x) = 0 + x + 0 = x$

$\qquad$ $\tan(.1) \approx p_2(.1) = .1$

9. a. $f(x) = x^{1/2};\ f'(x) = \dfrac{1}{2}x^{-1/2};\ f''(x) = -\dfrac{1}{4}x^{-3/2}$

$\qquad$ $p_2(x) = 3 + \dfrac{1}{6}(x-9) - \dfrac{1}{216}(x-9)^2$

b. $p_2(8.7) = 3 + \dfrac{1}{6}(-.3) - \dfrac{1}{216}(-.3)^2 \approx 2.949583$

c. $f(x) = x^2 - 8.7;\ f'(x) = 2x$

$\qquad$ $x_0 = 3,\ x_1 = 3 - \dfrac{.3}{6} = 2.95,\ x_2 = 2.95 - \dfrac{(2.95)^2 - 8.7}{2(2.95)} \approx 2.949576$

10. a. $f(x) = \ln(1-x);\ f'(x) = -\dfrac{1}{1-x};\ f''(x) = -\dfrac{1}{(1-x)^2};\ f'''(x) = -\dfrac{2}{(1-x)^3}$

$\qquad$ $p_3(x) = 0 - (1)x - \dfrac{1}{2!}x^2 - \dfrac{2}{3!}x^3 = -x - \dfrac{1}{2}x^2 - \dfrac{1}{3}x^3$

$\qquad$ $\ln(1.3) = f(-.3) \approx p_3(-.3) = -(-.3) - \dfrac{1}{2}(-.3)^2 - \dfrac{1}{3}(-.3)^3 = .264$

b. $f(x) = e^x - 1.3;\ f'(x) = e^x;\ x_0 = 0$

$$x_1 = 0 - \frac{e^0 - 1.3}{e^0} = 0.3$$

$$x_2 = 0.3 - \frac{e^{0.3} - 1.3}{e^{0.3}} \approx 0.2631$$

11. $f(x) = x^2 - 3x - 2,\ f'(x) = 2x - 3$
$x_0 = 4$

$$x_1 = 4 - \frac{4^2 - 3(4) - 2}{2(4) - 3} = \frac{18}{5} = 3.6$$

$$x_2 = 3.6 - \frac{3.6^2 - 3(3.6) - 2}{2(3.6) - 3} = \frac{374}{105} \approx 3.5619$$

12. $f(x) = e^{2x} - e^{-x} - 1,\ f'(x) = 2e^{2x} + e^{-x}$
$x_0 = 0$

$$x_1 = 0 - \frac{e^{2(0)} - e^{-0} - 1}{2e^{2(0)} + e^{-0}} = \frac{1}{3}$$

$$x_2 \approx .2832$$

13. The series is a geometric series with $a = 1$ and $r = -\frac{3}{4}$. Since $\left|-\frac{3}{4}\right| < 1$, the series is convergent and converges to $\dfrac{1}{1 + \frac{3}{4}} = \dfrac{4}{7}$.

14. The series is geometric with $a = \dfrac{5^2}{6} = \dfrac{25}{6}, r = \dfrac{5}{6}$. The sum is $\dfrac{\frac{25}{6}}{1 - \frac{5}{6}} = 25$.

15. The series is geometric with $a = \dfrac{1}{8}, r = \dfrac{1}{8}$. The sum is $\dfrac{\frac{1}{8}}{1 - \frac{1}{8}} = \dfrac{1}{7}$.

16. The series is geometric with $a = \dfrac{4}{7}, r = -\dfrac{8}{7}$, so it diverges.

17. The series is geometric with $a = \dfrac{1}{m+1}, r = \dfrac{m}{m+1}$, so (since $m > 0$) it converges to

$$\frac{\frac{1}{m+1}}{1 - \frac{m}{m+1}} = \frac{1}{m+1}\left(\frac{m+1}{1}\right) = 1.$$

18. The series is geometric with $a = \dfrac{1}{m}, r = -\dfrac{1}{m}$, so it converges if $m > 1$. In this case, the sum is

$$\frac{\frac{1}{m}}{1 + \frac{1}{m}} = \frac{1}{m+1}.$$ It diverges if $m \le 1$.

19. Recall that

$$e^x = 1 + \frac{1}{1!}x + \frac{1}{2!}x^2 + \frac{1}{3!}x^3 + \frac{1}{4!}x^4 + \cdots.$$

Therefore

$$1 + 2 + \frac{2^2}{2!} + \frac{2^3}{3!} + \frac{2^4}{4!} + \cdots$$

$$= 1 + \frac{1}{1!}(2) + \frac{1}{2!}(2)^2 + \frac{1}{3!}(2)^3 + \frac{1}{4!}(2)^4 + \cdots = e^2.$$

20. This is the Taylor series for e^x with $x = \frac{1}{3}$. Thus the sum is $e^{1/3}$.

21. Since $\displaystyle\sum_{k=0}^{\infty} \frac{1}{3^k}$ converges to $\dfrac{1}{1-\frac{1}{3}} = \dfrac{3}{2}$ and $\displaystyle\sum_{k=0}^{\infty} \left(\frac{2}{3}\right)^k$ converges to $\dfrac{1}{1-\frac{2}{3}} = 3$,

$$\sum_{k=0}^{\infty} \left[\frac{1}{3^k} + \left(\frac{2}{3}\right)^k \right] = \sum_{k=0}^{\infty} \frac{1 + 2^k}{3^k} = \frac{3}{2} + 3 = \frac{9}{2}.$$

22. $\displaystyle\sum_{k=0}^{\infty} \frac{3^k + 5^k}{7^k} = \sum_{k=0}^{\infty} \left(\frac{3}{7}\right)^k + \sum_{k=0}^{\infty} \left(\frac{5}{7}\right)^k = \frac{1}{1-\frac{3}{7}} + \frac{1}{1-\frac{5}{7}} = \frac{7}{4} + \frac{7}{2} = \frac{21}{4}$

23. Since $\displaystyle\int_1^{\infty} \frac{1}{x^3}\,dx = \lim_{b\to\infty} \left[-\frac{1}{2}x^{-2} \Big|_1^b \right] = \lim_{b\to\infty} \left[\frac{1}{2} - \frac{1}{2b^2} \right] = \frac{1}{2}$, the given series converges by the integral test.

24. The series is geometric with $a = \dfrac{1}{3}$ and $r = \dfrac{1}{3}$, so it converges.

25. Use the integral test, and consider $\int_1^\infty \frac{\ln x}{x}\,dx$. First, let $u = \ln x$ and $du = \frac{1}{x}dx$. Then

$$\int \frac{\ln x}{x}\,dx = \int u\,du = \frac{1}{2}u^2 + C = \frac{1}{2}(\ln x)^2 + C.$$

Next,

$$\int_1^b \frac{\ln x}{x}\,dx = \frac{1}{2}(\ln x)^2\Big|_1^b = \frac{1}{2}(\ln b)^2 - 0.$$

This quantity grows arbitrarily large as $b \to \infty$, so the improper integral above is divergent. By the integral test, the infinite series

$$\sum_{k=1}^\infty \frac{\ln k}{k}$$

is also divergent.

26. $\frac{k^3}{(k^4+1)^2} = \frac{k^3}{k^8 + 2k^4 + 1} \le \frac{k^3}{k^8} = \frac{1}{k^5}$ for $k \ge 1$. Thus since $\sum_{k=0}^\infty \frac{1}{k^5}$ converges by the integral test,

$\sum_{k=1}^\infty \frac{k^3}{(k^4+1)^2}$ converges by the comparison test.

27. The series converges if $\int_1^\infty \frac{1}{x^p}\,dx$ converges (by the integral test).

$$\int_1^\infty \frac{1}{x^p}\,dx = \lim_{b\to\infty}\left[\frac{1}{-p+1}x^{-p+1}\Big|_1^b\right] = \lim_{b\to\infty}\left[\frac{1}{1-p}(b^{1-p}-1)\right]$$

This limit is finite if $p > 1$.

28. This is a geometric series with $r = \frac{1}{p}$. Thus it converges when $\left|\frac{1}{p}\right| < 1$ or $|p| > 1$.

29. Replacing x by $-x^3$ in the series for $\frac{1}{1-x}$ gives $\frac{1}{1+x^3} = 1 - x^3 + x^6 - x^9 + x^{12} - \cdots$

30. $\frac{d}{dx}[\ln(1+x^3)] = \frac{3x^2}{1+x^3}$, so

$$\ln(1+x^3) = \int\left[3x^2 - 3x^5 + 3x^8 - 3x^{11} + 3x^{14} - \cdots\right]dx = \left[x^3 - \frac{3}{6}x^6 + \frac{3}{9}x^9 - \frac{3}{12}x^{12} + \cdots\right] + C$$

$\left(\text{using the expansion of } \frac{1}{1+x^3} \text{ in Exercise 29.}\right)$

$\ln(1) = 0 = 0 + 0 + \cdots + C$; so $C = 0$.

$\ln(1+x^3) = x^3 - \frac{1}{2}x^6 + \frac{1}{3}x^9 - \frac{1}{4}x^{12} + \ldots, |x| < 1$

31.
$$\frac{1}{(1-3x)^2} = \frac{1}{3}\frac{d}{dx}\left[\frac{1}{1-3x}\right]$$

$$= \frac{1}{3}\frac{d}{dx}[1+3x+3^2x^2+3^3x^3+\cdots]$$

$$= 1+6x+27x^2+108x^3+\cdots.$$

32.
$$\frac{e^x-1}{x} = \frac{1}{x}\left[x+\frac{x^2}{2!}+\frac{x^3}{3!}+\cdots\right] = 1+\frac{1}{2!}x+\frac{1}{3!}x^2+\frac{1}{4!}x^3+\cdots$$

33. a.
$$\cos 2x = 1-\frac{1}{2!}(2x)^2+\frac{1}{4!}(2x)^4-\frac{1}{6!}(2x)^6+\cdots = 1-\frac{2^2}{2!}x^2+\frac{2^4}{4!}x^4-\frac{2^6}{6!}x^6+\cdots$$

b.
$$\sin^2 x = \frac{1}{2}(1-\cos 2x) = \frac{1}{2}-\frac{1}{2}\cos 2x = \frac{1}{2}-\frac{1}{2}\left[1-\frac{2^2}{2!}x^2+\frac{2^4}{4!}x^4-\frac{2^6}{6!}x^6+\cdots\right]$$

$$= \frac{2}{2!}x^2-\frac{2^3}{4!}x^4+\frac{2^5}{6!}x^6-\cdots$$

34. a.
$$\cos 3x = 1-\frac{1}{2!}(3x)^2+\frac{1}{4!}(3x)^4-\frac{1}{6!}(3x)^6+\cdots = 1-\frac{3^2}{2!}x^2+\frac{3^4}{4!}x^4-\frac{3^6}{6!}x^6+\cdots$$

b. Adding the first three terms above to the corresponding terms in the expansion of $3\cos x$ and

multiplying by $\frac{1}{4}$ gives $p_4(x) = \frac{1}{4}\left[(1+3)-\left(\frac{3^2}{2!}+\frac{3}{2!}\right)x^2+\left(\frac{3^4}{4!}+\frac{3}{4!}\right)x^4\right] = 1-\frac{3}{2}x^2+\frac{7}{8}x^4.$

35.
$$\frac{1+x}{1-x} = \frac{1}{1-x}+\frac{x}{1-x} = [1+x+x^2+x^3+\cdots]+[x+x^2+x^3+x^4+\cdots] = 1+2x+2x^2+2x^3+\cdots$$

36. Using Exercise 32,
$$\int_0^{1/2}\frac{e^x-1}{x}dx = \int_0^{1/2}\left[1+\frac{1}{2}x+\frac{1}{3!}x^2+\frac{1}{4!}x^3+\cdots\right]dx = \left[x+\frac{1}{4}x^2+\frac{1}{3!\cdot3}x^3+\frac{1}{4!\cdot4}x^4+\cdots\right]\Bigg|_0^{1/2}$$

$$= \frac{1}{2}+\frac{1}{4\cdot2^2}+\frac{1}{3!\cdot3\cdot2^3}+\frac{1}{4!\cdot4\cdot2^4}+\cdots$$

37. (a) The fifth Taylor polynomial consists of all terms in the sixth Taylor polynomial with degree
≤ 5. Therefore, $p_5(x) = x^2.$

(b) Since the coefficient of x^3 in $p_6(x)$ is 0, $\dfrac{f^{(3)}(0)}{3!} = 0$. Therefore, $f^{(3)}(0) = 0.$

(c) $\displaystyle\int_0^1\sin x^2 dx \approx \int_0^1\left(x^2-\frac{1}{6}x^6\right)dx = \left(\frac{x^3}{3}-\frac{1}{42}x^7\right)\Bigg|_0^1 = \frac{1}{3}-\frac{1}{42}$

$$\approx .3095 \quad (\text{exact value: } .3103).$$

38. $p_4(x) = x + \dfrac{1}{3!}x^3$

39. a. $f'(x) = 2x + 4x^3 + 6x^5 + \cdots$

b. The series given for *f(x)* is the Taylor series of $\dfrac{1}{1-x^2}$. Thus $f(x) = \dfrac{1}{1-x^2}$ and

$f'(x) = \dfrac{2x}{(1-x^2)^2}$.

40. a. $\displaystyle \int f(x)\,dx = \int \left[x - 2x^3 + 4x^5 - 8x^7 + 16x^9 - \cdots \right] dx = \left[\dfrac{1}{2}x^2 - \dfrac{1}{2}x^4 + \dfrac{2}{3}x^6 - x^8 + \dfrac{8}{5}x^{10} - \cdots \right] + C$

b. The series given for $f(x) = x[1 - 2x^2 + 2^2x^4 - 2^3x^6 + 2^4x^8 - \cdots]$ is the Taylor expansion of

$\dfrac{x}{1+2x^2}$. Thus $f(x) = \dfrac{x}{1+2x^2}$ and $\displaystyle \int f(x)\,dx = \dfrac{1}{4}\ln(1+2x^2) + C$.

41. $100 + 100(.85) + 100(.85)^2 + 100(.85)^3 + \cdots = \dfrac{100}{1-.85} \approx 666.666667$

The amount beyond the original 100 million dollars is \$566,666,667.

42. $100 + (.85)(100) + (.80)(.85)^2(100) + (.80)^2(.85)^3(100) + \cdots$

$= 100 + 85 + (.80)(.85)(85) + (.80)^2(.85)^2(85) + \cdots$

$= 100 + \dfrac{85}{1-(.80)(.85)} = 365.625$ million dollars

43. $\displaystyle \sum_{k=1}^{\infty} 10,000e^{-.08k} = \sum_{k=1}^{\infty} 10,000(e^{-.08})^k = \dfrac{10,000(e^{-.08})}{1-e^{-.08}} \approx \$120,066.66.$

44. $\displaystyle \sum_{k=1}^{\infty} 10,000(.9)^k e^{-.08k} = \sum_{k=1}^{\infty} 10,000(.9e^{-.08})^k = \dfrac{10,000(.9)e^{-.08}}{1-.9e^{-.08}} \approx \$49,103.30$

45. $\displaystyle \sum_{k=1}^{\infty} 10,000(1.08)^k e^{-.08k} = \sum_{k=1}^{\infty} 10,000(1.08e^{-.08})^k = \dfrac{10,000(1.08)e^{-.08}}{1-1.08e^{-.08}} \approx \$3,285,603.18$

Chapter 12
Probability and Calculus

12.1 Discrete Random Variables

Although this section is provided mainly as a background for the sections that follow, there are interesting and important problems here. The main concepts are: relative frequency table, probability, expected value, variance, and (discrete) random variable. The probability density histograms on pages 556–557 will be used in Section 12.2 to motivate the definition of a probability density function.

1. There are two possible outcomes, 0 and 1, with probabilities $\frac{1}{5}$ and $\frac{4}{5}$, respectively. Thus

$$E(X) = 0 \cdot \frac{1}{5} + 1 \cdot \frac{4}{5} = \frac{4}{5}.$$

$$\text{Var}(X) = \left(0 - \frac{4}{5}\right)^2 \cdot \frac{1}{5} + \left(1 - \frac{4}{5}\right)^2 \cdot \frac{4}{5} = .128 + .032 = .16.$$

Standard deviation of $X = \sqrt{.16} = .4$.

7. (a) Recall that the area of a circle of radius r is given by $[\text{Area}] = \pi r^2$. Thus the area of a circle of radius 1 is π, and the area of a circle of radius $\frac{1}{2}$ is $\pi\left(\frac{1}{2}\right)^2 = \frac{1}{4}\pi$. To find the percentage of points lying in the circle of radius $\frac{1}{2}$, set up the ratio

$$\frac{[\text{Area of circle of radius }\frac{1}{2}]}{[\text{Area of circle of radius 1}]} = \frac{\frac{1}{4}\pi}{\pi} = \frac{1}{4} = .25 = 25\%.$$

Thus 25% of the points lie within $\frac{1}{2}$ unit of the center.

(b) The area of a circle of radius c is: $[\text{Area}] = \pi c^2$. To find the percentage of points lying within this circle set up the ratio

$$\frac{[\text{Area of circle of radius } c]}{[\text{Area of circle of radius 1}]} = \frac{\pi c^2}{\pi} = c^2 = 100c^2\%.$$

12.2 Continuous Random Variables

We recommend that you study the theoretical parts of Sections 6.1, 6.3, and 6.5 (particularly the Fundamental Theorem of Calculus) as background for the discussion here of the cumulative distribution function $F(x)$ and the probability density function $f(x)$. Given either function, you should be able to find the other with no difficulty. Observe that $F'(x) = f(x)$ and that $F(x)$ is the unique antiderivative of $f(x)$ for which $F(A) = 0$, when X is a random variable on $A \leq x \leq B$. You should also be able to use either $F(x)$ or $f(x)$ to compute various probabilities of the form $\Pr(a \leq X \leq b)$.

Pay attention to Example 6 because it provides a nice connection with earlier material on improper integrals and it introduces ideas that are needed in Section 12.4. For these reasons, questions similar to Example 6 often appear on exams.

1. Clearly, $f(x) \geq 0$, since $\frac{1}{18}x \geq 0$ when $0 \leq x \leq 6$. Thus Property I is satisfied. For Property II, check that

    $$\int_0^6 f(x)dx = \int_0^6 \frac{1}{18} x dx = \frac{1}{36} x^2 \Big|_0^6 = \frac{1}{36}(6)^2 - 0 = 1.$$

    Thus Property II is also satisfied, and $f(x)$ is indeed a probability density function.

7. In order to satisfy Property I, k must be nonnegative, for if $k < 0$ and $1 \leq x \leq 3$, then $f(x) = kx < 0$. For Property II, set

    $$\int_1^3 kx dx = 1,$$

    and above for k.

    $$\int_1^3 kx dx = \frac{k}{2} x^2 \Big|_1^3 = \frac{k}{2}(9) - \frac{k}{2} = 4k = 1.$$

    Therefore the value $k = \frac{1}{4}$ satisfies both Properties I and II, and $f(x) = \frac{1}{4}x, 1 \leq x \leq 3$, is a probability density function.

13. The set of possible values of X is determined by the domain of the density function $f(x)$. In this problem the domain is $0 \leq x \leq 4$, so the values of X range between 0 and 4. Since X can't be less than 0, the probability that X is less than or equal to 1 is the same as the probability that X is between 0 and 1. Thus $\Pr(X \leq 1)$ is an abbreviation for $\Pr(0 \leq X \leq 1)$. Similarly, since X cannot be larger than 4, $\Pr(3.5 \leq X)$ is an abbreviation for $\Pr(3.5 \leq X \leq 4)$.

    (a) $\Pr(X \leq 1)$ is represented by the area under the graph of $f(x) = \frac{1}{8}x$ where $0 \leq x \leq 1$.

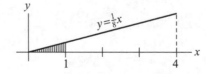

(b) $\Pr(2 \le X \le 2.5)$ is represented by the area under the graph of $f(x) = \frac{1}{8}x$ where $2 \le x \le 2.5$.

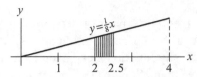

(c) $\Pr(3.5 \le X)$ is represented by the area under the graph of $f(x) = \frac{1}{8}x$ where $3.5 \le x \le 4$.

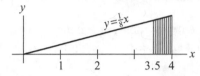

19. The probability that the lifetime X of a battery is *at least* 35 hours is given by $\Pr(35 \le X)$. However, we are told in this problem that the values of X range between 30 and 50. So $\Pr(35 \le X) = \Pr(35 \le X \le 50)$, and

$$\Pr(35 \le X) = \int_{35}^{50} \frac{1}{20} \, dx = \frac{1}{20}x \Big|_{35}^{50} = \frac{50}{20} - \frac{35}{20} = \frac{15}{20} = \frac{3}{4}.$$

Thus the probability is $\frac{3}{4}$ that the battery will last longer than 35 hours.

25. (a) $\Pr(2 \le X \le 3) = \int_{2}^{3} \frac{1}{21}x^2 \, dx = \frac{1}{63}x^3 \Big|_{2}^{3} = \frac{27}{63} - \frac{8}{63} = \frac{19}{63}.$

(b) Since $F(x)$ is an antiderivative of $f(x) = \frac{1}{21}x^2$,

$$F(x) = \int \frac{1}{21}x^2 \, dx = \frac{1}{63}x^3 + C.$$

Since X has values greater than or equal to 1, $F(1) = 0$. Set $F(1) = \frac{1}{63} + C = 0$ to find that $C = -\frac{1}{63}$, so

$$F(x) = \frac{1}{63}x^3 - \frac{1}{63} = \frac{1}{63}(x^3 - 1).$$

(c) $\Pr(2 \le X \le 3) = F(3) - F(2)$

$$= \frac{1}{63}(3^3 - 1) - \frac{1}{63}(2^3 - 1) = \frac{26}{63} - \frac{7}{63} = \frac{19}{63}.$$

31. $\Pr(0 \le X \le 5) = \int_{0}^{5} 2ke^{-kx} \, dx = \frac{2k}{-k}e^{-kx} \Big|_{0}^{5} = -2e^{-kx} \Big|_{0}^{5}.$

Since $k = (\ln 2)/10 \approx .0693$,

$$\Pr(0 \le X \le 5) = -2e^{-.0693x} \Big|_{0}^{5} = -2e^{(-.0693)(5)} - [-2e^{0}]$$

$$= -1.4142 + 2 = .5858.$$

Alternate calculation:

$$-2e^{-kx}\Big|_0^5 = -2e^{-k(5)} - (-2e^0) = -2e^{-5(\ln 2)/10} + 2.$$

Observe that

$$-2e^{-5(\ln 2)/10} = -2e^{-(1/2)\ln 2} = -2e^{(\ln 2)^{-1/2}}$$
$$= -2\cdot 2^{-1/2} = -2^{1/2} = -\sqrt{2}.$$

Thus

$$\Pr(0 \le X \le 5) = -\sqrt{2} + 2 \approx .59.$$

37. **(a)** Clearly, $f(x) = 4x^{-5} \ge 0$ for all values of x greater than or equal to 1, so Property I is satisfied. For Property II, check that

$$\int_1^\infty 4x^{-5}dx = 1.$$

If $b \ge 1$, then

$$\int_1^b 4x^{-5}dx = -x^{-4}\Big|_1^b = -(b)^{-4} - [-(1)^{-4}] = 1 - \frac{1}{b^4}.$$

Since as $b \to \infty$, the number b^4 gets arbitrarily large and $\frac{1}{b^4}$ approaches 0. Thus

$$\int_1^\infty 4x^{-5}dx = \lim_{b\to\infty}\int_1^b 4x^{-5}dx = \lim_{b\to\infty}\left(1 - \frac{1}{b^4}\right) = 1.$$

(b) Since $F(x)$ is an antiderivative of $f(x) = 4x^{-5}$,

$$F(x) = \int 4x^{-5}dx = -x^{-4} + C.$$

Since X has values greater than or equal to 1, $F(1) = 0$. Set $F(1) = -1 + C = 0$ to find that $C = 1$, so

$$F(x) = -x^{-4} + 1 = 1 - x^{-4}.$$

(c) $\Pr(1 \le X \le 2) = F(2) - F(1) = \frac{15}{16} - 0 = \frac{15}{16}$. Observe that $\Pr(1 \le X \le 2) + \Pr(2 \le X) = 1$, since *all* values of X are greater than or equal to 1. Hence,

$$\Pr(2 \le X) = 1 - \Pr(1 \le X \le 2) = 1 - \frac{15}{16} = \frac{1}{16}.$$

Helpful Hint: Decide now what you would do on an exam if you were given Exercise 37 with no mention of parts (a), (b), and $\Pr(1 \le x \le 2)$. Certainly there would be no need to find the cumulative distribution function. Would you want to compute $\Pr(1 \le X \le 2)$? Compute (a) and (b) below, and decide which approach is easier for you.

(a) $\Pr(2 \le X) = 1 - \Pr(1 \le X \le 2) = 1 - \int_1^2 4x^{-5}dx.$

(b) $\Pr(2 \le X) = \int_2^\infty 4x^{-5}dx.$

12.3 Expected Value and Variance

The expected value of a random variable X is easier to understand and compute than the variance. However, both concepts are important in statistics. An explanation of how $E(X)$ and $\text{Var}(X)$ are related to the graph of the probability density function for X will be given in Section 12.4, in the discussion of normal random variables. Check with your instructor to see if you need to know the definition of $\text{Var}(X)$, or if the alternate formula for Var (X) will suffice. The alternate formula is easier to compute.

Calculations of $E(X)$ and $\text{Var}(X)$ sometimes seem to require integration by parts, but this can be avoided when the probability density function is a polynomial. Observe how the integrals

$$\int_0^1 xf(x)dx \quad \text{and} \quad \int_0^1 x^2 f(x)dx$$

in Practice Problem 2 are simplified by expanding $f(x)$ so that $xf(x)$ and $x^2 f(x)$ are simple polynomials that are easily integrated.

1. Compute

$$E(X)=\int_0^6 x\cdot\frac{1}{18}xdx=\int_0^6 \frac{1}{18}x^2dx=\frac{1}{54}x^3\Big|_0^6=4.$$

To compute $\text{Var}(X)$ by the alternate formula, first compute

$$\int_0^6 x^2\cdot\frac{1}{18}xdx=\int_0^6 \frac{1}{18}x^3dx=\frac{1}{72}x^4\Big|_0^6=\frac{1296}{72}=18.$$

Then use this value and $E(x)$ to obtain

$$\text{Var}(X)=\int_0^6 x^2\cdot\frac{1}{18}xdx-E(X)^2=18-(4)^2=2.$$

7. Compute

$$E(X)=\int_0^1 x\cdot 12x(1-x)^2\,dx=\int_0^1 12x^2(1-2x+x^2)dx$$

$$=\int_0^1 (12x^2-24x^3+12x^4)dx$$

$$=\left(4x^3-6x^4+\frac{12}{5}x^5\right)\Big|_0^1=\frac{2}{5}.$$

$$\text{Var}(X)=\int_0^1 x^2\cdot 12x(1-x)^2\,dx-E(X)^2$$

$$=\int_0^1 (12x^3-24x^4+12x^5)dx-\left(\frac{2}{5}\right)^2$$

$$=\left(3x^4-\frac{24}{5}x^5+2x^6\right)\Big|_0^1-\frac{4}{25}$$

$$=\frac{1}{5}-\frac{4}{25}=\frac{1}{25}.$$

13. The average time is the expected value of X.

$$E(X) = \int_0^{12} \frac{1}{72} x^2 dx = \frac{1}{216} x^3 \Big|_0^{12} = \frac{1728}{216} = 8.$$

Therefore, the average time spent reading the editorial page is 8 minutes.

19. Set $\int_0^M \frac{1}{18} x dx = \frac{1}{2}$ and solve for M.

$$\int_0^M \frac{1}{18} x dx = \frac{1}{36} x^2 \Big|_0^M = \frac{1}{36} M^2 = \frac{1}{2}.$$

So $M^2 = 18$, and $M = \sqrt{18} = 3\sqrt{2}$. Note that $-3\sqrt{2}$ is not a solution since $0 \le M \le 6$.

25. By definition $E(X) = \int_A^B x f(x) dx$. Integrate by parts, setting

$$h(x) = x, \quad f(x) \quad [= g(x) \text{ in integration by parts formula}]$$
$$h'(x) = 1, \quad F(x) \quad [= \text{antiderivative of } f(x)]$$

where $f(x)$ is any probability density function on $A \le x \le B$ and $F(x)$ is the cumulative distribution function corresponding to $f(x)$. Then

$$E(X) = \int_A^B x f(x) dx = x F(x) \Big|_A^B - \int_A^B 1 \cdot F(x) dx$$

$$= B \cdot F(B) - A \cdot F(A) - \int_A^B F(x) dx$$

$$= B \cdot 1 - A \cdot 0 - \int_A^B F(x) dx \qquad [F(B) = 1, \ F(A) = 0]$$

$$= B - \int_A^B F(x) dx,$$

12.4 Exponential and Normal Random Variables

This section deserves your attention because exponential and normal random variables arise so often in applications. In a situation when an exponential density function ke^{-kx} is appropriate for a random variable X, the value of $E(X)$ is often estimated from experimental data, and then the constant k is obtained from the relation

$$E(X) = \frac{1}{k}, \quad \text{or equivalently,} \quad k = \frac{1}{E(x)}.$$

You need to know this relation in order to work Exercises 5–14.

The text's discussion of arbitrary normal random variables provides an opportunity to review changes of variable in a definite integral. Check with your instructor about how much detail you should show when you make a substitution $z = (x - \mu)/\sigma$.

1. Here $k = 3$, so

$$E(X) = \frac{1}{3}, \quad \text{and} \quad \text{Var}(X) = \frac{1}{9}.$$

7. The mean is given by $E(X) = \frac{1}{k} = 3$. Hence $k = \frac{1}{3}$, and the probability density function is $f(x) = \frac{1}{3}e^{-(1/3)x}$. The probability that a customer is served in less than 2 minutes is given by

$$\Pr(0 \le X \le 2) = \int_0^2 \frac{1}{3}e^{-(1/3)x}dx = -e^{-(1/3)x}\Big|_0^2$$
$$= -e^{-2/3} - (-e^0)$$
$$= 1 - e^{-2/3}.$$

13. The average life span (or mean) is given by $E(X) = \frac{1}{k} = 72$. Hence $k = \frac{1}{72}$, and the probability density function is $f(x) = \frac{1}{72}e^{-(1/72)x}$.

    (a) The probability that a component lasts for more than 24 months is given by

    $$\Pr(24 \le X) = \int_{24}^{\infty} \frac{1}{72}e^{-(1/72)x}dx$$
    $$= \lim_{b \to \infty} \int_{24}^{b} \frac{1}{72}e^{-(1/72)x}dx$$
    $$= \lim_{b \to \infty} \left[-e^{-(1/72)x}\Big|_{24}^{b} \right]$$
    $$= \lim_{b \to \infty} \left[e^{-1/3} - e^{-b/72} \right]$$

    Note that as $b \to \infty$, the number $-\frac{b}{72}$ approaches $-\infty$, so $e^{-b/72}$ approaches 0. Therefore,

    $$\Pr(24 \le X) = \lim_{b \to \infty} [e^{-1/3} - e^{-b/72}] = e^{-1/3}.$$

    (b) To find the reliability function, repeat the calculations in (a) with the number 24 replaced by the variable t.

    $$r(t) = \Pr(t \le X) = \int_t^{\infty} \frac{1}{72}e^{-(1/72)x}dx$$
    $$= \lim_{b \to \infty} \int_t^{b} \frac{1}{72}e^{-(1/72)x}dx$$
    $$= \lim_{b \to \infty} \left[-e^{-(1/72)x}\Big|_t^{b} \right]$$
    $$= \lim_{b \to \infty} \left[e^{-t/72} - e^{-b/72} \right]$$
    $$= e^{-t/72}.$$

19. To find a relative maximum, first set $f'(x) = 0$ and solve for x.

$$f'(x) = -xe^{-x^2/2} = 0.$$

Since $e^{-x^2/2}$ is strictly positive for all values of x, you can see that $f'(x) = 0$ if and only if $x = 0$. To check that $f(x)$ has a maximum at $x = 0$, check $f''(0)$.

$$f''(x) = x^2 e^{-x^2/2} - e^{-x^2/2}$$
$$f''(0) = 0 - 1 = -1 < 0.$$

Thus $f(x)$ is indeed concave down at $x = 0$. Therefore $f(x)$ has a relative maximum at $x = 0$.

25. (a) We have $\mu = 6$ and $\sigma = \frac{1}{2}$. Note that $\dfrac{x - \mu}{\sigma} = \dfrac{x - 6}{1/2} = 2x - 12$ and

$$\frac{1}{\sigma\sqrt{2\pi}} = \frac{1}{(1/2)\sqrt{2\pi}} = \frac{2}{\sqrt{2\pi}}.$$

Thus, the normal density function for X, the gestation period, is

$$f(x) = \frac{2}{\sqrt{2\pi}} e^{-(1/2)(2x-12)^2}$$

So,

$$\Pr(6 \le X \le 7) = \int_6^7 \frac{2}{\sqrt{2\pi}} e^{-(1/2)(2x-12)^2}\, dx.$$

Use the substitution $z = 2x - 12$, $dz = 2dx$. If $x = 6$, then $z = 0$, and if $x = 7$, then $z = 2$. So

$$\int_6^7 \frac{2}{\sqrt{2\pi}} e^{-(1/2)(2x-12)^2}\, dx = \int_0^2 \frac{1}{\sqrt{2\pi}} e^{-(1/2)z^2}\, dz.$$

(Do not forget to change the limits of integration.) The value of this integral is the area under the standard normal curve form 0 to 2. Use Table 1 in the Appendix of the text to find

$$\Pr(6 \le X \le 7) = A(2) = .4772.$$

Therefore, about 47.72% of births occur after a gestation period of between 6 and 7 months.

(b) Proceeding as in part (a), use the same substitution to evaluate

$$\int_5^6 \frac{2}{\sqrt{2\pi}} e^{-(1/2)(2x-12)^2}\, dx,$$

noting that if $x = 5$, then $z = 2x - 12 = -2$, and if $x = 6$, then $z = 0$. Therefore,

$$\Pr(5 \le X \le 6) = \int_{-2}^0 \frac{2}{\sqrt{2\pi}} e^{-(1/2)z^2}\, dz$$
$$= A(-2) = A(2) = .4772.$$

Therefore, about 47.72% of births occur after a gestation period of between 5 and 6 months.

Warning: You can use the TI-83/84 calculator to check your calculations of areas under normal curves, but do not use it to replace all of your written work. Many instructors use Section 12.4 to review the change of variable in a definite integral, and they may wish you to show these calculations and just use the area table $A(z)$ or the TI-83/84 to find areas under the *standard* normal curve. For example, the probability computed in exercise 25(a) is given by each of the commands

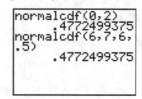

Your instructor may tell you to use only the first form of normalcdf on exams. To put normalcdf on the TI-83/84 home screen, press [2nd][DIST] 2. See page 581 in the text or a TI-83/84 manual for more information about the normal cumuluative distribution function (cdf).

31. The normal density function for this exercise is

$$f(x) = \frac{1}{(.8)\sqrt{2\pi}} e^{-(1/2)[(x-18.2)/.8]^2} .$$

Let X be the diameter of a bolt selected at random from the supply of bolts. Then

$$\Pr(20 \le X) = \int_{20}^{\infty} \frac{1}{(.8)\sqrt{2\pi}} e^{-(1/2)[(x-18.2)/.8]^2} dx.$$

Use the substitution $z = (x-18.2)/.8$, $dz = (1/.8)dx$, and note that if $x = 20$, then $z = 1.8/.8 = 2.25$, and if $x \to \infty$, then $z \to \infty$ too. Hence

$$\Pr(20 \le X) = \int_{2.25}^{\infty} \frac{1}{\sqrt{2\pi}} e^{-(1/2)z^2} dz.$$

The value of this integral is the area under the standard normal curve from 2.25 to ∞. If we consider areas under the standard normal curve, we see that

$$\begin{bmatrix} \text{area to} \\ \text{right of 2.25} \end{bmatrix} = \begin{bmatrix} \text{area to} \\ \text{right of 0} \end{bmatrix} - \begin{bmatrix} \text{area between 0} \\ \text{and 2.25} \end{bmatrix}$$

That is,

$$\Pr(20 \le X) = .5 - A(2.25)$$
$$= .5 - .4878$$
$$= .0122.$$

Therefore, about 1.22% of the bolts will be discarded.

37. Set $Y_1 = e^\wedge(-X^\wedge 2/2)/\sqrt{(2\pi)}$. On a TI-83/84 calculator, you can set $Y_1 = \text{normalpdf}(X)$. The command is found in the DIST menu. To graph Y_1, use the window $[-5, 5]$ by $[-.2, .5]$, with $\text{Yscl} = .1$. For the definite integral of $x^2 f(x)$, use fnInt $(X^\wedge 2 * Y_1, X, -8, 8)$.

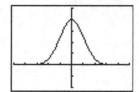
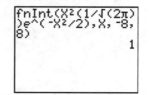

The fnInt command is on the MATH menu for the TI-82 and TI-83/84, and on the CALC menu for the TI-85 and TI-86. Repeat the command (use the [2nd] ENTRY key to save time), with ± 8 replaced by $\pm N$, for $N = 4$ and 6. Report your calculations, which should suggest that the improper integral of $x^2 f(x)$ from $-\infty$ to ∞ is 1. Since $E(Z) = 0$, this calculation shows that $\text{Var}(Z) = 1$ and hence the standard deviation of Z is 1, too.

12.5 Poisson and Geometric Random Variables

This last section of the text combines the probability theory in this chapter with the Taylor series for e^x (from Section 11.5) and the geometric series (Section 11.3). For convenience, state all probabilities to four decimal places.

1. From Example 1, $p_5 = .1008$. Since $p_n = (\lambda/n) p_{n-1}$,

$$p_6 = \left(\frac{3}{6}\right) p_5 = 3(.1008/6) = .0504$$

$$p_7 = \left(\frac{3}{7}\right) p_6 = 3(.0504/7) = .0216$$

$$p_8 = \left(\frac{3}{8}\right) p_7 = 3(.0216/8) = .0081.$$

7. Since the average of a Poisson random variable equals the parameter for the variable, $\lambda = 1.5$ for this problem.

   (a) $\Pr(X = 0) = p_0 = e^{-\lambda} = e^{-1.5} = .2231.$

   (b) First, compute $p_1 = (1.5/1) p_0 = .3347$, $p_2 = (1.5/2) p_3 = .2510$, and $p_3 = (1.5/3) p_2 = .1255$. Then
   $$p_2 + p_3 = .2510 + .1255 = .3765.$$

   (c) $\Pr(X \geq 4) = 1 - \Pr(X < 4)$
   $$= 1 - (p_0 + p_1 + p_2 + p_3)$$
   $$= 1 - (.2231 + .3347 + .2510 + .1255)$$
   $$= 1 - .9343$$
   $$= .0657.$$

13. The number X of Red taxis that appear before the first Blue taxi appears is a geometric random variable, with "success" corresponding to the arrival of a Red taxi. At any given time, the probability of success is $p = 3/4$, because three out of every four cabs are Red.

   (a) For $n \geq 1$, the probability of n successes (Red taxis) before the first failure (a Blue taxi) is

   $$p_n = p^n(1-p) = \left(\frac{3}{4}\right)^n \left(\frac{1}{4}\right).$$

(b) The probability of observing at least three Red taxis in a row is 1 minus the probability of observing less than three Red taxis, namely, $1 - (p_0 + p_1 + p_2)$. This probability is

$$1 - \left[\frac{1}{4} + \left(\frac{3}{4}\right)\left(\frac{1}{4}\right) + \left(\frac{3}{4}\right)^2\left(\frac{1}{4}\right) \right] = 1 - (.25 + .1875 + .1406) \approx .4219.$$

(c) The average number of Red taxis observed in a row is given by $E(X) = p/(1-p) = \frac{3}{4} \div \frac{1}{4} = 3$.

19. It can be shown that the number of defective fuses found in a box selected at random is a Poisson random variable. Let $f(\lambda) = (\lambda^2/2)e^{-\lambda}$, the probability of selecting a box with two defective fuses. Then,

$$f'(\lambda) = \left(\frac{\lambda^2}{2}\right)\left(-e^{-\lambda}\right) + \lambda e^{-\lambda} \quad \text{(by the product rule)}$$

$$= \left(\lambda - \frac{\lambda^2}{2}\right)e^{-\lambda}$$

$$f''(\lambda) = \left(\lambda - \frac{\lambda^2}{2}\right)\left(-e^{-\lambda}\right) + (1-\lambda)e^{-\lambda}$$

A possible maximum occurs when

$$\left(\lambda - \frac{\lambda^2}{2}\right)e^{-\lambda} = 0, \quad \text{or} \quad \lambda\left(1 - \frac{\lambda}{2}\right)e^{-\lambda} = 0.$$

Since $e^{-\lambda} \neq 0$ and $\lambda \neq 0$ (because there are some defective fuses in the box sampled), we conclude that $\lambda = 2$. Checking $f''(2) = 0 + (1-2)e^{-2}$, we see that $f''(2)$ is negative, so $f(\lambda)$ has a maximum at $\lambda = 2$. Thus, given one sample which has two defective fuses, the maximum likelihood estimate of the Poisson parameter is $\lambda = 2$.

25. (a) The formula $p_n = (\lambda/n)p_{n-1}$ shows that $p_7 = (6.9/7)p_6 < p_6$. So the answer is "no," seven babies are less likely to be born than six babies.

(b) The answer requires computing $p_0 + \cdots + p_{15}$. Use the basic formula for p_n, and the **sum(seq**(...)) command, (discussed on pages 532 and 588–589 in the text):

$$\text{sum(seq}(\lambda^n/n! * p_0, n, 0, 15, 1))$$

Here $\lambda = 6.9$ and $p_0 = e^{-6.9}$. Note X is more convenient to use than n. The sum is

$$\text{sum(seq}(6.9\^X/X! * e\^ - 6.9, X, 0, 15, 1)) \approx .9979061$$

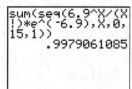

For the TI-83/84, you can use the command **poissonpdf**, which is entry "C" on the DISTR menu. Place this inside the sum(seq (. . .)) command, as follows:

$$\text{sum(seq(poissonpdf}(6.9, X), X, 0, 15, 1)).$$

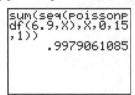

Warning: The TI-83/84 calculator has commands for both Poisson probabilities and geometric probabilities, on the DIST menu. However, the TI-83 definition of geometric probabilities is different from the one used in the text. ("Success" and "failure" are interchanged and the numbering of the probabilities p_0, p_1,..., is different.) You can safely use poissonpdf and poissoncdf to check your work, but ask your instructor about any limitations on using these functions on an exam. See the text, pages 583–584, for details about the Poisson functions.

Review of Chapter 12

Work through the Review of the Fundamental Concepts at the end of the chapter and determine which concepts you must know and which terms and formulas you must memorize. Summarized below are the main skills you should have, listed by the various types of random variables.

(a) *Discrete random variable.* Compute $E(X)$, Var(X), and the standard deviation of X.

(b) *Continuous random variable on a finite interval.* Compute $E(X)$, Var(X); test if $f(x)$ is a probability density function; given one of $f(x)$ and $F(x)$, find the other; use $f(x)$ or $F(x)$ to compute probabilities. Remember that $\Pr(X \le b)$ stands for $\Pr(A \le X \le b)$, etc. Harder questions are found in exercises 32–36 of Section 12.2.

(c) *Continuous random variable on an infinite interval.* A typical probability density function is $f(x) = kx^{-(k+1)}$ for $x \ge 1$. Same skills as for (b), plus use the formula $\Pr(X \ge b) = 1 - \Pr(X \le b)$ to simplify calculations.

(d) *Exponential random variable.* The probability density function is $f(x) = ke^{-kx}$ for $x \ge 0$. Show that $f(x)$ *is* a probability density functions. Compute various probabilities and the cumulative distribution function. You probably will not have to compute $E(X)$ and Var(X), unless your instructor gives you the following limits to learn:

$$\lim_{b \to \infty} be^{-kb} = 0 \quad \text{and} \quad \lim_{b \to \infty} b^2 e^{-kb} = 0 \quad \text{for } k > 0.$$

However, you should know that $E(X) = \dfrac{1}{k}$ and Var(X) $= \dfrac{1}{k^2}$.

(e) *Normal random variable.* Use a table to compute probabilities for a standard normal random variable, and make a change of variable to compute probabilities for an arbitrary normal random variable. Your instructor may set up guidelines for using a calculator to compute areas under normal curves.

(f) *Poisson random variable.* If X is a Poisson random variable with parameter λ, then you should know that $p_0 = e^{-\lambda}$,

$$p_n = \frac{\lambda^n}{n!} e^{-\lambda} \, (n = 1, 2, \ldots), \quad \text{and} \quad E(X) = \lambda.$$

(g) *Geometric random variable.* If X is a geometric random variable and the probability of success on one trial is p, then you should know that

$$p_n = p^n(1-p) \quad (n = 0, 1, 2, \ldots), \quad \text{and} \quad E(X) = p/(1-p).$$

Chapter 12 Review Exercises

While solutions for all review exercises are included, expanded explanations are included for every sixth exercise.

1. **(a)** $\Pr(X \le 1) = \int_0^1 \frac{3}{8} x^2 \, dx = \frac{1}{8} x^3 \Big|_0^1 = \frac{1}{8} - 0 = \frac{1}{8}.$

$\Pr(1 \le X \le 1.5) = \int_1^{1.5} \frac{3}{8} x^2 \, dx = \frac{1}{8} x^3 \Big|_1^{1.5} = \frac{1}{8}\left[\left(\frac{3}{2}\right)^3 - 1^3 \right] = \frac{27}{64} - \frac{1}{8} = \frac{19}{64}.$

(b) $E(X) = \int_0^2 x f(x) \, dx = \int_0^2 x \cdot \frac{3}{8} x^2 \, dx = \frac{3}{8} \cdot \frac{1}{4} x^4 \Big|_0^2 = \frac{3}{2}.$

$\text{Var}(X) = \int_0^2 x^2 f(x) \, dx - E(X)^2 = \int_0^2 x^2 \cdot \frac{3}{8} x^2 \, dx - \left(\frac{3}{2}\right)^2$

$= \frac{3}{8} \int_0^2 x^4 \, dx - \frac{9}{4} = \frac{3}{8} \cdot \frac{1}{5} x^5 \Big|_0^2 - \frac{9}{4} = \frac{12}{5} - \frac{9}{4} = \frac{3}{20}.$

2. $f(x) = 2(x-3) = 2x - 6, \ 3 \le x \le 4$

(a) $\Pr(3.2 \le X) = \int_{3.2}^4 (2x - 6) \, dx = \left(x^2 - 6x \right) \Big|_{3.2}^4 = .96$

$\Pr(3 \le x) = 1$ since the random variable is defined for $3 \le x \le 4$.

(b) $E(X) = \int_3^4 (2x^2 - 6x) \, dx = \left(\frac{2}{3} x^3 - 3x^2 \right) \Big|_3^4 = \frac{11}{3}$

$\int_3^4 (2x^3 - 6x^2) \, dx = \left(\frac{1}{2} x^4 - 2x^3 \right) \Big|_3^4 = \frac{27}{2}$

$V(X) = \frac{27}{2} - \left(\frac{11}{3}\right)^2 = \frac{1}{18}$

3. **I.** $e^{A-x} \geq 0$ for all x.

II. $\displaystyle\int_A^\infty e^{A-x}dx = \lim_{b\to\infty}\left[-e^{A-x}\Big|_A^b\right]$

$\displaystyle = \lim_{b\to\infty}[1 - e^{A-b}] = 1$

Thus $f(x) = e^{A-x}$, $x \geq A$ is a density function.

$\displaystyle F(x) = \int_A^x e^{A-t}dt = -e^{A-t}\Big|_A^x = 1 - e^{A-x}$

4. $f(x) = \dfrac{kA^k}{x^{k+1}}$, $k > 0, A > 0, x \geq A$

I.   Since k and A are > 0, $f(x) \geq 0$ for all $x \geq A$.

II. $\displaystyle\int_A^\infty\left(\frac{kA^k}{x^{k+1}}\right)dx = \lim_{b\to\infty}\left[\frac{-A^k}{x^k}\Big|_A^b\right] = \lim_{b\to\infty}\left[1 - \frac{A^k}{b^k}\right] = 1$

Thus $f(x)$ is a density function.

$\displaystyle F(X) = \int_A^x\left(\frac{kA^k}{t^{k+1}}\right)dt = -\frac{A^k}{t^k}\Big|_A^x = 1 - \frac{A^k}{x^k}$

5.   For $n \geq 2$, any choice of $c_n > 0$ will ensure $f_n(x) \geq 0$ for all $x \geq 0$. Thus we need only

$\displaystyle\int_0^\infty c_n x^{(n-2)/2}e^{-x/2}dx = 1$. If $n = 2$ this becomes $c_2\displaystyle\int_0^\infty e^{-x/2}dx = 1$

$\displaystyle c_2\lim_{b\to\infty}\left[-2e^{-x/2}\Big|_0^b\right] = 1 \Rightarrow 2c_2 = 1 \Rightarrow c_2 = \frac{1}{2}$.

For $n = 4$, we have $c_4\displaystyle\int_0^\infty xe^{-x/2}dx = 1$. Integrating by parts twice gives

$\displaystyle\int_0^b xe^{-x/2}dx = e^{-x/2}(-4 - 2x)\Big|_0^b$.

Therefore, $c_4\displaystyle\lim_{b\to\infty}\left[e^{-x/2}(-4-2x)\Big|_0^b\right] = 1 \Rightarrow 4c_4 = 1 \Rightarrow c_4 = \frac{1}{4}$.

6.   **I.**   If $k > 0$, $\dfrac{1}{2k^3}x^2 e^{-x/k} \geq 0$ for all x.

II. Integrating by parts twice, $\displaystyle\int_0^b x^2 e^{-x/k}dx = -e^{x/k}[kx^2 + 2k^2 x + 2k^3]\Big|_0^b$. Thus

$\displaystyle\int_0^\infty\frac{1}{2k^3}x^2 e^{-x/k}dx = \frac{1}{2k^3}(2k^3) = 1$.

7.   **(a)** From the definition of expected value,

$$E(X) = 1(.599) + 11(.401)$$
$$= 5.01.$$

This can be interpreted as saying that if a large number of samples are tested in batches of ten samples per batch, then the average number of tests per batch will be close to 5.01.

(b) The 200 samples will be divided into 20 batches of ten samples. In part (a) we found that, on average, 5.01 tests must be run on each batch. So the laboratory should expect to run about $(20)(5.01) \approx 100$ tests altogether.

8. (a) $E(X) = 1(.774) + 6(.226) = 2.13$

(b) 200 samples is 40 batches of 5. Thus they can expect to run $40(2.13) \approx 85$ tests.

9. $F(x) = 1 - \dfrac{1}{4}(2-x)^2,\ 0 \le x \le 2$

(a) $\Pr(X \le 1.6) = F(1.6) = .96$

(b) $\Pr(X \le t) = 1 - \dfrac{1}{4}(2-t)^2 = .99$
$t = 1.8$ (thousand gal.)

(c) $f(x) = F'(x) = \dfrac{1}{2}(2-x),\ 0 \le x \le 2$

10. $E(X) = \dfrac{1}{625}\displaystyle\int_0^5 x(x-5)^4\,dx = \dfrac{1}{625}\left[\dfrac{x}{5}(x-5)^5 - \dfrac{1}{30}(x-5)^6\right]\Bigg|_0^5$ (Integration by parts)

$= \dfrac{(-5)^6}{30 \cdot 5^4} = \dfrac{5}{6} = .8333$ (hundred dollars)

Thus on average the manufacturer can expect to make $100 - 83.33 = \$16.67$ on each service contract sold.

11. (a) $E(X) = \displaystyle\int_{20}^{25} \dfrac{1}{5}x\,dx = \dfrac{1}{10}x^2\Bigg|_{20}^{25} = 22.5$

$V(X) = \displaystyle\int_{20}^{25} \dfrac{1}{5}x^2\,dx - 22.5^2$

$= \dfrac{x^3}{15}\Bigg|_{20}^{25} - 22.5^2 \approx 2.0833$

(b) $\Pr(X \le b) = \displaystyle\int_{20}^{b} \dfrac{1}{5}\,dx = .3 \Rightarrow$

$\dfrac{1}{5}b - 4 = .3 \Rightarrow b = 21.5$

12. $F(x) = \dfrac{(x^2-9)}{16},\ 3 \le x \le 5$

(a) $f(x) = F'(x) = \dfrac{x}{8},\ 3 \le x \le 5$

(b) $\Pr(a \le X) = \dfrac{1}{4}$, so $F(a) = \dfrac{(a^2-9)}{16} = \dfrac{3}{4} \Rightarrow a = \sqrt{21}$

13. **(a)** Find the value of k that satisfied the equation

$$\int_5^{25} kx\,dx = 1.$$

So compute

$$\int_5^{25} kx\,dx = \frac{k}{2}x^2\Big|_5^{25} = \frac{k}{2}625 - \frac{k}{2}25 = 300k = 1.$$

Therefore $k = \frac{1}{300}$.

(b) $\Pr(20 \le X \le 25) = \int_{20}^{25} \frac{1}{300}x\,dx = \frac{1}{600}x^2\Big|_{20}^{25}$

$$= \frac{1}{600}625 - \frac{1}{600}400 = \frac{1}{600}(625 - 400) = .375.$$

(c) $E(X) = \int_5^{25} \frac{1}{300}x^2\,dx = \frac{1}{900}x^3\Big|_5^{25} = \frac{1}{900}(15,625 - 125) = 17.222.$

Therefore, the mean annual income is \$17,222.

14. **(a)** $F(x) = \Pr(3 \le X \le x)$

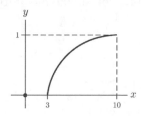

(b) $F(7) - F(5) = \Pr(5 \le X \le 7)$

(c) $\Pr(5 \le X \le 7) = \int_5^7 f(x)\,dx$

15. Points (θ, y) satisfying the given condition are precisely those points under the curve

$y = \sin\theta, 0 \le \theta \le \pi$. This region has area $\int_0^\pi \sin\theta\,d\theta = -\cos\theta\Big|_0^\pi = 2$. The area of the rectangle is π,

so the probability that a randomly selected point falls in the region $y \le \sin\theta$ is $\dfrac{2}{\pi}$.

16. Since the length of the needle is 1 unit, $\sin\theta$ is the difference in the y-coordinates of the base and the end of the needle.

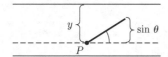

The needle will touch a ruled line if and only if this difference exceeds y, the vertical distance from the base to the next ruled line. To compute $\Pr(y \leq \sin\theta)$, view dropping the needle as a random choice of a point (θ, y) from the square $0 \leq \theta \leq \pi, 0 \leq y \leq 1$. Then Exercise 15 applies.

17. Let X be the lifetime of the computer monitor. Then

$$\Pr(Y = 0) = \Pr(X \leq 3) = \int_0^3 \frac{1}{5} e^{-(1/5)x} dx = -e^{-(1/5)x}\Big|_0^3 = 1 - e^{-3/5} \approx .45119$$

Thus $\Pr(Y = 100) \approx .54881$ and $E(Y) \approx \$54.88$.

18. Let Y be as in the hint and let X be the life span of the motor. Then

$$\Pr(Y = 300) = \Pr(X \leq 1) = \int_0^1 \frac{1}{10} e^{-(1/10)x} dx = -e^{-(1/10)x}\Big|_0^1 = 1 - e^{-1/10} \approx .09516$$

Thus $E(Y) \approx 300(.09516) \approx \$28.55$. Since the insurance costs $25 to buy, you should buy it for the first year.

19. To show that $\Pr(X \leq 4) = 1 - e^{-4k}$, compute

$$\Pr(X \leq 4) = \int_0^4 ke^{-kx} dx = -e^{-kx}\Big|_0^4 = -e^{-4k} - (-e^0) = 1 - e^{-4k}.$$

To make $\Pr(X \leq 4) = .75$, the parameter k must satisfy

$$1 - e^{-4x} = .75$$
$$e^{-4x} = .25$$
$$-4x = \ln .25 \approx -1.386.$$

Thus

$$k = \frac{-1.386}{-4} \approx .35.$$

20. $E(X) = .01 \int_0^\infty x^2 e^{-x/10} dx$

Integrating by parts twice, $\int_0^b x^2 e^{-x/10} dx = -e^{-x/10}(10x^2 + 200x + 2000)\Big|_0^b$. So $E(X) = 2000(.01) = 20$ (thousand hours) and the expected additional earnings from the machine are $20(5000) = \$100,000$. Since this amount exceeds the price, the machine should be purchased.

21. $f(x)$ is the density of a normal random variable X with $\mu = 50$, $\sigma = 8$. Thus

$$\Pr(30 \leq X \leq 50) = A\left(\frac{50 - 30}{8}\right) = A(2.5) = .4938$$

22. Let X be the length of a randomly selected part. Then

$$\text{Pr}(79.95 \leq X \leq 80.05) = A\left(\frac{79.99 - 79.95}{.02}\right) + A\left(\frac{80.05 - 79.99}{.02}\right) = A(2) + A(3)$$

$$= .4772 + .4987 = .9759$$

Hence out of a lot of 1000 parts, $1000(.9759) = 975.9$ should be within the tolerance limits, leaving about 24 defective parts.

23. Let X be the height of a randomly selected man in the city.

$$\text{Pr}(X \geq 69) = .5 + A\left(\frac{70 - 69}{2}\right) = .5 + A(.5) \qquad = .6915$$

Thus about 69.15% of the men in the city are eligible.

24. Let Y be the height of a randomly selected woman from the city.

$$\text{Pr}(Y \geq 69) = .5 - A\left(\frac{69 - 65}{1.6}\right) = .5 - A(2.5) = .0062$$

So only about .62% of the women are eligible.

25. Recall that $\text{Pr}(a \leq Z)$ is the area under the standard normal curve to the right of a. Since $\text{Pr}(a \leq Z) = .4$, and $\text{Pr}(0 \leq Z) = .5$, the number a lies to the right of 0. That is, a is positive. Thus,

$$\begin{bmatrix} \text{area to} \\ \text{right of } a \end{bmatrix} = \begin{bmatrix} \text{area to} \\ \text{right of } 0 \end{bmatrix} - \begin{bmatrix} \text{area between } 0 \\ \text{and } a \end{bmatrix}$$

$$\text{Pr}(a \leq Z) = \text{Pr}(0 \leq Z) - \text{Pr}(0 \leq Z \leq a)$$

$$.4 = .5 - \text{Pr}(0 \leq Z \leq a).$$

So $\text{Pr}(0 \leq Z \leq a) = A(a) = .1$. From the $A(z)$ table, look for the z that makes $A(z)$ as close to .1 as possible. Since $A(.25) = 0.987 \approx .1000$, we conclude that $a \approx .25$.

26. Using the result of exercise 25, the cutoff grade t must satisfy $\dfrac{t - 500}{100} = .25 \Rightarrow t = 525$.

27. (a) $\text{Pr}(-1 \leq Z \leq 1) = 2A(1) = .6826$

    (b) $\text{Pr}(\mu - \sigma < X < \mu + \sigma) = \text{Pr}(-1 < Z < 1) = .6826$

28. (a) $\text{Pr}(-2 \leq Z \leq 2) = 2A(2) = .9544$

    (b) $\text{Pr}(\mu - 2\sigma < X < \mu + 2\sigma) = \text{Pr}(-2 < Z < 2) = .9544$

29. (a) Let X be an exponential random variable with density $f(x) = ke^{-kx}$. Then $E(X) = \mu = \dfrac{1}{k}$ and

    $V(X) = \sigma^2 = \dfrac{1}{k^2}$. Applying the inequality with $n = 2$ gives

    $$\text{Pr}\left(\frac{1}{k} - \frac{2}{k} \leq X \leq \frac{1}{k} + \frac{2}{k}\right) = \text{Pr}\left(-\frac{1}{k} \leq X \leq \frac{3}{k}\right) = \text{Pr}\left(0 \leq X \leq \frac{3}{k}\right) \geq 1 - \frac{1}{2^2} = \frac{3}{4}$$

    (b) $\text{Pr}\left(0 \leq X \leq \dfrac{3}{k}\right) = \displaystyle\int_0^{3/k} ke^{-kx}dx = -e^{-kx}\Big|_0^{3/k} = 1 - e^{-3} \approx .9502$

30. Let X be a normal random variable with $E(X) = \mu$ and $V(X) = \sigma^2$. Applying the inequality with

$n = 2$ gives $\Pr(\mu - 2\sigma \le X \le \mu + 2\sigma) \ge 1 - \dfrac{1}{2^2} = \dfrac{3}{4}$.

The exact value is $\Pr(\mu - 2\sigma \le X \le \mu + 2\sigma) = 2A(2) = .9544$

31. For $\lambda = 4$, $p_4 = \lambda^4 \dfrac{e^{-4}}{4!} = (4)^4 \dfrac{e^{-4}}{24} \approx .1954$.

32. $1 - [p_0 + p_1 + p_2 + p_3 + p_4 + p_5 + p_6 + p_7] \approx .0511336$

33. $E(X) = 4$

34. $\left(\dfrac{2}{9}\right)\left(\dfrac{7}{9}\right)^n$

35. $E(X) = \dfrac{\dfrac{7}{9}}{1 - \dfrac{7}{9}} = \dfrac{7}{2}$

36. $1 - (p_0 + p_1 + p_2) \approx .4705075$

DATE DUE

APR 4 1979			
JUN 7 1979			
APR - 1995			

The Interpretation of
Geological Phase Diagrams

A SERIES OF BOOKS IN GEOLOGY
Editors:
James Gilluly
A. O. Woodford
Thane H. McCulloh

The Interpretation of Geological Phase Diagrams

ERNEST G. EHLERS

The Ohio State University

W. H. FREEMAN AND COMPANY
San Francisco

Printed in the United States of America

Library of Congress Catalog Card Number: 75–182129

International Standard Book Number: 0–7167–0254–1

2 3 4 5 6 7 8 9

Contents

Preface

Most textbooks currently used in geology courses include rather sketchy presentations of phase equilibria, and the student is often expected to memorize a few phase diagrams without gaining any real understanding of the nature of the various mineralogical reactions that such diagrams record. There is also a large gap between the conventional petrology text and the phase equilibrium data in the literature. This gap has widened in recent years with the proliferation of information on high-pressure, hydrothermal, and various gas-containing systems.

This book is intended for supplementary use in graduate and undergraduate courses in petrology and for reference use. Emphasis is on the interpretation and understanding of simple and moderately complex phase diagrams; the treatment of the subject is not intended to be exhaustive. Most of the examples chosen for discussion are of pertinent geological systems; others, some of which are hypothetical, are included to demonstrate particular types of diagrams or reactions. Much attention is given to systems that are under either fluid pressure or confining pressure. Bibliographic references are given at the back of the book to enable the reader to consult original works for more extensive descriptions of experimental techniques or for information on the geological ramifications of laboratory studies.

The approach used here is essentially the traditional method of the petrologist—the description of the continuous changes that take place in the kind and number of phases along paths of equilibrium crystallization or melting, as heat and/or pressure is applied or withdrawn from the system. The thermodynamics and crystal chemistry behind the phase relations presented are not discussed. References to these topics do, however, appear in the bibliography.

I wish to thank the staff and students of the Department of Mineralogy of The Ohio State University and of the Vening Meinesz Laboratory of The University of Utrecht, Netherlands, for their many discussions and helpful assistance. In particular, I wish to thank Dr. Charles Shultz for his critical reading of the manuscript, Mrs. Kathleen Wuichner for her assistance in typing, and my wife, Diane, for her patience and encouragement.

January 1972   *Ernest G. Ehlers*

1

Pertinent Definitions and the Phase Rule

In order to read the geological literature on phase equilibria and understand the implications of the Phase Rule (the basis of the classification of equilibrium relationships), one must be familiar with some of the standard terminology. The few terms defined here will be used throughout the book, and should be understood before the reader proceeds further.

System

A system is any part of the universe that has been isolated for the purpose of considering changes that take place within it in response to differing conditions. A system may be a liquid within a beaker, a magma chamber, or even an entire planet. Usually we shall regard a system as a particular chemical substance or group of substances that exists independent of quantity or location.* Thus one may consider as a system any mixture of the three oxides CaO, Al_2O_3, and SiO_2. The oxides may exist together on the moon or in a submarine in the depths of the Sea of Okhotsk. Systems may be further subdivided into closed or open. A *closed system* is one that changes only by receiving energy from the external environment or by yielding energy to it; an *open system* may exchange both matter and energy with the external environment.

---

*Normally such variables as gravity or magnetism are not taken into consideration.

Equilibrium

A system may be either at equilibrium or nonequilibrium. A system at equilibrium is in its lowest energy state consistent with the imposed conditions; it has no tendency to change spontaneously. A nonequilibrium system is one that is either changing or has a tendency to change. Similarly, a system may be defined as stable, metastable, or unstable. A *stable system* is one that is at equilibrium, whereas a *metastable system* is one that may appear to be at equilibrium but in fact is not at its lowest energy state (Fig. 1). Many of the

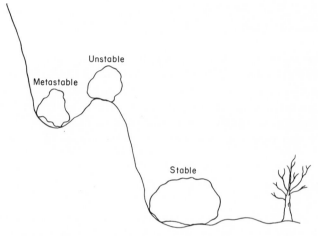

FIGURE 1
The various states of equilibrium, assuming ground level to
represent the most stable state.

denser minerals, such as diamond, kyanite, jadeite, or coesite, are truly stable only under pressures higher than one atmosphere; they persist metastably because reaction rates at low pressures and temperatures are extremely slow. Similarly, many minerals formed at high temperature, such as sanidine or cristobalite, are metastable at room temperature. An *unstable* mineral or mineral assemblage is one that is either in the process of changing to a more stable arrangement or just on the verge of doing so.

Phase

A phase is any part of a system that is physically distinct and mechanically separable from other parts of the system. There exists a boundary surface between different phases. Phases may exist in the solid, the liquid, or the gaseous state. Gases are completely miscible with each other, regardless of composition; consequently there can be only a single gas phase in a system at equilibrium. Although many liquids can be mixed in any proportion to form a single phase, some are partially or completely immiscible (such as oil and water); such liquids remain as separate phases with a distinct boundary surface between them. Some solids have very strict compositional limits (like quartz, which can exist only as relatively pure SiO_2); others show a wide variation in chemical composition: the plagioclase feldspars may have compositions ranging from $NaAlSi_3O_8$ to $CaAl_2Si_2O_8$. Because of the compositional limits of many solid phases, there is commonly more than one solid phase in a system.

Phase Diagram

A phase diagram is a graphic representation of the assemblage of phases that exist in a system as a function of the imposed conditions. The conditions that describe the system are usually taken as pressure, temperature, and composition, although other variables may be used. The phase assemblages indicated on the diagram are normally equilibrium (minimum energy) assemblages, but occasionally the diagrams are used to show nonequilibrium relationships.

Phase Rule

A postulate of basic importance in the classification and use of phase diagrams is the Phase Rule, derived by J. Willard Gibbs in the 1870's (see Gibbs, 1961) and later explored in more detail by many others. The thermodynamic basis of this rule is discussed in most standard textbooks on physical chemistry. The usual statement of the Phase Rule is:

$$P + F = C + 2$$

where P = the number of phases, F = degrees of freedom, and C = minimum number of components. The term C, *component*, refers to the minimum number of chemical constituents that are necessary and sufficient to describe the composition of all phases within the system. This immediately allows any system to be classified according to the number of chemical constituents present. A system that, under various conditions, may consist of water, ice, or steam, or various combinations of these, must be classified as a one-component system, since all of the phases consist of the same composition, H_2O.

Consider a system that, under various conditions, contains quartz (SiO_2), tridymite (SiO_2), cristobalite (SiO_2), and the pyroxene enstatite ($MgSiO_3$). There are two different phase compositions represented, and the system would be considered a two-component system. The choice of components in this example would normally be SiO_2 and $MgSiO_3$. If an additional phase, such as the olivine forsterite (Mg_2SiO_4) were possible in this system, the system must

4

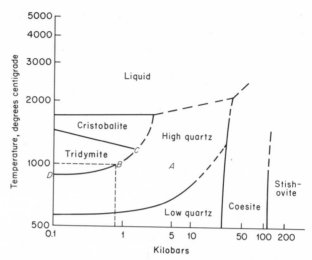

FIGURE 2
The one-component system SiO_2. [After Ostrovsky (1966); the coesite-stishovite boundary line has been revised in accordance with new data (personal communication from Ostrovsky).]

still be regarded as two-component, because all possible phases could be made up of SiO_2, Mg_2SiO_4, or a mixture of these (e.g., $SiO_2 + Mg_2SiO_4 = 2MgSiO_3$). The addition of the mineral periclase (MgO) to the system would not change the number of components, as all phases would be regarded as consisting of SiO_2, MgO, or a mixture of these.

If one were considering a system in which calcite ($CaCO_3$) undergoes a thermal decomposition to lime (CaO) and CO_2, it would be incorrect to regard this as a three-component system (Ca-C-O), because all of the phases can be represented in composition by two components, CaO, CO_2, or a mixture of these. The Phase Rule does not stipulate the nature of the particular components chosen, but merely the minimum necessary number. For convenience in graphical representation certain obvious combinations are chosen over others.

The term F, degrees of freedom (variance), can best be discussed with reference to a phase diagram. Figure 2 shows the one-component system SiO_2. Temperature in degrees centigrade is shown increasing logarithmically on the vertical axis, and pressure in kilobars is shown increasing to the right logarithmically on the horizontal axis. *One kilobar is equal to 1000 bars, and 1 bar is equal to 0.9869 atmosphere.* Every point within the diagram represents a particular pressure and temperature. Furthermore, the interior of the diagram is subdivided to indicate the stability regions of the various SiO_2 phases. Thus the diagram can be used to determine the phase or phases that exist at equilibrium for any pressure and temperature. In addition to being able to exist as gas or liquid, SiO_2 is able to exist in six different solid structural modifications (polymorphs). At point A (1000°C and 5 kb.), high quartz is the stable phase; at point B, high quartz and tridymite exist together; and at C, high quartz, tridymite, and cristobalite exist together.

The term F, *degrees of freedom*, can be defined as the minimum number of variables that must be fixed in order to define perfectly a particular condition of the system. To do this it is sometimes sufficient to indicate the number of phases present, but often the temperature and pressure must also be stated.

If one stipulates that the conditions are such as to allow the three phases cristobalite, tridymite, and high-quartz to exist together, the phase diagram shows that there is only one point, C, at which this is possible. The coexistence of these three phases is possible only at a particular pressure and temperature. The pressure and temperature are found from the coordinates of the diagram. As it is not necessary to state PT values to define the condition of the system, the number of degrees of freedom is zero. The condition of the system is perfectly defined by the statement that the three phases coexist. This may be derived from the Phase Rule by substituting the proper numbers:

$$P + F = C + 2$$
$$3 + F = 1 + 2$$
$$F = 0$$

Three phases coexisting in a one-component system lead to a condition where $F = 0$, a condition of invariance. The point C is an invariant point. Neither P nor T may vary (no freedom) without causing one or two of the phases to be eliminated.

Conversely, although both tridymite and high-quartz coexist at the point B, the coexistence of these two phases does not uniquely define the P and T of the assemblage, as there is a variety of combinations of P and T (along the line CBD) at which these two phases coexist. To define the condition of the system perfectly, one must be able to determine the pressure and temperature of the phase assemblage. This can be achieved by stipulating either the pressure or temperature of the two-phase assemblage. If tridymite and high-quartz coexist stably at 1000°C, it can be deduced from the diagram that the pressure on the system is somewhat less than 1 kb. There is one independent and one dependent variable. One independent variable is another way of stating that this condition has one degree of freedom or one degree of variance. For this reason, such transition lines are referred to as univariant curves. This condition is easily derived from the Phase Rule by substituting the proper values for the terms:

$$P + F = C + 2$$
$$2 + F = 1 + 2$$
$$F = 1$$

The various transition curves throughout the diagram are all univariant curves, and each possesses one degree of freedom.

To state merely that the condition of the system is such that high quartz exists alone (such as at point A) is not sufficient to define the condition of the system completely. High quartz can exist in a wide variety of P and T combinations. Pressure and temperature can be varied independently without changing the phase assemblage. The phase field of high quartz, as well as those of the other single phases, is a divariant region. Use of the Phase Rule indicates two degrees of freedom.

The Phase Rule not only allows classification of systems, but is a great aid in interpreting the diagram of the system. In the diagram for the SiO_2 system (Fig. 2), it can be noted that there is no region or point at which more than three phases can stably coexist. If a four-phase assemblage were postulated for this system, the Phase Rule would indicate that this was an impossible condition for a stable assemblage, since negative degrees of freedom are indicated:

$$P + F = C + 2$$
$$4 + F = 1 + 2$$
$$F = -1$$

Because the term F gives the minimum number of variables that must be fixed to define the condition of the system, the number must be either positive or zero. The maximum number of stable coexistent phases in a system will be present when $F = 0$. This argument can be applied to multicomponent systems; for example, it can be immediately deduced that the maximum number of stably coexisting phases in a 20-component system is 22. If additional phases are present, the system is not in equilibrium. This criterion has been used in metamorphic petrology to determine whether a particular mineral assemblage is in equilibrium (see Turner, p. 55, 1968). If the number of phases present exceeds that stipulated by the Phase Rule, nonequilibrium is proven, but the converse is not necessarily true. Although the number of phases present may be consistent with equilibrium conditions, some phases may be unstable under the set conditions. Applying the Phase Rule to metamorphic assemblages is often difficult because of lack of precise knowledge of the number of components present.

In summary then, it is possible to classify systems in terms of components. From the Phase Rule, and the number of phases present, the degrees of freedom of any phase assemblage can be determined. A divariant phase assemblage (such as the high-quartz field, Fig. 2) may be subject to change in two variables—pressure and temperature (or concentration, in systems of more than one component); these changes may take place independently. A univariant phase assemblage (curve CBD, Fig. 2) may be maintained if a change of one variable is accompanied by a dependent change in a second variable. An invariant phase assemblage (point C, Fig. 2) can be maintained only if pressure, temperature, composition, or any other dependent variables are not allowed to vary. It is the task of the experimental investigator to specify the conditions of invariancy and univariancy.

The reader should be cautioned that although many experimentally determined reactions satisfy requirements of the Phase Rule, they may not represent true equilibrium relations. The mere fact that an investigator has repeatedly synthesized a phase under particular conditions of pressure and temperature does not necessarily mean that that phase is the most stable phase under those conditions. It may merely indicate that the phase is more stable than the original starting materials. Given sufficient time (perhaps years rather than days), the mineral may convert to a still more stable form. The question of metastable reactions is particularly important in relatively low-temperature hydrothermal reactions. In an excellent discussion on the determination of equilibrium in experimental systems, Fyfe (1960, p. 565) concludes with the following statement: "Where, as has commonly happened, experimental results conflict with inferences based on geological observations, the experimentalist has a special responsibility to scrutinize and state clearly the limitations of his laboratory procedures." Similarly, the man in the field should remind himself of these limitations when applying experimental data to field problems.

2

Binary Systems

Because one-component (unary) systems are usually described in terms of variation in both temperature and pressure, they will not be discussed here; instead, they are taken up in Chapter 6, "Systems Under Confining Pressure." We therefore begin with the study of two-component (binary) systems at atmospheric pressure. For systems under atmospheric pressure, the Phase Rule (normally written $P + F = C + 2$) must be altered to $P + F = C + 1$, as pressure is not a variable. In this form it is called the Condensed Phase Rule, since any gas phase may be regarded as condensed to a liquid or as present in negligible quantities.

The simplest type of equilibrium relationship is exemplified by the behavior of the two-component system $CaAl_2Si_2O_8$ (anorthite) and $CaSiTiO_5$ (sphene), shown in Figure 3. The abscissa indicates composition, with pure $CaSiTiO_5$ at the left and pure $CaAl_2Si_2O_8$ at the right. The percentage of the component $CaAl_2Si_2O_8$ increases from left to right, as indicated by the numbered scale. The percentage of the other component increases from right to left, and is not indicated by a numbered scale. In most phase diagrams the percentages of components are given in weight percent; in some they are given in mole percent. The ordinate indicates temperature, which in most diagrams is stated in degrees Centigrade; diagrams prepared for use in ceramics, however, may state the temperature in degrees Fahrenheit.

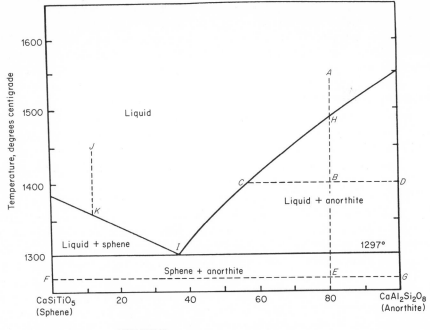

FIGURE 3
The binary system $CaSiTiO_5$—$CaAl_2Si_2O_8$.
[After Prince (1943).]

Any point within a phase diagram specifies both a composition and a temperature. The solid lines within the diagram outline regions, or fields, of temperature and composition, and the phase assemblages that exist in equilibrium within these fields are indicated by labels. As we are here considering only condensed systems, these regions are limited to assemblages of four kinds —liquid, liquid and solids, solids, and, rarely, liquid and liquid. Depending on the complexity of the system, a phase diagram may contain more than one of each kind of region. Phase diagrams are prepared from data obtained by heating samples of various mixtures of minerals, chemicals, or glasses to particular temperatures and then determining the phase assemblages that exist at those temperatures. For silicate systems this is usually accomplished by rapidly chilling (quenching) heated samples, thus preserving the high-temperature phase assemblages metastably at lower temperatures.

Observe first that in Figure 3 the various regions of the diagram are labeled as consisting of either one or two phases—liquid, liquid and sphene, liquid and anorthite, and sphene and anorthite. Any point in a one-phase field consists of a single phase with a composition indicated directly below, on the abscissa. A sample corresponding to point A within the diagram is a liquid phase with a composition of 80 weight percent $CaAl_2Si_2O_8$ and 20 weight percent $CaSiTiO_5$. A sample of the same bulk composition, but at a lower temperature (point B), lies within a two-phase region; the compositions of the two phases in that region are found by drawing a horizontal line through the point in question to the extremes of the region—in this example, the points C and D. Such a line is referred to as a *tie line*, or less commonly as a *conode*. This line indicates the compositions of the two phases that are in equilibrium with each other. The composition of the liquid phase, indicated at C, is found by reference to the abscissa to be about 57 weight percent $CaAl_2Si_2O_8$ and 43 weight percent $CaSiTiO_5$. The composition of the solid phase, indicated at point D, is that of pure anorthite, $CaAl_2Si_2O_8$. If this same bulk composition were cooled to point E within the two-phase field sphene and anorthite, it would consist of these two crystalline phases, whose compositions are read at the extreme left and right of the diagram, at points F and G; these points are connected by a tie line through E.

The tie line not only indicates the individual compositions of coexisting phases at any point within two-phase fields, but also indicates the relative *amounts* of each that exist under equilibrium conditions. Consider again the tie-line that joins liquid C with coexisting anorthite at D. It can be seen that if the bulk composition of the sample (indicated by B) is closer to the composition of one of the two phase, then the one to which it is closest must be present in the largest amount. If the bulk composition falls exactly at the midpoint between the compositions of the two phases, the phases are present in equal amounts. It is possible, therefore, to measure the distances along the tie line from the bulk composition point to each of the phase composition points and determine the relative amount of each phase present. In this example, the length CB is a measure of the relative amount of anorthite present, BD the amount of the liquid phase, and CD the total. Putting this another way we can say that

$$\frac{BC}{CD}(100) = \frac{2.25 \text{ cm}}{4.15 \text{ cm}}(100) = 54.2 \approx 54\% \text{ anorthite}$$

$$\frac{BD}{CD}(100) = \frac{1.90 \text{ cm}}{4.15 \text{ cm}}(100) = 45.8 \approx 46\% \text{ liquid}$$

This simple but extremely useful relationship has been graced with the name *Lever Rule*.

Figure 3 indicates that the mineral sphene melts at 1382°C and anorthite at 1550°C. Melts of pure sphene or anorthite will completely freeze or melt at these temperatures. Liquids formed by melting mixtures of these two minerals are completely miscible, as is shown by the single continuous liquid region in the diagram. Melts of intermediate composition have a complex freezing or melting behavior. The melt labeled A in Figure 3 consists of 80 percent of the $CaAl_2Si_2O_8$ component and 20 percent of the $CaSiTiO_5$ component. Although this melt consists mainly of $CaAl_2Si_2O_8$, it will not freeze at 1550°C, as does pure $CaAl_2Si_2O_8$. The effect of admixing $CaSiTiO_5$ is to depress the freezing point of the melt below that of pure $CaAl_2Si_2O_8$. The liquid can be cooled to a temperature of about 1490°C before freezing begins (point H). At temperatures somewhat below 1490°C the diagram shows a field of liquid + anorthite. Crystallization therefore begins with the precipitation of anorthite crystals. As the $CaAl_2Si_2O_8$ component is removed from the melt by crystallization of anorthite, the remaining melt becomes richer in $CaSiTiO_5$; its composition shifts toward the left side of the diagram, and its freezing point decreases. These changes in composition and freezing temperature of the residual liquid can be traced along the path HCI, which shows graphically what happens to the liquid phase as heat is removed from the system and anorthite continues to precipitate. By using the Lever Rule, one can determine the phase compositions and the relative proportions of each phase for all temperatures in the two-phase region.

Continued removal of heat and precipitation of anorthite eventually brings the composition of the liquid phase to the point I, which is the minimum temperature point of the liquid field, or the *eutectic point*. No liquid can be present below the eutectic temperature. The horizontal line drawn through I is called the *solidus*; it defines the upper limits of those fields that include only solid phases. The two sloping curves that mark the lower limits of the one-phase liquid field are called *liquidus lines*.

At the eutectic, both solid phases begin to crystallize simultaneously from the remaining melt. The relative amounts of the two solid phases precipitating at the eutectic point are determined by measuring the two legs of the tie line drawn through I, which gives 37 percent anorthite and 63 percent sphene. As these two phases crystallize in a fixed ratio, the composition of the liquid phase remains constant. The freezing temperature also remains constant until the last drop of liquid is consumed, after which further removal of heat allows the temperature to fall. No matter how much further the system is cooled, sphene and anorthite will continue to exist in stable equilibrium. The final phase assemblage will consist of 80 percent anorthite and 20 percent sphene, as can be seen from the lengths of the two legs of a tie line such as FEG. The final percentage of each solid phase is a combination of the amount that formed in fixed proportions at the eutectic plus the amount that formed earlier above the eutectic temperature (in this case during the temperature interval HI). A final crystallization mixture that is in eutectic proportions can only result if the original melt was of eutectic composition.

It has been shown that the relative amounts of *two* coexisting phases can always be determined readily by means of the Lever Rule (as for points B and E). For reactions that occur at invariant points (such as I), where three phases coexist, phase proportions or percentages *cannot* be determined by means of the Lever Rule because the relative amounts of phases present depend upon the amount of heat that has been removed from the system. Proportions can, however, be determined at the beginning or end of such reactions.

The crystallization history of a liquid, such as A, may be deduced during experimentation by observing the variation in cooling rate. Assume that the melt A (Fig. 3) is allowed to cool naturally and is monitored by a temperature-sensing device, such as a thermocouple, and a continuous strip chart recorder.

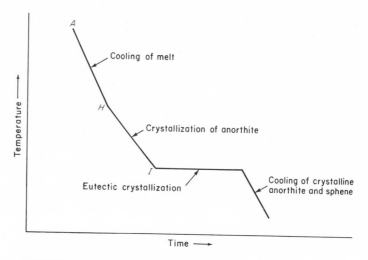

FIGURE 4
Schematic sketch of cooling rates for sample A in the binary system
$CaSiTiO_5$—$CaAl_2Si_2O_8$. Lettered points correspond to those in
Figure 3.

The time-versus-temperature plot, or *cooling history,* would resemble that
shown in Figure 4. Initially the melt will cool quickly. When anorthite begins
to crystallize (at H), the rate of cooling is slowed because of the heat of crystallization. When the melt has cooled to the eutectic temperature, at I, anorthite
and sphene crystallize together, and the temperature remains fixed until the
liquid is exhausted (as can be seen by applying the Condensed Phase Rule,
$P + F = C + 1, 3 + F = 2 + 1, F = 0$). When crystallization is complete the
system cools again at a rapid rate.

A melt of any randomly chosen composition would crystallize according
to the same general pattern as that shown by melt A. For example, crystallization of a melt of composition J (Fig. 3), when cooled to the liquidus line
at K, begins with the precipitation of sphene. With continuous removal of
heat, additional sphene forms, and the remaining liquid becomes enriched
in $CaAl_2Si_2O_8$. The liquid changes in composition, as shown by the liquidus
line, until the system has cooled to the eutectic temperature at I. Sphene
and anorthite then precipitate simultaneously at a fixed temperature and in
a fixed ratio until crystallization is complete.

When a liquid whose composition corresponds to that of the eutectic point
cools to the eutectic temperature, all crystallization takes place at a fixed
temperature, as has been verified by using the Condensed Phase Rule; a three-
phase assemblage has no degrees of freedom and is therefore *invariant.* The
temperature and phase compositions of a three-phase assemblage must remain
fixed until at least one of the phases is eliminated.

Melting in the system $CaSiTiO_5$-$CaAl_2Si_2O_8$ (Fig. 3) proceeds in a manner
exactly opposite to that of freezing. Assuming that the solid is an intimate
mixture of sphene and anorthite, both minerals will begin melting simulta-

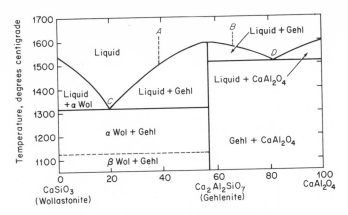

FIGURE 5
The binary system CaSiO₃—CaAl₂O₄. [After Rankin and
Wright (1915); modified by Osborn and Schairer (1941).]

neously when heated to the eutectic temperature, and the liquid will be of
eutectic composition. With additional heating, one of the solid phases is eventually eliminated, and the melt then changes in composition and temperature
as indicated by the appropriate liquidus line (depending upon whether sphene
or anorthite is in excess) until the last crystal has been melted.

Very rapid cooling (quenching) may result in the metastable preservation
of high-temperature equilibrium phase assemblages. Thus it is possible to
quench melt *A* (Fig. 3) and preserve it as a glass. Similarly, an assemblage
of melt and anorthite (point *B*) can be quenched to an assemblage of anorthite
crystals in a glass matrix.

Removal of crystals from a cooling melt in a simple binary system has no
effect on the course of crystallization; a cooling melt from which crystals
are removed will continue to change in composition as shown by the liquidus
until it has cooled to the eutectic temperature. The reason is that the crystals,
once formed, do not chemically react with the remaining liquid. In many
other systems, however, liquid-crystal reactions do occur, and the presence
of crystals in the melt or their removal will significantly affect the course of
crystallization.

Nonequilibrium effects of rapid cooling during crystallization in systems of
this type are discussed in detail by Buckley (1951) and Turner and Verhoogen
(1960, pp. 47–49, 97, 98). A method of producing layered igneous bodies by
undercooling is discussed by Taubenick and Poldervaart (1960, p. 1317).

INTERMEDIATE CONGRUENTLY MELTING COMPOUNDS

A slightly more complicated example of a simple binary system is that of
CaSiO₃—CaAl₂O₄ (Fig. 5), a system that is of primary interest to the cement
industries rather than geology. Between the two end-members, wollastonite

(CaSiO$_3$) and CaAl$_2$O$_4$, there exists an intermediate compound – the mineral gehlenite (Ca$_2$Al$_2$SiO$_7$). When heated, gehlenite melts completely at 1595°C to form a liquid that has the same composition as the solid. Melting of this type, which might be regarded as "normal," is termed *congruent*. The end-member compounds melt congruently as well.

This system can be treated as two simple binaries, CaSiO$_3$—Ca$_2$Al$_2$SiO$_7$ and CaAl$_2$O$_4$—Ca$_2$Al$_2$SiO$_7$. As liquids more silica-rich than gehlenite (such as *A*) crystallize, they change in composition until they reach the binary eutectic composition at *C*, whereas liquids more alumina-rich than gehlenite (such as *B*) change in composition until they reach the binary eutectic at *D*. Depending upon the ratio of the two components in the initial melt, the final crystalline solid may be a mixture of wollastonite and gehlenite, a mixture of gehlenite and CaAl$_2$O$_4$, or pure gehlenite, wollastonite, or CaAl$_2$O$_4$.

Another feature of interest in this diagram is the polymorphism of CaSiO$_3$. This compound undergoes a structural change at about 1125°C. The lower left portion of the phase diagram is subdivided to indicate the fields in which gehlenite coexists either with the low-temperature β form or with the high-temperature α form of CaSiO$_3$. The transformation from one polymorph to the other occurs in the solid state and has no effect on the crystallization or melting relations.

INCONGRUENT MELTING

The binary system KAlSi$_2$O$_6$—SiO$_2$ (Fig. 6) is of particular interest because of the behavior of the intermediate compound, potash feldspar. Unlike the intermediate compound in the system shown in Figure 5, when this mineral is heated to about 1150°C it reacts to form a mixture of liquid and crystals, and neither phase has the composition of the original solid. This type of reaction is known as *incongruent melting*. The end members in this system melt congruently. The compositions of the phases formed at the incongruent melting point *A* are found at the ends of the horizontal tie line drawn through that point; one phase is leucite (at point *B*) and the other a liquid of composition *C*. The relative proportions of these two phases are found by applying the Lever Rule (see p. 10).

The incongruent behavior of this feldspar is observed in both freezing and melting. Consider a liquid *D*, identical in composition to potash feldspar. When this is cooled to the liquidus at *E*, crystallization begins with the formation of leucite. With continued cooling, the liquid becomes increasingly enriched in the SiO$_2$ component, as is indicated by the slope of the liquidus, and eventually attains the composition shown at *C*. At this temperature potash feldspar is stable, and the following reaction occurs:

$$\text{Liquid} + \text{Leucite} \longrightarrow \text{Potash feldspar}$$

The reaction is invariant because it involves three phases, and therefore proceeds to completion at a fixed temperature, as is obvious from the diagram.

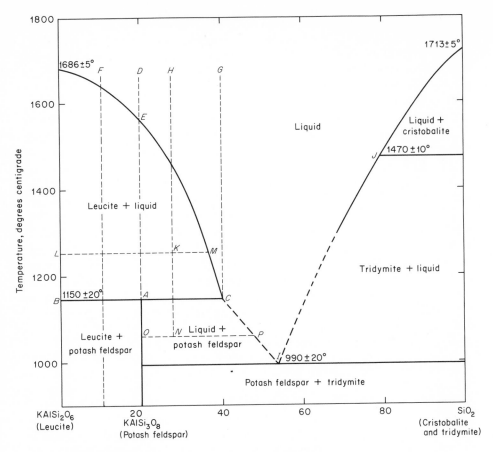

FIGURE 6
The binary system KAlSi$_2$O$_6$—SiO$_2$. [After Schairer and Bowen (1948).]

Once the feldspar forms, it remains stable throughout cooling to room temperature. The reaction that occurs when the liquid reaches C is said to be *peritectic* rather than eutectic, and the point C is called a *peritectic* or *reaction point.*

It is always possible, of course, to determine the correct phase sequence for either heating or cooling merely by drawing a vertical line (such a line of constant composition is called an *isopleth*) through the composition of interest and noting the fields that must be encountered during heating or cooling. This does not, however, directly reveal the particular reactions that occur, but rather the sequence of phase assemblages, which can be read from the fields that the line passes through. For example, it can be seen by inspection that a liquid of composition F (Fig. 6) forms leucite + liquid when cooled, and finally leucite + potash feldspar. Any cooling liquid whose composition lies between leucite and potash feldspar eventually forms (under equilibrium conditions) a mixture of leucite and potash feldspar. During cooling, liquids in this range of compositions initially precipitate leucite and shift in composition to the peritectic, *C*, at which point the reaction to produce potash feldspar takes place. As is indicated by the original composition of the liquid, there is an excess of leucite over liquid. Consequently, not all of the leucite is consumed by the liquid during the reaction to produce feldspar. The final crystalline mixture consists of feldspar formed by the peritectic reaction plus the excess leucite not consumed in that reaction.

Similarly, liquids with compositions between points *D* and *G* yield mixtures of potash feldspar and silica when cooled to room temperature; this is, in fact, true of any composition richer in SiO_2 than the one indicated by point *D*. A liquid at *H*, when cooled to the liquidus, initially precipitates leucite. With continued precipitation, the liquid shifts in composition to the peritectic, *C*. Since liquid is present in excess, all of the leucite and some of the liquid react to produce potash feldspar. When the reaction is completed at this temperature (point *C*) and all of the leucite is consumed, an excess of liquid remains. With further cooling, some of the liquid crystallizes to form more feldspar, and the liquid changes in composition as indicated by the feldspar liquidus until the eutectic composition *I* is reached. At the eutectic, both feldspar and tridymite crystallize in eutectic proportions until the batch is solidified.

Examination of the silica-rich part of the diagram reveals that silica exists in two different structural states, or polymorphs, within the temperature range of the diagram. Above approximately 1470°C cristobalite is the stable phase, and below that temperature, tridymite. Any silica precipitated above 1470°C is cristobalite. The diagram indicates that, upon cooling (under equilibrium conditions), a structural rearrangement of silica will convert cristobalite to tridymite. Actually, however, this change takes place extremely slowly, and cristobalite not uncommonly persists metastably to room temperature. Any silica precipitated below 1470°C should crystallize as tridymite. Although the change is not indicated on the diagram, both of these polymorphs should convert to quartz before reaching room temperature.

The liquidus surfaces for both cristobalite and tridymite in this system are seen as a continuous curved line from the melting point of pure silica to the eutectic at *I*. In most systems in which a polymorphic transition takes place at liquidus temperatures, there is a slight change in the slope of the liquidus at the conversion temperature. This may be seen at point *D* in Figure 9.

From an examination of the stability fields at the bottom of the phase diagram shown in Figure 6, it can be seen that the possible end products of reactions that proceed under equilibrium conditions are pure leucite, pure potash feldspar pure tridymite, or mixtures of either potash feldspar and leucite, or potash feldspar and tridymite. The diagram shows that leucite and a silica polymorph should never exist stably together. But in nature, not all reactions proceed under equilibrium conditions, and occasionally nonequilibrium mineral assemblages are produced—particularly when cooling proceeds too rapidly to maintain equilibrium. Assume that a liquid of composition *D* has cooled to the peritectic. At this temperature it consists of leucite and liquid. Under equilibrium conditions the liquid and leucite should react completely at this temperature to form potash feldpar. But if the system is cooled too quickly at this stage, the reaction may not go to completion, in which case only the outer parts of the leucite crystals might be converted to feldspar. The liquid-crystal reaction begins at the surfaces of the leucite crystals, and a reaction rim of feldspar is formed. In order for complete reaction to proceed, it is necessary for the liquid to diffuse through this outer layer to the unaffected central part of the crystal. Given sufficient time, that is just what happens, and the original leucite crystals are completely converted to potash feldspar. But if cooling takes place too rapidly, the process remains incomplete, producing armored crystals. The remaining liquid might quench to a glass or continue precipitating feldspar along the liquidus from C to I until final crystallization of potash feldspar and tridymite takes place at the eutectic. Should the liquid follow the second of these two possible nonequilibrium paths, the final assemblage would consist of potash feldspar, leucite, and tridymite. The proportion of the phases produced under nonequilibrium conditions cannot be calculated from the Lever Rule, and will vary as a function of the cooling rate. Similarly, if the heating rate is extremely rapid it is possible to convert potash feldspar directly to a liquid of similar composition without the intermediate mineral leucite ever forming.

COMPLETE SOLID SOLUTION

Some minerals are fixed in composition, and others variable. A mineral of variable composition is said to show *solid solution*. The plagioclase feldspars ($NaAlSi_3O_8$—$CaAl_2Si_2O_8$) show complete solid solution between the pure sodic and pure calcic end-members. In fact, the most common plagioclases are of intermediate composition rather than pure end-members. The compositions are usually given in an abbreviated form, such as An_{67}, which denotes a plagioclase consisting of 67 weight percent of the $CaAl_2Si_2O_8$ component (anorthite) and 33 weight percent of the $NaAlSi_3O_8$ component (albite). The system $NaAlSi_3O_8$—$CaAl_2Si_2O_8$ (Fig. 7) was first investigated by Bowen in 1913, and was one of the earliest to be examined for geological and ceramic interpretation.

The first thing to notice about a system of this kind is that a liquid of any composition can crystallize to produce a plagioclase of the same composition. Only under nonequilibrium conditions can plagioclase of two different compositions be produced from a single liquid.

In their simplest form, phase diagrams of systems that show solid solution are characterized by two curves. Above the upper curve, the liquidus, is a field consisting of a liquid phase; below the lower curve. the solidus, is a field consisting of a solid phase. Between the liquidus and solidus curves, liquids and solids coexist in various mixtures. The compositions and relative amounts of coexisting liquid and solid phases are found by means of horizontal tie lines. A sample of the composition and temperature of A (Fig. 7) consists of liquid of composition B and crystals of composition C. We know from the Lever Rule that the length of AC is proportional to the amount of liquid

18

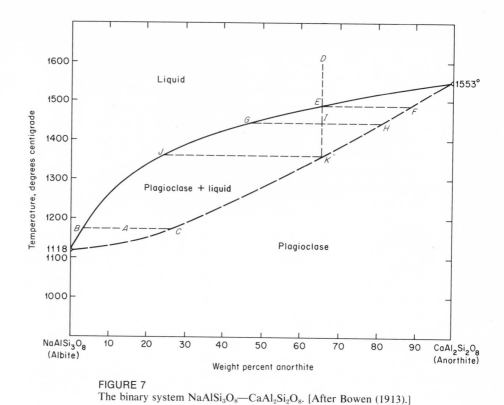

FIGURE 7
The binary system $NaAlSi_3O_8$—$CaAl_2Si_2O_8$. [After Bowen (1913).]

present and that the length of BA is proportional to the amount of crystals present.

The arrangement of liquidus and solidus lines in systems of this type leads to an interesting type of crystallization. A liquid D, when cooled to the liquidus at E, begins to crystallize; the first-formed crystals are of the composition F. As the crystals are richer in the $CaAl_2Si_2O_8$ component than the original liquid, this causes the composition of the remaining liquid to become enriched in the $NaAlSi_3O_8$ component. A consequence of this change in liquid composition is a depression in the freezing point of the liquid, which follows the path shown by the liquidus line. But as can be seen by examining the horizontal tie lines, the liquid cannot exist in equilibrium with crystals that have previously precipitated; that is, a liquid of composition G can coexist stably only with crystals of the composition H, rather than the earlier-formed crystals of composition F. The liquidus line shows how the liquid changes in composition as it cools; at each point on the liquidus, the liquid precipitates only crystals that are in equilibrium (i.e., liquid G may precipitate H, liquid J may precipitate K, etc). Simultaneously, the liquid reacts with previously formed crystals by diffusion processes and changes their composition. Thus, as the liquid continuously changes in composition from E to J along the liquidus, the composition of all precipitating and previously precipitated crystals simultaneously and continuously changes from F to K. By the time the crystals

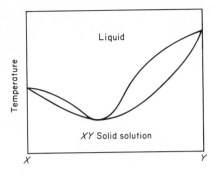 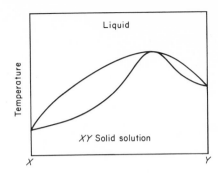

FIGURE 8
Hypothetical binary systems with complete solid solution and minimum and maximum melting points at intermediate compositions.

are of composition K, they have the same composition as the original liquid (D), and the last of the liquid is consumed. The plagioclase remains stable throughout cooling to room temperature (aside from various low-temperature structural changes not indicated on the diagram).

In systems of this type, where continuous reaction takes place between a changing melt and previously formed crystals, there is always a possibility of incomplete reaction due to nonequilibrium conditions. Incomplete reactions are common in both natural and synthetic plagioclase-containing systems as a result of variation in the rate of crystallization. The end-products of incomplete reaction between such a cooling melt and earlier-formed, calcium-rich crystals are compositionally zoned plagioclases with rims more sodic than the core. Such compositional zoning may be sharply defined or gradational, depending on the cooling history. If the cores of the plagioclase crystals are more calcic than the bulk composition of the system, it follows that the outer rims are more sodic than the bulk composition. This requires that the final melt composition change further in the direction of the sodic component than would be expected under equilibrium conditions.

Two variations on this type of diagram are seen in Figure 8. Both are of complete binary solid-solution systems, one with a temperature maximum and one with a temperature minimum. The melting or crystallization sequence is the same as that shown by the plagioclase system, with the exception that a liquid having the composition of a temperature maximum or minimum will crystallize directly to a solid phase of identical composition. Note that such a minimum point is not a eutectic, as only a single solid is produced upon cooling. Two solids form at a binary eutectic.

LIMITED SOLID SOLUTION

Minerals that show complete solid solution (such as the plagioclases, olivines, and melilites) are fairly common, but even more common are those that show limited solid solution. An example is the binary system $NaAlSiO_4$—$NaAlSi_3O_8$

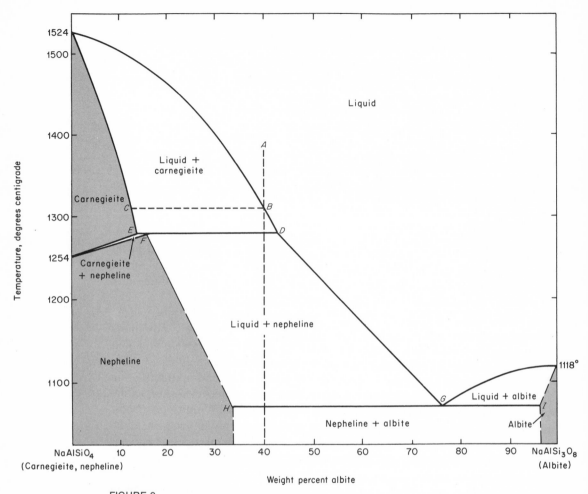

FIGURE 9

The binary system $NaAlSiO_4$—$NaAlSi_3O_8$. [After Greig and Barth (1938).]

(Figure 9). Note that there are three shaded areas on the diagram. Each is a composition-temperature field in which a single solid phase exists, the three solids being carnegieite, nepheline, and albite. At low temperatures, nepheline may exist as a single phase with as much as about 34 percent $NaAlSi_3O_8$ in solid solution; at higher temperatures, the compositional variation of nepheline becomes much more limited. At 1254°C pure nepheline ($NaAlSiO_4$) changes to carnegieite, a phase stable at higher temperatures; the degree of solid solution shown by carnegieite first increases and then decreases with increasing temperature. Albite has a rather limited compositional range; it exhibits solid solution with a maximum of about 4 percent $NaAlSiO_4$ in its structure.

If the limit of solid solution of a solid phase is exceeded by the addition of an excess of one or more components, the component in excess will react to form a second phase. For example, nepheline at 1200°C allows up to 23 weight percent $NaAlSi_3O_8$ in its structure. If additional $NaAlSi_3O_8$ is added, reaction will take place to produce a two-phase assemblage of liquid and nepheline (containing 23 weight percent $NaAlSi_3O_8$). If the limit of solid solution of nepheline is exceeded at low temperatures by the addition of $NaAlSi_3O_8$, a two-phase assemblage of nepheline and albite will result.

Let us examine what would happen if one attempted to crystallize nepheline from a melt containing more than 34 percent $NaAlSi_3O_8$. Consider the cooling path that begins with the melt A, which contains 40 percent of the $NaAlSi_3O_8$ component. The isopleth at 40 percent $NaAlSi_3O_8$ passes through the following phase fields: liquid, liquid and carnegieite, liquid and nepheline, and finally nepheline and albite. When the liquid A cools to the liquidus at B, carnegieite of composition C begins to precipitate. With continued cooling, reaction, and precipitation, the liquid shifts in composition to D on the liquidus as the co-existing carnegieite changes in composition to E. Below this temperature, carnegieite of that composition cannot exist, and the following reaction must take place under invariant conditions:

$$\text{Liquid } (D) + \text{Carnegieite } (E) \rightarrow \text{Nepheline } (F)$$

During the reaction, liquid, present in excess, decreases in amount, as can be seen from the Lever Rule. Once the system contains only nepheline of composition F and liquid of composition D, the temperature again decreases, with direct precipitation of nepheline from the liquid. During this next stage, continued precipitation of nepheline causes the liquid to change continuously in composition from D to G on the liquidus. Similarly, the nepheline changes in composition from F to H. With decrease in temperature, compositional changes take place in both the nepheline previously precipitated and that being precipitated in equilibrium with the liquid. As with the plagioclases, the earlier-formed crystals change in composition by reaction with the liquid of changed composition. The compositions and percentages of phases can be determined at any temperature by constructing tie lines between the nepheline solidus and liquidus lines. When the liquid phase has reached eutectic composition at G, another invariant point is encountered, and a new reaction takes place. All of the liquid in the system is of composition G; it crystallizes to produce albite of composition I and additional nepheline of composition H. The temperature can then decrease with nepheline and albite of essentially fixed composition existing stably together.

A quite different type of cooling behavior can be expected from a $NaAlSiO_4$-rich liquid, such as composition J. Refer to the enlarged portion of the diagram in Figure 10. The phase fields encountered along the isopleth during cooling are seen to be liquid, liquid and carnegieite, carnegieite, carnegieite and nepheline, and nepheline. In detail, the liquid J, when cooled to the liquidus surface, first begins to crystallize carnegieite of composition K. Continued cooling causes the liquid to become enriched in the $NaAlSi_3O_8$ component, as is shown by the liquidus. Simultaneously the earlier-formed carnegieite

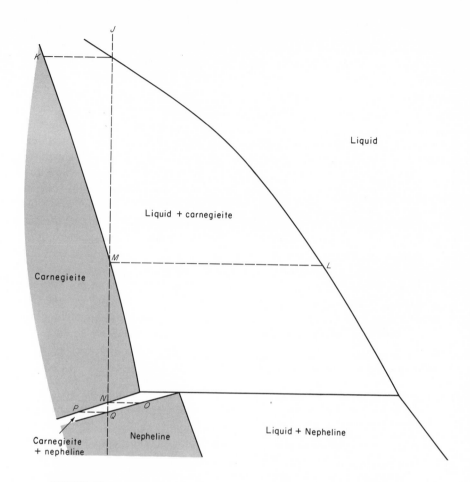

FIGURE 10
Enlargement of the NaAlSiO$_4$—rich part of the system NaAlSiO$_4$—NaAlSi$_3$O$_8$.

is changed in composition by reaction with the liquid, and newly-precipitated carnegieite is crystallized with an increasingly NaAlSi$_3$O$_8$-rich composition. This process continues until the liquid is completely consumed at L and all of the carnegieite crystals are of the composition M. The carnegieite crystals exist stably until the system reaches the temperature of point N, where they begin to convert in the solid state to the more stable low-temperature phase nepheline. Observe that there is here a very narrow two-phase region where carnegieite crystals of one composition coexist with nepheline crystals of a slightly different composition. The reason for the two-phase region is that the carnegieite and nepheline structures at this temperature do not have the same capability for incorporating the NaAlSi$_3$O$_8$ component in their structure.

As the carnegieite begins to convert to nepheline that is slightly enriched in NaAlSi$_3$O$_8$, the composition of the remaining carnegieite becomes slightly depleted in NaAlSi$_3$O$_8$. With decreasing temperature, the carnegieite shifts in composition to P and decreases in amount while simultaneously converting

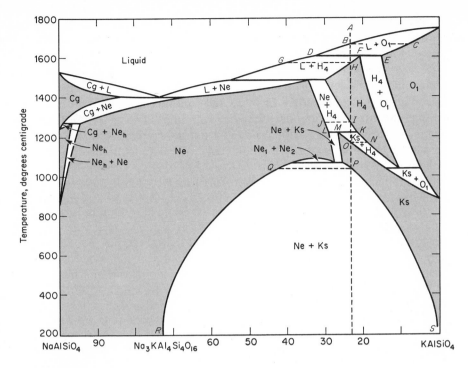

FIGURE 11
The binary system NaAlSiO$_4$—KAlSiO$_4$ (Cg = Carnegieite, L = liquid, Ne$_h$ = high-temperature nepheline, Ne = low-temperature nepheline, O$_1$ = orthorhombic KAlSiO$_4$, Ks = kalsilite, H$_4$ = tetrakalsilite). [After Tuttle and Smith (1958). All tentative (broken) lines have been made solid for clarification.]

to nepheline, which changes in composition from O to Q. The composition of coexisting carnegieite and nepheline phases can be found at intermediate stages by means of the usual two-phase tie lines. This process continues with the gradual elimination of carnegieite until the entire batch has been transformed to nepheline of the composition Q. Nepheline remains stable during additional cooling.

COMPLEX BINARY SYSTEMS

The simple binary systems covered so far illustrate most of the common features to be found in more complex binary systems. It can be seen by a perusal of the literature, as well as by the following example, that the more complicated-looking systems merely consist of various combinations of the features that have just been discussed.

An example of the complexities of polymorphism and solid solution is offered by the system NaAlSiO$_4$—KAlSiO$_4$ (Fig. 11). Many of the interior phase boundaries are inferred rather than precisely known. Nevertheless, the reader will be able to spend many happy hours studying the various complicated reactions. The shaded areas of the diagram indicate fields in which single solid phases exist. The composition $NaAlSiO_4$ is seen to consist of both high- and low-temperature varieties of nepheline of somewhat similar structure, in addition to the still higher-temperature modification, carnegieite. Potassium-rich kalsilite converts at high temperatures to orthorhombic $KAlSiO_4$, whereas soda-rich kalsilite converts at higher temperature to tetrakalsilite. The compositions of phases that exist at any temperature and composition point within the two-phase fields are, as usual, determined by means of tie lines.

The crystallization sequence of a melt A follows a complex pattern but contains no new features. When cooled to the liquidus at B, the liquid precipitates orthorhombic $KAlSiO_4$ of composition C. With further cooling, the liquid and coexisting crystals shift in composition to D and E, respectively. Here an invariant point is reached, and at a fixed temperature the following reaction occurs:

$$\text{Liquid } (D) + \text{Orthorhombic } KAlSiO_4 \ (E) \rightarrow \text{ Tetrakalsilite } (F)$$

Liquid D is present in excess, as the bulk composition falls to the left of point F (i.e., closer to liquid D). With additional cooling and precipitation of more tetrakalsilite, the liquid D changes in composition to G, where it is consumed. The tetrakalsilite is of composition H. The tetrakalsilite cools stably until the next two-phase field is reached at I, where tetrakalsilite begins to convert in the solid state to nepheline of composition J. Continued cooling and conversion brings the tetrakalsilite to composition K with coexisting nepheline of composition L. Here, where a third phase forms, is another invariant point, and the following reaction takes place:

$$\text{Nepheline } (L) + \text{Tetrakalsilite } (K) \rightarrow \text{ Kalsilite } (M)$$

Tetrakalsilite (K) is present in excess. When the nepheline is consumed, additional cooling converts the remaining tetrakalsilite to kalsilite. The final tetrakalsilite composition is at N, as the total batch is converted to kalsilite of composition O. The kalsilite remains stable until a temperature corresponding to point P is reached, at which time nepheline of composition Q exsolves (unmixes) from kalsilite. The two phases continue to unmix, and approach the compositions R and S respectively, as the system continues cooling. It is of interest to note that nepheline is formed during the cooling interval IK, is consumed, and forms again at temperature P.

LIQUID IMMISCIBILITY

In addition to several of the phenomena previously discussed, the phase diagram of the binary system $MgO—SiO_2$ (Fig. 12) includes a region of liquid immiscibility. Two intermediate binary compounds are present. Forsterite

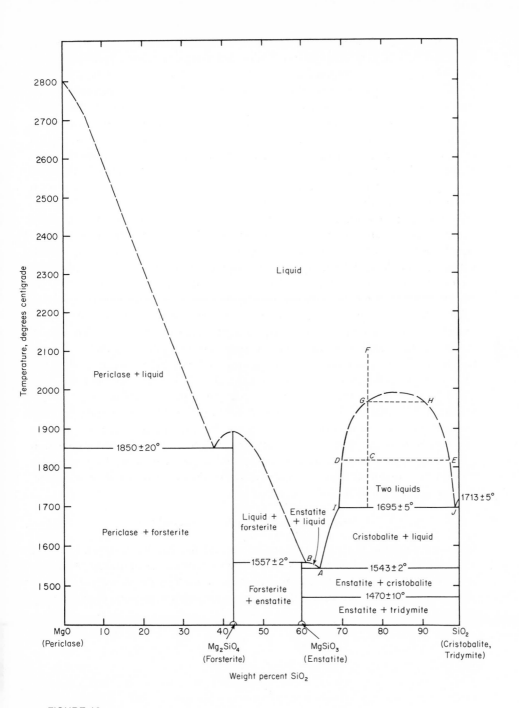

FIGURE 12
The binary system MgO—SiO₂. The exact position of the liquid immiscibility region is unknown, and is consequently indicated by a broken line in this diagram. [After Bowen and Andersen (1914), and Greig (1927).]

(Mg_2SiO_4) melts congruently at 1890°C. Between it and periclase (MgO) there exists a binary eutectic. The pyroxene enstatite with a composition of $MgSiO_3$ melts incongruently at 1557°C to yield forsterite and liquid. Point *A* is a eutectic between enstatite and cristobalite; Point *B* is a peritectic. The melting and freezing relationships of these intermediate compounds are essentially identical to those previously discussed for other intermediate compounds.

The unique feature of this diagram, however, is the presence of the region of liquid immiscibility in the silica-rich portion. Compositions within this area of the diagram consist of two liquid phases. For example, point *C* represents two liquids whose composition are given by *D* and *E*; the percentages of each can be determined by the Lever Rule.

A homogeneous liquid *F*, when cooled to *G*, begins to separate into two liquids: a second liquid, of composition *H*, begins to separate from the original liquid. Ordinarily, the second liquid separates initially as tiny droplets, which either sink or float, depending upon the difference in the specific gravities of the two liquids. Given sufficient time, a horizontal boundary develops between the two liquids. Such boundaries may be evident as a result of differences in color, refractive index, or some other physical property or feature. With continued removal of heat, the SiO_2-rich liquid phase increases in amount and in silica content; the original liquid becomes enriched in MgO due to the depletion of its silica-rich fraction. It changes in composition toward *I*, while the second liquid changes in composition toward *J*. Below the temperature indicated by the line *IJ*, the region of liquid immiscibility ceases to exist. By extending the isopleth downward from *C*, we can see that it encounters a field in which cristobalite and liquid exist stably. By constructing tie lines, we can see further that the liquid with which cristobalite coexists is of composition *I*. Therefore, at this temperature the second liquid, which is of composition *J*, must be consumed if equilibrium crystallization is to proceed. The following reaction takes place:

$$\text{Liquid } (J) \rightarrow \text{ Cristobalite} + \text{Liquid } (I)$$

When the reaction has gone to completion and the batch consists entirely of cristobalite and liquid of composition *I*, the system can cool further. That this reaction does occur at a fixed temperature is verified by the Condensed Phase Rule. When cristobalite first appears, three phases are present in this binary system.

$$P + F = C + 1$$
$$3 + F = 2 + 1$$
$$F = 0$$

The three phases can coexist only at an invariant point—namely, *J*.

Continued cooling results in further precipitation of cristobalite. The liquid changes in composition until the system has cooled to the eutectic, *A*, where both enstatite and cristobalite crystallize in eutectic proportions until all of the liquid is consumed.

3

Ternary Systems

A ternary system consists of three components. The presence of a third component requires a change in the method of graphical presentation of both temperature and composition. The three components are represented at the corners of a triangular coordinate system, each corner representing 100 percent of one of the three components. Consider a system X-Y-Z (Fig. 13). A binary composition such as A, plotted on the side of the triangle between X and Y, represents a mixture of X and Y. As the point A is halfway between X and Y, it represents 50 percent X and 50 percent Y. The composition at B represents about 80 percent Z and 20 percent X.

Ternary compositions, such as C, may be determined by erecting perpendiculars from the point of interest to each of the sides of the triangle. Each

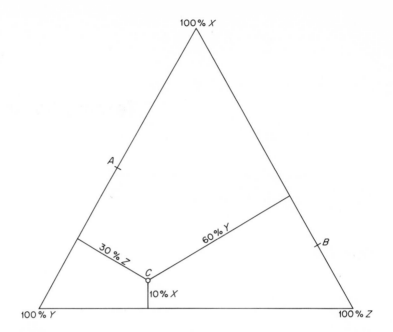

FIGURE 13
Compositional triangle showing method of determining or plotting composition *C* within the three-component system *X-Y-Z*.

perpendicular is proportional in length to the percentage of the component represented at the corner opposite the side that it intersects. Point *C* represents 10 percent *X*, 60 percent *Y*, and 30 percent *Z*. An easier way to arrive at the proportions is illustrated in Figure 14. Two lines, each parallel to one of two side of the compositional triangle, are drawn through the point in question, *C*. These two lines both intersect the third side (*XZ*), dividing it into three parts. The relative lengths of the three parts are seen to correspond to the relative amounts of the three components represented by point *C*.

The simplest method of all, and the one that everyone uses when they can, is to go to the bookstore and obtain some internally ruled triangular graph paper. The ruled lines represent the loci of points that have equal amounts of one of the components. As illustrated in Figure 15, the mixture *C* lies on a line *DE* parallel to the base. A composition point anywhere on this line contains 10 percent of *X*. This same point lies also on a reference line *FG* parallel to the right side of the triangle, which indicates a content of 60 percent *Y*. The shrewd observer will immediately deduce, without further perusal of the diagram, that the *Z* content must be 30 percent, but as it never hurts to check on these things, refer to the line *HI* for verification.

The temperature coordinate is perpendicular to the triangular compositional grid, that is, perpendicular to the surface of the paper. When liquidus surfaces

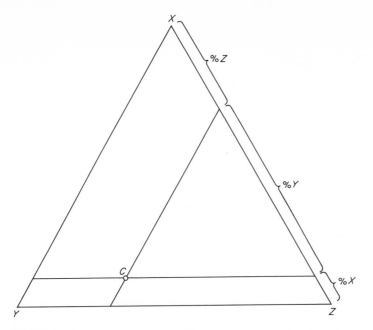

FIGURE 14
Determination of point C by the two-line method. Two lines are drawn through C
parallel to any two of the sides of the triangle (here XY and YZ). The
intersection of these two lines with the third side (XZ) divides that side into
three parts whose lengths are proportional to the relative amounts of components
X, Y, and Z at point C.

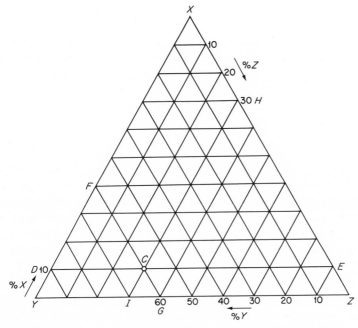

FIGURE 15
The point C on triangular graph paper. The percentages of components X, Y, and
Z are read directly from the numbered coordinate axes.

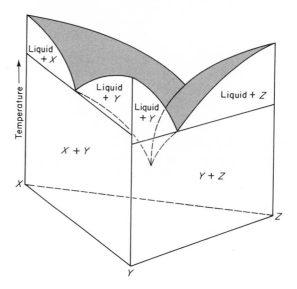

FIGURE 16
The temperature-composition model of the ternary system
X-Y-Z. Each of the three vertical sides is a binary system.
The top shows the liquidus surfaces.

are plotted, a three-dimensional object is created, as seen in perspective in Figure 16. Although models of ternary systems can actually be constructed of wood or plaster (an interesting challenge to the eager student), the construction of a model does not lend itself to detailed examination of the hundreds of ternary systems that have been determined. The usual procedure is to indicate the compositions on the triangular grid and to show the configuration of the various liquidus surfaces by means of lines of equal temperature (isotherms), as in a topographic map. In addition, the boundaries between the liquidus surfaces of the various solid phases are delineated by means of heavy lines called *boundary curves*. Such a representation is seen in the phase diagram of the ternary system $CaSiO_3$—$CaSiTiO_5$—$CaAl_2Si_2O_8$ (Fig. 17). As there is no solid solution in this system, it is thus a combination of three simple binary systems. The isotherms increase in temperature toward the melting points of the three components. The heavy lines that delineate the boundaries between liquidus surfaces slope to a minimum temperature at their point of intersection. The thermal slope of the boundary curves can be determined by noting that, as in a topographic map, the contours always point "upstream" as they cross the curves.

The boundary curves subdivide the diagram into areas known as *primary fields,* which are merely liquidus surfaces for the minerals in the system. For example, the area labeled sphene represents the range of liquid compositions which will precipitate sphene as the first, or primary, phase. The boundary curves can be regarded as extensions of binary eutectics into the ternary

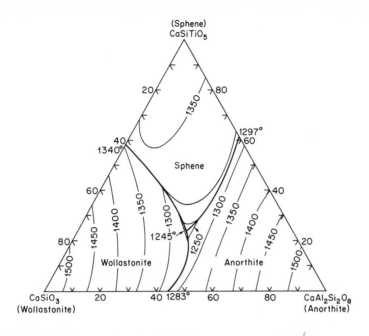

FIGURE 17

The temperature-contoured liquidus surface of the ternary system $CaSiO_3$—$CaSiTiO_5$—$CaAl_2Si_2O_8$. The three primary phase fields are separated by boundary curves (heavy lines), which intersect at a point of minimum temperature (ternary eutectic). [After Nishioka (1935).]

system. The three boundary curves intersect at a temperature minimum called a ternary eutectic. For simplicity in considering the relations that exist during crystallization in ternary systems, the isotherms will not be shown on most of the diagrams that follow; arrows on the boundary curves will indicate the direction of decreasing temperature. Figure 18 shows a phase diagram of the same ternary system with only boundary curves and primary fields indicated. A liquid A cools until it intersects the liquidus surface in the primary field of anorthite. As anorthite begins to crystallize, the liquid becomes depleted in anorthite, causing a change in composition directly away from the anorthite composition point. Continued precipitation of anorthite with decreasing temperature changes the composition of the residual liquid from A to B. The primary field of wollastonite is encountered at B, and wollastonite begins to precipitate along with anorthite. As three phases are now present, use of the Condensed Phase Rule ($P + F = C + 1$) indicates that invariant equilibrium is established, since $3 + F = 3 + 1$, which gives $F = 1$. The three phases can only exist along a line that defines the compositional limits of the liquid— that is, along the boundary curve. Since both wollastonite and anorthite are being precipitated as the liquid cools, the liquid changes in composition away from the wollastonite—anorthite base line. Simultaneous crystallization of wollastonite and anorthite continues until the primary field of sphene is en-

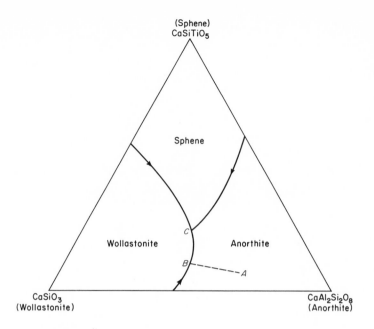

FIGURE 18
Crystallization path of liquid A within the ternary system
$CaSiO_3$—$CaSiTiO_5$—$CaAl_2Si_2O_8$.

countered at C. Here, sphene precipitates along with anorthite and wollastonite. As is seen from the Condensed Phase Rule, the coexistence of four phases
(three solid and one liquid) in a ternary system defines an invariant point:

$$P + F = C + 1$$
$$4 + F = 3 + 1$$
$$F = 0$$

Consequently, although the phases may vary in relative amount, their compositions and temperature remain defined at this point — the ternary eutectic —
until at least one of the phases is eliminated. With continued loss of heat from
the system, crystallization of the three minerals proceeds (in proportions determined by the position of the eutectic) until the liquid is consumed. This
diagram tells nothing about changes that may take place in the solid state.
Diagrams that show changes that take place in the solid state are of a different
type and will be discussed later.

It is possible to determine the amounts and compositions of the various
phases that exist at most stages of crystallization or melting by means of the
Lever Rule. Consider the liquid A in Figure 19. It is located within the primary
field of wollastonite. With cooling, the liquidus surface is encountered, wollastonite crystallizes, and the liquid shifts in composition directly away from

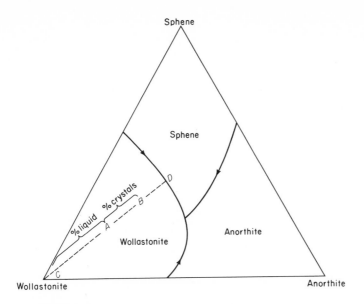

FIGURE 19
Use of the Lever Rule to determine crystal-liquid proportions
within a primary field.

the wollastonite composition point. The amount by which the liquid changes in
composition is necessarily related to the amount of wollastonite that has been
precipitated. When the liquid has shifted in composition to B, the distance
AB represents the proportion of wollastonite crystals that have formed, and
the distance between the wollastonite composition point C and the original
liquid A represents the proportion of remaining liquid. Percentages of these
phases are found by the following:

$$\frac{AB}{CB}(100) = 34\% \text{ Wollastonite Crystals}$$

$$\frac{CA}{CB}(100) = 66\% \text{ Liquid}$$

With additional cooling, wollastonite continues to precipitate until the primary field of sphene is encountered at D. Wollastonite and sphene then precipitate simultaneously, and the liquid changes in composition along the boundary
curve, toward the eutectic. Let us take stock of what has happened by the
time the system has cooled to point E (Fig. 20). In order to determine liquid-
crystal percentages existing at this time, a line is extended from the point of
interest, E, through the original liquid composition point, A, to the side of the
compositional triangle at F. The fraction $AE/FE \times 100$ represents the percentage of crystals, and $AF/FE \times 100$ indicates the percentage of remaining
liquid. The amounts are 62 percent crystals and 38 percent liquid.

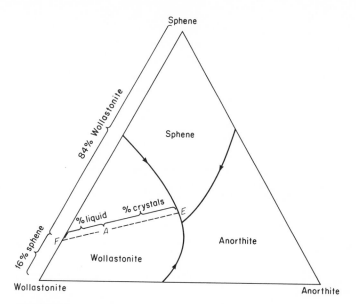

FIGURE 20
Use of the Lever Rule to determine crystal-crystal and crystal-liquid
proportions on a boundary curve.

In addition to determining the total amount of crystals that have formed, we
can also determine the percentage of each solid phase that has formed by considering the position of point F at the left side of the triangle. Since this point
is closest to the wollastonite composition, it is clear that more wollastonite has
formed than sphene. The percentages of the two solid phases that have been
precipitated are read directly from the left side of the triangle, as seen in
Figure 20.*

Knowing the relative percentages of the solid phases that *have been* precipitated during the previous cooling history is of course not the same as knowing
the relative percentages of the two solid phases that *are being* precipitated at
any particular instant. To obtain this information a different procedure is required. A liquid of composition A (Fig. 21) will, when cooled to the boundary
curve between the primary fields of wollastonite and anorthite, precipitate both
of those phases. Since the liquid must change in composition directly away
from the bulk composition of the phases being precipitated, a tangential extension of the boundary curve from A to the compositional line between wollastonite and anorthite reveals the relative percentages of these two phases

*Alternatively, the percentages of sphene, wollastonite, and liquid E can be determined by the
method shown in Figure 14. Wollastonite, sphene, and liquid E can be taken as the corners of a
compositional triangle with point A as the unknown interior point. After constructing the compositional triangle, lines are drawn through point A, parallel to two of the sides of this triangle.
The side of this triangle that is intersected by these two lines is subdivided into three lengths,
which represent the relative amounts of wollastonite, sphene, and liquid E.

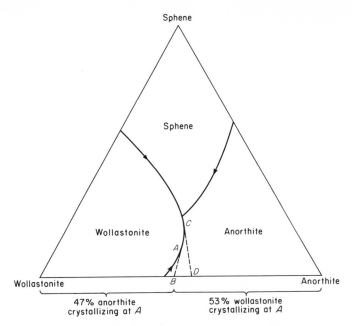

FIGURE 21

Use of the Tangent Rule to determine the ratio of phases that precipitate along a boundary curve. Point B indicates the proportion of phases being precipitated from liquid A. Point D indicates the proportion of phases being precipitated from liquid C.

being precipitated at A. The percentages of precipitating wollastonite and anorthite may be read directly at point B (about 47 percent anorthite and 53 percent wollastonite). Thus the direction of the boundary curve at any liquid composition point is an expression of the relative percentages of the phases precipitating at that point on the boundary curve.

Since boundary curves are usually not straight lines, it can readily be seen that the relative percentages of the precipitating phases will change as the liquid cools toward the ternary eutectic. The tangent from point C (Fig. 21) indicates that the percentages of wollastonite and anorthite will have changed as shown at point D (a higher proportion of anorthite than earlier at B).

The hypothetical system X-Y-Z* (Fig. 22) is presented to illustrate an example of extreme variation along a boundary curve. A liquid at A on the boundary curve between the primary fields of Y and Z will precipitate phases Y and Z in the proportions shown by the tangent from A to B (about $\frac{1}{3}Y$ and $\frac{2}{3}Z$). When the liquid has changed in composition to point C, the proportions of the precipitating phases are shown at D (more of Z, less of Y). When the liquid has cooled to point E on the boundary curve, it is changing directly

*A hypothetical system was chosen because no similar, simple, geologically applicable system could be found.

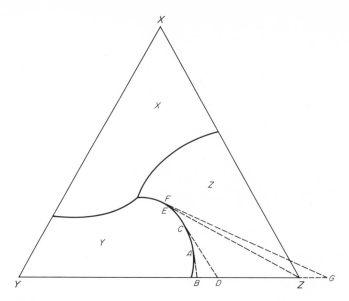

FIGURE 22
Subtraction curves and reaction curves can be distinguished by means of the Tangent Rule. That part of a boundary curve whose tangent intersects a composition line is a subtraction curve. If the tangent intersects an extension of the composition line, that part of the boundary curve is a reaction curve.

away from the corner of the triangle that represents pure phase Z, as seen from the tangent; consequently, it is precipitating only phase Z and no Y at all. Further along on the boundary curve, at point F, the tangent extension at G does not encounter any part of the composition line between the phases Y and Z. Intersection of the tangent with an extension of the compositional line indicates that while phase Z is precipitating, some of the previously precipitated crystals of Y must now be dissolving back into the liquid. Thus this boundary curve indicates not only when material is being subtracted from a liquid, but also when material is being both added to and subtracted from a liquid.* It is now possible to distinguish boundary curves of two types: (1) a *subtraction curve,* that part of a boundary curve along which crystalline materials precipitate from the liquid, and (2) a *reaction curve,* that part of a boundary curve along which crystalline materials both precipitate and dissolve.

---

*It might be argued here that the tangent, if extended in the opposite direction from F, encounters a composition point between compounds X and Y and that the liquid must be now precipitating these two phases. This is an impossibility, as the liquid composition at F is not in contact with the primary field of X, and therefore phase X cannot exist as a solid at the temperature and composition shown by point F. Similarly, the tangent drawn from F to G crosses the side XZ, which might perhaps imply that X and Z are precipitating; this, too, is impossible as the liquid is only in contact with the primary fields of Y and Z, and therefore only phases Y and Z can coexist with the liquid.

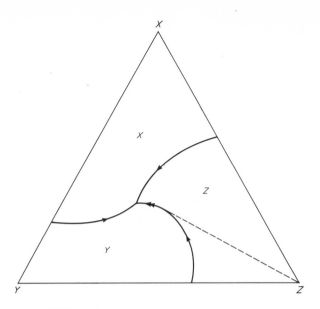

FIGURE 23
The convention of using single arrows to denote
subtraction curves, and double arrows to denote
reaction curves.

A subtraction curve is often indicated by means of a single arrow in the
direction of decreasing temperature, and a reaction curve by means of double
arrows in the direction of decreasing temperature. These curves are so indicated in Figure 23 for the hypothetical system X-Y-Z. The general rule on
distinguishing between these two types of curves can be stated as follows:
if the tangent drawn from the boundary curve intersects the side of the composition triangle that represents the two coexisting solid phases, then that
portion of the boundary curve is a subtraction curve; if the tangent from the
boundary curve encounters the extension of the pertinent side of the compositional triangle, then that portion of the boundary curve is a reaction curve.

Let us return to the crystallization history of the system X-Y-Z. As a liquid
changes in composition along the reaction curve it will precipitate Z while
simultaneously dissolving Y. The gradual shift in proportions of the *previously*
crystallized phases can be observed by the method of extending a line back
through the original composition point. The ratio of Z increases over Y as the
liquid composition changes along the reaction curve. We are faced with the
question as to what occurs at the minimum point. Is this point a ternary eutectic
or something else? Certainly X must begin to precipitate when its primary field
is reached; but do crystals of Y continue to dissolve, or does Y begin to precipitate again at this minimum point?

Observe first that as three solids and one liquid phase coexist at the minimum point, this must be a ternary invariant point; the liquid cannot change

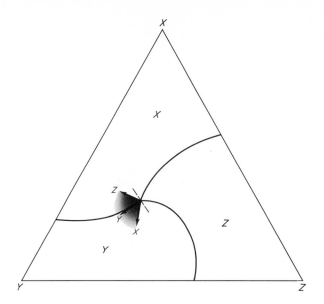

FIGURE 24

The use of vectors at an invariant point to deduce what reaction will take place. The vectors show possible directions of change of liquid composition at the invariant point. A vector pointing away from a solid-phase composition point indicates that the liquid is being depleted in that phase due to crystallization (X and Z). A vector pointing toward a solid phase composition point (Y) indicates that the phase in question is dissolving. Each vector indicates that a phase is either dissolving or crystallizing. Addition of the vectors in any ratio indicates that the composition of the liquid must change in some direction within the shaded area. As the liquid is of fixed composition at the invariant point, this arrangement, which implies a change in liquid composition, is impossible.

in composition and still remain with the three solid phases. The problem can be solved by drawing vectors from the minimum point to indicate possible directions in which the liquid might tend to change in composition.

Figure 24 shows the situation that would have to prevail if X and Z were precipitating and Y were dissolving. Vectors X and Z show that if X and Z are precipitating, there must be tendencies for the liquid to shift in composition away from pure X and pure Z. Vector Y shows that if Y is dissolving at this minimum point, there must be a tendency for the liquid to become enriched in Y or to change in composition toward Y. Since the vectors that show these tendencies toward change in the liquid composition all lie on the same side of the broken line drawn through the minimum point, the liquid must change in composition into the shaded area, depending upon the proportions of X, Y, and Z dissolving or precipitating. As was previously noted, however, the minimum point is invariant, hence the liquid cannot change in composition. This possibility is therefore ruled out.

If, on the other hand, it is considered that the three phases X, Y, and Z precipitate simultaneously, the three vectors would be drawn as shown in

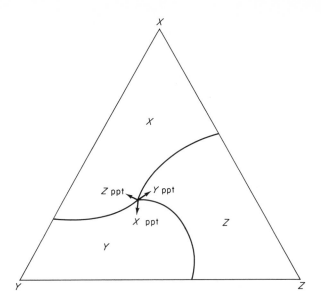

FIGURE 25
Vector arrangement indicating that X, Y, and Z are precipitating at the invariant
point. As these three vectors are balanced (so as to produce no change in
liquid composition), the reaction is the correct one for this invariant point.

Figure 25 to indicate that the liquid tends to change in composition away
from all three composition points X, Y, and Z. Vectors so oriented can be
balanced in proportions that would allow the liquid composition to remain
unchanged, which is what is required by the Phase Rule. Hence it is possible
in this system for crystals of Y to precipitate, partially dissolve along the
reaction curve under equilibrium crystallization, and then finally reprecipitate
at what is now seen to be a ternary eutectic. If this type of reaction occurred
during magmatic crystallization, examination of the final products in a conventional thin section could lead an unwary investigator to conclude that a
temperature rise took place at some time during the "normal" cooling sequence. Reaction curves are fairly common in more complex systems, and will
be discussed again later.

A final point that emerges in connection with this system is the question of
possible thermal slopes of liquidus surfaces. In Figure 26, hypothetical isotherms are drawn for the system X-Y-Z. The liquidus surfaces adjacent to
subtraction curves rise on both sides (see cross section AA′). But cross section BB′ reveals that the reaction curve in the diagram is not situated along
a temperature minimum—that is, the liquidus fields on both sides of the reaction curve slope in the same direction.

With this type of configuration it is possible for a crystallizing liquid to leave
the reaction curve—that is, the liquid may migrate in composition along part
of the reaction curve, and then leave it to follow a new compositional path

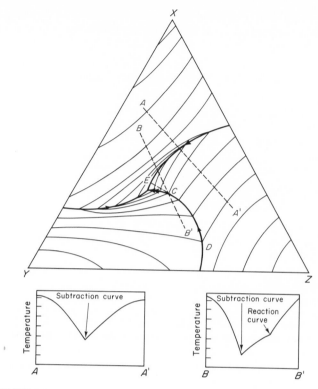

FIGURE 26

The ternary system *X-Y-Z* with isotherms. The two small inserts show the thermal cross sections along the lines *AA'* and *BB'*. The subtraction curve represents a thermal valley, whereas the reaction curve indicates a break in slope on the liquidus surface.

(across a primary field). A liquid remains on the reaction curve only when it contains crystals of those solid phases whose primary fields adjoin the reaction curve; to turn the argument around we can say that when three phases are coexistent in a condensed binary system, the Phase Rule requires that they must exist in univariant equilibrium—namely in a univariant relationship, such as a boundary curve. When one or more of these three phases is eliminated, there is no requirement for the remaining assemblage to remain in univariant equilibrium—that is, to remain on the boundary curve.

This can be demonstrated by considering the cooling of a liquid *D* (Fig. 26). The liquid cools to the liquidus surface, where it begins precipitating *Z*. The liquid changes in composition directly away from *Z* to the boundary curve between the primary fields of *Z* and *Y*. Continued cooling causes precipitation of *Z* and *Y*. A situation of univariant equilibrium has been established, and the liquid remains on the boundary curve as long as both *Z* and *Y* are present.

Suppose that at C, where the boundary curve becomes a reaction curve, all of the crystals are removed from the liquid by a nonequilibrium process such as separation by decantation. Univariant equilibrium is destroyed, and the cooling liquid is prevented from reacting with the previously formed crystals of Y. It is as though the liquid had an original composition of C, and merely cooled to the reaction curve at C. At C, the tangent of the reaction curve indicates that only phase Z can crystallize. As no Y is present to go back into the melt, the liquid only precipitates Z and moves directly away from the Z composition point, across the primary field of Z to E. At E the primary field of X is encountered; Z and X crystallize together, as the liquid moves along the boundary curve to the ternary eutectic, where X, Y, and Z precipitate together. As was pointed out, the process of decantation is one of nonequilibrium. It is also possible, however, for a liquid to leave a reaction curve while crystallizing under conditions of equilibrium, as will be demonstrated in later examples.

CONGRUENTLY MELTING COMPOUNDS

Most geologically important ternary systems are not of the simple type just discussed, but contain several intermediate compounds with diverse melting relationships. The system $NaF—LiF—MgF_2$, although of no geological significance, illustrates somewhat more complex behavior. In addition to the ternary end-members, it includes an intermediate binary compound of the composition $NaMgF_3$. Figure 27 shows the binary system $NaF—MgF_2$ and the intermediate phase $NaMgF_3$ as a congruently melting compound. The iso-

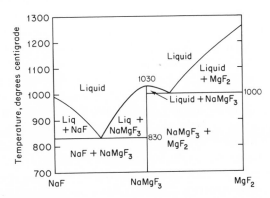

FIGURE 27
The binary system $NaF—MgF_2$. The intermediate compound $NaMgF_3$ melts congruently to produce a liquid of identical composition. [After Bermann and Dergunov (1941).]

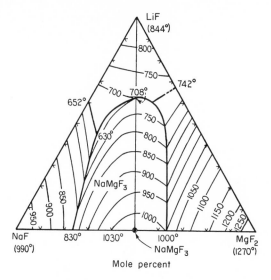

FIGURE 28
The ternary system LiF—NaF—MgF$_2$
[After Bermann and Dergunov (1941).]

therms of the ternary system are shown in Figure 28 so that the liquidus surfaces can easily be visualized. The first thing to note about a complex ternary system is the nature of the boundary curves—their type and slope. When a diagram is presented without any isotherms, as in Figure 29, this can be done by means of Alkemade's Theorem. An *Alkemade line* is a straight line connecting the composition points of two phases whose primary phase fields share a common boundary curve; Alkemade lines join the composition points of those phases which can exist stably together, and are commonly called "compatibility lines" or, less commonly, "equilibrium thermal divides" (Yoder and Tilley, 1962, p. 398). For example, the primary fields of LiF and NaF share the boundary curve *AB*; the edge of the composition triangle between the points LiF and NaF is therefore an Alkemade line. Similarly, since the primary fields of LiF and NaMgF$_3$ intersect at a common boundary curve, there is an Alkemade line through the triangle from the composition points LiF to NaMgF$_3$. It follows that since there are five boundary curves within the compositional triangle, there must also be five Alkemade lines related to these. They are the lines LiF—NaF, LiF—NaMgF$_3$, LiF—MgF$_2$, NaF—NaMgF$_3$, and NaMgF$_3$—MgF$_2$. We may regard particular boundary curves and Alkemade lines as being *pertinent* to each other when they both relate to the same phases; that is, the boundary curve between the primary phase fields of LiF and MgF$_2$ is pertinent to the Alkemade line between the composition points LiF and MgF$_2$. *Alkemade's Theorem* states that the intersection of an Alkemade line with its pertinent boundary curve represents a temperature maximum for the boundary curve. With the aid of

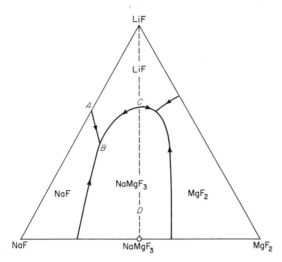

FIGURE 29
Primary fields and labeled boundary curves
within the ternary system LiF—NaF—MgF$_2$.

this simple theorem we can now show the direction of decreasing temperature
of the boundary curves. The composition points of LiF and NaF are joined
by an Alkemade line; the pertinent boundary curve for this Alkemade line is
the curve AB, which connects the primary phase fields of LiF and NaF. According to Alkemade's Theorem, the intersection of this boundary curve with
its pertinent Alkemade line represents a temperature maximum for the boundary curve; the curve AB is now known to have its temperature maximum at
A. Arrows placed on the curve would have to point toward B, indicating a
downward thermal slope from A to B.

In a similar way, arrows can be placed on the other boundary curves to indicate their directions of decreasing temperature. The interesting part of the
diagram is, of course, the point C, where the interior Alkemade line LiF—
NaMgF$_3$ crosses its pertinent boundary curve between the primary fields of
LiF and NaMgF$_3$. As this point is a maximum, the boundary curve must slope
off in two directions from C, the point of intersection. Thus, the seemingly
complex diagram actually turns out to be composed of two small subternary
systems NaF—LiF—NaMgF$_3$ and LiF—MgF$_2$—NaMgF$_3$. The paths of
crystallization and melting within these subternaries are similar to those in
the previously described simple ternary systems. The only unique path of
crystallization and melting is along the Alkemade line LiF—NaMgF$_3$. A
liquid D when cooled, intersects the liquidus surface of the primary field of
NaMgF$_3$ and begins to crystallize. The liquid moves directly away from the
composition point of NaMgF$_3$ until it reaches the saddle-point on the boundary curve at C. Here LiF and NaMgF$_3$ crystallize simultaneously. Inasmuch
as the original liquid composition fell exactly on the Alkemade line between

these compounds, there is no tendency for the liquid to move off along the boundary curve to either one of the two ternary eutectics in the subternaries. All further crystallization takes place at the temperature of C until the mass consists of solidified LiF and $NaMgF_3$. That the saddle-point C acts as an invariant point seems at first incompatible with the Phase Rule, which requires four phases in a ternary system for invariancy. But this difficulty is resolved as soon as it is noted that compositions from LiF to $NaMgF_3$ actually make up a simple binary system within the larger ternary (Fig. 30); the eutectic in this binary is simply the saddle-point C within the larger ternary system.

INCONGRUENT MELTING

The system $CaAl_2Si_2O_8$—$KAlSi_2O_6$—SiO_2 (anorthite-leucite-silica), originally described by Schairer and Bowen in 1947, is shown in Figure 31. This system has significant geological implications, which are discussed by the original investigators, by Yoder and Tilley (1962), and by others. The binary system $KAlSi_2O_6$—SiO_2, shown in Figure 6, was covered earlier under incongruent melting. The side $KAlSi_2O_6$—$CaAl_2Si_2O_8$ is a simple binary eutectic; the side $CaAl_2Si_2O_8$—SiO_2 is the same, with the exception of the polymorphic transition between cristobalite and tridymite.

The first step is to draw the arrows on the boundary curves to indicate the direction of decreasing temperature. One can then investigate cooling and melting paths. Notice immediately that the boundary curve between cristobalite and tridymite is isothermal. Such a line is never followed during any stage of crystallization, as its presence in the diagram merely indicates a polymorphic transition (without solid solution in either phase).

In order to label the boundary curves it is necessary to use Alkemade's Theorem. Here we find a slight complication. Observe that the short boundary curve between the primary fields of leucite and potassium feldspar does not intersect the Alkemade Line between the composition points of leucite and potassium feldspar. In order to obtain an intersection point, the Alkemade Line must be extended to A. The tangent rule described earlier (p. 34) can now be reworded to state that *if the boundary curve or any tangent drawn from the boundary curve does not intersect the pertinent Alkemade line, but merely an extension of that line, then that portion of the boundary curve is a reaction curve rather than a subtraction curve.* Since any tangent drawn from this boundary curve can only intersect an extension of the Alkemade line, the entire length of this boundary curve is a reaction curve. Any boundary curve that extends into a ternary system from a binary peritectic point, such as A, is always initially a reaction curve, and may or may not become a subtraction curve with a change in direction.

A second complication in this system is the boundary curve between anorthite and potash feldspar. This boundary curve does not intersect the Alkemade line between potash feldspar and anorthite. But an intersection can be accomplished by extending the boundary curve (rather than the Alkemade

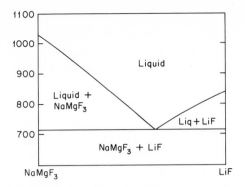

FIGURE 30
The binary system $NaMgF_3$—LiF constructed from isotherms in the ternary system LiF—NaF—MgF_2 (Figure 28).

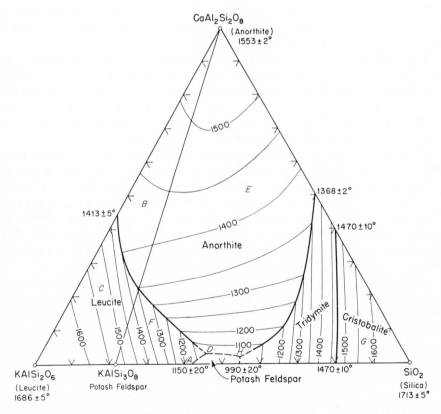

FIGURE 31
The system $CaAl_2Si_2O_8$—$KAlSi_2O_6$—SiO_2.
[After Schairer and Bowen (1948).]

line). The point of intersection is a temperature maximum and the boundary curve decreases in temperature away from it. Because all tangents drawn from this boundary curve intersect the pertinent Alkemade line, it is a subtraction curve, and is indicated as such (Fig. 33) by placing a single arrow in the proper direction.

To summarize the situation in which the Alkemade line and its pertinent boundary curve do not intersect:

1. If the Alkemade line must be extended to obtain an intersection with lines drawn tangent to the pertinent boundary curve, then those portions of the boundary curve are reaction curves. Either all or part of a boundary curve may be a reaction curve. The point of intersection is a temperature maximum.

2. If the pertinent boundary curve or its tangents must be extended to intersect the Alkemade line, the point of intersection represents a thermal maximum for the boundary curve. This portion of the boundary curve is a subtraction curve.

The other three boundary curves (and their tangents) — between the primary fields of leucite and anorthite, anorthite and tridymite, and potash feldspar and tridymite — intersect their Alkemade lines without complications and are subtraction curves throughout their length; each is labeled with a single arrow in the direction of decreasing temperature.

Knowing the positions of the Alkemade lines and the nature of the boundary curves, it is now possible to trace paths of crystallization within the system. But before tracing these in detail, it would be profitable to determine the final phase assemblages that are formed from the various liquid compositions — that is, determine the answers without actually working out the problems in detail. This is easily done by means of Alkemade lines.

There are five internal boundary curves between the various primary phase fields. Each boundary curve joins the primary fields of two solid phases that are compatible with each other in the presence of liquid. We can therefore construct the five pertinent Alkemade (compatibility) lines as in Figure 32; these lines subdivide the ternary system into two smaller compatibility triangles. Knowing this relation, it is now possible to state the final solid phase assemblage that crystallizes from any liquid composition within the system without tracing the detailed paths of crystallization.

A liquid whose original composition falls on an Alkemade line will crystallize to form the two solid phases indicated at the ends of that line. A liquid such as A, which is on the anorthite-silica Alkemade line, will crystallize to form a mixture of anorthite and tridymite. Similarly, any liquid that falls on the anorthite-potash feldspar Alkemade line, such as B, will crystallize to form a mixture of those two phases. A liquid C will crystallize potash feldspar and tridymite, and D will yield leucite and potash feldspar; the association of leucite and a silica polymorph is impossible under equilibrium conditions, as there is no Alkemade Line between these phases.

The Alkemade Lines form two triangles indicating the three solid phases that may coexist with liquid. Any liquid, such as E, located within the Alkemade

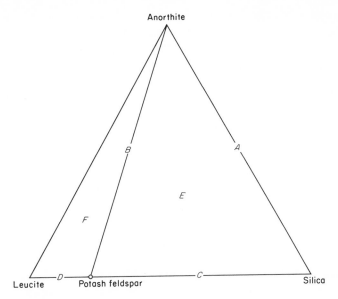

FIGURE 32
The system $CaAl_2Si_2O_8$—$KAlSi_2O_6$—SiO_2, showing alkemade lines
and triangles, which are used to determine the nature of the
crystalline phases that will form from any melt within the diagram.

triangle anorthite-potash feldspar-silica will crystallize to form those three
solids. Liquid of the composition F will crystallize anorthite, leucite, and
potash feldspar.

Returning now to the liquidus surfaces of this ternary system (Fig. 31),
we can say with confidence that liquids B and C will each crystallize to form a
mixture of leucite, anorthite, and potash feldspar. The liquids will change in
composition during crystallization in a manner similar to those of previously
discussed systems. Examination of the compatibility relations of the system
reveals the final composition that these liquids will have. Since each of the
liquids will precipitate anorthite. leucite, and potash feldspar as their final
products, they must change in composition so as to be compatible with these
three solids. There is only one point on the diagram at which a liquid can exist
with these three phases. This is the invariant point D, where the three primary
fields of these solid phases intersect. This point represents the final composition of any liquid whose original bulk composition is located within the Alkemade triangle anorthite-leucite-potash feldspar. Similarly, any liquid within
the anorthite-potash feldspar-silica Alkemade triangle (such as E, F, or G)
must migrate to the invariant point H. For each Alkemade triangle in a ternary
system there is an invariant point to which all liquid compositions migrate by
crystallization processes. Clearly, knowledge of Alkemade lines and their
implications is extremely important in understanding phase relations of typical
extremely complex systems.

Keeping the generalities of the system in mind, we can proceed to a detailed examination of the paths of crystallization of various liquids (see Fig. 33). The silica-rich liquid A falls within the Alkemade triangle anorthite-potash feldspar-silica. We can immediately predict, therefore, that these three phases are the final crystallization products. Upon cooling, the liquid reaches the liquidus surface of the primary field of cristobalite. Cristobalite precipitates, and the liquid migrates directly away from the SiO_2 composition point, and at 1470°C encounters the stability field of the lower-temperature polymorph tridymite. Under equilibrium conditions (actually a fairly uncommon occurrence in this system), all of the previously precipitated cristobalite converts to tridymite, and additional tridymite crystallizes as the liquid changes in composition away from the SiO_2 corner, eventually reaching the primary field of anorthite at B. Here, both tridymite and anorthite precipitate simultaneously, and the liquid changes in composition along the boundary curve to C, where simultaneous precipitation of anorthite, tridymite, and potash feldspar occurs in eutectic proportions until the liquid is consumed. Note that these are the three phases initially predicted. The final amounts of the three solid phases can be determined by the position of the original composition A within the triangle anorthite-potash feldspar-silica, and is independent of the position of the ternary eutectic point.

Because all liquids whose compositions lie within the Alkemade triangle leucite-anorthite-potash feldspar migrate to the invariant point D, the function of this point should be determined before examining any paths of crystallization in detail. This point is not a minimum (as is the eutectic at C), as the boundary curve DC slopes thermally downward from it. Because the boundary curves, by analogy with a drainage system, form a tributary arrangement, the point is logically called a *tributary reaction point*. The reaction point will be seen to be a ternary analog to the binary peritectic point. The reaction that occurs there can be deduced by the position of the point within the system. First observe that four phases must coexist at the point, as the primary phase fields of leucite, anorthite, and potash feldspar adjoin D, the composition of the liquid. As this is an invariant point, we know that any crystallization reaction that occurs here must be such as to maintain the composition of the liquid at this point; recall that the only reactions permitted at invariant points are those that change the quantities of the phases, but not their compositions or numbers. What type of reaction could occur that would not affect the composition of the liquid D? The first obvious possibility is to consider that this point is similar to a eutectic, and postulate that the three solid phases crystallize together. This possibility can be checked by using the vector approach (as was done earlier, on page 38). If the three solids crystallize, the liquid will have tendencies to change in composition away from the three composition points of these phases, as is illustrated in Figure 34. Note that there is no possible way to balance the vectors so as to allow the liquid to remain at D. Since all of the vectors are on the right side of the broken line, the liquid would have to move into the shaded area. This possibility must therefore be eliminated.

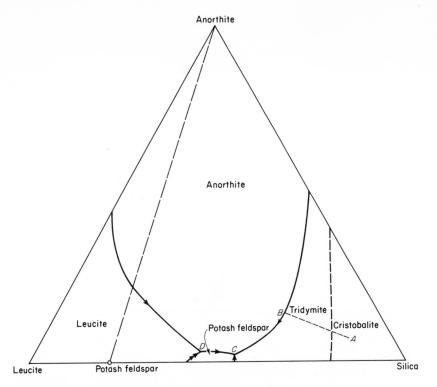

FIGURE 33
Labeled boundary curves, invariant points, and crystallization path of liquid *A*, which produces a mixture of anorthite, potash feldspar, and tridymite.

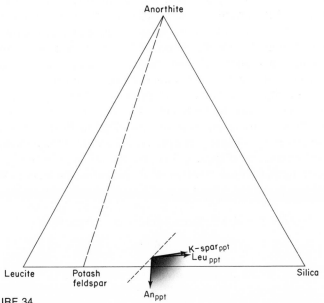

FIGURE 34
Vectors illustrating an impossible crystallization arrangement for leucite, anorthite, and potash feldspar at invariant point *D* (Fig. 33). Since the vectors cannot be balanced about the invariant point, the liquid cannot maintain a constant composition, which is required by the Phase Rule (ppt = phase precipitating).

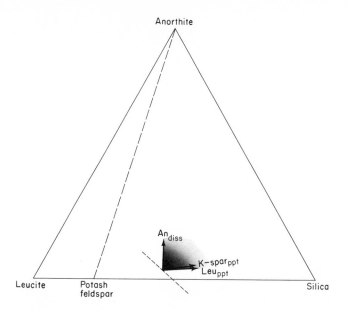

FIGURE 35
Vectors illustrating another impossible arrangement, with anorthite
melting and leucite and potash feldspar crystallizing. The liquid
cannot maintain a constant composition, as is required by the
Phase Rule (diss = phase dissolving).

Another possibility to consider (Fig. 35) is that leucite and potash feldspar precipitate while anorthite goes back into solution (the anorthite vector must now point toward the anorthite composition point). Once again, however, we see that the vectors cannot be balanced in such a way that the liquid composition remains at point D. Another impossible arrangement is obtained (Fig. 36) if we postulate that anorthite and leucite precipitate and potash feldspar dissolves. A possible answer is obtained when we postulate that leucite dissolves while anorthite and potash feldspar crystallize (Fig. 37), as this possibility allows the vectors to be balanced when the rates are in the proper proportions. If we pursue this line of investigation further—for example, to postulate whether two phases can dissolve while one precipitates—we obtain another possible answer (Fig. 38); it is possible to balance the vectors by dissolving anorthite and potash feldspar while precipitating leucite. Which of the two possibilities is correct? Both are, as they are opposites:*

(1) Leucite ↑, Potash feldspar ↓, Anorthite ↓

(2) Leucite ↓, Potash feldspar ↑, Anorthite ↑

The first reaction occurs during cooling, the second during heating.

---

*Upward-pointing arrows indicate melting of the adjacent phase, and downward-pointing ones, precipitation.

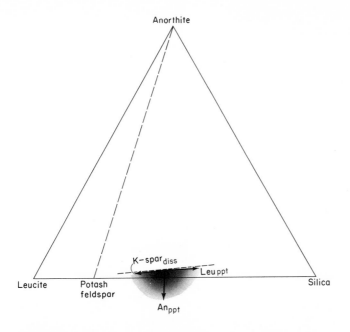

FIGURE 36
Vectors illustrating still another impossible arrangement, with potash
feldspar dissolving and anorthite and leucite crystallizing. Again, the
liquid cannot maintain a constant composition, as required by the
Phase Rule.

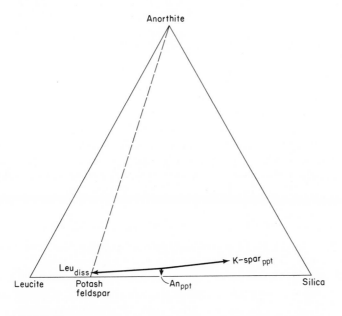

FIGURE 37
Vectors indicating the correct arrangement for melting of leucite and
crystallization of potash feldspar and anorthite. Since these vectors
can be balanced about the invariant point, they thus indicate the
reaction during cooling.

52

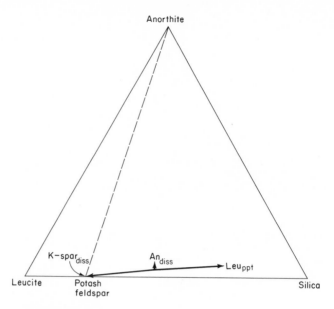

FIGURE 38
Correct balanced vector arrangement, but opposite to that in
Figure 37, indicating the reaction that occurs during heating.

It is fortunately unnecessary to go through such detail to determine the type
of reaction that occurs at a reaction point, as a general rule is available that
enables one to decide the proper sequence at a glance. Shown in Figure 39
are a reaction point and three adjacent phase fields, *X, Y,* and *Z.* The rule is

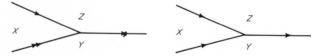

FIGURE 39
Tributary reaction points with labeled adjacent primary fields *X, Y,*
and *Z.* The phase *X* between the two tributary boundary curves is
either eliminated or decreased in amount during cooling at the
reaction point as *Y* and *Z* precipitate.

this: during cooling, the phase that is partly or completely lost by dissolution
is the one whose primary field falls between the two tributary boundary curves
—in this example, *X.* During heating, the situation is reversed.

A liquid *A* (Fig. 40), on the Alkemade line between anorthite and potash
feldspar, should crystallize to produce those two phases. Cooling of liquid *A*
first causes crystallization of anorthite, and the remaining liquid changes in
composition directly away from the anorthite composition point to the subtraction curve at *B.* Here, leucite and anorthite precipitate together, and the

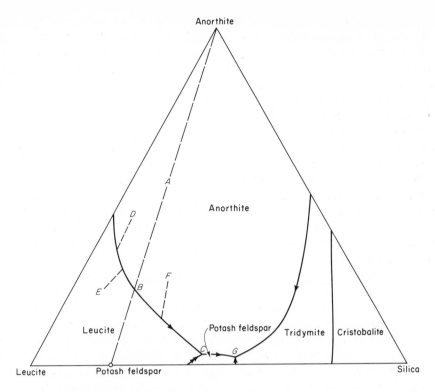

FIGURE 40
Paths of crystallization in the system $CaAl_2Si_2O_8$—$KAlSi_2O_6$—SiO_2.
[After Schairer and Bowen (1948).]

liquid composition changes with decreasing temperature along the boundary curve to point C, the tributary reaction point. Arrival at this invariant point necessitates the presence of a fourth phase, potash feldspar. The reaction previously discussed (leucite ↑, potash feldspar ↓, anorthite ↓) must now occur at a fixed temperature while the system loses heat. Since we know from the position of the original liquid composition A that the final products must be anorthite and potash feldspar, it can be predicted that the reaction at the invariant point will proceed until all of the leucite is consumed; as the last crystal of the leucite is lost, the liquid (which has been decreasing in amount during the reaction) is also used up. This leaves a final assemblage of anorthite and potash feldspar, with no trace of the earlier-formed leucite.

Other liquids, such as D or E, will arrive at the reaction point after having crystallized anorthite and leucite. At first the usual reaction will occur (leucite ↑, anorthite ↓, potash feldspar ↓), but not all of the leucite will be dissolved before the liquid is consumed. Thus the final assemblage that results from liquids D or E consists of leucite, anorthite, and potash feldspar, as could be predicted from the positions of the composition points, D and E, of the original liquids.

Consider a liquid F (Fig. 40) that falls within the Alkemade triangle anorthite-potash feldspar-silica; it can be predicted immediately that this liquid will complete crystallization at the invariant point associated with these three phases—namely, the eutectic G. As the liquid C cools and arrives at the reaction point C with its previously precipitated crystals of anorthite and leucite, the usual reaction begins—anorthite and potash feldspar precipitate as leucite dissolves by reaction with the liquid. But some of this liquid will still remain with the anorthite and potash feldspar after the leucite is completely consumed. Thus the liquid must leave the invariant point, as the absence of leucite prevents further reaction there. With further cooling, more anorthite and potash feldspar precipitate, and the liquid shifts in composition along the boundary curve to G, the ternary eutectic, where tridymite precipitates with anorthite and potash feldspar at a fixed temperature until the liquid is exhausted. It is thus possible to predict, from the original composition of the liquid whether the intermediate liquid composition C will leave the reaction point and proceed to the eutectic at G.

The sequence of events that occur along the reaction curve will be discussed for a different system—the system Mg_2SiO_4—$CaAl_2Si_2O_8$—SiO_2 (Fig. 41). This is not a true ternary system, as it contains a field of spinel, $MgAl_2O_4$, whose composition cannot be represented by any combination of the three components of the triangle. This system is but one slice of a more complex one (CaO—MgO—Al_2O_3—SiO_2). Situations of this sort will be discussed later. Ignoring for now the primary field of spinel (and its boundary curves), observe the labeling of the boundary curves in accordance with Alkemade's Rule. Observe also that the pyroxene enstatite ($MgSiO_3$) melts incongruently to forsterite and liquid, as the binary insert shows, and that the boundary curve between the forsterite and enstatite fields is therefore a reaction curve rather than subtraction curve.

Any liquid whose original composition lies within the Alkemade triangle forsterite-enstatite-anorthite must, under equilibrium conditions, crystallize those three phases. We can therefore predict that a liquid of original composition A must complete crystallization at the tributary reaction point B, where these three phases coexist stably with liquid. When the liquid A cools to the liquidus surface of forsterite, forsterite begins crystallizing and the liquid moves directly away from the Mg_2SiO_4 corner until it reaches the reaction curve at C. Here enstatite is formed as forsterite dissolves by reacting with the liquid. Some of the enstatite is formed as a direct precipitate from the melt, and some is formed by reaction of the liquid with previously formed forsterite. This process continues as the liquid changes composition along the reaction curve to B, the tributary reaction point. Here, forsterite continues to be lost by reaction as enstatite and anorthite are precipitated, but the liquid is completely used up before all of the forsterite is consumed, leaving three solid phases.

Any liquid whose original composition lies within the Alkemade triangle enstatite-silica-anorthite must complete crystallization at the ternary eutectic point E and produce all three as solids. We will use the crystallization of

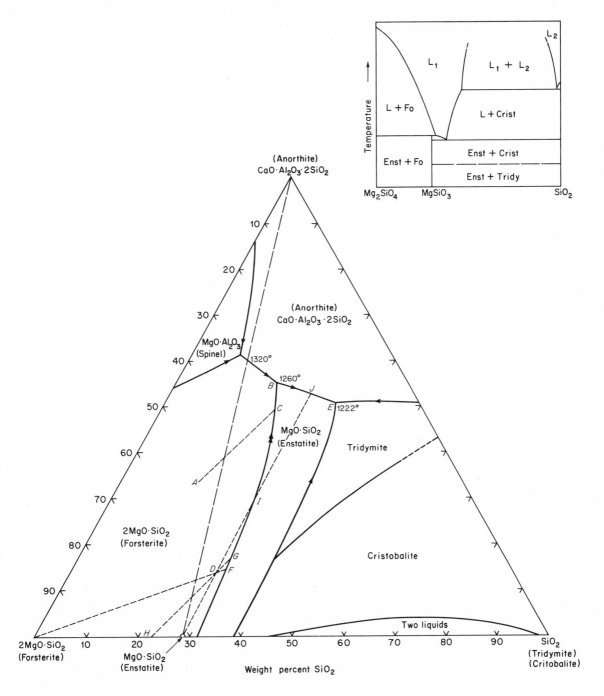

FIGURE 41

The pseudo-ternary system Mg$_2$SiO$_4$—CaAl$_2$Si$_2$O$_8$—SiO$_2$, showing crystallization paths. The Mg$_2$SiO$_4$—SiO$_2$ binary insert illustrates the incongruent melting of MgSiO$_3$ (L = liquid, Fo = forsterite, Enst = enstatite, Crist = cristobalite, Tridy = tridymite). [From Andersen (1915).]

liquid D as an example, first considering the sequence of crystallization, and then re-examining the sequence to demonstrate those aspects that can be predicted by analysis of the diagram.

The liquid D cools and precipitates forsterite until it reaches the reaction curve at F. The liquid now changes in composition along the reaction curve, while simultaneously crystallizing enstatite, and decreasing in forsterite content by reaction and dissolution. As only a small amount of forsterite was precipitated initially, (the interval DF), this reaction goes to completion with the total elimination of forsterite. This occurs at point I; here the liquid contains only crystals of enstatite. As the reaction relation is eliminated, and univariant equilibrium is destroyed, the liquid may leave the reaction curve. The liquid continues to crystallize enstatite, and migrates across the primary field of enstatite, directly away from the enstatite composition point, $MgSiO_3$. This continues until the subtraction curve at J is encountered. Here anorthite and enstatite precipitate simultaneously, and the liquid migrates to the ternary eutectic at E. The three phases enstatite, anorthite, and tridymite crystallize simultaneously at this invariant point until the liquid is consumed.

Let us re-examine the crystallization sequence in greater detail. Initial crystallization of liquid D yields forsterite, and the liquid migrates to the reaction curve at F. As only forsterite crystallizes, it is possible to draw a straight line from the liquid at F back through the original composition at D and extend it to the forsterite composition point at Mg_2SiO_4. As the liquid migrates down the reaction curve, it contains both forsterite and enstatite. The proportion of these two solids can be determined by the technique discussed earlier on p. 34. For a liquid composition such as G on the reaction curve, a line is extended back through the original liquid composition point at D; the line is continued until it encounters the pertinent Alkemade line of the two solids involved. This is the point H on the Alkemade line between forsterite and enstatite. The position of point H on the pertinent Alkemade line provides a method of determining the composition of the precipitated solids. Since point H falls closer to enstatite than to forsterite, enstatite constitutes the major portion of the precipitated phases. The liquid moves along the reaction curve with constantly increasing amounts of enstatite and constant loss of forsterite; at I the forsterite is completely consumed, as can be verified by observing that a straight line drawn from the liquid at I through the original composition at D will intersect the composition point of enstatite, $MgSiO_3$. As the forsterite is eliminated at I, continued crystallization of enstatite causes the liquid to change in composition directly away from the composition point $MgSiO_3$. The liquid migrates in composition to the subtraction curve at J, and then to the ternary eutectic at E.

The point at which the liquid may leave the reaction curve can be determined in advance. This is accomplished by constructing a straight line from the composition point of the incongruently melting compound (here $MgSiO_3$) through the original liquid composition (D), and observing the point of intersection (I) with the reaction curve. If a line constructed in this manner does not intersect the reaction curve, then the liquid will remain on that curve until the reaction point is encountered.*

THE DISTRIBUTARY REACTION POINT

Suppose that in the hypothetical system X-Y (Fig. 42) there is a compound XY_2 that is stable at low temperatures but breaks down upon heating to form two other phases by means of a solid-state reaction. In the example, XY_2

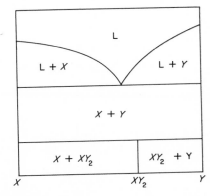

FIGURE 42
The hypothetical binary system X-Y, showing an intermediate compound XY_2, which decomposes to X and Y without melting.

reacts to form X and Y. With continued heating, compounds X and Y melt in the normal manner.

   With the addition of a third component, such as Z (Fig. 43), liquidus temperatures within the ternary system may be lowered below the upper temperature limit of XY_2, so that the compound XY_2 may exist stably in the presence of liquid. Labeling of the boundary curves in the standard manner shows that there are three different types of invariant points in the system. We have already considered the kind of reaction that takes place at a ternary eutectic (A) and at a tributary reaction point (B). We now do the same for a point such

---

*Another common method for finding the point at which the liquid leaves the reaction curve makes use of three-phase triangles. Assume that the liquid has arrived at G during the course of crystallization. The three phases present (liquid G, forsterite, and enstatite) can be considered to form the corners of a triangle, within which lies the bulk composition D. The bulk composition D must lie within a triangle made up of the three coexistent phases. As the liquid composition changes along the reaction curve, the bulk composition D remains within the three-phase triangle until I is reached. Here, composition D lies along the side of the triangle between enstatite and liquid I. This indicates that all of the forsterite has been consumed, and the liquid composition must leave the reaction curve.

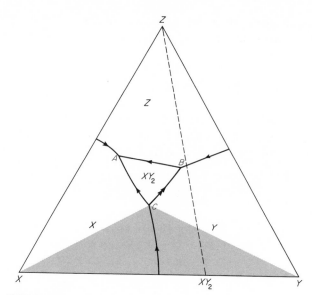

FIGURE 43
The hypothetical system X-Y-Z, showing a ternary eutectic (A), a tributary reaction point (B), and a distributary point (C).

as C, which represents the temperature maximum of XY_2 and is called a *distributary reaction point* by analogy with a river that divides, or branches. Note that at C, one boundary curve approaches the point in the direction of cooling, and two leave.

Upon cooling, a liquid can approach the distributary reaction point only if its original composition lies within the shaded area of Figure 43. Such a liquid will first precipitate a single phase and then shift in composition along the subtraction curve, crystallizing both X and Y. When the composition of the liquid reaches the distributary point C, precipitation of XY_2 must begin, as point C is adjacent to the primary field of XY_2. Since C is an invariant point, the liquid cannot change in composition until one or more of the phases are eliminated. Thus there appear to be several possible reactions that could occur here:

(a) $X\downarrow \quad Y\downarrow \quad XY_2\downarrow$

(b) $X\downarrow \quad Y\uparrow \quad XY_2\downarrow$

(c) $X\uparrow \quad Y\downarrow \quad XY_2\downarrow$

(d) $X\uparrow \quad Y\uparrow \quad XY_2\downarrow$

The vectors that originate at point C in Figure 44 indicate the direction of liquid compositional change as a result of particular reactions of each compound. If either or both X and Y precipitate with XY_2, it is impossible to

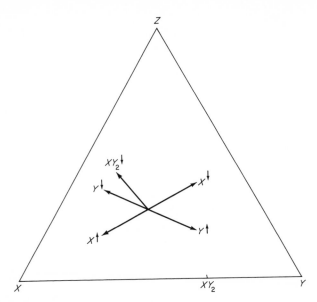

FIGURE 44
Composite vector diagram about the distributary point C (Fig. 43), showing all possible reactions and their effect upon liquid composition.

balance the various tendencies so as to maintain a fixed liquid composition.

The only reaction that allows the composition of the liquid to remain at the invariant point until the reaction is completed is one in which some of both X and Y is consumed by the liquid during precipitation of XY_2 (Fig. 45). Additional consideration of the migration vectors will show that the relative amounts of each of the phases being crystallized or melted are fixed (as a function of their composition and the position of the distributary reaction point within the system); that is, the relative proportions of the solid phases absorbed or precipitated at the distributary reaction point are independent of the amounts of the previously crystallized phases. It follows from this that the relative amounts of crystallized phases present in the liquid when it arrives at the distributary reaction point will determine whether X will be completely consumed before Y or vice versa.

A brief examination of the system as a whole (Fig. 46) reveals that it is divided into two compatibility triangles. A liquid within the triangle X-Z-XY_2 will cool and crystallize X, Z, and XY_2; the final liquid must consist of composition A (the point of intersection of those three primary phase fields). Similarly, a liquid within the triangle Y-Z-XY_2 will produce those three phases and have a final liquid composition of B. The solid phases at the distributary reaction point C are X, Y, and XY_2. As there is no Alkemade triangle with X, Y, and XY_2 as composition points, it follows that no final crystalline product can consist of those three phases. Thus we see that the distributary reaction point cannot indicate a final liquid composition within this system. A liquid

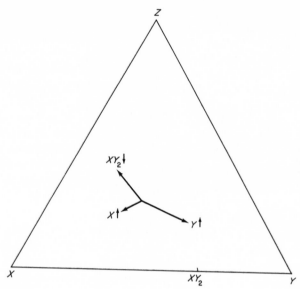

FIGURE 45
Balanced vectors showing the correct reaction about the distributary point.

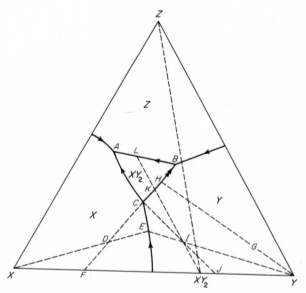

FIGURE 46
Crystallization paths within the system X-Y-Z.

may shift in composition to the distributary point during crystallization, but will then continue to change, reaching its final composition either at the ternary eutectic A or at the tributary reaction point B.

It should be pointed out that in some systems, it is possible for the composition of the final liquid to be that indicated at the distributary point, as will be seen in the system $FeO-Al_2O_3-SiO_2$ (pp. 72–73).

A liquid of original composition D cools to precipitate phase X and shifts in composition to the boundary curve at E, where both X and Y precipitate simultaneously; as crystallization proceeds, the liquid shifts in composition to the distributary point, where X and Y decrease in amount, and XY_2 precipitates. The relative proportions of the previously precipitated phases is indicated by the Lever Rule (point F, between corners X and Y). The lengths X-F and F-Y are proportional to the relative amounts of Y and X that have been precipitated. But the point F also lies between the composition points for compounds X and XY_2; the lengths X-F and F-XY_2 indicate the relative proportions of XY_2 and X after reaction at the distributary point has gone to completion. Knowing that the reaction products consist of X and XY_2, it follows that the liquid will migrate along boundary curve CA by simultaneously precipitating X and XY_2. At the ternary eutectic, X, XY_2, and Z precipitate together until the liquid is exhausted.

A liquid G cools and precipitates crystals of Y. It then shifts in composition to the reaction curve at H. Along this curve XY_2 is precipitated and some Y is lost by reaction with the liquid. This reaction continues at the tributary point B, where Z also precipitates until the liquid is consumed.

Upon cooling, a liquid I within the Alkemade triangle X-Z-XY_2 initially precipitates crystals of Y and moves to the subtraction curve at E. Both X and Y are precipitated until the liquid arrives at the distributary point C. Here X and Y react with the liquid and decrease in amount as XY_2 precipitates. The question of which crystals (X or Y) are eliminated first, is again answered by applying the Lever Rule. A lever constructed from the distributary reaction point C back through the original composition point I intersects the side of the triangle at J. Point J lies between composition points X and Y and also between composition points XY_2 and Y; when we consider its position with reference to X and Y, we see that it indicates the proportions of X and Y that have been precipitated while the liquid has migrated to the reaction point (the distances JY and XJ). When we consider the position of J with reference to XY_2 and Y, we see that it indicates the proportions of XY_2 and Y that exist at the end of the reaction that takes place at the distributary point (the distances JY and JXY_2). Thus it is possible to use the Lever Rule to determine the relative proportions of solid phases that exist at the beginning and end of the reaction that takes place at the distributary reaction point.

When this reaction is complete, the liquid contains Y and XY_2; the liquid must now migrate along the reaction curve toward B while precipitating XY_2 and absorbing Y by reaction. Knowing from the original composition I that the liquid must shift to the ternary eutectic A, one can guess that the liquid might leave the reaction curve. The liquid does, in fact, leave the curve when all of the Y phase is eliminated. The point at which the liquid has absorbed all remaining crystals of Y is found by drawing a line from XY_2 through the original composition I to the reaction curve (K), or by the three-phase triangle method discussed on pages 56–57. When the liquid is at point K, it continues to precipitate XY_2 and moves directly away from composition point XY_2, across the primary field of XY_2. The liquid reaches the subtraction curve at L, where Z and XY_2 precipitate together. It then shifts to the ternary eutectic at A, where X, Z, and XY_2 precipitate simultaneously at a fixed temperature until all of the liquid is consumed.

Returning again to the question of the direction in which a liquid will migrate from the distributary reaction point, we can summarize the procedure as follows: Draw a lever from the distributary point back through the original liquid composition to the pertinent Alkemade line (as was done for CDF and CIJ in Figure 46). The intersection point on the Alkemade line (F or J) indicates the relative proportion of the solid phases (X and Y) that exist at the outset of the reaction as well as the relative proportions of the solids that exist after the reaction is completed (X and XY_2, and Y and XY_2). The final assemblage indicates which of the two boundary curves the liquid will follow. From Figure 47 we can see that for liquids with original compositions in the lightly shaded area, the crystalline phases produced at the distributary point will be mixtures of X and XY_2; therefore, the boundary curve between X and XY_2 must be followed. For liquids whose original compositions fall on the boundary between the light and the dark gray areas, the lever indicates that both X and Y are simultaneously absorbed leaving only crystals of XY_2. Here a consideration of levers drawn from liquid compositions on either possible path indicates that the boundary curve between X and XY_2 must be followed.* All liquid compositions within the dark grey area follow the right-hand path, as indicated by levers intersecting the Alkemade line between XY_2 and Y.

Observe that one cannot state, in general, that a liquid whose original composition lies within Alkemade triangle $X—Z—XY_2$ will always follow the left boundary curve or that a liquid whose original composition lies within the Alkemade triangle $Y—Z—XY_2$ must follow the right boundary curve. The path of each liquid composition must be determined by means of the Lever Rule.

COMPLEX TERNARY SYSTEMS WITHOUT SOLID SOLUTION

In the examination of systems of greater complexity than those so far discussed, one encounters the problem of deciding which possible arrangement of a combination of Alkemade lines is the correct one. In the simplest ternary system, with no intermediate compounds, each of the three end-member minerals is compatible with the other two. The sides of the compositional triangle

---

*The right hand path, being a reaction curve, requires loss of Y during precipitation of XY_2. As all of the Y has been eliminated at the distributary point, this path is impossible.

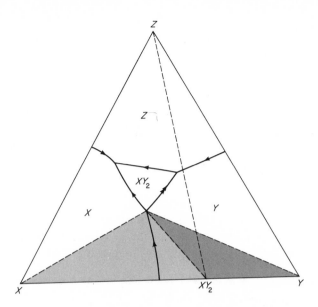

FIGURE 47
The use of the Lever Rule to indicate the path of liquid migration through the distributary
point. Original liquids within the dark gray area follow the right-hand path, and the original
liquids in the light gray area follow the left-hand path. Levers drawn from the distributary point
through the bulk composition to the binary edge XY indicate the cooling, solid-phase
composition upon arrival and after reaction at the distributary point.

are, in fact, Alkemade lines or joins. When a single binary compound such as
XY_2 (Fig. 48) is present, the larger system (X—Y—Z) breaks down into two
small ternary subsystems (X—Z—XY_2 and Y—Z—XY_2). But when two binary
compounds, such as XZ and YZ, are present in the system (Fig. 49), it becomes
necessary to look at the concept of Alkemade lines a bit more closely. Recall
that an Alkemade line is said to exist when two primary fields intersect along
a boundary curve; that is, the existence of an Alkemade line means that two
compounds exist stably together over a range of temperatures and liquid
compositions. The maximum number of solid phases that can exist in the presence of liquid in a condensed ternary system is given by the Phase Rule as
three. From Figure 49, it is immediately obvious that two Alkemade lines
cannot cross (as at A) in the presence of liquid, as this would imply that four
compounds, X, Y, XZ, and YZ, could all exist with liquid, which is impossible.
Under the conditions stipulated, only one of the two intersecting Alkemade
lines can exist. For this reason, we can make the general rule that two Alkemade lines cannot cross in the presence of liquid.* Nevertheless, this still
leaves us with the problem of deciding which of the two possible arrangements
shown in Figure 49 is correct. This cannot be solved, however, unless we know

*Note that this rule only applies to condensed systems of the type discussed, and not those in
which pressure is also a variable.

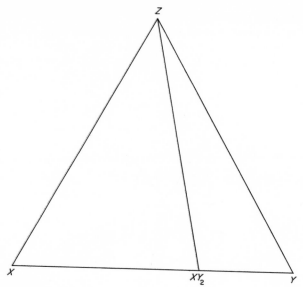

FIGURE 48
Compatibility arrangement in the system X-Y-Z, which has a single intermediate compound XY_2. Only one Alkemade triangle arrangement is possible.

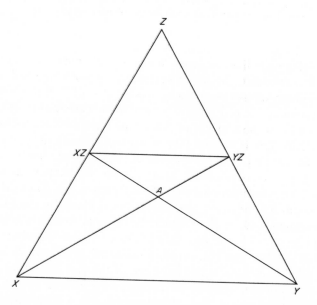

FIGURE 49
Two binary compounds XZ and YZ in the system X-Y-Z. The compatibility arrangement shown is impossible at a randomly chosen pressure, as the intersection of two Alkemade lines at point A indicates coexistence of four solids with a liquid phase. Either line XZ-Y or line YZ-X is correct, but not both.

which crystalline phases are stable together. That is, we need to know which boundary curves are present. For a system whose primary fields of X and YZ share a common boundary curve (Fig. 50), the arrangement of Alkemade lines shown in Figure 51 is correct. If, on the other hand, the primary fields of XZ

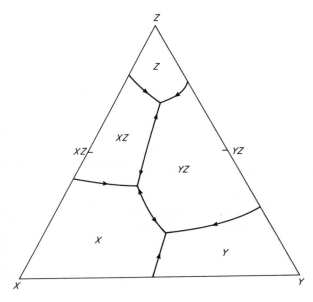

FIGURE 50
A possible arrangement of primary fields within the system X-Y-Z.

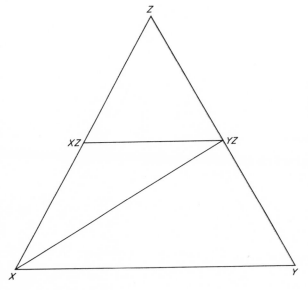

FIGURE 51
Compatibility relations consistent with Figure 50, as indicated by Alkemade triangles.

and Y share a common boundary curve like that shown in Figure 52, then the arrangement shown in Figure 53 is correct.

The possible arrangements of Alkemade lines become much more numerous as the number of compounds in the system increases, as in Figure 54, where a third binary compound XY is introduced. In addition to binary compounds, a system may contain one or more ternary compounds. Such compounds make the arrangement of Alkemade lines even more difficult to determine.

FIGURE 52
A second possible arrangement of primary fields within the system
X-Y-Z.

FIGURE 53.
Alkemade triangles that are consistent with Figure 52.

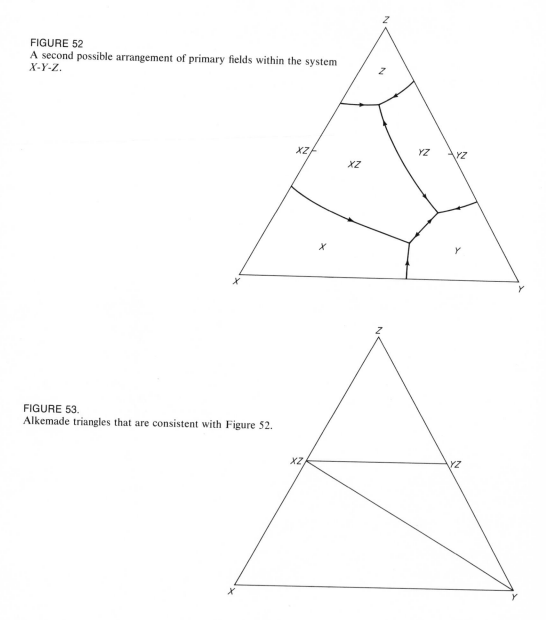

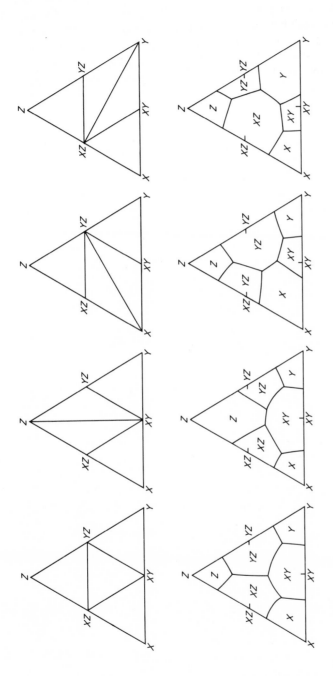

FIGURE 54
The system X-Y-Z and its three intermediate binary compounds.

The system FeO—Al₂O₃—SiO₂ (Fig. 55) contains several interesting features. This figure has been redrawn from a more detailed diagram in order to show the various primary fields, and has been simplified by the elimination of isotherms. We shall ignore, for present purposes, the limited degree of solid solution exhibited by wüstite, hercynite, corundum, and mullite. From such a diagram it is possible to deduce not only the courses of crystallization and final phase assemblages, but also the relative temperature values of some of the invariant points.

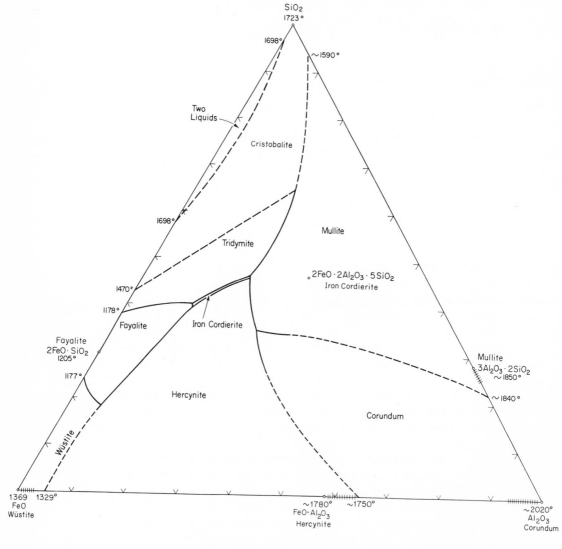

FIGURE 55
Liquidus surface for the system FeO—Al₂O₃—SiO₂.
[After Osborn and Muan (1960a).]

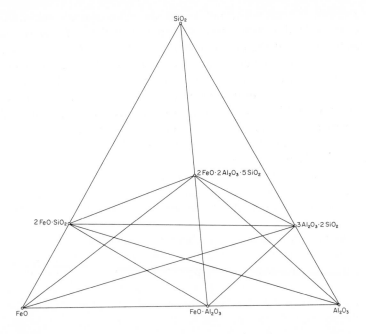

FIGURE 56
The system FeO—Al$_2$O$_3$—SiO$_2$, showing every possible compatibility line. The correct arrangement of Alkemade triangles can be deduced from consideration of the liquidus surfaces shown in Figure 55. Solid-solution relations have been eliminated for simplification.

When confronted with such a diagram, the first step to take is to determine the positions of the Alkemade lines; the correct arrangement of the lines must be deduced from all of the possibilities shown in Figure 56. This is done by observing the boundary curves. The two solid phases that coexist with a liquid on the boundary curve are compatible, and their composition points must therefore be joined by a compatibility, or Alkemade, line. The tiny primary field of iron cordierite shares boundary curves with the primary fields of tridymite, mullite, hercynite, and fayalite. Thus it follows that Alkemade lines must join the cordierite composition point with the composition points of each of those four phases. Since fayalite and hercynite have a common boundary curve, their composition points must be joined by an Alkemade line. Similarly, the composition points of mullite and hercynite must be joined by an Alkemade line, as they too share a common boundary curve. The remaining boundary curves indicate Alkemade lines along the outer edges of the triangle. If this procedure has been followed correctly, the large triangle should now be subdivided into smaller triangles. If the Alkemade lines form a four-sided figure, this means that one of the boundary curves has been overlooked. Another way of obtaining the Alkemade line configuration is to make use of the invariant points; for every ternary invariant point (with the exception of some distributary reaction points), an Alkemade triangle must exist whose corners are located at the composition points of the solid phases whose primary fields coexist at the invariant point.

The next step is to label the boundary curves. The first consideration here is to determine the direction of decreasing temperature. This is found by employing Alkemade's Theorem—namely, projecting (if necessary) the pertinent Alkemade line and boundary curve to an intersection point; the intersection point represents a temperature maximum for the boundary curve. Next it is necessary to determine whether the curve is of the reaction or subtraction type. This is accomplished by using the tangent rule (p. 36). If the tangents drawn from the boundary curve intersect the pertinent Alkemade line, that part of the boundary curve is a subtraction curve; but if the tangents encounter an extension of the pertinent Alkemade line, then that part of the boundary curve is a reaction curve.

After the boundary curves have been properly labeled (Fig. 57), the nature of the ternary invariant points becomes obvious: A and B are ternary eutectics; C, D, and E are tributary reaction points; and F is a distributary reaction point.

With the correct compatibility triangles established, it is possible to state the compositions of the final liquids and final solid phase assemblages for any liquid composition within the system.

Original liquid within Alkemade triangles	
Solid phase assemblages	Final liquid composition
Silica-Cordierite-Fayalite	A
Fayalite-Hercynite-Wüstite	B
Fayalite-Cordierite-Hercynite	C
Cordierite-Mullite-Hercynite	D
Mullite-Hercynite-Corundum	E
Silica-Mullite-Cordierite	F

Original liquid on Alkemade line	
Solid phase assemblages	Final liquid composition
Silica-Cordierite	F
Fayalite-Cordierite	C
Fayalite-Hercynite	G
Cordierite-Hercynite	D
Cordierite-Mullite	F
Hercynite-Mullite	E

Original liquid of cordierite composition	
Solid phase assemblages	Final liquid composition
Cordierite	F

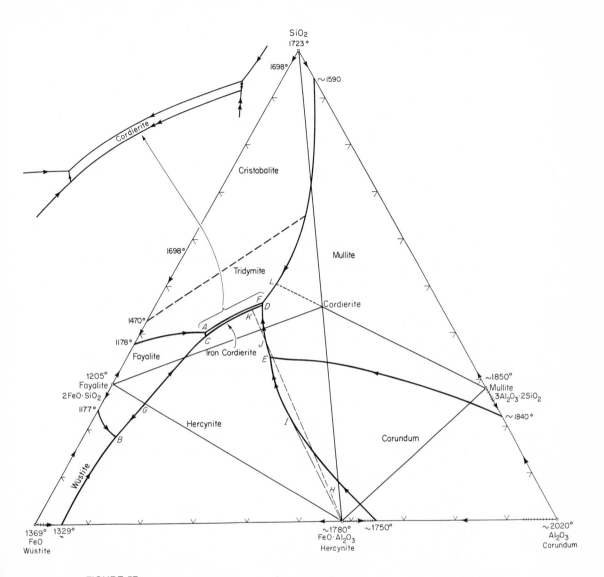

FIGURE 57
Crystallization paths within the system FeO—Al$_2$O$_3$—SiO$_2$. Boundary curves are labeled, and the correct Alkemade triangle arrangement is indicated. Solid solution relations have been eliminated for simplification.

The final crystalline products correspond to the particular Alkemade triangle or line on which the original liquid composition lies; if the original liquid is of the same composition as a ternary solid, that phase alone will constitute the final crystallized assemblage.

With the boundary curves properly labeled, paths of crystallization (or melting) can now be considered. Of interest because of its "devious" reactions is the liquid of original composition H. Upon cooling, the liquid first precipitates hercynite. The liquid then shifts in composition to the subtraction curve, where hercynite and corundum precipitate simultaneously. As a result of the curvature of this boundary curve, less and less corundum precipitates as cooling progresses until, at I, only hercynite is precipitated from the liquid. Beyond this point, the boundary curve is a reaction curve, indicating that some corundum is lost by reaction with the liquid as the hercynite continues to crystallize. This continues until the liquid reaches the tributary point, at E. Here the corundum is absorbed at a fixed temperature as hercynite and mullite precipitate. When the corundum is consumed, the liquid can again change in composition. As the liquid migrates along the reaction curve from E toward D, hercynite crystallizes and some mullite is absorbed. But the liquid never reaches D, because the remaining mullite is absorbed by reaction (at J), causing the liquid to leave the reaction curve. This point is determined by drawing a line from the hercynite composition point through the original composition and extending it until it encounters the reaction curve. Examination of this line reveals that, at J, the liquid contains only hercynite and must leave the reaction curve with continued precipitation of hercynite. This continues until another reaction curve is encountered at K. Cordierite now crystallizes and some of the previously formed hercynite is lost by reaction with the liquid. This process continues as the liquid changes in composition from K to C. At C, another tributary reaction point, cordierite and fayalite precipitate as additional hercynite is lost by reaction. This continues until the liquid is consumed, leaving a final assemblage of hercynite, cordierite, and fayalite. This final phase assemblage is, as usual, predictable from the composition of the original liquid.

The role of the distributary reaction point can be further elaborated by considering the crystallization of several cordierite-rich compositions. A liquid whose original composition corresponds to that of cordierite must at some temperature yield a final product of pure solid cordierite. This happens in the following way. The cooling liquid initially precipitates mullite (since cordierite can be seen to be an incongruently melting compound). This brings the liquid composition to a subtraction curve at L, where mullite and tridymite precipitate simultaneously. The liquid then migrates to the distributary reaction point, at F, where tridymite and mullite decrease in amount by reaction with the liquid while cordierite is precipitated. As the original liquid composition coincided with that of cordierite, this reaction will go to completion; all of the liquid, tridymite, and mullite will be consumed to yield pure cordierite. In this example, the distributary reaction point corresponds to the final liquid composition.

Liquids whose original compositions lie on either the silica-cordierite or the mullite-cordierite Alkemade lines also arrive at the distributary reaction point, but must produce two-phase crystalline assemblages. The reactions will be

$$\text{Liquid} + \text{Tridymite} + \text{Mullite} \rightarrow \text{Tridymite} + \text{Cordierite}$$

and

$$\text{Liquid} + \text{Tridymite} + \text{Mullite} \rightarrow \text{Mullite} + \text{Cordierite}$$

Numerous other interesting crystallization paths exist in this system, and it is suggested that the reader trace a number of them as well as several melting sequences. Particularly complex crystallization sequences occur for liquids whose compositions fall within Alkemade triangles near the composition point of cordierite.

The very complex system $CaO-Al_2O_3-SiO_2$ (Fig. 58), which may at first appear to be a hopeless maze, actually contains nothing new. A consistent use of the principles so far discussed will, it is hoped, see the reader through the most complex parts of the diagram.

Figure 59 is a diagram of the same system, but with only the composition points and boundary curves shown so that the reader may have the pleasure of working out the details. Again, slight solid solution effects are ignored for the sake of simplification.

The procedure is the same as that used earlier.

1. Identify the Alkemade lines. For every boundary curve there exists an Alkemade line.

2. Consider the relation between a boundary curve and its pertinent Alkemade line.
   a. If tangents drawn from the boundary curve intersect the pertinent Alkemade line, then that part of the boundary curve is a subtraction curve.
   b. If tangents drawn from the boundary curve intersect an extension of the pertinent Alkemade line, then that part of the boundary curve is a reaction curve.
   c. The actual or projected intersections represent a temperature maximum for the boundary curve.

3. After the thermal slopes of the boundary curves are determined, the various ternary invariant points can be identified as either eutectic, tributary, or distributary reaction points.

For every boundary curve between two primary phase fields, there must be an Alkemade line between the composition points that correspond to those primary phase fields. For example, the primary phase field of anorthite, $CaO \cdot Al_2O_3 \cdot 2SiO_2$, is surrounded by six boundary curves joining it to the primary fields of SiO_2, $3Al_2O_3 \cdot 2SiO_2$, Al_2O_3, $CaO \cdot 6Al_2O_3$, $2CaO \cdot Al_2O_3 \cdot SiO_2$, and $CaO \cdot SiO_2$. Consequently, there must be six Alkemade lines drawn from the $CaO \cdot Al_2O_3 \cdot 2SiO_2$ composition point to the composition points of these six

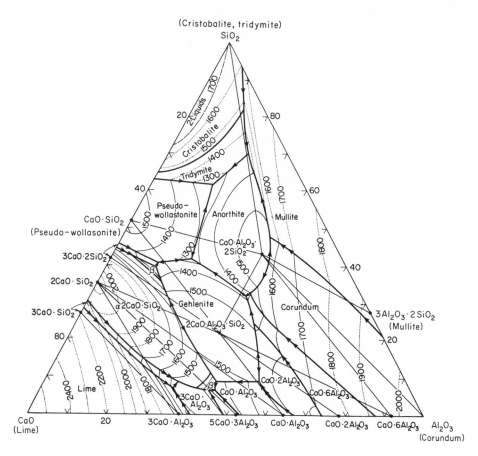

FIGURE 58

The system $CaO-Al_2O_3-SiO_2$. Compositions are in weight percent. [After Osborn and Muan (1960c) and Gentile and Foster (1963).]

phases. Similarly, the primary field of gehlenite, $2CaO \cdot Al_2O_3 \cdot SiO_2$, is deline-ated by seven boundary curves, hence there must be seven Alkemade lines drawn from the composition point of $2CaO \cdot Al_2O_3 \cdot SiO_2$ to the composition points of all seven compatible phases.

When the large triangle has been subdivided into Alkemade triangles, the "secrets" of the system stand revealed. For any compositional point within the system, it becomes possible to state the final equilibrium products of crystallization. The crystallization of any melt whose composition falls with-in an Alkemade triangle will produce a solid phase assemblage that consists of the three compounds that make up the corners of that Alkemade triangle. Any liquid whose original composition falls on an Alkemade line will produce a solid phase assemblage composed of the two compounds whose composi-tion points are connected by that line. Finally, a liquid whose original compo-

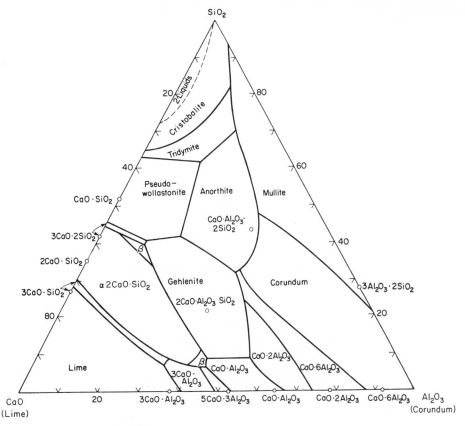

FIGURE 59
Primary fields within the system CaO—Al$_2$O$_3$—SiO$_2$.

sition falls directly on the composition point of a compound will yield that compound.

The crystallization sequences are identical in principle to those discussed earlier. The effect of small differences in initial liquid composition is shown in Figure 60 by the sequence of points A, B, C, D, E, F about the anorthite (CaO·Al$_2$O$_3$·2SiO$_2$) composition point. Each of these compositions differs only slightly from the others, but because each one is within a different Alkemade triangle, the final liquids will conclude their crystallization at different invariant points (a, b, c, d, e, and f, respectively), and the final solid phase assemblages will each be different. Melting sequences are exactly the opposite under equilibrium conditions. Solid assemblages A, B, C, D, E, and F would commence melting at the temperatures and compositions indicated by the corresponding invariant points a, b, c, d, e, and f. By way of contrast, note that large compositional changes *within* a single Alkemade triangle cause no changes whatsoever in the initial temperature of melting, as the initial compo-

76

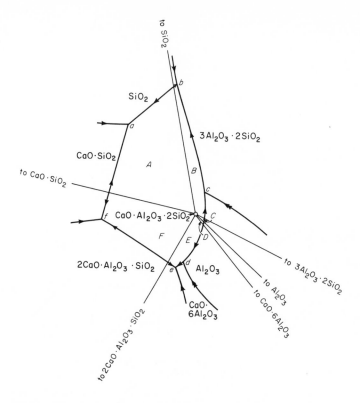

FIGURE 60

Enlarged part of the region about CaO·Al$_2$O$_3$·2SiO$_2$ within the system CaO—Al$_2$O$_3$—SiO$_2$. Liquids *A*, *B*, *C*, *D*, *E* and *F* complete crystallization at invariant points *a*, *b*, *c*, *d*, *e*, and *f*, respectively, as can be deduced by the various labeled Alkemade lines.

sition of the first liquid formed is fixed by a particular invariant point, and is the same for any composition within the Alkemade triangle.

The effect of minor compositional changes on the crystallization of magmas can, in some instances, have a very drastic effect on the course of crystallization and differentiation. This point is made clear by Yoder and Tilley (1962, especially pp. 398–401) in their discussion of the differentiation of simplified basalt compositions to produce either silica-rich or silica-poor derivatives.

It is instructive to consider a few selected courses of crystallization within the system CaO—Al$_2$O$_3$—SiO$_2$. Composition *A* (Fig. 61) lies within the Alkemade triangle 3Al$_2$O$_3$·2SiO$_2$—CaO·Al$_2$O$_3$·2SiO$_2$—SiO$_2$. Crystallization must therefore be completed at the invariant point *B*, where these three solid phases coexist with liquid of composition *B*. Since the composition point *A* of the original liquid lies in the primary field of Al$_2$O$_3$, this solid begins precipitating when the liquid cools to the liquidus. The liquid moves directly away from the Al$_2$O$_3$ composition point until it reaches the reaction curve at *C*. At

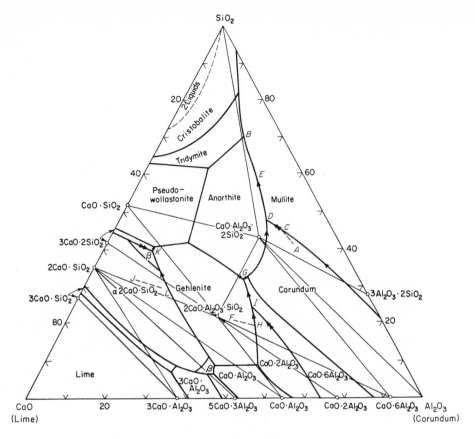

FIGURE 61
Crystallization paths within the system CaO—Al$_2$O$_3$—SiO$_2$.

the reaction curve, some of the Al$_2$O$_3$ is lost by reaction with the liquid, and 3Al$_2$O$_3$·2SiO$_2$ is formed by reaction and precipitation. This reaction continues as the liquid composition shifts along the reaction curve to point D. At the reaction point, D, all of the originally precipitated Al$_2$O$_3$ is consumed, while mullite and anorthite precipitate simultaneously. The remaining liquid of composition D now continues to precipitate crystals of mullite and anorthite and changes in composition along the subtraction curve DEB. At the ternary eutectic, precipitation of CaO·Al$_2$O$_3$·2SiO$_2$, 3Al$_2$O$_3$·2SiO$_2$, and SiO$_2$ takes place at a fixed temperature until all of the liquid is consumed.

The liquid F (near the composition point 2CaO·Al$_2$O$_3$·SiO$_2$) undergoes a similar crystallization sequence. Since this liquid falls within the Alkemade triangle 2CaO·Al$_2$O$_3$·SiO$_2$—CaO·Al$_2$O$_3$·2SiO$_2$—CaO·6Al$_2$O$_3$, it must eventually precipitate those three phases from a liquid at the invariant point G. As it is within the primary field of 2CaO·Al$_2$O$_3$·SiO$_2$, it first precipitates 2CaO·Al$_2$O$_3$·SiO$_2$ upon cooling to the liquidus, and changes in composition directly away from the composition point toward the subtraction curve at H, where $2CaO \cdot Al_2O_3 \cdot SiO_2$ and $CaO \cdot 2Al_2O_3$ precipitate simultaneously as the liquid migrates to the tributary reaction point at I. We know from the original bulk composition, however, that the liquid must somehow change in composition from I to G along the boundary curve between the primary fields of $2CaO \cdot Al_2O_3 \cdot SiO_2$ and $CaO \cdot 6Al_2O_3$. Since no $CaO \cdot 2Al_2O_3$ can exist along this boundary curve, we can therefore deduce that the reaction that takes place at the tributary reaction point is:

$$CaO \cdot 2Al_2O_3 + Liquid \rightarrow 2CaO \cdot Al_2O_3 \cdot SiO_2 + CaO \cdot 6Al_2O_3$$

The $CaO \cdot 2Al_2O_3$ is eliminated by this reaction, and liquid is in excess. The remaining liquid leaves the invariant point, shifting in composition from I to G while simultaneously precipitating additional $2CaO \cdot Al_2O_3 \cdot SiO_2$ and $CaO \cdot 6Al_2O_3$. At G, the ternary eutectic, $2CaO \cdot Al_2O_3 \cdot SiO_2$, $CaO \cdot 6Al_2O_3$, and $CaO \cdot Al_2O_3 \cdot 2SiO_2$ precipitate together until the liquid is exhausted.

Consider the crystallization of a liquid of composition J. We can see immediately that the final crystalline products must consist of $2CaO \cdot SiO_2$, $2CaO \cdot Al_2O_3 \cdot SiO_2$, and $3CaO \cdot 2SiO_2$, and that the final liquid must have a composition K. First to precipitate in the actual cooling sequence is $2CaO \cdot SiO_2$. As the liquid becomes partly depleted in $2CaO \cdot SiO_2$, it changes in composition and encounters the boundary curve at L. Here, both $2CaO \cdot SiO_2$ and $2CaO \cdot Al_2O_3 \cdot SiO_2$ precipitate, causing the liquid to shift in composition to the tributary reaction point at K. We know from the original composition that the reaction at the tributary reaction point must be:

$$2CaO \cdot SiO_2 + 2CaO \cdot Al_2O_3 \cdot SiO_2 + Liquid \longrightarrow$$

$$2CaO \cdot SiO_2 + 2CaO \cdot Al_2O_3 \cdot SiO_2 + 3CaO \cdot 2SiO_2$$

The final proportions of the three solid phases can be deduced from the position of the original liquid composition J within its Alkemade triangle, and are independent of the position of the invariant point.

In addition to the usual considerations of melting and freezing relations it is occasionally useful to attempt to derive, by examination of a ternary diagram, the binary systems that exist within the ternary. For the ternary system shown in Figure 61, the first attempts should be directed toward the limiting binary systems CaO—SiO_2, Al_2O_3—SiO_2, and CaO—Al_2O_3. (The derivation of binary and pseudo-binary systems from within ternary systems is discussed on pages 100–105.)

The derivation of the binary system Al_2O_3—SiO_2 presents no special problems, as it has only a single intermediate congruently melting compound $3Al_2O_3 \cdot 2SiO_2$, and is similar to the system shown earlier in Figure 5. The binary CaO—Al_2O_3 presents both congruently and incongruently melting compounds. The problem here is simplified if we recall that a binary eutectic extends into a ternary system initially as a subtraction curve; furthermore, a binary peritectic point is merely a point on a ternary reaction curve. Distinguish first between congruent and incongruently melting compounds. Between

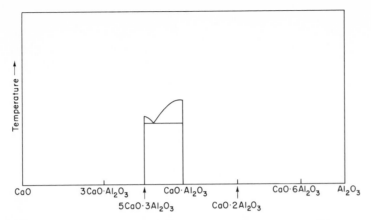

FIGURE 62
Beginning construction of a tentative binary system (CaO—Al_2O_3) from examination of the ternary CaO—Al_2O_3—SiO_2. Congruent melting of 5 $CaO \cdot 3 Al_2O_3$ and $CaO \cdot Al_2O_3$ requires an intermediate binary eutectic. Precise temperatures are not considered.

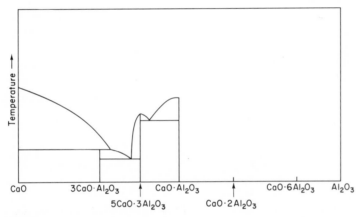

FIGURE 63
Incongruent melting of $3CaO \cdot Al_2O_3$ requires a binary peritectic and eutectic between the composition points $3CaO \cdot Al_2O_3$ and $5CaO \cdot 3Al_2O_3$.

two adjacent congruently melting compounds there must be a binary eutectic. This is true between the congruently melting compounds $5CaO \cdot 3Al_2O_3$ and $CaO \cdot Al_2O_3$ (Fig. 62). Observe next that $3CaO \cdot Al_2O_3$ melts incongruently to produce CaO + Liq; therefore CaO has the higher melting temperature, and this is indicated as in Figure 63; in addition, a eutectic exists between $5CaO \cdot 3Al_2O_3$ and $3CaO \cdot Al_2O_3$, and is necessarily lower than the temperature at which $3CaO \cdot Al_2O_3$ melts incongruently. On the Al_2O_3-rich side, observe that $CaO \cdot 2Al_2O_3$ and $CaO \cdot 6Al_2O_3$ melt incongruently. The compound $CaO \cdot 2Al_2O_3$ melts to yield $CaO \cdot 6Al_2O_3$ + Liq, and $CaO \cdot 6Al_2O_3$ melts to produce Al_2O_3 + Liq. As $CaO \cdot 2Al_2O_3$ melts to yield $CaO \cdot 6Al_2O_3$ + Liq, it follows that $CaO \cdot 6Al_2O_3$ must have a higher melting temperature than $CaO \cdot 2Al_2O_3$. Similarly, the conversion of $CaO \cdot 6Al_2O_3$ to Al_2O_3 + Liq with increasing temperature indicates that Al_2O_3 melts at a still higher temperature. These relationships

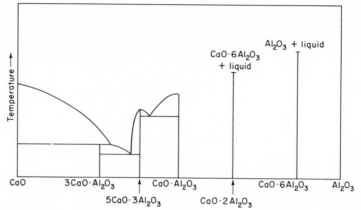

FIGURE 64
The composition of liquids produced by incongruent melting of $CaO \cdot 2Al_2O_3$ and $CaO \cdot 6Al_2O_3$ requires that Al_2O_3 has a higher melting point than $CaO \cdot 6Al_2O_3$.

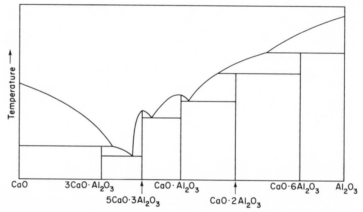

FIGURE 65
The tentative binary system $CaO \cdot Al_2O_3$. Precise temperatures are not considered.

are indicated in Figure 64. Finally, it can be noted that the ternary subtraction curve indicates the presence of a eutectic between $CaO \cdot Al_2O_3$ and $CaO \cdot 2Al_2O_3$; this must be lower than the melting points of both compounds. Putting all of this information together, we emerge with the schematic diagram shown in Figure 65. Using the temperature data from the CaO—Al_2O_3 binary (taken from the ternary CaO—Al_2O_3—SiO_2, Fig. 58), the correct binary liquidus surface can be redrawn as in Figure 66.

Using the same general approach on the CaO—SiO_2 binary (and ignoring temperature), we can conclude that the liquidus portion of the diagram must look something like the schematic shown in Figure 67. The correct diagram for this system (Fig. 68) shows that it includes a variety of subsolidus reactions that are of course not apparent from examination of the ternary liquidus; but it can be seen that the essential features of the liquidus region are shown clearly in the schematic diagram.

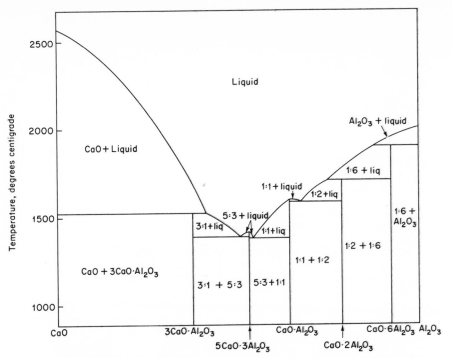

FIGURE 66
The system CaO—Al₂O₃ corrected with temperatures taken from the ternary
system CaO—Al₂O₃—SiO₂. Subliquidus reactions cannot be deduced from the
ternary liquidus surfaces.

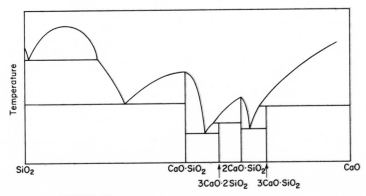

FIGURE 67
The tentative binary CaO—SiO₂, deduced from the ternary system
CaO—Al₂O₃—SiO₂.

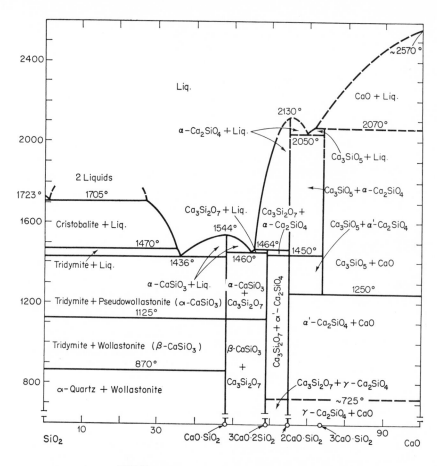

FIGURE 68
Correct diagram of the binary CaO—SiO$_2$.
[After Phillips and Muan (1959).]

SOLID SOLUTION

Solid solution is of geological importance because many of the major mineral groups show partial or complete solid-solution relationships. The system CaMg(SiO$_3$)$_2$—NaAlSi$_3$O$_8$—CaAl$_2$Si$_2$O$_8$, for example, shows complete solid solution of plagioclase between the Na (albite) and Ca (anorthite) end-members, as was discussed earlier under binary systems (p. 18). Plagioclase compositions are thus represented by a compositional line on the edge of the composition triangle (Fig. 69) rather than by points, as are the compounds in the previously described ternary systems.

This system is not a true ternary, but will be considered so for the sake of discussion. Liquid crystallizing in this system produces a two-phase assem-

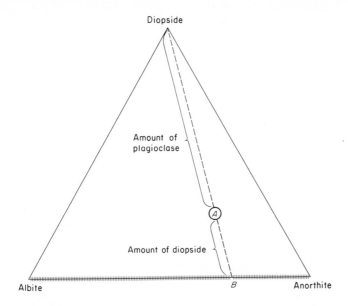

FIGURE 69
Possible solid-phase compositions in the system CaMg(SiO$_3$)$_2$—NaAlSi$_3$O$_8$—
CaAl$_2$Si$_2$O$_8$. Complete solid solution (hachured lines) is present between albite
and anorthite.

blage consisting of diopside and plagioclase, rather than the usual three solid
phases. For example, a liquid of composition A will, upon cooling, produce
plagioclase of composition B and diopside. The composition of the plagioclase
is determined by extending a straight line from the diopside composition point
at CaMg(SiO$_3$)$_2$ through the bulk composition A to the plagioclase composition line; the bulk composition must of course lie on a straight line between
the two phases that have crystallized. The relative amounts of the two crystalline phases are determined by use of the Lever Rule.

Figure 70 shows that this system has a single boundary curve between two
liquidus surfaces, one for diopside of fixed composition and one for plagioclase
of variable composition. It has no ternary invariant point. In contrast to previously discussed ternary systems, which contained invariant points that were
characterized by the coexistence of three solid phases with liquid, only two
solid phases can form in this system, and they can coexist with a liquid only
on a boundary curve rather than at an invariant point (a three-phase assemblage
is not enough to define an invariant point in a ternary system). The boundary
curve* indicates the liquid compositions that can coexist with both diopside
and plagioclase. A particular liquid composition on the curve can coexist only
with a specific plagioclase composition. Thus a liquid of composition A on

---

*This type of boundary curve has been called a eutectic curve (Eitel, 1951), but this usage is
uncommon. As both reaction and subtraction occur along this curve, it is often given the special
name of cotectic curve.

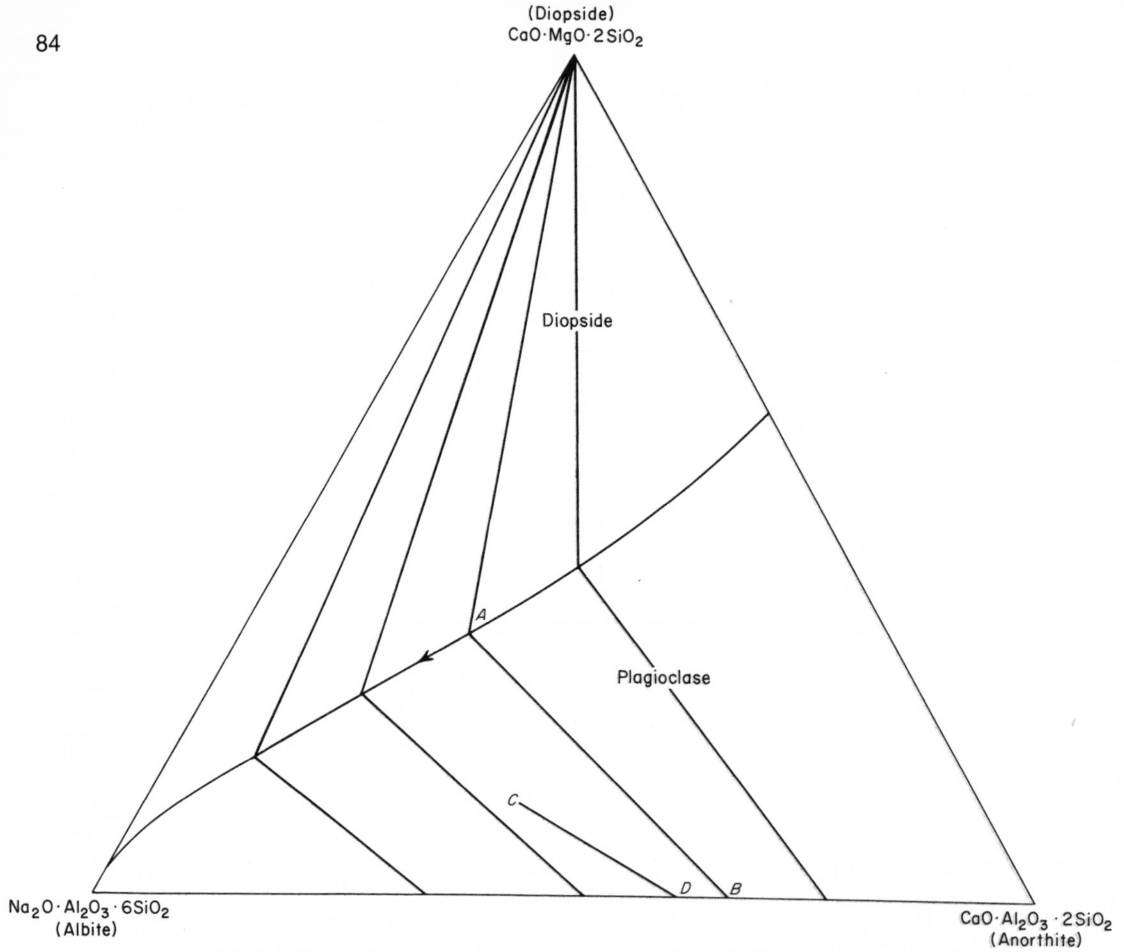

FIGURE 70

The system $CaMg(SiO_3)_2$—$NaAlSi_3O_8$—$CaAl_2Si_2O_8$, showing primary fields and tie (conjugation) lines. A liquid A on the boundary curve can exist with diopside, and plagioclase of composition B. Liquid C can exist with plagioclase D. [After Bowen (1915).]

the boundary curve can coexist only with a plagioclase of composition B and diopside. The compositions of some coexisting liquid-plagioclase pairs are shown in Figure 70 by a series of two-phase tie lines. These tie lines must be determined by detailed experimental work, and are not obvious from inspection of the liquidus configuration of the diagram. Note carefully that these tie lines apply only to coexistence of plagioclase with liquids *on* the boundary curve; liquids within the primary field of plagioclase coexist with plagioclases of other compositions.*

As diopside can occur with any composition of plagioclase, it follows that there must then be an infinite number of Alkemade lines between diopside and

*Tie lines that join liquid compositions with coexistent plagioclase (such as AB or CD) are occasionally referred to as *conjugation lines*.

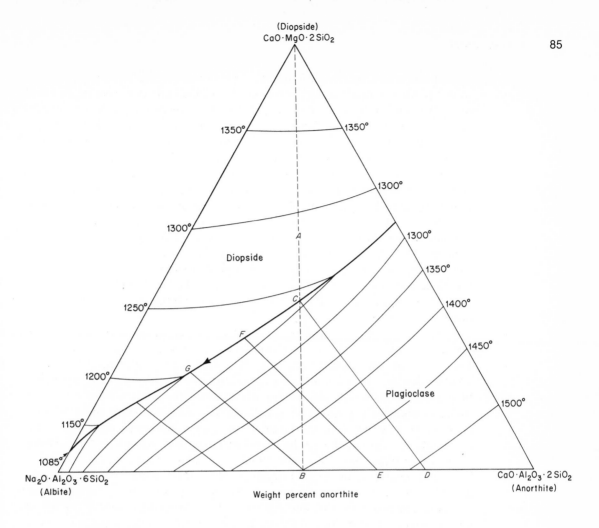

FIGURE 71
Crystallization of a liquid of composition A, above the primary field of diopside. The initial plagioclase composition is at D. The final plagioclase is at B, and the final liquid at G.

plagioclase. Therefore, Alkemade's Theorem, from which we obtain the thermal slope of the pertinent boundary curve cannot be used; the thermal slope must be determined experimentally. The situation is not entirely hopeless, however, since the general rule for such systems is that the thermal slope of the boundary curve is usually in the same direction as that of the liquidus curve or surface of the phases that possess solid solution. In this particular example, the general rule holds; the boundary curve slopes toward the albite composition point, as does the liquidus curve of the binary system albite-anorthite.

Consider now the cooling of a liquid within this system (Fig. 71). A liquid of composition A must crystallize to form plagioclase of composition B and diopside, as can be seen by its position within the system. Since the liquid A lies in the primary field of diopside, cooling will first cause the precipitation of diopside when the liquidus surface is encountered. The liquid then changes in composition directly away from diopside and encounters the boundary curve at C. We can see from the tie line CD that the composition of the first plagioclase to precipitate is D. Simultaneous precipitation of diopside and plagioclase causes the liquid to shift in composition down the boundary curve. Liquid compositions at lower positions on the boundary curve can coexist only with plagioclase more sodic than D (as can be seen from the tie lines). Equilibrium between plagioclase and liquid is maintained by two processes. As the liquid migrates down the boundary curve, it continuously reacts with the previously formed crystals, causing them to become more sodic. At the same time, the liquid continuously precipitates plagioclase of increasingly higher soda content. At F the liquid coexists with plagioclase of composition E and diopside. When the plagioclase is of the final predicted composition B, the last of the liquid is consumed at G. The composition of the final liquid G can be determined from the tie-line GB, since B is known to be the final plagioclase composition.

A more complex situation is encountered in tracing the crystallization sequence of liquids within the primary field of plagioclase, such as liquid A in Figure 72. Melt A initially precipitates a very Ca-rich plagioclase, such as composition D. As the liquid migrates down the liquidus surface, precipitation of more soda-rich plagioclase occurs, and the previously precipitated (more calcium-rich) plagioclase simultaneously becomes more sodic in composition by reaction with the liquid. This double reaction causes the liquid to migrate in a curved path to the boundary curve at E. A liquid at E must have a coexisting plagioclase of composition F (as seen from the liquid-crystal tie line). The migration of original liquid A to E on the boundary curve is predictable, since the bulk composition (A) must lie between the composition points of the two phases (E and F), which have formed from it.* Each liquid whose original composition falls on the tie line EF follows a unique curved path that intersects the boundary curve at E. The specific path can be determined *only* by experiment.

The liquid of original composition A would, upon cooling to the boundary curve at E, precipitate both diopside, and plagioclase of composition F. A gradual shift in the compositions of both liquid and plagioclase would occur with continuous crystallization of plagioclase and diopside. At G, for example, the liquid contains diopside and plagioclase of composition H. The liquid would be used up as it reached composition I; the plagioclase would simultaneously reach its final composition B (which was known from the original liquid composition A). Note that in the two examples described above the liquid does not reach the minimum point on the boundary curve.

In nonequilibrium crystallization, the plagioclase would react incompletely with the liquid, causing the liquid to migrate to a lower point on the boundary

---

*The generality here is derived from the fact that any original liquid composition in the primary field of plagioclase coincides with some point on a two-phase tie line that joins a liquid on the boundary curve with a plagioclase composition. The composition of the liquid as it arrives at the boundary curve is one end of this tie line, and the coexistent plagioclase the other.

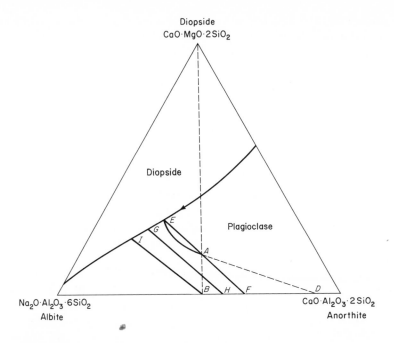

FIGURE 72
Crystallization of liquid *A* located above the primary field of plagioclase. The
first plagioclase to crystallize is *D* (which cannot be deduced from this diagram).
Final plagioclase and liquid are at points *B* and *I*.

curve than that predicted by its bulk composition. Examination of the feldspars in igneous rocks indicate that such shifts are common in nature; in many
rocks that cooled rapidly, the plagioclase crystals are not homogeneous in
composition, but exhibit calcium-rich centers and soda-rich rims.

Another example of solid solution in ternary systems is given by the system
$NaAlSi_3O_8$—$KAlSi_3O_8$—SiO_2 (Fig. 73). This is not a true ternary system, as
it contains the mineral leucite, $KAlSi_2O_6$, the composition of which cannot be
represented in terms of the three ternary end-members. Such a system is
termed pseudoternary. The reason for the presence of leucite is (as discussed
on p. 15) that potash feldspar melts incongruently to produce leucite and liquid
(see Fig. 74). The true ternary system is in fact SiO_2—$KAlSiO_4$—$NaAlSiO_4$.
Discussion will be limited here to the silica-rich part of the system, as it is of
more interest geologically.

At the higher temperatures shown in the pseudo-binary system $NaAlSi_3O_8$—
$KAlSi_3O_8$ (Fig. 75), the feldspars show complete solid solution. At low
temperatures, intermediate alkali feldspar compositions unmix, as is indicated
by the two-feldspar area. The liquidus portion of this pseudo-binary is characterized by the presence of a field of leucite (which becomes unstable with cooling), as well as a minimum point on the solid solution curves; such a minimum
point is characterized by precipitation of a single solid phase. Near the center

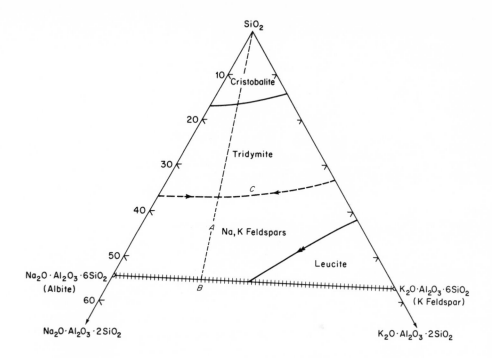

FIGURE 73

The pseudo-ternary system NaAlSi₃O₈—KAlSi₃O₈—SiO₂, which is a part of the ternary system NaAlSiO₄—KAlSiO₄—SiO₂. Liquid A cools to produce an alkali feldspar of composition B and tridymite. Complete solid solution between the alkali feldspars is indicated by hachured lines. [After Schairer (1950).]

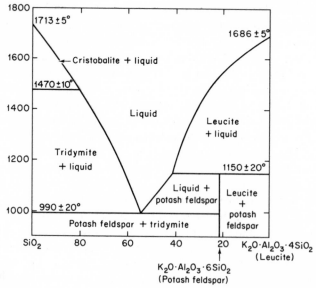

FIGURE 74

The binary system KAlSi₂O₆—SiO₂. [After Schairer and Bowen (1948).]

of the ternary system (Fig. 73), a boundary curve separates the feldspar field from the field of tridymite. This curve has a minimum point at about C for intermediate compositions.*

It can be seen from Figure 73, that a liquid of composition A will cool to produce tridymite and a feldspar of composition B. Under conditions of rapid cooling, the feldspar might remain as a single phase with the composition B, but under conditions of slower cooling, the feldspar will unmix to some extent upon encountering the two-phase area (see Fig. 75). The unmixing or exsolution produces the well-known perthitic texture often observed in igneous rocks of acidic composition.

The cooling sequence is indicated to some extent by the hypothetical tie-lines (shown as broken lines in Figure 76). A liquid A upon cooling to the liquidus first produces crystals of a rather soda-rich feldspar of a composition such as C. Reaction occurs between these crystals and the liquid, in addition to precipitation of more potassium-rich feldspars. The liquid follows a curved path to the boundary curve at D as the feldspar approaches composition E. At the boundary curve tridymite simultaneously precipitates with alkali feldspar, and the liquid migrates in the direction of the minimum point F. But the liquid (as in the examples given for the system $CaMg(SiO_3)_2$—$NaAlSi_3O_8$—$CaAl_2Si_2O_8$) does not necessarily reach the minimum point. From the original composition, it can be predicted that the final feldspar will be of composition B. The composition of the liquid that coexists with the feldspar B is given by a tie line to G on the boundary curve. Under equilibrium conditions, therefore, the liquid must be consumed as it reaches the composition G on the boundary curve. Only if the reaction between liquid and feldspar is incomplete does the liquid pass this point.

A more interesting situation develops if the original liquid composition happens to lie within the primary field of leucite. Cooling of liquid of composition H (Fig. 76) causes initial precipitation of leucite. The remaining liquid must migrate away from the leucite composition point until it encounters the reaction curve at I. It then moves along the reaction curve with precipitation of a potash-rich feldspar and absorption of leucite. The leucite is entirely consumed at point J, as is verified by extending a line from the coexistent incongruently melting potash-rich feldspar (at a point such as K†) through the bulk composition H of the original batch to the boundary curve. The line intersects the curve at J. Additional precipitation of potash-rich feldspar continues as the reaction proceeds, and the liquid migrates in a curved path to the boundary curve at L. During this time the potash feldspar has gradually become more soda-rich, and has a composition M when the liquid reaches L. Here simultaneous precipitation of feldspar and tridymite takes place. The liquid is consumed at N, and the final crystalline products are tridymite and

---

*Minimum points of this kind may fall anywhere on such a boundary curve; that is, their position is not necessarily near the center as in this diagram.

†As the bulk composition falls within the ternary system, the feldspar produced by reaction of liquid with leucite will not be pure K-feldspar, but rather a Na-containing K-rich feldspar (such as point K).

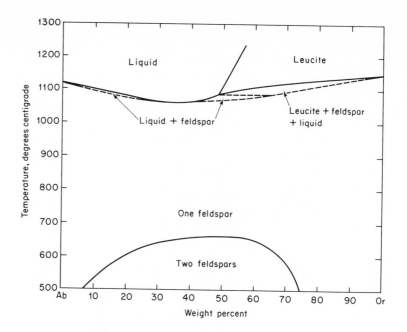

FIGURE 75
The pseudo-binary system $NaAlSi_3O_8$—$KAlSi_3O_8$.
[After Tuttle and Bowen (1958).]

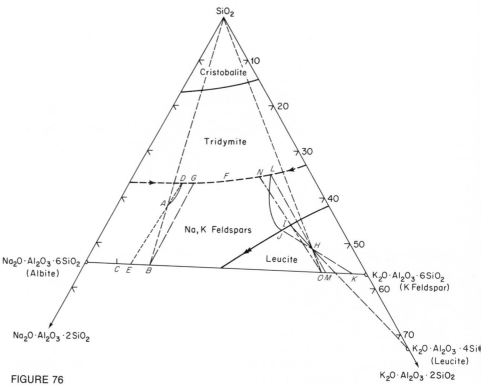

FIGURE 76
Hypothetical crystallization paths in the pseudo-ternary system $NaAlSi_3O_8$—$KAlSi_3O_8$—SiO_2.
[After Schairer (1950).]

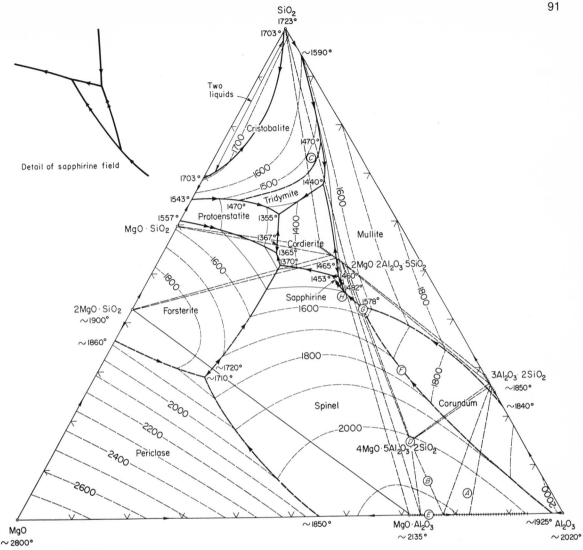

FIGURE 77
The system MgO—Al$_2$O$_3$—SiO$_2$, showing paths of crystallization. [After Osborn and Muan
(1960b), and Eitel (1965).]

feldspar of composition O (as could be predicted from the original liquid
composition).

Another important petrologic system that exhibits limited solid solution is
MgO—Al$_2$O$_3$—SiO$_2$ (Figure 77), a system that has been examined and revised to a great extent over the years. The diagram in Figure 77 can be analyzed by the methods described earlier. The boundary curves indicate the
existence of Alkemade lines, and in turn the positions of the Alkemade lines
indicate the nature of the boundary curves, from which the nature of the invariant points can be determined.

The solid solution shown by this system complicates the placement of

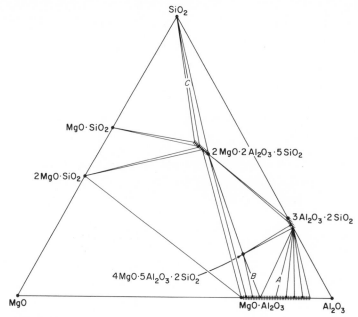

FIGURE 78

The effect of limited solid solution in the system MgO—Al$_2$O$_3$—SiO$_2$ is to create areas that consist of swarms of two-phase tie lines. The phase compositions for points within these areas are found at the extremities of the tie lines. The percentage of each of these phases is found by applying the Lever Rule to the tie line.

Alkemade lines for the minerals spinel, cordierite, and mullite (Fig. 78). Spinel, in particular, shows considerable solid solution with Al$_2$O$_3$; the compositional limits are indicated by a hatched line rather than a point. The Alkemade lines occur here in swarms rather than as single lines. These swarms indicate that each spinel composition exists with a particular composition of mullite, cordierite, or sapphirine (4MgO·5Al$_2$O$_3$·2SiO$_2$). Hence alumina-rich spinels may coexist with an alumina-rich mullite, and magnesian-rich spinel may coexist with cordierite. Spinels of intermediate composition may exist either with sapphirine or with sapphirine plus cordierite or mullite, depending upon the composition. The positions of these compatibility lines must be determined experimentally, and cannot be deduced from the configuration of the liquidus surfaces.

The presence of such swarms of Alkemade lines increase the possibility that the final assemblages will consist of two phases. For example, a liquid at A (Figs. 77 and 78) will crystallize to produce mullite and spinel; the final liquid will be located in composition not on an invariant point, but rather somewhere on the mullite-spinel boundary curve (its exact position depending upon the experimentally determined tie lines). A liquid B will crystallize to yield sapphirine and spinel; the final liquid composition will be adjacent to point *H* (Fig. 77), on the tiny boundary curve between the primary fields of sapphirine and spinel. The liquid *C* will crystallize to produce tridymite and cordierite, and crystallization will culminate on the boundary curve between the primary fields of these two phases.

There is nothing unique or new in this diagram, but it is instructive to consider the complex crystallization sequence that is shown by a liquid with the composition of sapphirine, point *D* (Fig. 77). We can state immediately that this liquid must complete its crystallization with the production of a single phase, sapphirine. Spinel will crystallize as soon as the liquid cools to the primary phase field of spinel. As the tie-lines are not yet determined for the spinel phase in the ternary system, we can predict only that the liquid will first precipitate relatively Al-rich spinels, such as *E*, and then follow a curved path of crystallization to the boundary curve, perhaps encountering it at *F*. Here precipitation of both corundum and spinel takes place as the liquid shifts in composition. During this time, the spinel is constantly changing in composition by reaction with the liquid. When the liquid reaches the tributary point at *G*, the corundum must be absorbed. The reaction is:

$$\text{Spinel} + \text{Corundum} + \text{Liquid} \rightarrow \text{Spinel} + \text{Mullite} + \text{Liquid}$$

When the last crystal of corundum is consumed, the liquid leaves the invariant point and migrates along the subtraction curve to *H* while precipitating mullite and spinel. At *H*, a distributary point, the following reaction occurs:

$$\text{Spinel} + \text{Mullite} + \text{Liquid} \nrightarrow \text{Sapphirine}$$

Because of the particular original composition chosen, all of the previous phases are consumed to yield sapphirine.

4

Quaternary Systems

A detailed discussion of systems of more than three components is beyond the scope of this book; this brief chapter is intended only to introduce the reader to four-component (quaternary) systems. Those interested in a thorough discussion of systems of four or more components are referred to Eitel (1951).

Quaternary systems must, of course, behave in accordance with the Phase Rule. The only new problem they pose is in their representation. Compositions within four component systems (Fig. 79) are ordinarily represented by means of a tetrahedron. The four sides of the tetrahedron are composed of ternary systems. Each point inside the tetrahedron represents some combination of all four components. The composition of any point within the tetrahedron can be determined by measuring the perpendicular distances from it to the various sides. The interior of the tetrahedron is subdivided into primary phase volumes—that is, regions in which particular liquid compositions may exist with a single solid phase. It is theoretically possible to contour these areas with isotherms to indicate the temperatures at which various liquids begin to crystallize, but this would be extremely difficult in a three-dimensional solid model and would be virtually useless in a two-dimensional drawing like the one shown in Figure 79. Customarily, only the temperatures of the invariant points are indicated.

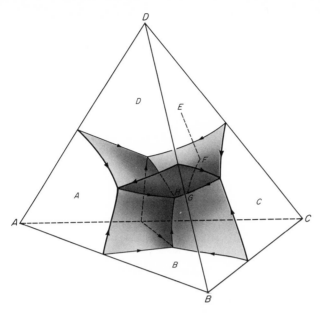

FIGURE 79
A tetrahedral projection of a hypothetical quaternary system *A-B-C-D*. Primary phase regions are now volumes rather than areas, as in ternary systems. Two solids and a liquid coexist along interior divariant surfaces. The intersections of the three divariant surfaces form univariant lines along which three solids and a liquid coexist. Four univariant lines intersect to form an invariant point, where four solids and a liquid coexist.

The primary phase volumes adjoin one another along curved surfaces. Along such divariant surfaces a liquid may, at the proper temperature, coexist with two solid phases (which make up the adjacent primary phase volumes). Three primary phase areas intersect along a univariant line (which is an extension of a ternary invariant point into the quaternary system). Finally, four univariant lines will intersect at an invariant point, where the liquid may exist with four solid phases. The invariant point may be a eutectic, a reaction point, or an inversion point, depending upon the phases involved and the thermal slopes of the univariant lines that meet at the point.

A quaternary system composed of four simple ternary systems and a quaternary eutectic point is shown in Figure 79. As a liquid of composition *E* in the primary phase volume of phase *D* cools, crystals of *D* are precipitated and the liquid moves directly away from *D*, until it encounters the primary field of *C* at the divariant surface *F*. With additional cooling, both *D* and *C* crystallize together and the liquid migrates away from both composition points along the divariant surface until the primary phase region of *B* is encountered at *G*. This is a univariant line along which *B*, *C*, and *D* precipitate. The liquid changes in composition along this line by depletion of these three phases until the quaternary eutectic is encountered at H, where four phases precipitate at a fixed temperature until the liquid is consumed.

The cooling sequence of a system that includes an intermediate binary compound, however, is more complex, as is shown in Figure 80. The intermediate compound illustrated, AB, melts congruently. Thus the quaternary $A - B - C - D$ can be subdivided into two simple subquaternaries $A - AB - C - D$, and $AB - B - C - D$, each with its own quaternary eutectic. Alkemade's Rule, which applies also to systems of this type, indicates that the temperature is a maximum where the univariant line between the primary phase volumes AB, C, and D crosses the pertinent Alkemade triangle $AB - C - D$. Bulk compositions within the tetrahedron $AB - B - C - D$ must, upon cooling, reach the invariant point E, where all four primary phase volumes intersect; liquids within the tetrahedron $A - AB - C - D$ cool to the invariant point F.

This arrangement must be modified if an intermediate binary compound, such as AB, melts incongruently, as seen in Figure 81; that AB does in fact melt incongruently can be verified by noting that the primary phase volume of AB does not include the composition point AB. The quaternary invariant point E becomes a reaction point rather than a eutectic; the point F is a eutectic. Directions of falling temperature for the univariant lines are, again, determined by means of Alkemade's Theorem. A liquid whose composition lies within a particular compositional tetrahedron must eventually reach the pertinent invariant point of this tetrahedron.

An example of a quaternary system that exhibits solid solution is shown in Figure 82. This system, $CaMg(SiO_3)_2 - Mg_2SiO_4 - NaAlSi_3O_8 - CaAl_2Si_2O_8$ (diopside-forsterite-albite-anorthite), approximates many aspects of the composition of a basaltic rock. The presence of a small field in which spinel crystallizes, however, indicates that the system is not the simple quaternary that it appears to be at first glance, but is actually a portion of the more complex system $Na_2O - CaO - MgO - Al_2O_3 - SiO_2$. Excluding from consideration the field of spinel, the system is divided into three primary phase volumes. Whether initial crystallization begins with plagioclase, diopside, or forsterite depends upon the bulk composition; continued cooling brings the liquid to a divariant surface, and then to the univariant line that extends between those points indicated with temperatures of 1270° and 1135°C. Cooling will cause the liquid to migrate some distance along this line. The final composition and crystallization temperature of the liquid will depend upon the bulk composition and the positioning of the two-phase tie lines between liquid and plagioclase compositions. As in the binary and ternary systems that have plagioclase compositions as components, there is necessarily a continuous reaction between liquid and coexisting plagioclase crystals, resulting in a continuous change in their compositions.

In many quaternary systems, large numbers of compounds are present, as in the system $MgO - Al_2O_3 - SiO_2 - ZrO_2$. The solid-state relationships of the various compounds in this system are shown in Figure 83. The Condensed Phase Rule indicates that in a quaternary system, the maximum number of solid phases that can coexist with a liquid is four. Thus a complex quaternary

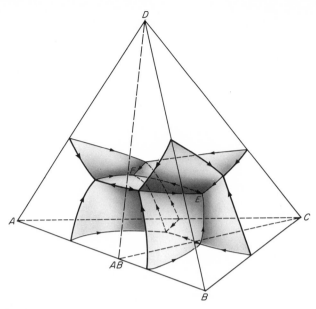

FIGURE 80
The quaternary system *A-B-C-D* and its congruently melting binary compound *AB*.

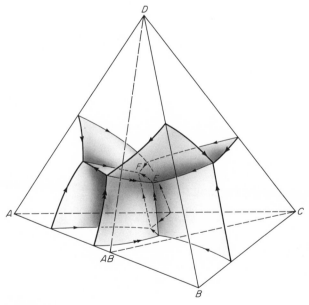

FIGURE 81
The quaternary system *A-B-C-D* and its incongruently melting binary compound *AB*.

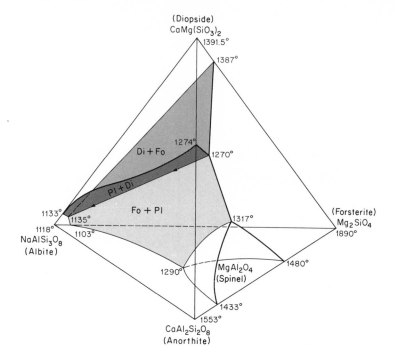

FIGURE 82

The pseudo-quaternary system NaAlSi$_3$O$_8$—CaAl$_2$Si$_2$O$_8$—Mg$_2$SiO$_4$—CaMg(SiO$_3$)$_2$. The system is pseudo-quaternary because of the presence of spinel, MgAl$_2$O$_4$, whose composition cannot be obtained by mixing any proportion of the tetrahedral apex compositions. Complete solid solution exists between albite and anorthite. This system is used to represent a simplified alkali basalt (Fo = forsterite, Pl = plagioclase, and Di = diopside). [After Schairer and Yoder (1967).]

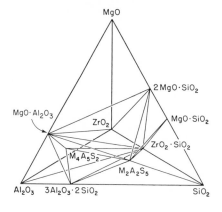

FIGURE 83

Subsolidus relations in the system MgO—Al$_2$O$_3$—SiO$_2$—ZrO$_2$. The presence of the fourth component allows the formation of compatibility pyramids as well as triangles (M = MgO, A = Al$_2$O$_3$, and S = SiO$_2$). [After Harold and Smothers (1954).]

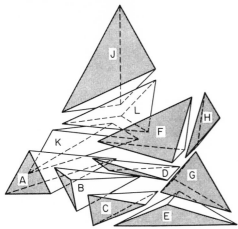

FIGURE 84
An exploded model of the system
MgO—Al_2O_3—SiO_2—ZrO_2 in which
the compatibility pyramids have been
slightly separated for greater clarification.
[After Harold and Smothers (1954).]

system may be broken down into smaller subquaternary systems in the same way that a complex ternary system is broken down into smaller subternaries. This requires an experimental determination of Alkemade lines within the system. Once the Alkemade lines are established, the large system can be broken down into compatibility triangles and tetrahedrons. Such an arrangement is shown in the exploded view (Fig. 84). Note that this type of diagram merely indicates which solid phases may coexist stably together, and does not show anything about the locations of the primary crystallization volumes as in the previous examples.

Other methods of representing quaternary systems include the use of isothermal surfaces and the imposition of various compositional restraints, such as holding the percentage of one component constant. The type of diagram used depends mainly upon the intended application of the information to be presented. More detailed discussions of the representation of multicomponent systems are to be found in Korzhinskii (1959), Ricci (1951), and Perel'man (1966).

5

Pseudo-Systems

The reader will encounter in the literature many systems that do not appear to follow the Phase Rule. Figure 85 is an example. The "eutectic" (just above the solidus field of $A_2B + C$) indicates the coexistence of A_2B, C, A, and liquid; a four-phase assemblage, however, is unreasonable in a condensed binary system; three should be the maximum number.

The compound A is not made up of a mixture of the two end-member components A_2B and C. This compound is therefore not a binary compound, and the representation is not that of a true binary system. This system can be termed pseudo-binary (or quasi-binary) because it superficially appears binary, but in fact represents a slice of a more complex system—the ternary system $A - B - C$, shown in Figure 86.*

The liquidus surfaces of the pseudo-binary are derived directly from a temperature-composition section of the ternary system. Primary fields and boundary curves along the Alkemade line $A_2B - C$ of the ternary system (Fig. 86) are seen as liquidus lines and liquidus minimums in the pseudo-binary system.

---

*A pseudo-binary system such as $A_2B - C$ is often referred to as a *join*. A pseudo-ternary system is occasionally referred to as a ternary join.

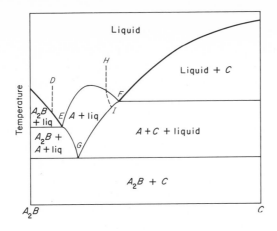

FIGURE 85
The hypothetical pseudo-binary join A_2B-C.

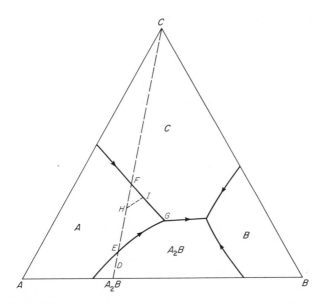

FIGURE 86
The hypothetical ternary system A-B-C and its binary congruently melting compound A_2B.

Boundary surfaces that exist below the liquidus of the pseudo-binary system can be derived by tracing crystallization sequences within the ternary system. Any liquid, such as D (Fig. 86), whose composition is between the composition point A_2B and the point E, will cool to precipitate A_2B; continued crystallization will bring such liquids to the boundary curve at E, where A_2B and A precipitate together. As any liquid of this type will arrive at the boundary curve at E, a horizontal field boundary exists between point E and the composition A_2B in the pseudo-binary system (Fig. 85). This line separate the two-phase area, liquid $+ A_2B$, from the three-phase area, liquid $+ A_2B + A$. The same sort of line can be drawn to the right of point F (Fig. 85) for all liquids whose compositions are between F and composition point C in the ternary system.

The liquid E (Fig. 86) at the boundary curve between the primary fields of A_2B and A, will cool to crystallize A_2B and A. The liquid composition changes in the direction of G, the invariant point. As the liquid migrates down the boundary curve from E, it leaves the Alkemade line $A_2B - C$; as the point G is merely a projection of the ternary invariant point on the pseudo-binary system (Fig. 85), the liquid composition also leaves the TX plane of the pseudo-binary system $A_2B - C$.* The path of crystallization (E to G) will be at an angle to the pseudo-binary TX plane; this angle is dependent upon the angle between the Alkemade line $A_2B - C$ and the boundary curve EG in the ternary system.

All liquids with an original composition on the Alkemade line $A_2B - C$ will migrate to the tributary reaction point G (where they react to form A_2B and C). The pertinent segments of the two boundary curves in the ternary system (FG and EG) are projected onto the pseudo-binary surface and are indicated with the same letter designations.

A liquid such as H cools to precipitate crystals of A (Fig. 86). This immediately drives the liquid off the Alkemade line and off the TX plane of the pseudo-binary system. The liquid migrates to I, where A and C precipitate together. This same sequence is seen in the pseudo-binary as a curved crystallization path across the field of Liq $+ A$ to the boundary curve at I. The curvature and direction of this path can be deduced only from examination of the thermal slope of the liquidus surface of A in the ternary system. From the point I, continued crystallization of A and C changes the remaining liquid to G, where A and liquid are consumed while converting the entire batch to solid A_2B and C.

A more complex pseudo-binary system can be derived from the section $A - BC$ in Figure 87 by the same procedure used to obtain the pseudo-binary in Figure 85. The general character of the liquidus surfaces can be derived by noting the configuration of the ternary system; various ternary crystallization paths reveal the internal features of the pseudo-binary (Fig. 88). Corresponding points are labeled in both figures to facilitate comparison. In addition, paths of crystallization for liquids I and J are shown on the ternary system by means of broken lines.

The phase diagram of the system $CaAl_2Si_2O_8$—Mg_2SiO_4—SiO_2 (Fig. 89) contains a primary field of spinel, $MgAl_2O_4$, situated between anorthite ($CaAl_2Si_2O_8$) and forsterite (Mg_2SiO_4). The pseudo-binary, seen in Figure 90, is actually a portion of the very complex system CaO—MgO—Al_2O_3—SiO_2. The presence of spinel indicates that the system $CaAl_2Si_2O_8$—Mg_2SiO_4—SiO_2 is in fact pseudo-ternary. The relationships within this system are discussed in some detail by Osborn and Tait (1952).

---

*The more intimidated readers may here feel obliged to hold the page away from their person to avoid the molten silicate melt.

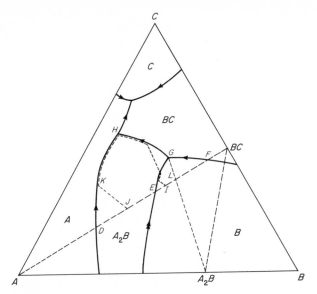

FIGURE 87
The hypothetical system *A-B-C* and its binary congruently and
incongruently melting compounds *BC* and A_2B.

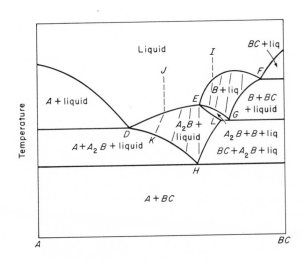

FIGURE 88
The pseudo-binary join *A-BC*, from the system
A-B-C of Figure 87.

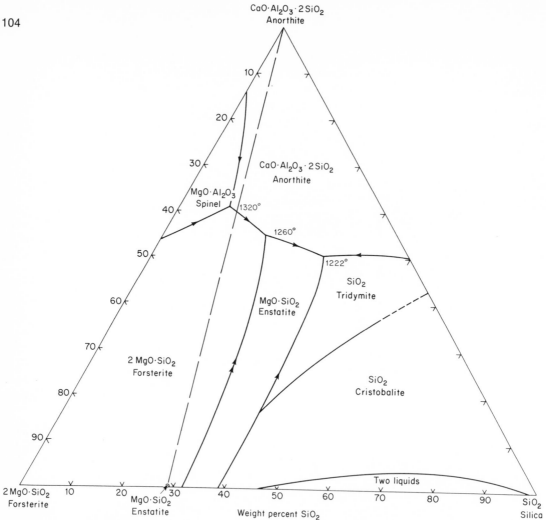

FIGURE 89

The pseudo-ternary system Mg$_2$SiO$_4$—CaAl$_2$Si$_2$O$_8$—SiO$_2$. [After Andersen (1915).]

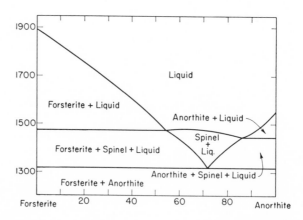

FIGURE 90

The pseudo-binary join Mg$_2$SiO$_4$—CaAl$_2$Si$_2$O$_8$. [After Osborn and Tait (1952).]

Studies of such pseudo-systems, although quite difficult to visualize when complex, often provide valuable insight into the even more complex systems of which they are a part. For example, Bowen (1928, pp. 277–281) was able to describe the crystallization of spinels on the basis of the information available in a diagram of the pseudo-ternary system (Fig. 89), described earlier by Andersen (1915). More recently, many different investigators have been able to obtain much useful information about the melting and crystallization behavior of peridotites and eclogites at high pressure by analyzing various pseudo-systems within the quaternary system $CaO-MgO-Al_2O_3-SiO_2$.

6

Systems under Confining Pressure

In general when systems are said to be under pressure, the pressure referred to (unless otherwise specified) is a confining pressure on the solids, liquids, or gas.* Before discussing such systems, however, some mention should be made of vapor pressure-temperature diagrams, as these are occasionally found in the geological literature. The thought behind such a diagram is straightforward — namely, that each solid or liquid possesses a definite vapor pressure, even if it is too small to be measured conveniently; furthermore, at any temperature the phase with the smaller vapor pressure is the stable phase. Thus, if we have the two polymorphs X_1 and X_2, and the vapor pressure of X_1 is greater than X_2, eventually all of the X_1 will convert to X_2. Relative stability of various phases can be indicated by plotting vapor pressure against temperature.† Diagrams of this sort are generally used to show either metastable

---

*Various units of pressure are in common use in the literature. Conversions between these units are given below. The bar and the kilobar (1,000 bars) have been adopted as the standard units of pressure (Kennedy, 1961), and it is expected that most future usage will conform to this.

1 bar = 0.986924 atmospheres = 1.019716 kilogram/cm² = 14.5038 pounds/in²

1 atmosphere = 1.013249 bars = 1.033226 kg/cm² = 14.6969 lbs/in²

In order to relate the pressures and temperatures of the experimental data to conditions that are expected to prevail within the earth, geothermal gradients are presented in Figure 91.
†The variation in vapor pressure with temperature is given by the relationship

$$\frac{d(\ln P_{vap})}{d(\frac{1}{T})} = \frac{-\Delta H_{vap}}{R}$$

where P_{vap} = vapor pressure, ΔH_{vap} = heat of vaporization, R = gas constant, T = absolute temperature. This relationship is discussed in Mahan (1964, p. 113), and in most other textbooks on physical chemistry. A plot of $\ln P_{vap}$ against reciprocal temperature yields a line of slope $-\Delta H_{vap}/R$. In most systems ΔH_{vap} varies slightly with temperature, and the lines show a gentle curvature.

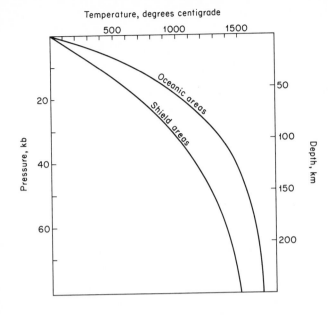

FIGURE 91
Oceanic and Precambrian
Shield geotherms.
[After Ringwood, MacGregor,
and Boyd (1964).]

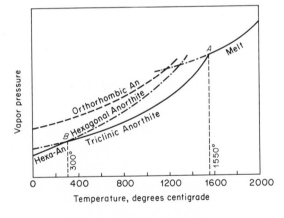

FIGURE 92
Hypothetical vapor-pressure–temperature
diagram for the system $CaAl_2Si_2O_8$.
[After Yoshiki, Koide, and Waki (1953).]

persistence or inferred relative temperature stabilities when precise data are
lacking. A vapor pressure diagram for the system $CaAl_2Si_2O_8$ is shown in
Figure 92; at any temperature, the stable phases occupy the minimum vapor
pressure curves (shown as solid lines); metastable curves are shown as broken
lines. The ambient pressure is one atmosphere. From these curves, one can
deduce that a liquid of composition $CaAl_2Si_2O_8$ is stable above 1550°C.
Below 1550°C the liquid has a higher vapor pressure than triclinic anorthite.
Therefore the triclinic anorthite is stable; the intersection of the two vapor
pressure curves at *A* represents the freezing point of the liquid. The vapor

pressure of triclinic anorthite is lower than that of the other solid phases from the melting point (at *A*) down to about 300°C (at *B*), below which the diagram indicates that hexagonal anorthite is the most stable phase. Orthorhombic anorthite has been synthesized from melts, but eventually converts to triclinic anorthite under prolonged heating at elevated temperatures. The implication of the diagram is that orthorhombic $CaAl_2Si_2O_8$ has no stable field of existence, as there is no temperature at which its vapor pressure curve is lower than that of any of the other phases.

UNARY SYSTEMS

An examination of the geological literature will reveal that a considerable amount of the research on high-pressure phase equilibrium is done on one and two-component systems. The reason for this is that most geological research on this subject has had to await recent technological improvements in equipment. Within the past decade, devices have become available that are capable of routinely attaining pressure of up to 200 kb and temperatures of 1300°C. Consequently, considerable geologically oriented work is now being done.

One perhaps obvious rule of thumb to keep in mind when working with pressure-temperature diagrams is that pressure on a system causes a decrease in volume, and temperature (in general) causes an increase in volume. Thus pressure and temperature can be thought of as generally opposing each other. The particular changes that take place in response to changing *P* and *T* are varied, but the simple principle remains the same.

The most well-known one-component system is certainly that of carbon. The original synthesis of diamond was described by Bundy, Hall, Strong, and Wentorf in 1955, and the diamond-graphite equilibrium curve (Fig. 93) was published by Bundy, Bovenkirk, Strong, and Wentorf in 1961. As can be seen, high pressure favors the stability of the denser polymorph diamond (mol. vol. = 3.42 cm³/mole), whereas high temperature favors the less dense polymorph graphite (mol. vol. = 5.33 cm³/mole). The transition curve derived on a thermodynamic basis by Berman and Simon (1955) shows very good agreement with the experimentally derived curve seen in Figure 93.

Bundy in 1962 explored the higher-temperature region of the carbon system and determined the melting curves for graphite and diamond (Fig. 94). Along the graphite-liquid curve (between *A* and *B*) it can be observed that increasing pressure on the liquid at a fixed temperature causes the formation of graphite, indicating that the graphite is denser than the liquid. At higher pressures (between *B* and *C*) the slope of the curve is reversed, and increasing pressure causes the graphite to convert to a denser liquid. The curve between diamond and liquid also has a negative slope (between *C* and *D*), indicating that increasing pressure causes the dense diamond to convert to a still denser liquid. By analogy with other systems, a solid phase (Solid III)

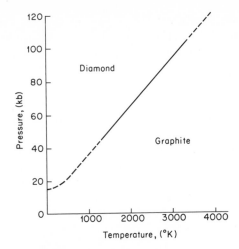

FIGURE 93
Pressure–temperature relations of the univariant line
diamond-graphite. [After Bundy, et al. (1961).]

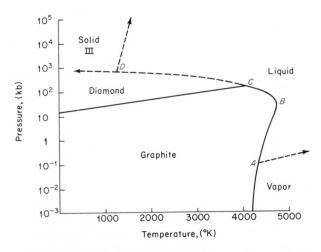

FIGURE 94
The carbon system. [Data after Bundy (1962); Figure
after Stewart (1967).]

was postulated to exist at pressures above that required for diamond stability.
It can further be observed from the diagram that diamond is not stable under
ordinary conditions of T and P. It does persist metastably, however, as do
many high-pressure polymorphs.

As far back as the early 1900's, P. W. Bridgman (1958) conducted large
numbers of high-pressure investigations, one of which was a detailed study of

the system H_2O at low temperatures (1912). Recent data on this system, as compiled by Kennedy and Holser (1966), are shown in Figure 95. Five polymorphs of ice exist in addition to normal ice (ice I). Because ice I is less dense than water, its melting temperature *decreases* with increasing pressure. The negative slope of the melting curve is often used to explain how masses of glacial ice can become mobile. At the base of a thick glacier, the melting point of ice might become sufficiently depressed to allow melting and consequent movement of the frozen mass.

As can be seen from Figure 95, ice I (at point A) melts when the pressure is increased to point B. With additional pressure the liquid will freeze at point C to form ice III. With still more pressure ice III converts at point D to ice V; ice V converts to ice VI at E, and to ice VII at F.

From diagrams of this type, a general idea of the volume differences may be obtained from the Clapeyron equation, the general equation that describes the slope of a PT curve:

$$\frac{dT}{dP} = \frac{T\Delta V}{\Delta H}$$

where dT/dP is the rate of change of the transition temperature with pressure, T is the temperature at which the transition curve is calculated, ΔH is the heat of transition, and ΔV the difference in volume between the high- and low-temperature phases. The volume difference may be $V_{gas} - V_{liquid}$, $V_{liquid} - V_{solid}$, or $V_{solid_I} - V_{solid_{II}}$. This equation is most effective for gas-liquid transitions, but has been applied also to liquid-solid and solid-solid transitions. Because of the decrease in ΔV with increasing pressure (as well as variation of ΔH), most transition curves show a decrease in slope (dT/dP) with increasing pressure: see McLachlan and Ehlers (1971).

In the normal density relationship between phases, the lower-temperature phase has the smaller volume and the ΔV is positive, since $V_{\text{High } T \text{ phase}} > V_{\text{Low } T \text{ phase}}$; the slope of the transition curve is also positive. The general density relationships between phases are summarized in Figure 96. If there is no volume change between the high- and low-temperature phases, the slope is neither positive nor negative and the transition is unaffected by a change in pressure. If the higher-temperature phase is the more dense, as in the transition from ice I to water, the curve will have a negative slope.

The relative densities of the various phases in the H_2O system are readily obtained from Figure 95. The phase sequences observed with increasing pressure from points A and G are shown below:

(Point A)   I $\rightarrow$ Water $\rightarrow$ III $\longrightarrow$ V $\rightarrow$ VI $\rightarrow$ VII

(Point G)   I $\longrightarrow$ III $\rightarrow$ II $\rightarrow$ V $\rightarrow$ VI $\rightarrow$ VII

By combining the two sequences, we can show that the order of increasing density is I, water, III, II, V, VI, and VII.

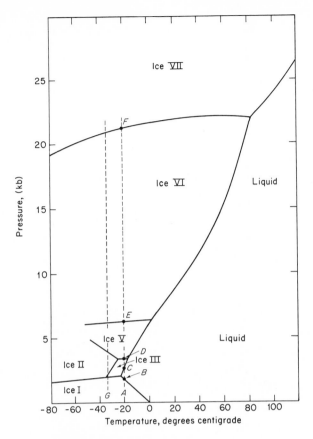

FIGURE 95
The system H_2O. [After Bridgman (1912, 1958) and
Kennedy and Holser (1966).]

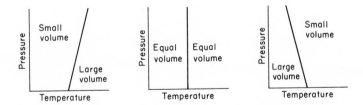

FIGURE 96
The effect on the *PT* curve of different molecular volumes of two
polymorphs. The stability field of the denser polymorph, which has
the smaller molecular volume, is increased with pressure at the
expense of the less-dense phase.

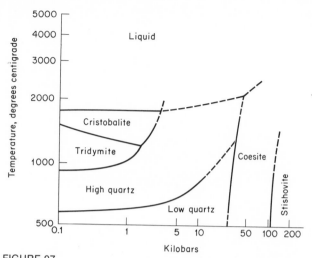

FIGURE 97

The system SiO_2. [After Ostrovsky (1966); the coesite-stishovite boundary has been revised in accordance with new data (personal communication from Ostrovsky).]

The PT diagram for the system SiO_2 is given in Figure 97. The slopes of the transition curves indicate that the densities of cristobalite, tridymite, and liquid are about equal, whereas a large density increase is evident from high and low quartz to coesite and from coesite to stishovite. Cristobalite and tridymite, with low densities, are eliminated in favor of quartz near 1 kb. The very dense coesite and stishovite have only been found under conditions requiring shock processes.

BINARY SYSTEMS

In the study of binary systems under pressure, it is usually necessary that pressure, temperature, and composition be represented simultaneously. One way to depict such information graphically is to prepare a three-dimensional diagram with P, T, and X as perpendicular coordinates. This means was used by Dachille and Roy (1960) for the system Mg_2SiO_4—Mg_2GeO_4. One of the main purposes of their study was to show, by extrapolation of known data, the pressure and temperature at which the Mg_2SiO_4 olivine structure would be converted to the denser spinel structure.

Such a diagram is perhaps best introduced by presenting several isobaric binary diagrams of the system. Figure 98,A shows the system at a pressure of 0.69 kilobars (10,000 p.s.i.).* The pure magnesium orthogermanate pos-

---

*Pressure units used in the original studies are used throughout this book, so as to familiarize the reader with the various systems of measurement.

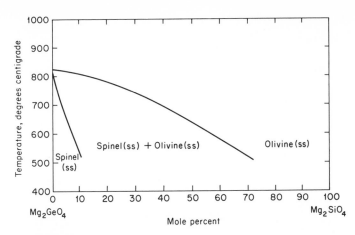

FIGURE 98A
Isobaric section at 10,000 p.s.i. for the system Mg_2GeO_4—Mg_2SiO_4.
[After Dachille and Roy (1960).]

sesses a spinel structure below 823°C and an olivine structure at higher temperatures. The higher-temperature olivine structure shows complete solid solution between Mg_2GeO_4 and Mg_2SiO_4, whereas the lower-temperature spinel structure allows only limited amounts of Mg_2SiO_4 to be substituted for Mg_2GeO_4. A large two-phase region exists in the central part of the diagram.

The lower-temperature spinel phase has a higher density than the corresponding olivine structure. Therefore at higher pressures the spinel field increases, as is seen in Figures 98, B, C, and D. A series of such isobaric TX diagrams can be arranged along a pressure coordinate to produce a three-dimensional PTX diagram, as was done by Dachille and Roy (Fig. 99). The four shaded isobaric sections labeled a, b, c, and d correspond to the four diagrams shown in Figure 98, A, B, C, D. Note that the size of the stability field of spinel increases with increase in pressure (to the right in Fig. 99). Extrapolation of these data indicated the possible existence of a spinel form of Mg_2SiO_4 at a pressure not too far above 100 kb; this is shown in an isothermal PX diagram (Fig. 100, A), which corresponds to the base of the PTX projection. The extrapolation, however, has since been demonstrated to be incorrect (Ringwood, 1969); in the transition from Mg-rich olivine to spinel, an intermediate spinel-like "β" phase forms, as shown in Figure 100, B.

Another geologically significant system is that of $NaAlSiO_4$—SiO_2 (nepheline-silica). This was investigated at atmospheric pressure by Greig and Barth (1938) and Schairer and Bowen (1947). The central region of the system was examined at high confining pressure by Robertson, Birch, and MacDonald (1957), and was subsequently studied in considerably more detail by Newton and Smith (1967), Newton and Kennedy (1968), and Bell and Roseboom

114

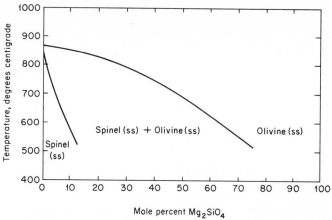

FIGURE 98B
Isobaric section at 55,000 p.s.i. for the system Mg_2GeO_4—Mg_2SiO_4.
[After Dachille and Roy (1960).]

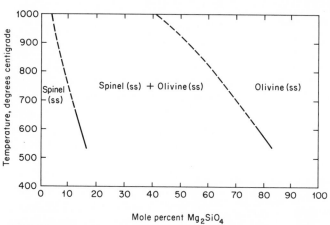

FIGURE 98C
Isobaric section at 200,000 p.s.i. for the system Mg_2GeO_4—Mg_2SiO_4.
[After Dachille and Roy (1960).]

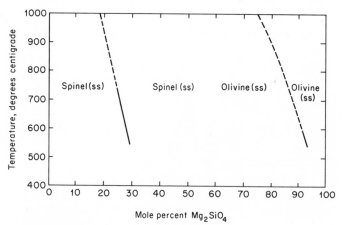

FIGURE 98D
Isobaric section at 580,000 p.s.i. for the system
Mg_2GeO_4—Mg_2SiO_4. [After Dachille and Roy (1960).]

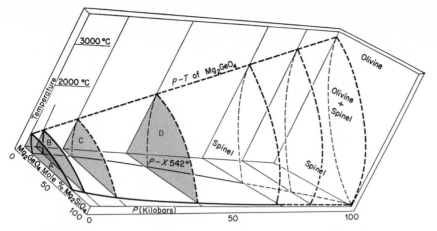

FIGURE 99
PTX projection of the system Mg$_2$GeO$_4$—Mg$_2$SiO$_4$. The four shaded *TX* areas labeled *A*, *B*, *C*, and *D* correspond to Figure 98A,B,C,D. [After Dachille and Roy (1960).]

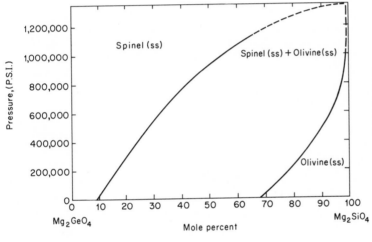

FIGURE 100A
Isothermal *PX* section of the system Mg$_2$GeO$_4$—Mg$_2$SiO$_4$, which corresponds to the base of the *PTX* projection of Figure 99. [After Dachille and Roy (1960).]

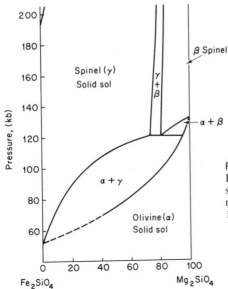

FIGURE 100B
Isothermal *PX* section of the system Mg$_2$SiO$_4$—Fe$_2$SiO$_4$ at 1000°C, showing the presence of a high-pressure phase, β spinel, above the magnesian olivine (α) structure. The spinel in Figures 98 to 100 is of the γ type. [After Ringwood (1969).]

116

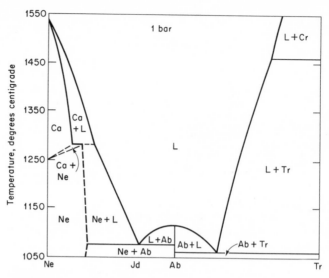

FIGURE 101

The system NaAlSiO$_4$—SiO$_2$ at 1 atmosphere (L = liquid, Ab = Albite, Ca = Carnegieite, Ne = Nepheline, Jd = Jadeite, Tr = tridymite, Cr = cristobalite). [After Greig and Barth (1938).]

(1969). Most of the data on the following diagrams was taken from Bell and Roseboom.

Figure 101 shows that at atmospheric pressure the system includes the congruently melting binary compound albite (NaAlSi$_3$O$_8$). Nepheline (NaAlSiO$_4$) exhibits considerable solid solution, and converts to the high-temperature polymorph carnegieite above 1250°C. A high-pressure phase, jadeite (NaAlSi$_2$O$_6$), is indicated on the compositional base of the diagram, but is not stable in this PT range.

With increasing pressure (Fig. 102) the liquidus surfaces rise as the eutectic between nepheline and albite shifts from the silica-rich to the silica-poor side of the jadeite composition point. At 22 kb (Fig. 103) the eutectic point between albite and quartz has risen to a higher temperature than the eutectic between albite and nepheline. In addition the denser phase jadeite is stable in the lower part of the indicated temperature range; the presence of jadeite eliminates the lower-temperature portion of the albite-nephline two-phase field.

The jadeite stability field continues to increase in size with pressure, and at a pressure just below 25 kb (Fig. 104) intersects the albite-nepheline eutectic. The temperature of this eutectic increases at a slower rate than the upper temperature limit of jadeite. The point at which the eutectic temperature is intersected by upper temperature limit of jadeite is labeled I_{Jd}. Here we have a unique situation, where four phases may coexist in a binary system—liquid, nepheline, jadeite, and albite. As a particular pressure (as well as temperature)

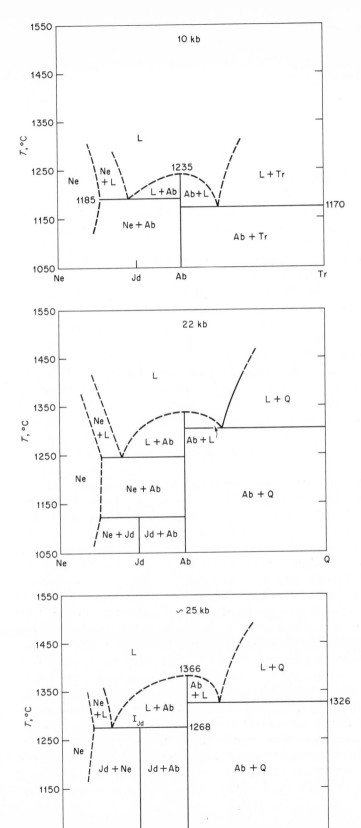

FIGURE 102
The system NaAlSiO₄—SiO₂ at 10 kb.
[After Bell and Roseboom (1969).]

FIGURE 103
The system NaAlSiO₄—SiO₂ at 22 kb.
[After Bell and Roseboom (1969).]

FIGURE 104
The system NaAlSiO₄—SiO₂ at about 25 kb.
[After Bell and Roseboom (1969).]

is required for this relationship, the condition of the system must be described in terms of the Standard Phase Rule, where

$$P + F = C + 2$$

$$4 + F = 2 + 2$$

$$F = 0$$

This point is thus an invariant point in the complete PTX description of the system. If the pressure were fixed and chosen *randomly* to describe a particular TX section (as was done in earlier chapters), such an invariant point would normally not be encountered and the TX section would be described in terms of the Condensed Phase Rule, $P + F = C + 1$; in such a description a three-phase assemblage would constitute an invariant point. Three-phase assemblages that are invariant in a static pressure situation become univariant lines when pressure is added as a variable to T and X. This is also demonstrated in these diagrams by the change in eutectic temperatures as a function of pressure.

Increasing the pressure to 26 kb (Fig. 105) brings the upper temperature limit of jadeite stability above the temperature of the albite-nepheline eutectic, causing jadeite to melt incongruently. Near 28 kb (Fig. 106) the melting temperature of jadeite reaches the liquidus surface between albite and the albite-nepheline eutectic. Here jadeite makes a transition from incongruent to congruent melting. This point of transition in melting behavior is known as a *singular point*, and is labeled S_{Jd} in the figure. In addition, albite is seen to break down at lower temperatures to a denser assemblage of jadeite and quartz.

Congruent melting of both jadeite and albite necessitates a eutectic point between the composition points of these two phases. This eutectic point rises with pressure until it is at the same temperature as the albite-quartz eutectic (Fig. 107). Increasing pressure further (Fig. 108) raises the jadeite-albite eutectic above that of the albite-quartz eutectic. The jadeite melting point rises with pressure at a greater rate than that of albite (Fig. 109), and at about 33 kb (Fig. 110) albite undergoes a transition from congruent to incongruent melting (labeled S_{Ab}). This is a singular point for albite. At pressures slightly above 33 kb albite melts incongruently (Fig. 111).

While the changes near the melting temperatures take place, the stability region of jadeite + quartz increases constantly at the expense of albite. This phenomenon results in a drastic decrease in the temperature stability range of albite at the higher pressures shown in Figures 110 and 111. The decrease in the stability field of albite continues until the phase is completely eliminated at about 34 kb (Fig. 112).

At pressures above about 33 kb, coesite, a high-pressure polymorph of SiO_2, first appears on the TX projection (Fig. 111). With increasing pressure the stability region of coesite expands at the expense of quartz (Fig. 112). At 40 kb (Fig. 113) quartz has been eliminated and its place taken by coesite. At considerably higher pressures (not shown) coesite is replaced by a still denser polymorph, stishovite.

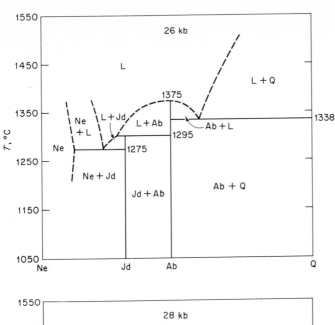

FIGURE 105
The system NaAlSiO₄—SiO₂ at 26 kb.
[After Bell and Roseboom (1969).]

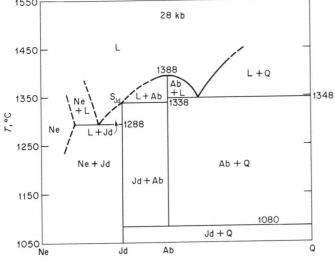

FIGURE 106
The system NaAlSiO₄—SiO₂ at 28 kb.
[After Bell and Roseboom (1969).]

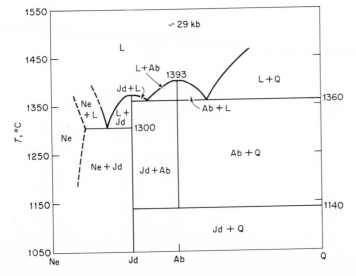

FIGURE 107
The system NaAlSiO₄—SiO₂ at about 29 kb.
[After Bell and Roseboom (1969).]

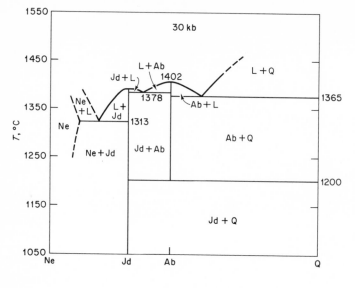

FIGURE 108
The system NaAlSiO₄—SiO₂ at 30 kb.
[After Bell and Roseboom (1969).]

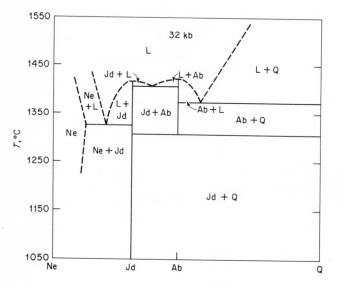

FIGURE 109
The system NaAlSiO₄—SiO₂ at 32 kb.
[After Bell and Roseboom (1969).]

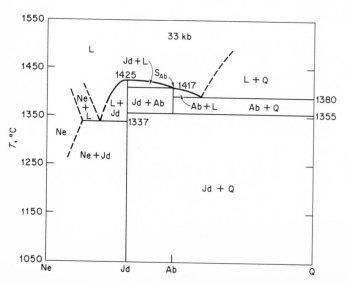

FIGURE 110
The system NaAlSiO₄—SiO₂ at 33 kb.
[After Bell and Roseboom (1969).]

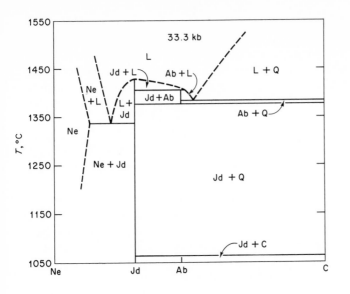

FIGURE 111
The system NaAlSiO$_4$—SiO$_2$ at 33.3 kb (C = coesite).
[After Bell and Roseboom (1969).]

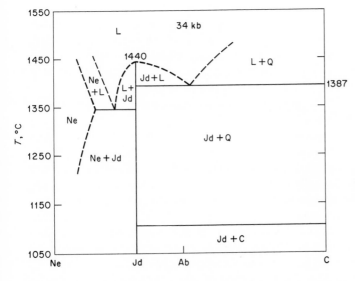

FIGURE 112
The system NaAlSiO$_4$—SiO$_2$ at 34 kb.
[After Bell and Roseboom (1969).]

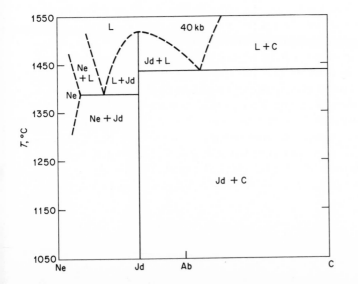

FIGURE 113
The system NaAlSiO$_4$—SiO$_2$ at 40 kb.
[After Bell and Roseboom (1969).]

The whole sequence of changes could be shown as a three-dimensional PTX projection, but the complex relationships would be lost amidst the maze of lines and planes. Furthermore, precise data could be extracted from such a projection only with the greatest difficulty. A better approach is to plot the univariant lines and invariant points on a PT projection. Such a plot, taken from Bell and Roseboom (1969), is shown in Figure 114; the diagram contains an insert showing the details of the phase relations in the vicinity of 33 kb and 1400°C. Four types of univariant curves are present. One of these shows the variation in the temperature of the melting point of a single phase, such as albite or jadeite, as a function of pressure. A second type indicates the PT conditions for a polymorphic change, such as the transition from quartz to coesite. A third type shows the temperature changes of the various eutectics as a function of pressure (e.g., $AB + Q \rightleftharpoons L$). A fourth type shows the combination of two solid phases to produce a third ($Ne + Ab \rightleftharpoons Jd$). The only thing not indicated on a PT diagram like the one in Figure 114 is the phase compositions; hence knowledge of liquid compositions at eutectics and peritectics, as well as solid phase compositions must be obtained by other means. Having all of the data for the various reactions plotted as a function of P and T allows one to construct an approximate TX diagram for any of the pressures indicated. Data extracted for a pressure of 15 kb are indicated by the broken line AA'. Albite at this pressure melts at temperature B; the albite-quartz eutectic is at temperature D, and the nepheline-albite eutectic is at temperature E.

Another use of PT diagrams is to show the limits of stability of various phases or phase assemblages. The few curves that appear in Figure 115 are extracted from the multitude shown in Figure 114. Figure 115 shows the limits of albite stability as a function of P and T. Albite is seen to have definite upper P and T limits. The stability limits of jadeite (Fig. 116), extracted again from Figure 114, show that jadeite is clearly favored by pressure, and is stable to the highest pressure shown in Figure 114. In addition, the presence of a second phase added to either albite or jadeite causes a decrease in the area of stability of these phases; Figure 117, A, B shows the effect of adding nepheline to albite and silica to jadeite. Detailed discussions of PT diagrams for binary systems are given by Bell and Roseboom (1969).

It is instructive to consider the utilization of raw experimental data in the preparation of a finished pressure-temperature diagram. The experimental work done by Bell and Roseboom (1969) on the system $NaAlSiO_4 - SiO_2$ (shown in Fig. 114) consisted of PT studies of only albite and jadeite compositions. The rest of the data for the system were either taken from earlier investigations or derived on the basis of general equilibrium relationships.

The use of a fixed composition in the preparation of PT diagrams is illustrated in part by Figure 118. The hypothetical system $A - B$ contains the intermediate compound A_3B. TX sections are shown at various pressures (with P increasing from P_1 to P_5). Assume that the aim is to get a pressure-temperature projection for the composition A_3B and that all experiments are done with that composition. At pressure P_1, the phase A_3B is stable at low temperatures; with increasing temperature it decomposes to a mixture of $A +$

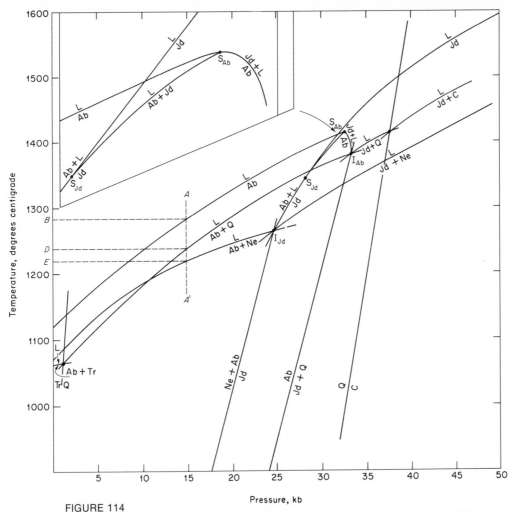

FIGURE 114

PT projection of the system NaAlSiO$_4$—SiO$_2$. Abbreviations correspond to those in Figures 101 to 113. [After Bell and Roseboom (1969).]

B. Since three phases in a two-component system are involved in the reaction $A_3B \rightleftharpoons 3A + B$, this decomposition takes place at a point on a univariant line in a complete *PTX* description of the system. The temperature of this reaction changes as a function of pressure, as is shown in the insert of the *PT* projection in Figure 118. At a somewhat higher temperature (also at pressure P_1) the bulk composition A_3B, which now consists of a mixture of *A* and *B*, converts to liquid (denoted by L) and *A* as $A + B \rightleftharpoons L + A$. Here we have a crossover from one divariant region to another by means of a divariant surface. Although only three phases are involved in the crossover, *A*, *B*, and liquid, the

124

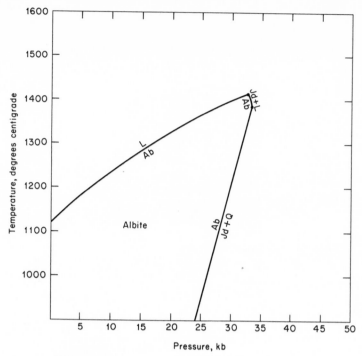

FIGURE 115
The stability field of albite, extracted from Figure 114.

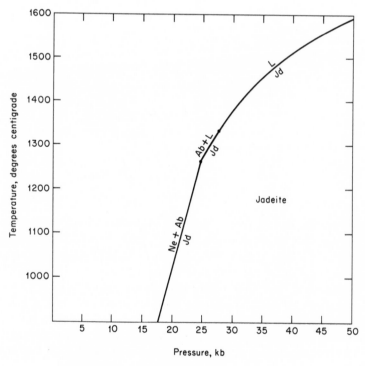

FIGURE 116
The stability field of jadeite, extracted from Figure 114.

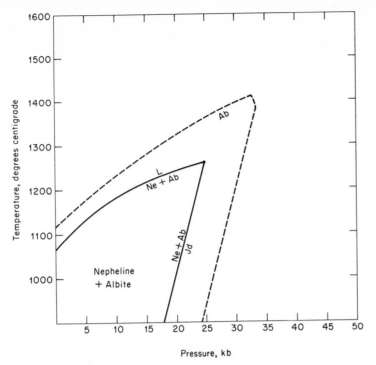

FIGURE 117A
Decrease in the field of stability of albite in the presence of
nepheline.

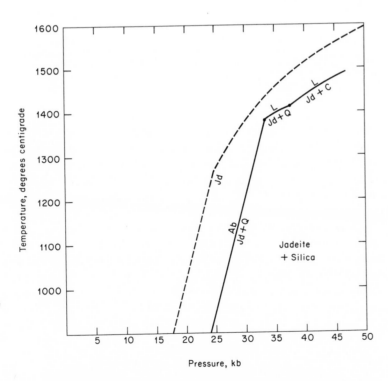

FIGURE 117B
Decrease in the stability field of jadeite in the presence of silica.

126

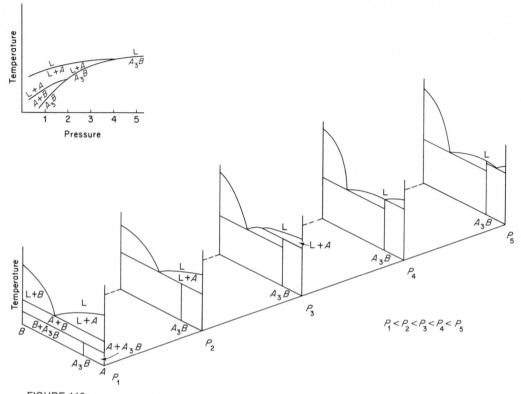

FIGURE 118

Temperature-composition sections at pressures increasing from P_1 through P_5 for the hypothetical system A-B. The effect of pressure is to increase the stability field of intermediate compound A_3B as well as the liquidus surfaces. The TP insert shows the various reactions that composition A_3B undergoes at different pressures.

same phase, A, is present as both a product and a reactant (it merely changes in amount). Thus this transition cannot be represented by a point on a univariant curve. The related univariant reaction occurs at the eutectic, where A and B in the eutectic ratio produce only liquid. The intersection of the two divariant regions $A + B$ and $L + A$ can also be plotted as a line on the PT projection (see insert). This line and the previously discussed univariant curve $A_3B \rightleftharpoons 3A + B$ converge with pressure and intersect at pressure P_2. Returning to the bulk composition A_3B at pressure P_1, note that the last reaction that takes place there is $L + A \rightleftharpoons L$, where the liquidus surface is encountered. This point is not an intersection with a univariant curve, as an insufficient number of phases are involved. Hence, of the three reactions observed for the bulk composition A_3B, only one takes place at an intersection with a univariant curve.

As the pressure is raised the upper stability limit of compound A_3B continues to rise. At P_2 a transition is made from a solid-solid decomposition relation to an incongruent melting relation (seen at P_3). The univariant reaction that shows incongruent melting is indicated in the insert as the curve $A_3B \rightleftharpoons L + A$. At a pressure P_4 the melting of A_3B changes from incongruent to congruent, which is indicated by beginning of the curve $A_3B \rightleftharpoons L$.

The usefulness of such PT diagrams can be shown by a study of Bell and Roseboom's diagram (Fig. 119) for the jadeite composition. In examining such figures it is first necessary to distinguish between univariant lines and intersections of divariant areas. The various curves and points may be distinguished with the aid of the Phase Rule. In a two-component system with P, T, and/or X as possible variables, the relation $P + F = C + 2$ indicates that the association of four phases constitutes an invariant point. In Figure 119 the phase boundaries intersect at points A, B, and C. The phase assemblages are as follows:

Point	Phases*
A	L, Ab, Ne, Jd
B	L, Ab, Jd
C	L, Ab, Ne

*L = liquid, Ab = albite, Ne = nepheline, Jd = jadeite.

Of the three intersections, only A indicates equilibrium of four phases, hence only A is an invariant point. Points B and C indicate three-phase equilibrium and are points on univariant curves.

It follows from the definition of an invariant point that if pressure, temperature, or composition is changed, one or more of the four coexisting phases will be eliminated. Loss of a single phase will reduce the condition to univariancy (as $P + F = C + 2, 3 + F = 2 + 2, F = 1$); a three-phase univariant assemblage exists along a univariant line in PTX space. As any one of the four phases may be eliminated, four univariant curves must be associated with each invariant point. Loss of two phases brings the condition to divariancy (as $2 + F = 2 + 2, F = 2$); divariant assemblages exist on PTX surfaces that join the various univariant curves. Finally, single-phase volumes ($F = 3$) will exist between the various divariant surfaces.

The invariant point A in Figure 119 indicates coexistence of the four phases liquid, albite, nepheline, and jadeite. Elimination of any one of these will leave the other three existing on a univariant curve. The possible univariant assemblages that exist about point A are shown in the table below. Note that not all of these are present in the experimental results shown in Figure 119. The phase that has been eliminated is indicated in parentheses. This same nota-

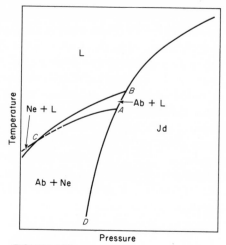

FIGURE 119
Experimental results for jadeite composition.
[After Bell and Roseboom (1969).]

tion will be used in figures; the phase that has been eliminated along a particular curve will be so indicated by a label in parentheses at the end of that curve.

Phase Eliminated	Phases Stable on Univariant Curve
(L)	Ab, Ne, Jd
(Ab)	L, Ne, Jd
(Ne)	L, Ab, Jd
(Jd)	L, Ab, Ne

If a change in conditions causes deviation from a univariant curve, either one or two phases of the univariant assemblage is lost. It is helpful to think of the univariant curve as representing the equilibrium conditions for a chemical reaction—the line in PTX space along which both products and reactants exist together. The reaction related to a particular curve can be deduced from the chemical compositions of the phases involved. The phase compositions are shown along a compositional line in Figure 120. Consider the univariant

FIGURE 120
Compositional bar, showing differences in composition of various phases between nepheline and quartz.

curve along which liquid is eliminated to leave albite, nepheline, and jadeite. The phase compositions in Figure 120 show that albite and nepheline could react to produce jadeite, but that other reactions (such as Ne + Jd ⇌ Ab and Jd + Ab ⇌ Ne) are impossible. The permissible reaction is seen in Figure 119 as the curve AD.

Returning again to the invariant point A in Figure 119, one can, by considering the composition limitations, deduce the reactions associated with each of the four related univariant curves.* These are shown below:

Curve Shown in Fig. 119	Phase Eliminated	Phases Stable on Univariant Curve	Reactions Related to Each Curve
AD	(L)	Ab, Ne, Jd	Ne + Ab ⇌ Jd
Not present	(Ab)	L, Ne, Jd	Ne + Jd ⇌ L
AB	(Ne)	L, Ab, Jd	L + Ab ⇌ Jd
Not present	(Jd)	L, Ne, Ab	Ne + Ab ⇌ L

*The liquid composition for these reactions is shown as L_1 in Figure 120, as can be deduced by consideration of the experimentally determined reaction Ab + L ⇌ Jd (Fig. 119).

When properly described, the invariant point A in Figure 119 should have associated with it the four univariant curves listed above. The curves, in turn, should show the four reactions listed above. Figure 119 shows three experimentally determined curves associated with point A. The curves AB and AD indicate the PT values for the reactions Ab + L ⇌ Jd, and Ab + Ne ⇌ Jd, as shown above in the table. The curve AC shows the conversion Ab + Ne ⇌ Ab + L and is therefore not one of the four univariant curves. Of the four necessary curves only two are determined experimentally; in spite of this, the relative positions of the four univariant curves can be deduced by means of the Morey-Schreinemakers Coincidence Theorem (usually known as the *Schreinemakers Rule*). The rule simply states that the reaction that occurs at a univariant curve can be also considered to occur on a metastable extension of the curve (drawn through the invariant point). From this rule, given one univariant curve and reaction, the relative positions of the other curves can be found.

Figure 121 shows one of the known experimental curves with its dotted metastable extension through the invariant point A. This is the liquid-absent curve (L)—the curve AD of Figure 119. Jadeite is present on the high-pressure side of the curve. Therefore, from the Schreinemakers Rule, we know the jadeite-absent curve (Jd) must lie on the low-pressure side, as shown schematically in Figure 122. Similarly, since Ab and Ne are on the low-pressure side of the curve (L), it follows that the albite- and nepheline-absent curves (Ab) and (Ne) lies on the high-pressure side. Before tentatively arriving at a placement of these two curves, it is necessary to label the reaction on the jade-

130

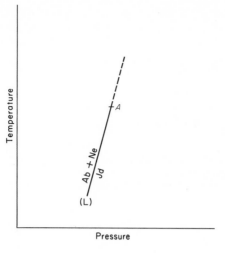

FIGURE 121
The univariant curve for the reaction Ab + Ne $\rightleftharpoons$ Jd. The symbol "(L)" indicates that liquid is absent. The invariant point corresponds to A.

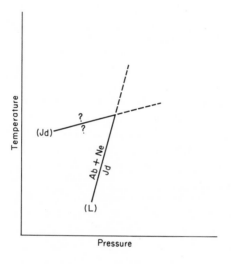

FIGURE 122
Possible position for the jadeite-absent univariant curve (Jd), which must fall on the low-pressure side of the liquid-absent curve (L).

ite-absent curve (Jd), as this will be seen later to show their correct relative positions. Two possibilities for labeling the curve (Jd) seem to exist, as seen in Figure 123, A, B. Figure 123, A is correct. This must be so because the curve (Jd) in Figure 123, A shows liquid existing on its high-temperature side; from this it follows that the univariant curve (L), which indicates an absence of liquid, must lie on the low temperature side of the (Jd) curve and its metastable extension. Figure 123, B is incorrect, as both the field of liquid (shown on the (Jd) curve) and the liquid-absent curve (L) lie on the same side of the (Jd) curve.

Figure 124 shows the correct relative placement of the (Ab) curve. This curve must lie on the high P side of the (L) curve and on the high T side of

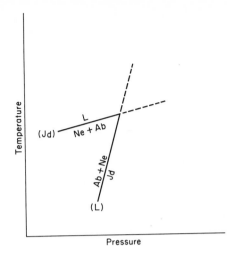

FIGURE 123A
The correct placement of phases about the jadeite-absent curve (Jd). Liquid L must be located on the high-temperature side of this curve as the liquid-absent curve (L) is located below it.

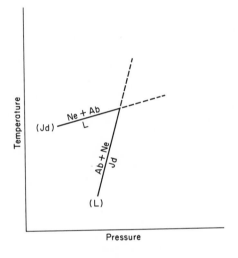

FIGURE 123B
Incorrect placement of phase assemblage about the jadeite-absent curve (Jd). The field of liquid L is inconsistent with the position of the liquid-absent curve (L).

the (Jd) curve. The only region that fulfills both of these conditions is the one between the metastable extensions of these two curves. The (Ab) curve must now be labeled according to the reaction Ne + Jd ⇌ L. The liquid field must be on the high-temperature side, as the curve (L) is on the low-temperature side. The assemblage Ne + Jd must be on the low-temperature side, as the curve (Jd) lies on the high-temperature side.

The final univariant curve (Ne) is placed as shown in Figure 125, as it must lie on the high-temperature side of the (Jd) and (Ab) curves, as well as on the high-pressure side of the (L) curve. The labeling of the curve with the reaction L + Ab ⇌ Jd conforms to the placement of the (Jd), (L), and (Ab) curves.

It is seen from the above exercise that a limited amount of data can be used

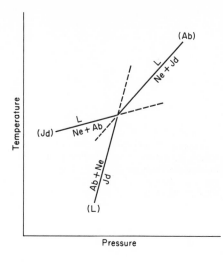

FIGURE 124
The correct placement of the albite-absent
curve (Ab) in relation to curves (Jd) and (L).

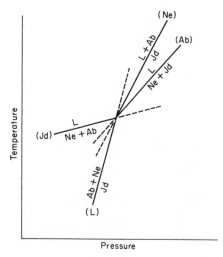

FIGURE 125
A complete and consistent arrangement of
four univariant lines about the central
invariant point.

to obtain a considerable amount of information. Note that the diagram in Figure
125 is in general conformity with the experimental data given in Figure 119
for jadeite, as well as with the data for the binary PT projection in Figure 114
(which shows the invariant point discussed as I_{Jd}. Experimentally derived
curves for albite compositions (Figure 126, A) allow a similar derivation for
invariant points A or B. The relative positions of the univariant curves and
reactions about point A are shown in Figure 126,B. In this construction, one
must know that the liquid composition falls between albite and quartz, as is
indicated by the compositional bar shown in Figure 120. The approximate
liquid composition can be derived from the experimental reaction $Ab \rightleftharpoons Jd + L$.

A

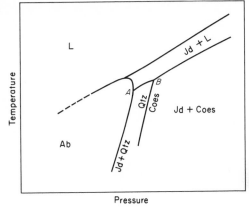

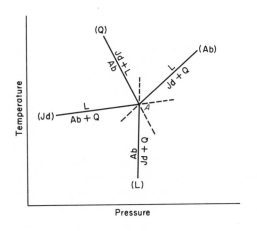

FIGURE 126A
Experimental data on the albite composition.
[After Bell and Roseboom (1969).]

FIGURE 126B
A possible arrangement of univariant lines about
the invariant point A (from Figure 126,A).

The placement of these curves according to known PT data is shown about the point I_{Ab} in Figure 114. More detailed discussions on the use of the Schreine-makers Rule are found in Korzhinskii (1959, pp. 96–103) and Niggli (1954, pp. 403–412).

TERNARY SYSTEMS

In the earlier discussions of complex ternary systems, it was pointed out that the study of liquidus relations allows one to delineate primary fields of crystal-lization, boundary curves, and invariant points. From such information it is possible to deduce the compatibility relations among the various mineral phases that exist in equilibrium with a liquid. Compatibility is indicated by

134

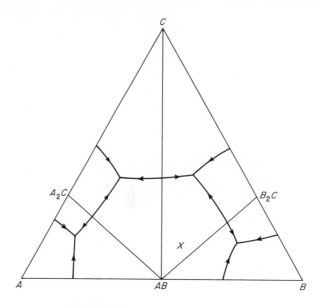

FIGURE 127
Compatibility relations in the hypothetical
system A-B-C at liquidus temperatures.

Alkemade lines and triangles, as in the hypothetical system $A-B-C$ shown in Figure 127. A melt of composition X will cool to produce the crystalline phases C, AB, and B_2C, rather than A, A_2C, and B or some other assemblage of compounds. Although the compatibility relations that are determined for liquidus temperatures may persist to room temperatures, it is quite common for new compatibility relations to develop via solid-state reactions at lower temperatures. For example, it might be possible that at some lower temperature the compounds A_2C and B_2C would be more stable together than C and AB, as would be indicated by the reaction

$$2AB + 2C \rightleftharpoons A_2C + B_2C$$

The sample X, which originally consisted of a mixture of AB, C, and B_2C at the liquidus, would have to undergo a solid-state reaction to form the mixture AB, A_2C, and B_2C in proper proportions (Fig. 128).

Similar types of changes in the compatibility arrangements occur in the solid state as a result of pressure changes. Dense compounds are developed at the expense of less dense, as has already been seen in the previous example of the system $NaAlSiO_4$—SiO_2. Such reactions necessitate changes in the Alkemade line arrangement.

Such a group of reactions is shown in Figures 129 to 133 for several isobaric isothermal sections in the system MgO—Al_2O_3—SiO_2, as compiled by Boyd and England (1963, p. 122) and revised by the data of Newton (1966). These

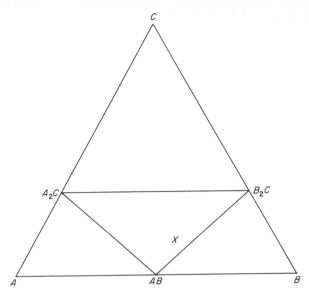

FIGURE 128
Subsolidus compatibility relations in
the hypothetical system *A-B-C*.

diagrams do not indicate the considerable solid solution shown earlier in Figures 77 and 78.

The Alkemade line arrangements for the system $MgO—Al_2O_3—SiO_2$ at 1300°C and one atmosphere are shown in Figure 129. At a pressure of 8 kb and 1100°C (Fig. 130), protoenstatite (PT) is no longer stable and exists as the denser orthorhombic enstatite (En). Enstatite shows some solid solution with Al_2O_3 under these conditions, and hence narrow swarms of two-phase tie-lines are shown rather than single Alkemade lines. In addition, enstatite exists in stable equilibrium with spinel, sapphirine, and sillimanite, causing significant rearrangement of the tie-lines.

As Boyd and England point out, this system illustrates that pressure favors structures with alumina in octahedral coordination (six oxygen atoms about each alumina) rather than tetrahedral (four-fold), as the higher coordination results in a denser and more tightly packed arrangement. Cordierite, typical of metamorphism at high temperature and low pressure, has all of its alumina in tetrahedral coordination; at atmospheric pressure (Fig. 129) it has a wide range of coexistence with many other minerals. At 8 kb (Fig. 130) the stable region of cordierite is decreased by formation of the enstatite-sillimanite join. At about 10 kb (Fig. 131) cordierite is completely eliminated. Pyrope, which has all of its alumina in octahedral coordination, is present at 21 kb (Fig. 132). This same diagram shows that kyanite, which contains only octahedral alumina, replaces sillimanite, which contains both tetrahedral and octahedral alumina. The system $MgO—Al_2O_3—SiO_2—H_2O$ has been discussed in greater detail

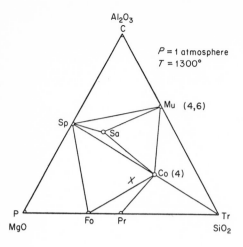

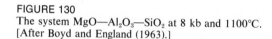

FIGURE 129
Compatibility relations in the system MgO—Al₂O₃—SiO₂ at 1 atmosphere and 1300°C. The numbers indicate alumina-oxygen co-ordination. For Figures 129 to 133, C = corundum, Co = cordierite, Cs = coesite, Fo = forsterite, Ky = kyanite, Mu = mullite, P = periclase, Pr = protoenstatite, Py = pyrope, Q = quartz, R En = rhombic enstatite, Sa = sapphirine, Si = sillimanite, Sp = Spinel; and Tr = tridymite. [After Boyd and England (1963).]

FIGURE 130
The system MgO—Al₂O₃—SiO₂ at 8 kb and 1100°C. [After Boyd and England (1963).]

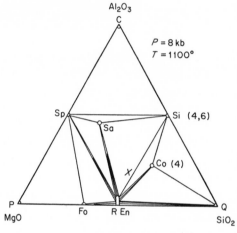

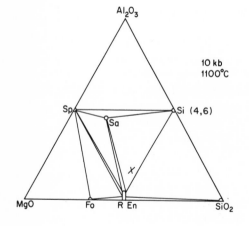

FIGURE 131
The system MgO—Al₂O₃—SiO₂ at 10 kb and 1100°C. [After Boyd and England (1963).]

FIGURE 132
The system MgO—Al₂O₃—SiO₂ at 21 kb and 1400°C. [After Boyd and England (1963).]

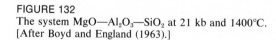

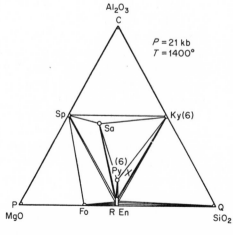

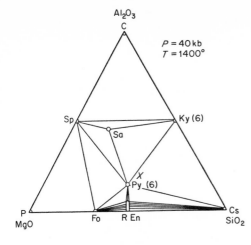

FIGURE 133
The system MgO—Al$_2$O$_3$—SiO$_2$ at 40 kb and 1400°C.
[After Boyd and England (1963).]

relative to coordination by Schreyer and Seifert (1970). Ringwood (1970) discusses mineral stability and coordination in the mantle relative to pressure-induced phase changes.

A typical series of mainly pressure-dependent reactions can be illustrated by considering a fixed composition, such as point X (Figures 129–133). The sequence of phase assemblages is listed below:

Pressure	Temperature*	Phase Assemblages Of Composition X
1 atm.	1300	Sp, Co, Fo
8 kb	1100	En, Sa, Sil
10 kb	1100	En, Sa, Sil
21 kb	1400	En, Ky, Py
40 kb	1400	Cs, Ky, Py

*Unfortunately the sections are not completely isothermal, as experimental data are not available. The temperature differences, however, are minor compared to the large pressure-induced changes.

It should be understood that the various phase assemblages shown are all subsolidus, and not necessarily stable in the presence of liquid.

There are, of course, univariant curves for each change of mineral compatibilities in the system; Figures 129 to 133 show only abbreviated versions of the system. For example, at 21 kb and 1400°C (Fig. 132), an Alkemade line exists between enstatite and spinel, whereas at 40 kb and 1400°C (Fig. 133) the same part of the diagram is crossed by a compatibility line between pyrope and forsterite. At the point where these two lines intersect, the reaction is:

$$4 \, MgSiO_3 + MgAl_2O_4 \rightleftharpoons Mg_2SiO_4 + Mg_3Al_2Si_3O_{12}$$
$$\text{(enstatite)} \quad \text{(spinel)} \quad \text{(forsterite)} \quad \text{(pyrope)}$$

138

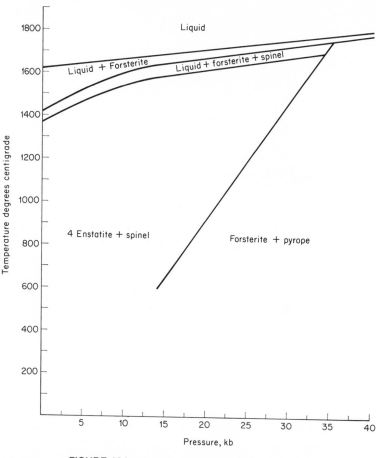

FIGURE 134
The reaction 4 Enstatite + Spinel ⇌ Forsterite + Pyrope.
[After Boyd and McGregor (1964).]

This reaction (Fig. 134) is both pressure and temperature dependent. Its position, along with liquidus data, was determined by Boyd and MacGregor (1964). The significance of this relationship is that it shows the pressures and temperatures, within the earth's mantle, at which the transformation takes place between peridotites containing spinel and those containing garnet.

The earlier-mentioned conversion of protoenstatite to rhombic enstatite with pressure is shown in Figure 135. This diagram is significant not only because it shows the breakdown of protoenstatite, but also because it indicates that clinoenstatite is stable at lower temperatures and higher pressures than enstatite—a reversal of the previously assumed stability relations of these two polymorphs. The high-pressure conversion of ortho- to clinoenstatite is regarded by Smith (1969) as a nonequilibrium process due to stress nucleation of clinoenstatite.

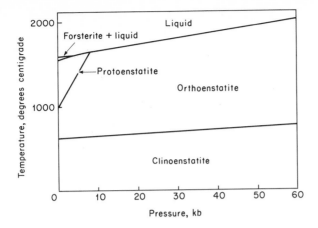

FIGURE 135
The system $MgSiO_3$. [After Boyd and England (1965).]

Figure 136 shows the liquidus surface of the system Mg_2SiO_4—SiO_2—$NaAlSiO_4$ at one atmosphere. The important thing to note here is the incongruent melting of $MgSiO_3$ to produce Mg_2SiO_4 and liquid. This causes the forsterite-enstatite boundary curve to be mainly a reaction curve rather than a subtraction curve. The curve ends at the ternary invariant point P_1, where the primary fields of forsterite, enstatite, and albite intersect. This point falls outside the pertinent Alkemade triangle and is a tributary reaction point. Any composition within the triangle forsterite-enstatite-albite will begin melting at P_1.

Kushiro (1968) discussed this system in connection with the partial melting of peridotites in the earth's mantle to produce liquids of basaltic composition. Quartz tholeiites are considered to have compositions equivalent to those in the silica-rich part of the diagram (the Alkemade triangle $MgSiO_3$—$NaAlSi_3O_8$—SiO_2), tholeiites to those in the intermediate area (Mg_2SiO_4—$MgSiO_3$—$NaAlSi_3O_8$), and alkali basalts to those in the nepheline-rich corner (Mg_2SiO_4—$NaAlSi_3O_8$—$NaAlSiO_4$). Figure 137 shows a general region indicated as X which can be considered representative of a typical peridotite composition. The invariant point between forsterite, enstatite, and albite is shown again as P_1. This point indicates the composition of the first liquid formed by partial melting of composition X at atmospheric pressure. With increasing confining pressure, $MgSiO_3$ melts congruently, as was seen earlier (Fig. 135), while the invariant point P_1 migrates to P_2, P_3, and P_4, causing a corresponding change in the composition of first liquids produced from melting of batch X. This is taken to indicate that when a peridotite (x) begins to melt at shallow depths in the earth's mantle, the first liquid produced is a quartz-normative tholeiitic melt. At greater depths and higher pressures, melting might first produce a tholeiitic liquid, such as P_3. At still greater depths, the same composition might produce a silica-

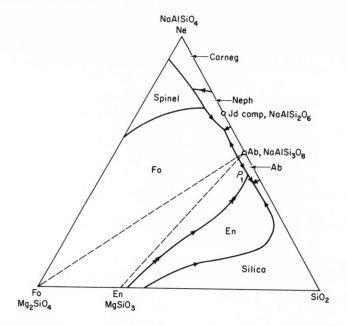

FIGURE 136
The system Mg_2SiO_4—SiO_2—$NaAlSiO_4$. [After Schairer and Yoder (1961).]

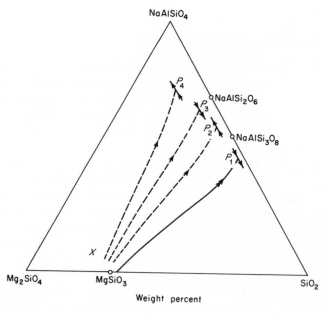

FIGURE 137
Shift in the forsterite-enstatite-liquidus boundary with pressure. The invariant
points P_1, P_2, P_3, P_4 correspond to pressures of 1 bar, and of 10, 20, and 30 kb.
[After Kushiro (1968).]

undersaturated liquid, such as P_4. Not only are the liquids produced at the various pressures different, but the solid phase assemblage changes as well — from forsterite, enstatite, albite at low pressures to forsterite, enstatite, jadeite at high pressures (Figs. 101 and 138, A,B). Although this sequence of first-formed liquids — a sequence proposed by Kushiro (1968) — is valid for this ternary system, O'Hara (1968) has pointed out that addition of a fourth component, $CaMg(SiO_3)_2$, changes the composition of these liquids, so that this simplified system cannot really be taken to be representative of melting in natural systems. It nevertheless illustrates the nature of the current approach being used by many investigators to describe partial melting processes at various depths within the mantle.

Another system of interest is that of $CaMg(SiO_3)_2$—$NaAlSi_3O_8$—$CaAl_2Si_2O_8$ (diopside-albite-anorthite), which was discussed earlier for atmospheric pressures (Figs. 69–71). Data are also available for this system at higher pressures. The join $NaAlSi_3O_8$—$CaAl_2Si_2O_8$ was investigated by Lindsley (1968), the join $CaMg(SiO_3)_2$—$CaAl_2Si_2O_8$ by Clark, Schairer, de Neufville (1962), and the area within the ternary by Lindsley and Emslie (1968). It should be emphasized, however, that the information on this ternary system is very tentative.

The system $NaAlSi_3O_8$—$CaAl_2Si_2O_8$ (Fig. 139) is binary at atmospheric pressure. At 10 kb both liquidus and solidus surfaces are raised to higher temperatures; the anorthite-rich compositions melt incongruently to liquid and corundum. Since corundum cannot be represented by any mixture of the end-member compositions, this part of the diagram is now pseudo-binary. At 20 kb the field of corundum and liquid increases greatly, and only plagioclase compositions more sodic than An_{35} demonstrate binary melting behavior. The range of incongruent melting continues to increase with pressure up to 32 kb, where even pure albite melts incongruently (see Figs. 110, 111). This means that at pressures above about 32 kb, no plagioclase can crystallize as a primary phase from a melt of plagioclase composition.

Addition of other components will cause a decrease in liquidus temperatures, and it may thus be possible for plagioclase to form as a primary precipitate in more complex systems. Figure 140 shows the liquidus surface of the system $CaMg(SiO_3)_2$—$CaAl_2Si_2O_8$—$NaAlSi_3O_8$ at one atmosphere and 15 kb. The plagioclase-diopside boundary curve shows a considerable shift in position toward anorthitic compositions with pressure. In addition, the field of corundum and liquid (not determined) overlies a large part of the primary field of plagioclase; as the former increases with pressure (Fig. 139) and albite melts incongruently, a point may be reached within the ternary system where all of the plagioclase primary field is covered, thus allowing only incongruent melting for all mixtures of plagioclase and diopside.

QUATERNARY SYSTEMS

A great variety of experimental data has been acquired within recent years on high-pressure equilibrium within the system CaO—MgO—Al_2O_3—SiO_2.

142

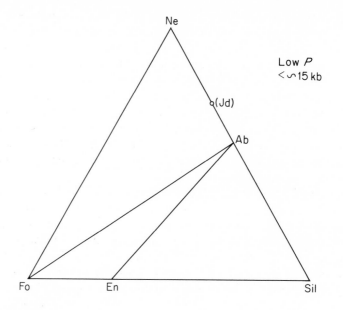

FIGURE 138A
Compatibility relations in the system Mg_2SiO_4—SiO_2—$NaAlSiO_4$ at low pressure.

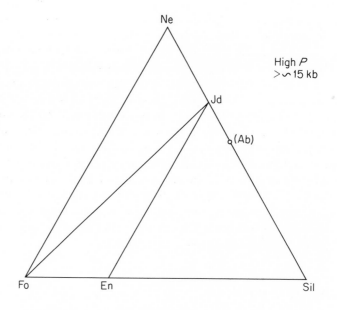

FIGURE 138B
Compatibility relations in the system Mg_2SiO_4—SiO_2—$NaAlSiO_4$ at high pressure. [After Yoder and Tilley (1962).]

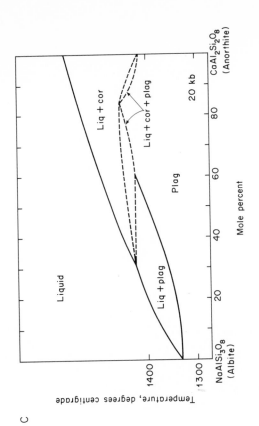

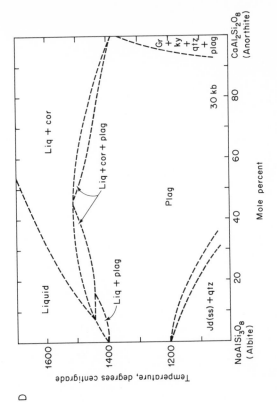

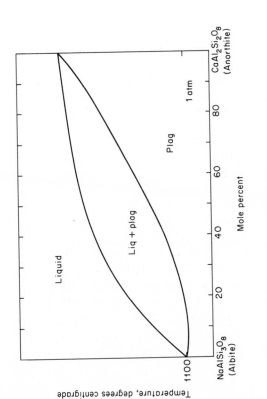

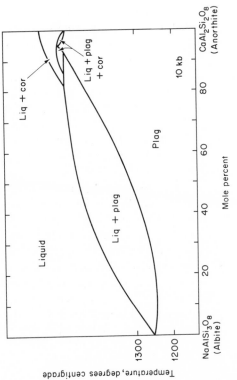

A

B

C

D

FIGURE 139.A,B,C,D
They join albite-anorthite at 1 bar, and at 10, 20, and 30 kb (Liq =
liquid, Plag = plagioclase, Cor = corundum, Jd = jadeite, Qtz =
quartz, Gr = garnet, Ky = kyanite). [After Lindsley (1968).]

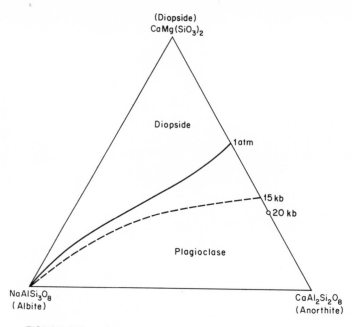

FIGURE 140
Variation in the diopside-plagioclase boundary curve at different
pressures. [After Lindsley and Emslie (1968).]

The major areas of interest within this quaternary system include the minerals
that make up the rocks peridotite and eclogite. The minerals are garnets,
pyroxenes, olivines, and spinel. Rather than systematically examining all
parts of the system, investigators have chosen to study only those compositional areas that would yield the greatest amount of information on the nature
of the upper mantle and the origin of basic and ultrabasic rocks. Some experiments were done with pure end-members, and others with various proportions
of naturally occurring "impure" minerals. The most detailed discussions of
this system and its geological applications are those of Ringwood and Green
(1966), Green and Ringwood (1967), O'Hara and Yoder (1967), and particularly O'Hara (1968).

The join $CaMg(SiO_3)_2$—$Mg_3Al_2Si_3O_{12}$ (diopside-pyrope) was studied by
O'Hara (1963) at 30 kilobars; the diagrams for this join are discussed in detail
by O'Hara and Yoder (1967). This join is related to the melting relationships
of eclogites, which are composed mainly of a diopsidic pyroxene and a pyropic
garnet. The join lies within the system CaO—MgO—Al_2O_3—SiO_2, and is on
the plane enstatite-wollastonite-corundum (as seen in Fig. 141). Figure 142
shows this plane removed from the tetrahedron; the join diopside-pyrope is
indicated by the dotted line. Experiments done at 30 kilobars produce the
diagram seen in Figure 143; the reason for the complexity is that the relationships are for the most part pseudo-binary rather than binary. Most phase

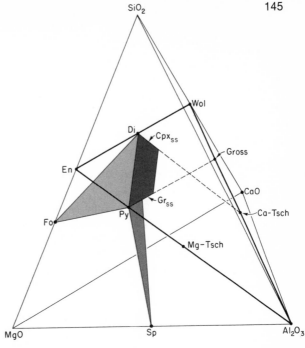

FIGURE 141
Phases in the system CaO—MgO—Al₂O₃—SiO₂.
The major minerals of the rock eclogite, diopside,
and pyrope, lie in the plane enstatite-wollastonite-
Al₂O₃ (En = enstatite, Di = diopside, Ca-Tsch =
calcium-tschermak's molecule, Mg-Tsch =
magnesian-tschermak's molecule, Py = pyrope,
Gross = grossular, Wol = wollastonite, Fo =
forsterite, Sp = spinel, Gr$_{ss}$ = garnet solid
solutions, and Cpx$_{ss}$ = clinopyroxene solid
solutions). [After O'Hara and Yoder (1967).]

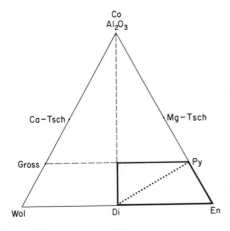

FIGURE 142
The plane enstatite-wollastonite-Al₂O₃, ex-
tracted from the tetrahedron CaO—MgO—
Al₂O₃—SiO₂. The join diopside-pyrope is
shown as a broken line. [After O'Hara and
Yoder (1967).]

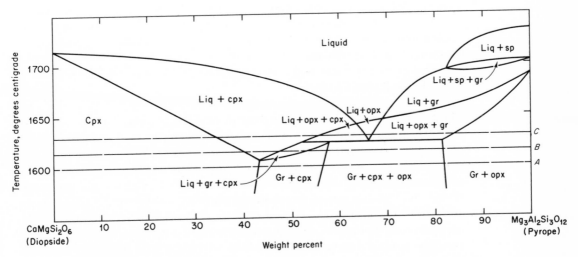

FIGURE 143
The join diopside-pyrope at 30 kb. Liq = liquid, cpx = clinopyroxene, gr = garnet, opx =
orthopyroxene, and sp = spinel. The 1600, 1615, and 1630°C levels are marked as A, B, and
C. [After O'Hara and Yoder (1967).]

compositions can be represented only on the wollastonite-enstatite-corundum plane (Fig. 142), and not on the diopside-pyrope join. This has been done by O'Hara (1963) for the three isothermal sections seen in Figure 144.

Figure 144, A shows the represented portion of the plane broken up into one-, two-, and three-phase regions. The one-phase regions labeled *CPX* and *OPX* illustrate the wide range of compositions available for clinopyroxene formation and a lesser compositional area of orthopyroxene. Pyropic garnet shows considerable solid solution with grossular (the Ca-rich garnet), as indicated by the hatched line on the top of the diagram. Several two-phase areas are present (i.e., *CPX* + *OPX*, *CPX* + Liq. *GR* + *OPX*, *GR* + Liq., and *GR* + *CPX*); these areas indicate coexistence of the two phases whose compositions are represented by adjacent one-phase areas. Although not determined by O'Hara, these areas could be crossed by swarms of two-phase tie-lines indicating the coexistence of particular compositions of the two co-existing phases. Thus a composition *A* in Figure 144, A might be shown to consist of a clinopyroxene of composition *B* and orthopyroxene of composition *C*; or composition *D* might consist of phases *E* and *F*. The three-phase areas are triangles; the compositions of the three coexisting phases are given by the corners of the triangle. A composition *G* consists of pyrope *H*, clinopyroxene *I*, and orthopyroxene *J*. The ratio of these three phases depends upon the position of *G* within the three-phase triangle.

At the temperature of 1600°C (Fig. 143) mixtures of diopside and pyrope yield the following phase assemblages:

Clinopyroxene;
Garnet and clinopyroxene;
Garnet, clinopyroxene, and orthopyroxene;
Garnet and orthopyroxene.

These same assemblages are represented in the isothermal (1600°C) section shown in Figure 144, A. The diopside-pyrope join crosses the diagram diagonally, from lower left to upper right (the line marked by solid circles). This join crosses the same phase assemblages listed above. The advantage of the isothermal section is that the compositions of the various phases can be represented. With the exception of clinopyroxene, none of the phase compositions lie on the diopside-pyrope join, thus once again indicating that the join is pseudo-binary for the most part. The relationship between Figures 143 and 144, A,B,C should be studied in detail.

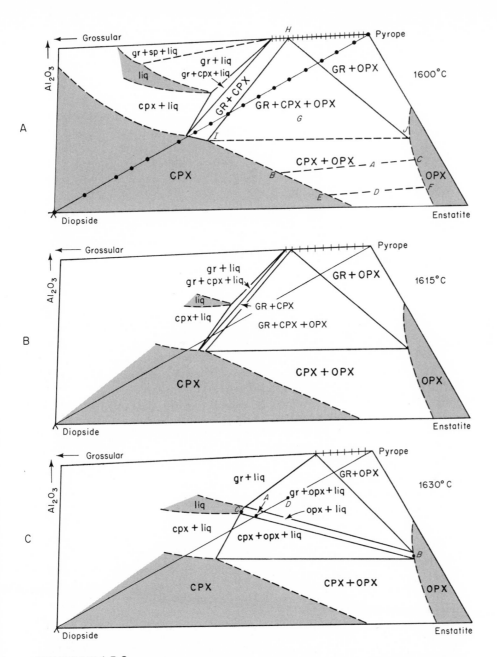

FIGURE 144,A,B,C
Isothermal sections at 1600, 1615, and 1620°C for a part of the plane enstatite-wollastonite-Al₂O₃ (outlines in Fig. 142). Phase compositions along the join pyrope-diopside correspond to data in the join shown in Figure 143. Single-phase areas are shaded. [After O'Hara and Yoder (1967).]

As a useful exercise, consider how an experimenter might be able to determine an approximate composition of liquid coexisting with the various solid phases; it is often important to know (when the liquid does not lie on the join) whether it is on one side or the other. From this sort of information magmatic fractionation trends can sometimes be deduced.

As can be seen from Figure 144, C, the easiest region to use as an example is the narrow two-phase field about the point A, where a diopside-pyrope mixture (A) produces orthopyroxene and liquid. If the composition of the orthopyroxene can be determined as B, and the bulk composition of the batch is known to be A, then, we know from the Lever Rule that the composition of the liquid must lie along the extension of the line AB drawn through A. The particular liquid composition could be determined precisely if the relative amounts of liquid and crystals are known. In the present example, the liquid composition C can be determined by noting that the quenched experimental batch consists of about 10 percent crystals and 90 percent glass.

A more complex situation develops when the experimenter finds that one of his runs has produced three phases. If he knows the starting bulk composition, the relative amounts of the three phases produced, and the composition of two of these, he can then determine (although with less precision) the composition of the third phase graphically. Recall that in a ternary system the proportions of the three components present can be determined by erecting perpendiculars from the unknown point (A) to the three sides of the triangle, as in Figure 145, A; the lengths of the three perpendiculars are proportional to the amounts of the three components. Another method of determining the composition of point A is shown in Figure 145, B. Lines are drawn through the unknown point A and parallel to any two of the sides of the triangle. The side of the triangle intersected by both lines (here the side XY) is divided into three segments, the relative lengths of which are proportional to the relative amounts of the three components, as indicated in the figure. These techniques work for triangles of any shape; see Korzhinskii (1959) for a discussion of various graphical techniques.

Assuming the validity of this technique, it now becomes possible to return to the original problem—that of deducing the composition of an unknown phase (in this case an orthopyroxene). Assume that the bulk composition D (in Fig. 144, C) reacts to form liquid, garnet, and orthopyroxene. This relationship is shown in Figure 146. The compositions of the garnet and liquid have

149

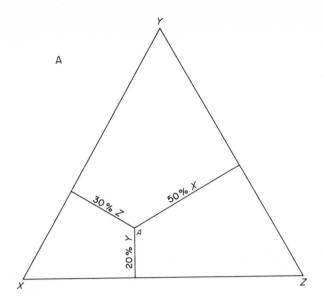

FIGURE 145A
The composition of point *A*, found by the method of erecting perpendiculars to the sides of the compositional triangle.

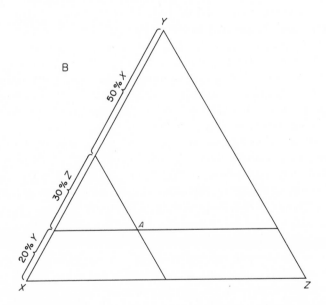

FIGURE 145B
The composition of point *A*, found by the method of drawing lines parallel to two sides of the compositional triangle.

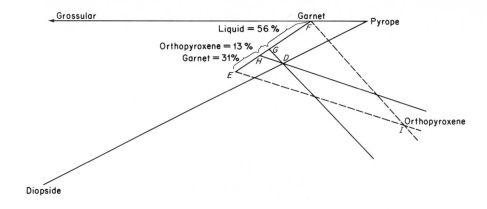

FIGURE 146
View of a part of Figure 144, C. Knowing the bulk composition at D, the composition
of two of the three phases produced (garnet F and liquid E), and the percentages of each
phase (garnet EH, liquid GF, orthopyroxene GH), the composition of the third phase,
orthopyroxene, can be found by use of the method shown in Figure 145, B.

been determined and plotted on the diagram at the ends of the line EF. The
relative proportions of the three synthesized phases are indicated along the
line EF (using the technique of Fig. 145, B) as 56 percent liquid, 13 percent
orthopyroxene, and 31 percent garnet. Lines are then drawn from the intersection points G and H on the line EF through the bulk composition point D.
These lines (GD and HD) are similar to the lines drawn in Figure 145, B and
are parallel to the sides of the compositional triangle. The sides of the compositional triangle can now be drawn from the corners E and F, parallel to GD
and HD. The point at which these two lines intersect (I) represents the composition of the coexisting orthopyroxene.

Although such techniques commonly make it possible to obtain approximate answers, they cannot always be used, as they are dependent upon the
precise determination of phase compositions and proportions. The samples
are usually quite small, very fine-grained, and compositionally segregated,
often making such procedures difficult or impossible.

7

Systems that Contain Water

The phase diagram for the system H_2O has been presented earlier (Fig. 95) in connection with various aspects of solid-solid and solid-liquid equilibria. Before commencing a detailed discussion of high-temperature, high-pressure relations in systems containing water, it is necessary to discuss some of the PVT relations of water. Figure 147 shows the region of coexistence of ice, water, and vapor. Point A defines the particular pressure and temperature at which these three phases coexist, and is usually referred to as a *triple point*. It is an invariant point, and the phase boundaries AB, AD, and AC are univariant lines. The sublimation curve for ice is AC; it shows the vapor pressure of ice at various temperatures. The curve AB is the vapor pressure curve for water; from this curve, one can see that the boiling point of water is a function of pressure. The curve AB terminates at the point B. At pressures and temperatures below B there is a distinct phase change between water and steam (E to F). At pressures and temperatures above B, no sharp phase change occurs between corresponding points such as G and H. At no stage between G and H is there more than one phase present. The point B is called the *critical point*; above this point a substance cannot be considered as either a liquid or a vapor, as it may exhibit the close molecular packing of a liquid (at high pressures) or the wide molecular spacing of a gas (at high temperatures). The temperature

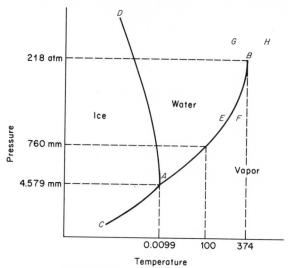

FIGURE 147
Schematic *PT* diagram, showing the relations
between ice, water, and vapor.

of the critical point, 374°C, is the critical temperature, and the pressure, 218 atmospheres, is the critical pressure. Most authors refer to the gas phase as *vapor* when its temperature and pressure are less than critical, and as *fluid*, or supercritical fluid, at higher pressures and temperatures. Relations at the critical point can also be shown on a *PV* plot (Fig. 148). The relative incompressibility of liquid, as compared to vapor, is shown by the steepness of the isothermal lines in the liquid region. The critical point is *G*, at the top of the two-phase region. At temperatures above critical—for example, along the isothermal curve *AB* at 450°C—an increase in the pressure of the fluid causes a continuous decrease in volume from *A* to *B*. This relationship is known as Boyle's Law, which states that the product *PV* equals a constant. At a temperature less than critical, such as point *C* at 300°C, an increase in pressure brings the vapor to *D*. Here the vapor liquefies along the path *DE* in the two-phase liquid-vapor region and decreases further in volume as a liquid from *E* to *F*. The large discontinuous volume change from *D* to *E* is equivalent to crossing the univariant curve from *F* to *E* in Figure 147. The discontinuous change decreases in amount as temperature increases, and is completely eliminated at supercritical conditions (the path *H* to *G* in Figure 147).

Most standard textbooks on physical chemistry include thorough discussions of critical behavior. Useful *PVT* data for H_2O and CO_2 are given by Kennedy and Holser (1966). A large number of geologically significant reactions in which H_2O is a component have been summarized in Eitel (1966) and in Levin, Robbins, and McMurdie (1964, 1969).

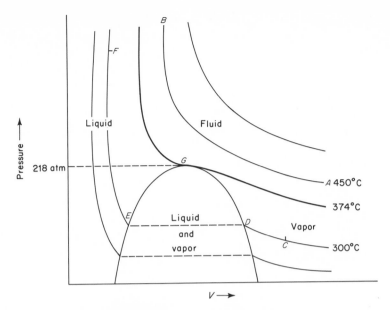

FIGURE 148
Schematic PV diagram for H_2O, illustrating
the nature of the critical point (G).

BINARY SYSTEMS

$NaAlSi_3O_8$—H_2O

The effects of water pressure on the stability of minerals differs greatly from those of simple confining pressure. Confining pressure merely changes mineral stability fields as a function of differences in density among the phases in the system, the more dense phases being favored at the expense of the less dense. Water, on the other hand, not only acts as a source of pressure on the system, but also becomes a component in chemical reactions. In addition, the partial or complete solution of many chemical substances in water greatly facilitates the attainment of equilibrium in geological and experimental environments.

In order to describe completely a two-component system that has water as one of its components, a three-dimensional PTX volume should be shown, with sufficient variation in temperature and pressure to embrace assemblages consisting of all combinations of phases. Such a representation was published by Wyllie and Tuttle (1960a), using data available at that time for the system $NaAlSi_3O_8$—H_2O (albite-water). Figure 149 shows four isobaric sections of this system. It must be mentioned that these schematic diagrams are now known to be incorrect. Boettcher and Wyllie (1967, 1969) have demonstrated

154

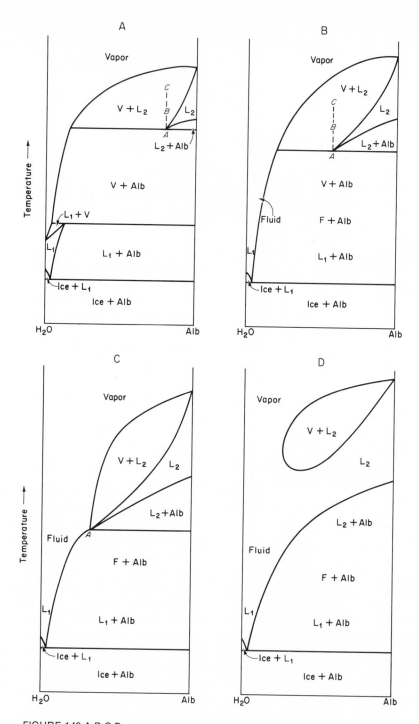

FIGURE 149,A,B,C,D
The system $NaAlSi_3O_8$—H_2O, showing P_{H_2O} increasing from A through D. The diagrams, when published, were tentative, and are now known to be incorrect (Alb = albite, L_2 = silica-rich liquid, L_1 = water-rich liquid, V = vapor, and F = fluid). [After Wyllie and Tuttle (1960a).]

that this system ceases to be binary above approximately 16 kb water pressure, as the reaction

$$\text{Albite} + \text{vapor} \rightleftharpoons \text{liquid}$$

changes to

$$\text{Albite} + \text{vapor} \rightleftharpoons \text{jadeite} + \text{liquid}$$

The ternary system, discussed in detail by Boettcher and Wyllie (1969), will not be treated here. We shall, for the sake of discussion, use the diagrams presented in Figure 149 as if they were correct.

The pressure P_A in the first diagram (Fig. 149, A) is below the critical point of H_2O; heating of ice yields water, and then vapor. Pure $NaAlSi_3O_8$ is similarly below its critical point and changes from solid to liquid to vapor with increasing temperature. At some as-yet-unknown pressure and temperature, the albite melt must also have its own critical point.

At a higher pressure, P_B (see Fig. 149, B), the critical point of water is exceeded, and a continuous gradation of properties exists in the supercritical fluid with variations in temperature.

Consider now the albite side of the four diagrams; the increase in pressure from P_A through P_D causes an increase in the melting temperature of albite. The increase is relatively rapid initially, but slows at higher pressures as the difference between the densities of liquid and solid phases becomes less marked. Similarly, the boiling point of the $NaAlSi_3O_8$ melt rises with pressure, as would be expected, because of the density difference between the vapor and liquid phases. It should be understood that for the pure $NaAlSi_3O_8$ composition, no H_2O is present, and an increase in pressure must be considered as an increase in confining pressure.

The presence of the field L_2 in Figure 149, A indicates that a small amount of water is soluble in albitic liquids. The presence of this second component in the melt causes the usual depression of melting temperature; the albite liquidus surface decreases to point A, where the amount of dissolved water reaches a maximum. At the higher pressure of Figure 149, B, the liquid field L_2 has expanded still more toward the H_2O component, and the minimum melting temperature is lower, indicating a further decrease in the stability of albite in the presence of water-saturated silicate melts. If the amount of water present exceeds the solubility limit of the melt, the excess H_2O will exist as a vapor phase, giving rise to the adjacent two-phase liquid-vapor region. Still higher pressures (Figs. 149, C,D) show a pinching-off of the base of the two-phase liquid-vapor field, indicating complete solubility between liquids of albitic composition and water vapor. Above the pressure indicated in Figure 149, C, any proportion of H_2O and $NaAlSi_3O_8$ can crystallize together at a binary eutectic to produce ice and albite.

Figure 150 shows the depression of the melting point of albite in the presence of water-saturated melts as a function of increasing water pressure. This is the depression of the point A in the isobaric sections of Figure 149. This

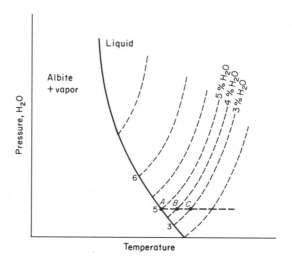

FIGURE 150

Water pressure-temperature diagram for the system NaAlSi$_3$O$_8$. The solid curve corresponds to the migration to point A in Figure 149. The broken (isohydrous) lines indicate the maximum amount of water that can be dissolved in the silicate melt at various values of T and P_{H_2O}.

curve can be extended to the pressure at which the liquid-vapor area pinches off at its base (Fig. 149, C). At pressures above this point the liquid and vapor fields become indistinguishable, hence the curve must terminate. Note again that these diagrams are schematic and that the curve does not terminate as indicated (Boettcher and Wyllie (1967)).

Figures 149, A,B,C show that the point A migrates to the left with increasing water pressure, indicating higher and higher water contents of the silicate melt. The depression curve in Figure 150 has been labeled to show the increase in water content for the transition Ab + V $\rightleftharpoons$ L. Returning to Figures 149, A and B, notice that if the temperature is increased above that of point A, the vapor-liquid field is encountered (points B and C) and the capacity of the melt to dissolve water is decreased. Thus in Figure 150 the water content of the melt can be shown as decreasing with increasing temperature (points A, B, and C). The maximum amount of dissolved water in the melt can be indicated and contoured in the liquid field of this diagram. These "water content contours," called *isohydrons*, demonstrate the melt's capacity for water saturation. Such lines terminate where the temperature exceeds that of the two-phase liquid-vapor region, as no region of melt exists.

Tracing isobaric crystallization sequences within this system along constant composition lines (isopleths) is done according to the same procedures used to trace crystallization sequences in nonhydrous binary systems.

Yoder (1965) discusses some interesting possibilities as to the effects on explosive vulcanism of systems of this type, particularly the system CaMg(SiO$_3$)$_2$—CaAl$_2$Si$_2$O$_8$—H$_2$O (diopside-anorthite-water).

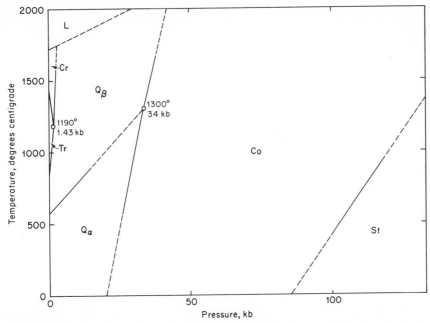

FIGURE 151
The system SiO_2, showing the stability of various phases as a function of temperature and confining pressure (L = liquid, Cr = cristobalite, Tr = tridymite, Q_β = β quartz, Q_α = α quartz, Co = coesite, St = stishovite). [After Ostrovsky, 1967; the position of the coesite-stishovite line has been shifted to the left (personal communication from Ostrovsky).]

SiO_2—H_2O

Another two-component system that has been investigated in detail is SiO_2—H_2O. The stability curves of the various polymorphs of SiO_2 are shown in Figure 151 as a function of temperature and *confining pressure*. The stability of these phases is related to the densities of the various polymorphs, which is typical; the less dense forms, tridymite and cristobalite, are eliminated with increasing pressure in favor of the denser quartz. In addition, the melting temperatures rise slightly with pressure, favoring the denser solid phases over the less dense liquid. When the total pressure on the system is water pressure, a vapor phase is present (Fig. 152); part of this vapor can dissolve in the silicate melt, resulting in a decrease in the melting temperature of the silica polymorphs. Because the water does not combine with any of the solid SiO_2 phases, the inversion curves between the various polymorphs are identical for both water pressure and confining pressure, for the water merely acts as a pressure medium.

Water pressure has a pronounced effect on melting temperatures, as can be seen in Figure 152. With a small increase in pressure, the melting temperature decreases at first to below the stability area of cristobalite, and then below

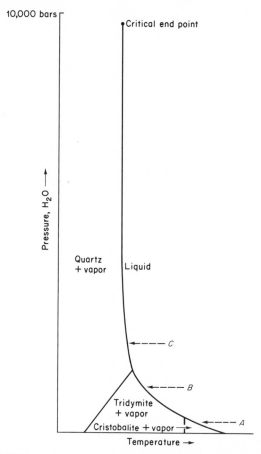

FIGURE 152

Maximum depression of the melting point of silica liquids as a function of P_{H_2O}. Transition curves between the various solid polymorphs are independent of the type of pressure applied, and correspond to the curves shown in Figure 151. [After Kennedy et al. (1962).]

that of tridymite. This means that at low water pressure, A, cristobalite may precipitate directly from the melt; at a higher water pressure, B, tridymite will be the primary precipitate, and at still higher pressure, C, quartz will crystallize directly from the melt.

The boundary curve between the solid-vapor and liquid regions continues to about 9.7 kb, where it reaches a critical end point (Kennedy et al., 1962). Above this point liquid and vapor show a continuous transition, and a single fluid phase exists in equilibrium with quartz. The composition of the fluid phase may vary from almost pure water to almost pure molten SiO_2.

This relationship may be also illustrated by means of isobaric temperature-composition sections (Fig. 153). The sections show the system at three different pressures, 1, 2, and 9.5 kb. As the water pressure is increased, the

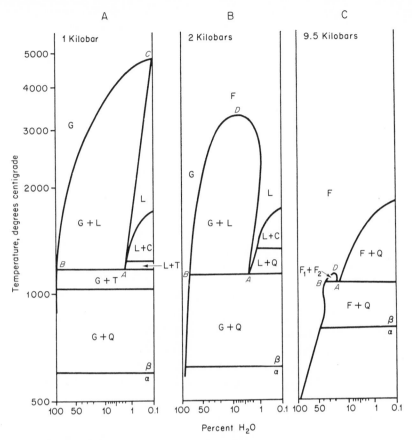

FIGURE 153,A,B,C
Isobaric *TX* sections of the system SiO₂—H₂O (G = gas, L = liquid, F = fluid, C = cristobalite, T = tridymite, and Q = quartz). The β–α line denotes the high- to low-temperature quartz transitions. [After Ostrovsky (1966).]

minimum temperature of the water-saturated silica melt, *A*, decreases. The vapor that coexists with liquid *A* is indicated as point *B* in Figure 153, A,B,C. Compositions *A* and *B* converge with increasing pressure. When the P_{H_2O} reaches 9.7 kb, *A* and *B* coincide. The two-phase region $F_1 + F_2$ disappears; this corresponds to the curve in Figure 152 that reaches the critical end point. Convergence of gas and liquid compositions with increase in water pressure is also shown in Figure 154.

The ideal way to represent such a system is to prepare a three-dimensional *PTX* projection. Short of this, a more commonly used approach is to represent important univariant reaction curves on a single *PT* surface, with all compositions on the same surface (Fig. 155). The solid and broken univariant curves in Figure 155 indicate the presumed transition curves of the anhydrous silicates with increasing *confining* pressure. At low pressures, cristobalite melts first

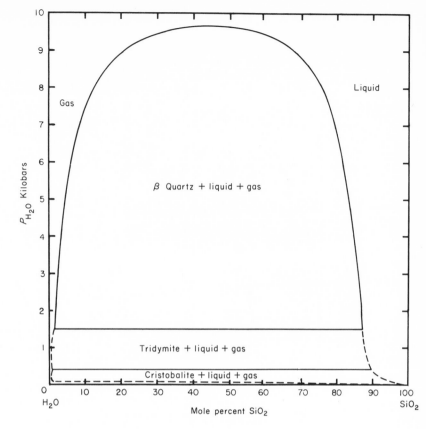

FIGURE 154
Compositions along the upper
three-phase boundary in the
system SiO_2—H_2O.
[After Kennedy et al. (1962).]

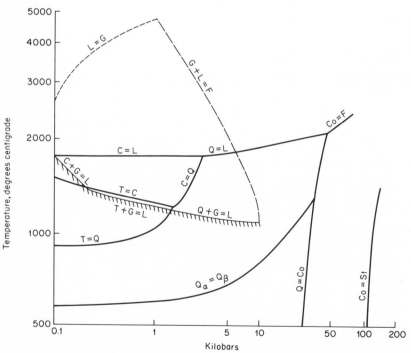

FIGURE 155
Univariant curves in the system SiO_2—H_2O. The heavier lines correspond to Figure 151. The
hachured lines correspond to the melting curve in Figure 152 (G = gas, L = liquid, F = fluid,
C = cristobalite, T = tridymite, Q = quartz, Co = coesite, and St = stishovite). [After Ostrovsky
(1966); the coesite-stishovite boundary has been revised in accordance with new data
(personal communication from Ostrovsky).]

to form a silica liquid; at higher pressures quartz melts, and at still higher pressures it is supposed that the denser coesite will melt directly to a fluid. Although it is not so indicated, the still denser phase stishovite probably melts directly to a fluid. The hatched curves correspond to the melting curve shown in Figure 152, and represent the depression of the melting points of the silica polymorphs in a hydrous environment. The other lower-temperature univariant curves represent transitions of the various solid polymorphs as a function of P and T. The curve $L \rightleftharpoons G$ shows the presumed variation of boiling point of silica liquids; this is the migration of the point C in Figure 153, A. At a pressure near 1 kb, the critical point of silica liquids is exceeded, and the gas to liquid transition changes to a fluid to liquid-gas transition. The final curve traces the position of the transition gas + liquid $\rightleftharpoons$ fluid; this is the point D in Figure 153, B,C.

$CaO - H_2O$

The effect of water pressure on hydrous compounds is quite different from that on anhydrous compounds. Figure 156 contrasts the melting curves of amphi-

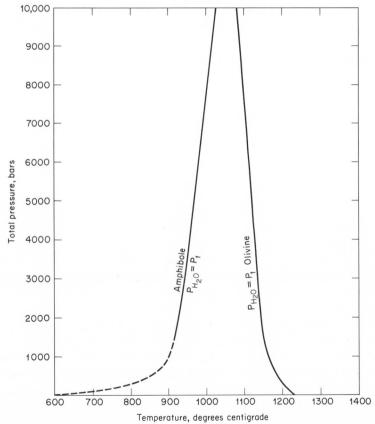

FIGURE 156
Melting curves of anhydrous (olivine) and hydrous (amphibole) phases as a function of P_{H_2O}. The batch composition is an olivine tholeiite. [After Yoder and Tilley (1962).]

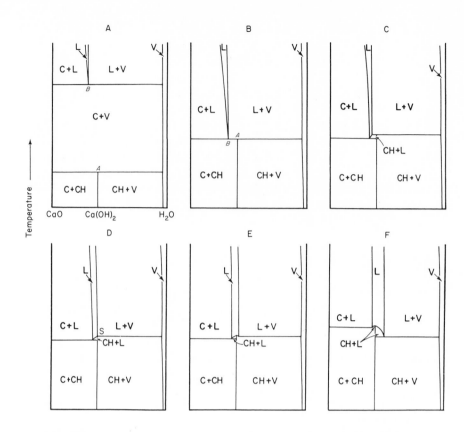

FIGURE 157
Schematic isobaric sections of the system $CaO-H_2O$. As water pressure increases from section A through section F, calcium hydroxide undergoes dissociation, and then melts incongruently and finally congruently (L = liquid, V = vapor, C = CaO, and CH = $Ca(OH)_2$). [After Wyllie and Tuttle (1960b).]

bole and olivine in a natural olivine tholeiite-water system; water pressure is the total pressure of the system. As is the case with all hydrated compounds, the melting curve of the amphibole initially rises with an increase in water pressure, in contrast to anhydrous compounds, whose melting curves decrease with increase in water pressure (see Figs. 150 and 152).

The situation can perhaps be better visualized in the series of schematic isobaric sections of the system $CaO-H_2O$ (Fig. 157). The P_{H_2O} rises from section A through F. Figure 157, A shows that $Ca(OH)_2$, portlandite, can coexist with either CaO or essentially pure water vapor. A rise in temperature causes the $Ca(OH)_2$ to dissociate at A to CaO and vapor. The melting temperature of the anhydrous CaO in the presence of water is seen to be depressed to point B; with an excess of water over that permitted by the melt, a two-phase region of liquid and vapor results. The nature of the higher-temperature regions

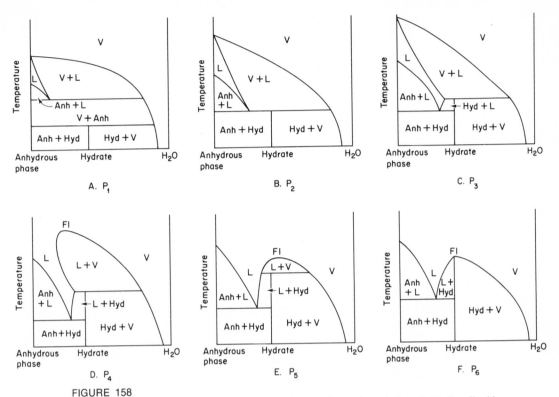

FIGURE 158
Hypothetical behavior of a hydrate wih increasing P_{H_2O} in sections A through F. (L = liquid, V = vapor, Anh = anhydrous solid phase, and Hyd = hydrated solid phase.)

of this binary are not indicated. Increase of P_{H_2O} causes a decrease in the minimum melting temperature of CaO (point B) and a rise in the decomposition temperature of Ca(OH)$_2$ (point A). At the pressure indicated in Figure 157, B, the points A and B have been brought to the same temperature, indicating a four-phase coexistence at an invariant point. At the higher pressure of Figure 157, C, the Ca(OH)$_2$ melts incongruently to produce both liquid and vapor phases. Still higher pressures cause a slight shift in the position of the liquid and vapor field, and in Figure 157, E,F, the Ca(OH)$_2$ melts congruently. The shift from incongruent to congruent melting is marked by the presence of a singular point, S, at the pressure of Figure 157, D. With each increase in the P_{H_2O} the upper stability limit of the hydrated phase Ca(OH)$_2$ is increased.

At sufficiently high pressures it would be expected that the liquid-vapor field could be eliminated. The hypothetical set of isobaric sections shown in Figure 158 illustrates another possible sequence that might develop with increasing P_{H_2O}. A difficulty in these types of sections is the relationship between water pressure and confining pressure. Consider a typical section such as Figure 158, A. The diagram is given for a particular water pressure. Yet if this is truly a binary system between H_2O and an anhydrous compound, then no water can exist on the anhydrous side of the diagram. The contradiction can be resolved only if the pressure on the anhydrous portion is in fact merely confining pressure (perhaps produced by an inert gas). Thus the changes in equilibrium conditions for the anhydrous portion of the diagram must be considered to result from a 'dry' pressure; with increase in the H_2O component, the partial pressure of water ($\bar{P}_{H_2O}$) increases until the $P_{H_2O} = P_{total}$.

The relationship between water pressure and confining pressure has been discussed in detail by Yoder (1955), Greenwood (1961), and Yoder and Tilley (1962, p. 466). Consider the melting of the anhydrous phase olivine (Fig. 159). The melting temperature of olivine decreases (curve AB) with increasing water pressure, because some of the H_2O dissolves in the melt and functions as a second component. If, on the other hand, the pressure is a dry confining pressure, then the melting temperature increases with pressure (as shown by curve AC), and is mainly a function of the difference between the density of olivine and that of its melt. Between these two extreme curves lie a swarm of intermediate melting curves (such as AD) whose positions depend upon the ratio of the two kinds of pressure. Other possible situations are those in which the P_{H_2O} reaches a maximum value (e.g., 3000 bars in Fig. 160) and the additional pressure is confining. The melting curve drops from A to B because of the increase of P_{H_2O} to its maximum, and then follows a rising slope from B to C as confining pressure increases.

The opposite situation is typically encountered for hydrated minerals (Fig. 161). An increase in P_{H_2O} causes a rise in melting temperature, and an increase in P_{conf} causes a slight decrease in the melting temperature (due to the slightly lower density of the hydrated phase as compared to the corresponding melt). When both confining pressure and water pressure are exerted, an intermediate melting curve is followed. A melting curve similar to Figure 160 could be developed for amphibole if the confining pressure were brought to a maximum, and then followed by the addition of water pressure.

Although the change of melting temperature just described is typical at low to moderate pressures, recent experiments have demonstrated that the situation can be reversed for very high water pressures. Boettcher and Wyllie (1969) examined the join $NaAlSi_2O_6$—H_2O (jadeite-water) at extremely high water pressures (Fig. 162). At low to moderate water pressures the melting curves

$$\text{Albite} + \text{Nepheline} + \text{Vapor} \rightleftharpoons L$$

and

$$\text{Albite} + \text{Analcite} + \text{Vapor} \rightleftharpoons L$$

decrease with increasing water pressure. Above a water pressure of about 11 kb, the very dense anhydrous jadeite becomes stable, and the melting curve has a positive slope. Another reverse tendency is shown for the melting curve of amphibole at very high P_{H_2O} (Fig. 163). The melting curve rises at moderate

165

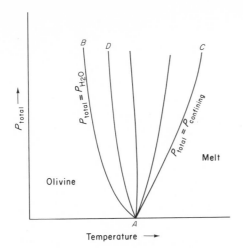

FIGURE 159
Schematic melting curves for olivine, showing different melting curves as a function of the type of pressure used. Intermediate curves indicate where the pressure is partly water pressure (P_{H_2O}) and partly confining (P_{conf}).

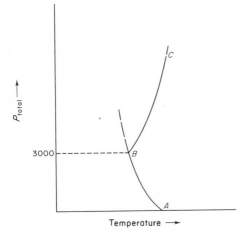

FIGURE 160
Melting curve for an anhydrous phase in which the P_{H_2O} reaches a maximum at point B and additional pressure is P_{conf}.

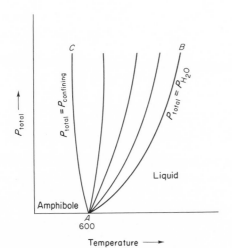

FIGURE 161
Schematic melting curves for a hydrous phase, showing differences as a function of the type of pressure used. Intermediate curves indicate where the pressure is partially water pressure.

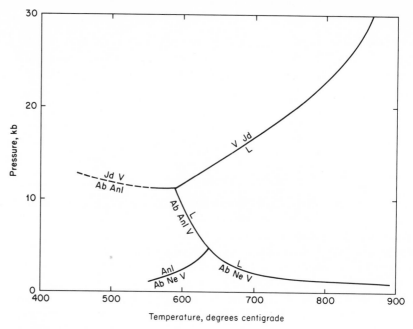

FIGURE 162
Univariant curves for jadeite–H$_2$O compositions. [After Boettcher and Wyllie (1969).]

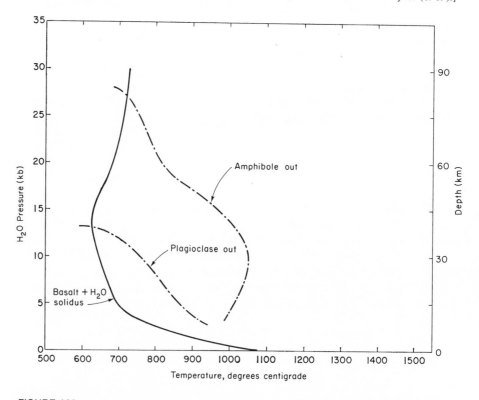

FIGURE 163
Upper stability limits of amphibole and plagioclase in a melt of basaltic composition. The curve labeled "basalt + H$_2$O" represents the minimum temperature at which liquid is present. [After Hill and Boettcher (1970).]

water pressures, but decreases at higher pressures. Because this curve was determined for crystallization of amphibole from a bulk composition of basalt, components other than those required to crystallize amphibole were also present.

TERNARY SYSTEMS

$MgO-SiO_2-H_2O$

Ternary systems that contain water require a special type of representation, as it is necessary to indicate changes in water pressure, as well as the usual variables of temperature and composition. Compositions of components and minerals can be plotted in the usual manner on a triangular graph (with H_2O as one of the components). The coexistence of the various phases within the system is given by Alkemade lines. A unique aspect here is the presence of compatibility lines between solids and a vapor or fluid consisting of essentially pure H_2O. The compatibility relations may exist over either a narrow or wide range of temperature and pressure. Hence it is common practice to show the compatibility arrangements in a triangular compositional diagram, with the applicable pressure-temperature conditions stipulated externally—either in the form of a PT graph or merely as a label.

In the system $MgO-SiO_2-H_2O$, the stability relations at low P_{H_2O} and temperatures greater than 800°C are shown (Fig. 164) as compatibility triangles. The four solid phases present—periclase, forsterite, enstatite, and quartz—are all shown on the triangular compositional base $MgO-SiO_2$. The compatibility lines indicate that each of these may exist in the presence of water vapor without combining to form a hydrate. With decreasing temperature, mixtures of enstatite and quartz react with water vapor to form anthophyllite* (Fig. 165):

$$\text{Enstatite} + \text{Quartz} + \text{Water Vapor} \rightleftharpoons \text{Anthophyllite}$$

The coexistence of four phases in a three-component system represents a condition of univariance (as $P + F = C + 2$, $4 + F = 3 + 2$, $F = 1$). This univariant reaction varies as a function of P and T and may be plotted as a univariant line, as in Figure 166. Stability arrangements on either side of the line are shown graphically by the appropriate compatibility triangles.

With further lowering of temperature, talc becomes stable (Fig. 167) and a different compatibility arrangement results. The reaction is

$$\text{Anthrophyllite} + \text{Quartz} + \text{Water vapor} \rightleftharpoons \text{Talc}$$

A fairly complete description of the system is given in Figure 168, where all

---

*The synthesis of anthophyllite is extremely difficult because of nucleation, which leads to the persistence of metastable phases. Sophisticated techniques of synthesis are required. Greenwood's (1963) discussion of techniques, and his application of the Schreinemakers Principle, make his study extremely informative.

168

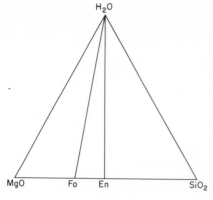

FIGURE 164
Compatibility relations in the system
MgO—SiO₂—H₂O above about 800°C (Fo =
forsterite, En = enstatite). [After Greenwood
(1963).]

FIGURE 165
Compatibility relations in the system
MgO—SiO₂—H₂O (at a lower T and/or
higher P_{H_2O} than for Figure 164), after the
reaction Enstatite + Quartz + H₂O →
Anthophyllite (Fo = forsterite, En =
enstatite, and An = anthophyllite). [After
Greenwood (1963).]

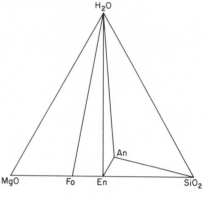

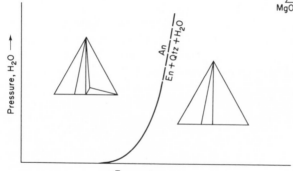

FIGURE 166
The univariant reaction curve for the decomposition of anthophyllite.
Compatibility relations on both sides of the reaction are shown in the
triangular composition diagrams. [After Greenwood (1963).]

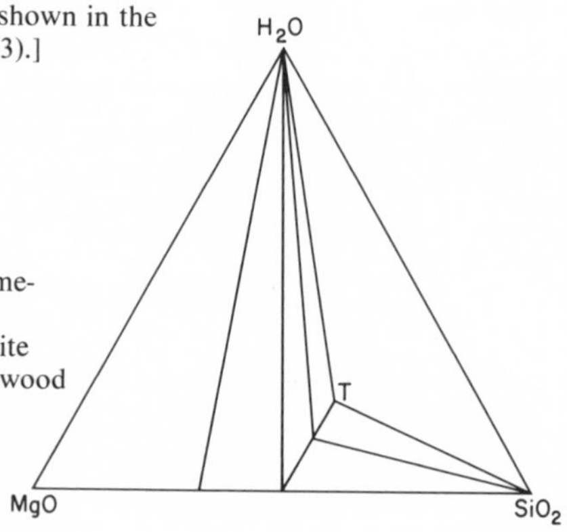

FIGURE 167
The system MgO—SiO₂—H₂O at a somewhat lower temperature than that for
figure 165, after the reaction Anthophyllite
+ Quartz + H₂O → Talc. [After Greenwood
(1963).]

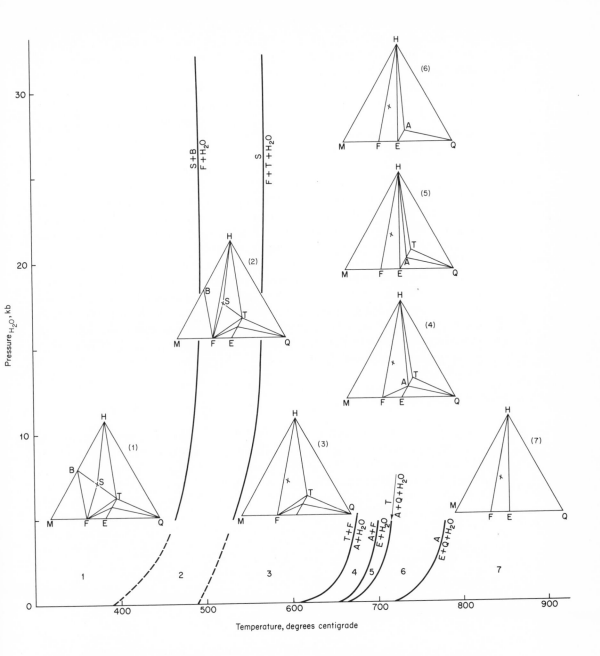

FIGURE 168

Composite T–P_{H_2O} diagram of reactions in the system MgO—SiO_2—H_2O. The compatibility triangles below indicate phase relations in each of the numbered areas. The serpentine composition is indicated as X ($H = H_2O$, $M = MgO$, $Q = SiO_2$, A = anthophyllite, B = brucite, E = enstatite, F = forsterite, S = serpentine, and T = talc). [After Greenwood (1963), Bowen and Tuttle (1949) and Kitahara, Takenouchi, and Kennedy (1966).]

of the known compatibility lines and univariant curves within the system have been plotted on the same PT coordinates. This figure is a composite of all of the univariant reactions of which Figure 166 is an example. Each area between univariant curves is characterized by a particular equilibrium assemblage and is so indicated by the small compatibility triangles. From this figure it is possible to determine the phase assemblage for any composition, within a wide range of pressure and temperature. Consider a serpentine composition. Serpentine is stable at low temperatures. With heating, it reacts to form an assemblage (point x) of forsterite, talc, and H_2O. With increasing temperature this same composition is converted to forsterite, anthophyllite, and H_2O; at still higher temperatures it forms forsterite, enstatite, and H_2O. Another method of representing phase stabilities in this system is shown in Figure 169; this mode of representation is unusual but very informative. It does, however, have the disadvantage of being limited to particular values of P_{H2O} and containing only those phases stable with excess H_2O.

Although univariant curves do define the maximum limits of stability of the various individual minerals or mineral groups, they should be applied to field problems only with great caution; the *entire* mineral assemblage must be taken into account. In Figure 168, the curve that originates at about 500°C defines the upper limit of stability of serpentine. The curve shows the maximum stability for serpentine in the pure state and under water pressure. The stability of serpentine is drastically decreased in the presence of brucite. The univariant reaction

$$\text{Brucite} + \text{Serpentine} \rightleftharpoons \text{Forsterite} + \text{Water vapor}$$

takes place at temperatures well below the breakdown temperature of serpentine alone (as is shown by the adjacent lower-temperature curve). Other phases present in the rock may cause additional decreases in the stability area of the mineral under consideration. It should also be re-emphasized that the natural assemblage has formed under a variety of different pressure regimes, rather than just P_{H2O} as the usual diagram would indicate; this also affects the mineral stability regions.

QUATERNARY SYSTEMS

The System MgO—Al$_2$O$_3$—SiO$_2$—H$_2$O

Four-component systems can be represented graphically by some three-dimensional scheme—ideally, a tetrahedron with each of the components at a corner. Compatibility relations at particular PT conditions are indicated by Alkemade lines joining the various phases. The Alkemade lines subdivide the tetrahedron into several smaller tetrahedrons that fit together in the manner of a three-dimensional jigsaw puzzle (Fig. 170, A). Besides using a single tetrahedron, subdivided by interior planes, an "exploded" diagram can

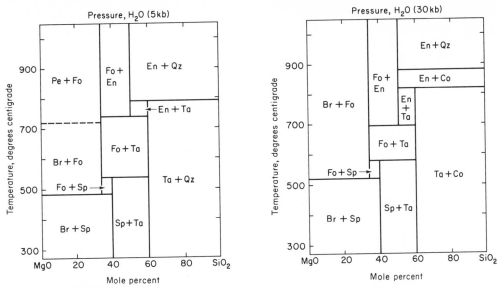

FIGURE 169
The system MgO—SiO₂—H₂O showing mineral assemblages at 5 kb (left) and 30 kb (right) water pressure (Br = brucite, Co = coesite, En = enstatite, Fo = forsterite, Pe = periclase, Qz = quartz, Sp = serpentine, and Ta = talc). [After Kitahara, Takenouchi and Kennedy (1966).]

also be used (Fig. 170, B). This mode of representation, although sometimes very useful, is often difficult to visualize and is not commonly used.

The problem of representation is not, however, as complex as might be expected, since most hydrothermal experimentation has been performed on systems that contain an excess of water. Therefore, the only phases that need be shown are those that are stable in the presence of a surplus of water. An additional simplification arises because these phases (actually at different distances from the H₂O apex) can be projected to the anhydrous base (Fig. 171, A). The base of the system can then be used in the usual manner (Fig. 171, B), with an excess of water assumed. Since the projection is made from the H₂O apex, the phase compositions can be plotted in the correct position relative to the three anhydrous components. This technique is used in the diagrams shown in Figure 172. Montmorillonite is present as an ill-defined region that gradually decreases and pinches out between 475° and 565°C. The heavy line between serpentine and the chlorite amesite indicates complete solid solution between these end-member compositions. Because of the large amount of compositional substitution in this phase (as well as the limited solid solution in talc), large areas of this sytem consist of two solids and vapor rather than three solids and vapor, as is indicated by the swarms of compatibility lines.

We can trace the series of reactions that will take place at constant water pressure as temperature is increased. Consider composition X. Initially, at

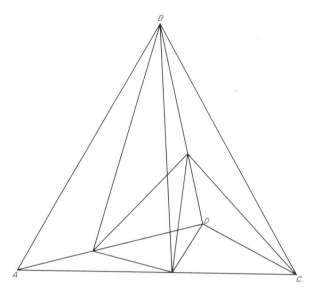

FIGURE 170A
Compatibility relations in the hypothetical quaternary system
A-B-C-D. The system is broken into Alkemade lines, triangles, and
pyramids.

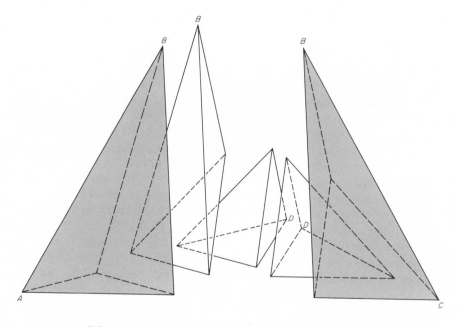

FIGURE 170B
The same system as shown in Figure 170, A, but with Alkemade
pyramids separated for greater clarification.

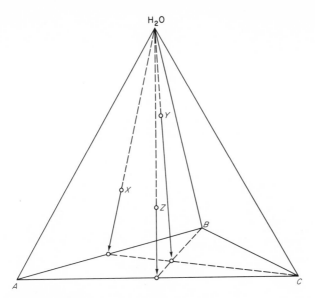

FIGURE 171A
Hypothetical system A-B-C-H_2O. The compounds X, Y, and Z are
projected from the H_2O apex onto the anhydrous base, allowing
correct geometric placement relative to the A, B, and C components.

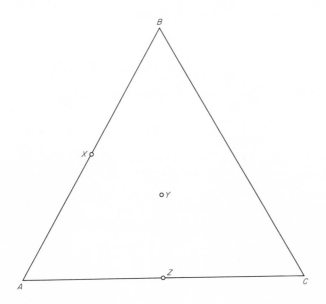

FIGURE 171B
The hydrous compounds X, Y, and Z of Figure 171A, plotted on the
anhydrous base A-B-C.

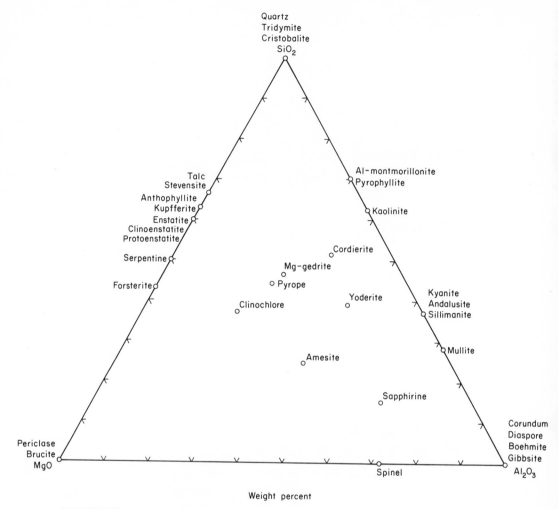

FIGURE 172
The system MgO—Al$_2$O$_3$—SiO$_2$ at a P_{H_2O} of 2 kb. The triangle above shows possible phases in the system. The six smaller triangles (facing page) are a series of isothermal isobaric sections (Q = quartz, B = brucite, D = diaspore, Cor = corundum, M = montmorillonite, S = serpentine, Am = amesite, And = andalusite, T = talc, S = serpentine, Al-M = Al-montmorillonite, K = kyanite, P = pyrophyllite, Sp = spinel, Fo = forsterite, Co = cordierite). [After Fawcett and Yoder (1966).]

350°C, X consists of montmorillonite. At 425°C it has reacted to form a mixture of quartz, montmorillonite, and chlorite; at 450°C and 475°C it consists of quartz and a chlorite. At 565°C the assemblage has changed to quartz, cordierite, and chlorite. Finally, because of the two-phase tie-line between talc and cordierite, the batch of original composition X reacts to form talc, cordierite, and chlorite as shown at 575°C. The reader interested in examining this system further will find additional information in Yoder (1952), Roy and Roy (1955), and Schreyer and Yoder (1964).

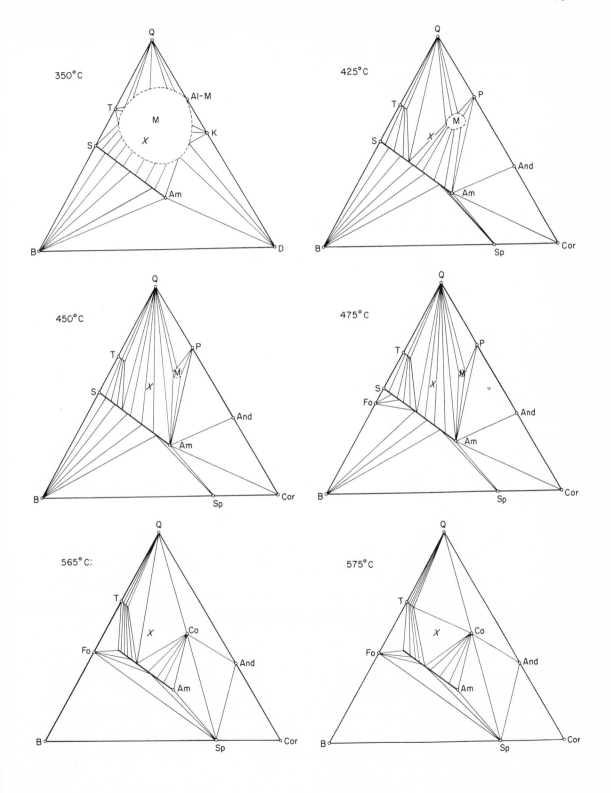

The System $NaAlSi_3O_8$—$KAlSi_3O_8$—$CaAl_2Si_2O_8$—H_2O

Because it contains the three major feldspar components, this system is quite important to the understanding of the genesis of igneous rocks. Although incompletely examined, considerable information is available, and the generalities are known. The liquidus surface of the anhydrous pseudo-ternary is shown in Figure 173. The shaded areas indicate regions of feldspar solid solution. Complete solid solution exists between the Ca-Na plagioclases and, at high temperatures, between the K-Na alkali feldspars. The K-feldspar field is complicated by the incongruent melting of orthoclase to yield leucite and liquid (as discussed on pp. 14–15); it is the presence of leucite ($KAlSi_2O_6$) that causes this system to be pseudo-ternary rather than ternary at one atmosphere.

Understanding of crystallization within this system under hydrous conditions is facilitated by observations that have been reported on the effect of water pressure on the boundary systems. According to Yoder, Stewart, and Smith (1957), the plagioclase solid solution loop shows an overall decrease in liquidus temperatures with increasing P_{H_2O} (Fig. 174). Moreover, the join $NaAlSi_3O_8$—$KAlSi_3O_8$ (Fig. 175, A,B) shows a continuous drop in liquidus temperatures and gradual elimination at high water pressures of the incongruent melting of orthoclase. Near 5 kb, where the solvus is intersected, continuous solid solution of the Na-K feldspars is no longer possible. At pressures above 5 kb (Fig. 175, B), two alkali feldspars can crystallize directly from the melt (as M and N), resulting in elimination of much of the perthitic exsolution texture characteristic of alkali feldspars that crystallize at low P_{H_2O} The effect of water pressure on the boundary $KAlSi_3O_8$—$CaAl_2Si_2O_8$ (Fig. 176, A,B) is to cause a decrease in liquidus temperatures and elimination of the fields containing leucite. With the elimination of leucite, the three-feldspar system becomes quaternary rather than pseudo-ternary.

At a water pressure of 5 kb, the ternary system (shown in Fig. 177) exhibits a considerable lowering of liquidus temperatures, as well as a shift in the boundary curve (Yoder, Stewart, and Smith (1957)). On the alkali feldspar boundary (Morse, 1970) the liquidus surface has been sufficiently depressed to encounter the solvus, and the former area of complete solid solution is now broken by the immiscibility gap (Fig. 175, B). The effects of the rather high degree of solid solution are indicated in Fig. 178, A, which shows the approximate composition of phases that precipitate from melts at a water pressure of 5 kb. Original liquid compositions that lie within the shaded areas will crystallize to form a single feldspar of like composition. Liquid compositions above the two-phase areas, such as point A, will crystallize to form a plagioclase of composition B and K-rich feldspar of composition C. The two-phase tie-lines (such as BC) vary in position as a function of P_{H_2O} (and consequently temperature of formation) and have been suggested for use as a geothermometer (Fig. 178, B) by Barth (1951, 1968).

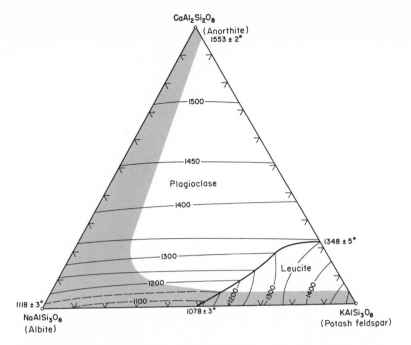

FIGURE 173
Liquidus surface of the system NaAlSi₃O₈—KAlSi₃O₈—CaAl₂Si₂O₈ at 1 atmosphere pressure. The shaded part indicates the area of feldspar solid solution. [After Franco and Schairer (1951).]

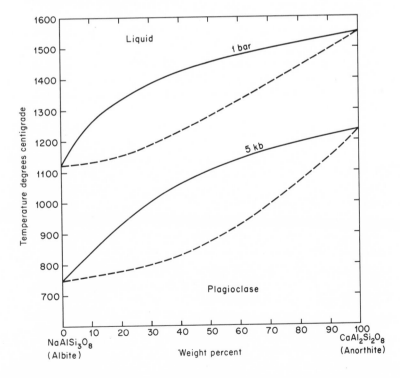

FIGURE 174
The system NaAlSi₃O₈—CaAl₂Si₂O₈—H₂O at 1 atmosphere and a P_{H_2O} of 5 kb., projected onto the Ab-An face of the temperature-composition prism. [After Bowen (1913) and Yoder, Stewart, and Smith (1957).]

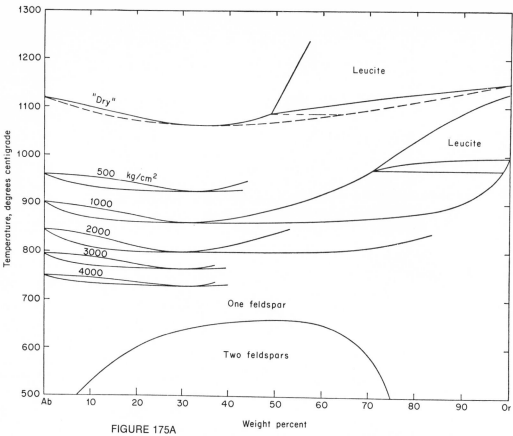

FIGURE 175A
Projection of isobaric equilibrium relations onto the anhydrous join of the system
NaAlSi₃O₈—KAlSi₃O₈—H₂O. Numbers on liquidus surfaces are P_{H_2O} in kg/cm².
[After Tuttle and Bowen (1958).]

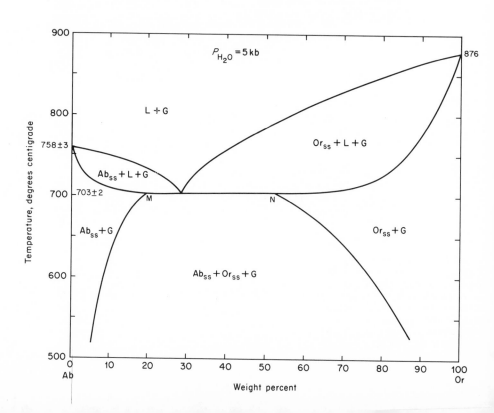

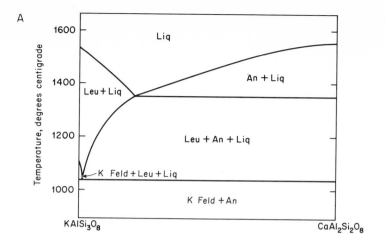

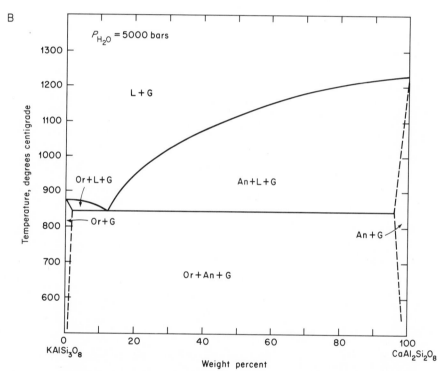

FIGURE 176,A,B
Projection of the system $KAlSi_3O_8$—$CaAl_2Si_2O_8$ at 1 atmosphere [after Schairer and Bowen (1948)] and projection onto the anhydrous join of isobaric equilibrium relations at P_{H_2O} of 5 kb. for the system $KAlSi_3O_8$—$CaAl_2Si_2O_8$—H_2O.

FIGURE 175B (opposite page)
Projection onto the anhydrous join of isobaric equilibrium relations at a P_{H_2O} of 5 kb for the system $NaAlSi_3O_8$—$KAlSi_3O_8$—H_2O. Ab_{ss} = albite solid solution, Or_{ss} = orthoclase solid solution, L = liquid, and G = gas. [After Yoder, Stewart, and Smith (1957) and Morse (1970).]

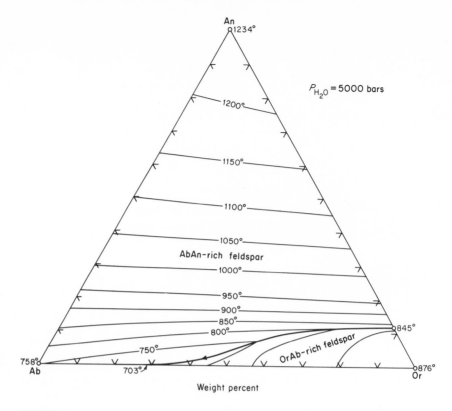

FIGURE 177
Projection of the system $NaAlSi_3O_8$—$KAlSi_3O_8$—$CaAl_2Si_2O_8$—H_2O at a P_{H_2O} of 5 kb.
[After Yoder, Stewart, and Smith (1957) and Morse (1970).]

Crystallization within this system can be described by use of the schematic diagram in Figure 179. The single boundary curve from O to L is a four-phase curve; liquid compositions on this curve coexist with vapor, plagioclase feldspar, and K-rich feldspar. The compositions of coexistent feldspars and liquids must be determined by experimentation. Such data is plotted as two-phase liquid-crystal tie-lines. For example, a liquid D coexists with feldspars B and C, while liquid G coexists with feldspars F and H. A triangle such as BCD, whose corners indicate the compositions of coexistent phases, can be useful in showing melting or crystallization relationships.* If an original liquid composition A is on the BC limb of triangle BCD, we can immediately state that the final feldspar compositions will be B and C, the composition of the last drop of cooling liquid will be at D, and the ratio of the final feldspar compositions B and C will be proportional to the lengths AC and AB respectively (from the Lever Rule).

Consider in detail the crystallization of a liquid of composition A. Cooling of liquid A initially produces a plagioclase, such as E, which is quite rich in

*This subject is discussed earlier on p. 84 for the system diopside-albite-anorthite.

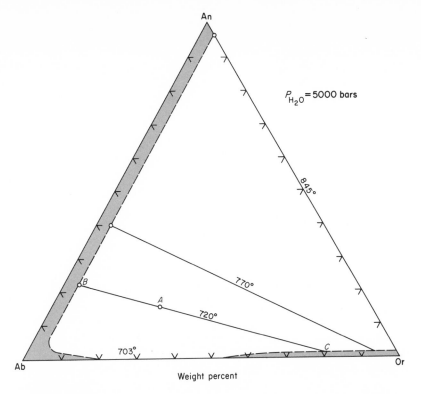

FIGURE 178A
Experimentally determined tie lines (at a P_{H_2O} of 5 kb), connecting coexisting feldspars in equilibrium with liquid and gas. [After Yoder, Stewart, and Smith (1957) and Morse (1970).]

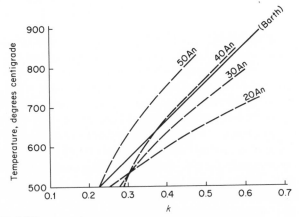

FIGURE 178B
The feldspar geological thermometer. The abscissa gives
$$k = \frac{\text{mole fraction of albite in the alkali feldspar}}{\text{mole fraction of albite in plagioclase}}$$
After determination of k, an intersection with the proper plagioclase curve yields a temperature of formation on the ordinate. The solid curve is an average for plagioclase up to An_{40} coexisting with alkali feldspar, and crystallized at a P_{H_2O} of 1000 bars. The geothermometer is not completely accurate, however, as more data are required. [After Barth (1968).]

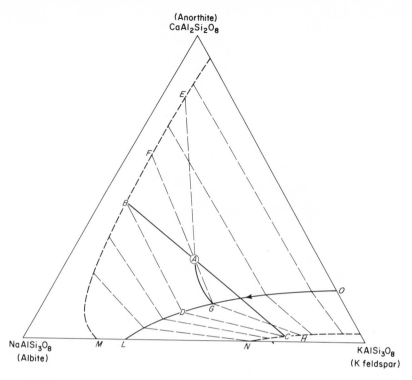

FIGURE 179
Hypothetical crystallization paths at high water pressure in the schematic system
$NaAlSi_3O_8$—$KAlSi_3O_8$—$CaAl_2Si_2O_8$—H_2O.

the anorthite component. With continued cooling, the liquid and previously precipitated crystals react and the crystal composition changes toward F; simultaneously, the melt continuously precipitates crystals that also change in composition toward F. The combination of reaction and precipitation causes the liquid to follow a curved rather than straight composition path toward the boundary curve. The liquid reaches the boundary curve at G as the coexisting plagioclase crystals arrive at F. The points F, A, and G must lie on a straight line; this line is the experimentally determined liquid-crystal tie-line that crosses the original liquid composition point A. Upon reaching the boundary curve at G, the liquid begins to precipitate alkali feldspar (composition H) with plagioclase. Simultaneous precipitation and reaction with the two feldspar phases causes the liquid to change gradually from G to D as the precipitating and coexisting feldspars change from F to B and H to C. Under equilibrium crystallization, the last drop of liquid is consumed at D, and the final feldspar compositions are B and C.* Incomplete (nonequilibrium) reaction of liquids and solids would cause the liquid to migrate past D and produce inhomogeneous feldspars whose compositions average out at B and C.

*Further slow cooling below the liquidus could allow unmixing of feldspar C to produce some Na-rich plagioclase.

The four-phase boundary curve *OL* terminates at *L* on the albite-orthoclase boundary system. Extreme fractionation processes may drive liquids (with a vapor and a coexisting pair of inhomogeneous feldspars) to this end-point, where the final feldspars produced are of composition *M* and *N* (indicated also in Figure 175, B).

The System $NaAlSi_3O_8$—$KAlSi_3O_8$—SiO_2—H_2O

The system $NaAlSi_3O_8$—$KAlSi_3O_8$—SiO_2—H_2O is the subject of a study that is of fundamental importance to igneous petrology—G.S.A. Memoir 74, by Tuttle and Bowen (1958). This work deals with the crystallization of the major minerals of granitic rocks; except for the Ca-rich feldspars and the minor ferromagnesian components, the minerals in this system make up essentially a whole rock composition. Furthermore, the work by Tuttle and Bowen includes an excellent treatment of experimental techniques, equilibrium relations, and interpretation, making it an extremely valuable reference for further study.

The alkali feldspar boundary of this system has already been discussed in the previous section, and is shown in Figure 175. The boundary $KAlSi_3O_8$—SiO_2 (Fig. 180) is part of the larger binary $KAlSi_2O_6$—SiO_2. Orthoclase ($KAlSi_3O_8$) melts incongruently to produce leucite and liquid. In the presence of increasing P_{H_2O} the temperature of the liquidus surfaces drops, and the field of leucite is eliminated. The third boundary surface, $NaAlSi_3O_8$—SiO_2 (Fig. 181), is a simple binary system whose liquidus surfaces decrease in temperature with increasing P_{H_2O}.

The anhydrous ternary system is seen in Figure 73, and crystallization within the system is discussed on p. 89. It is sufficient to say here that liquids migrate to the boundary curve to produce an alkali feldspar and a silica polymorph. Although liquids tend to approach the minimum point on the boundary curve, the final liquid composition is determined by the original composition. Under nonequilibrium conditions, final liquids will make a closer compositional approach to the minimum point than under equilibrium conditions.

Figure 182, which shows the liquidus surfaces, demonstrates that crystallization under increasing water pressure causes a continuous decrease in liquidus temperatures. At a pressure of 200 kg/cm², cristobalite, tridymite, and high quartz all appear at the liquidus. Crystallization of pure SiO_2 would result in initial precipitation of cristobalite from the melt; upon cooling (assuming equilibrium conditions) this would undergo solid-state conversion to tridymite, high quartz, and then low quartz. The addition of feldspar components to the system causes a drop in the liquidus temperatures; a silica-rich liquid of the composition *A* (Fig. 182, B) would begin to crystallize at a temperature below the stability range of cristobalite; initially tridymite would precipitate directly from the melt. A melt of the composition *B*, falling within the primary field of high quartz, lies below the stability field of tridymite and will precipitate high quartz as the primary phase. Similarly, a melt of composition *C* would precipitate a potash-rich feldspar without first precipitating leucite.

184

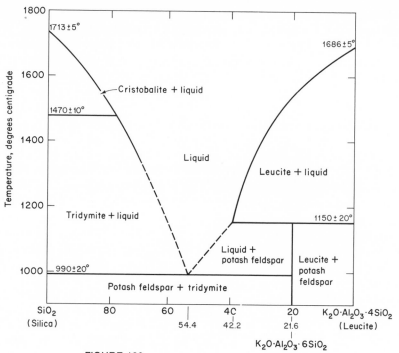

FIGURE 180
The system $KAlSi_2O_6$—SiO_2 at one atmosphere.
[After Schairer and Bowen (1948).]

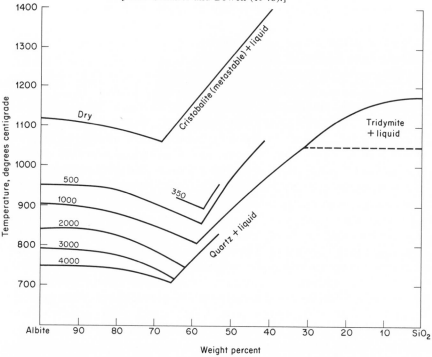

FIGURE 181
Isobaric liquidus equilibrium relations in the system $NaAlSi_3O_8$—SiO_2—H_2O projected onto the albite-quartz face of the temperature-composition prism. The numbers refer to P_{H_2O} in kg/cm². [After Tuttle and Bowen (1958).]

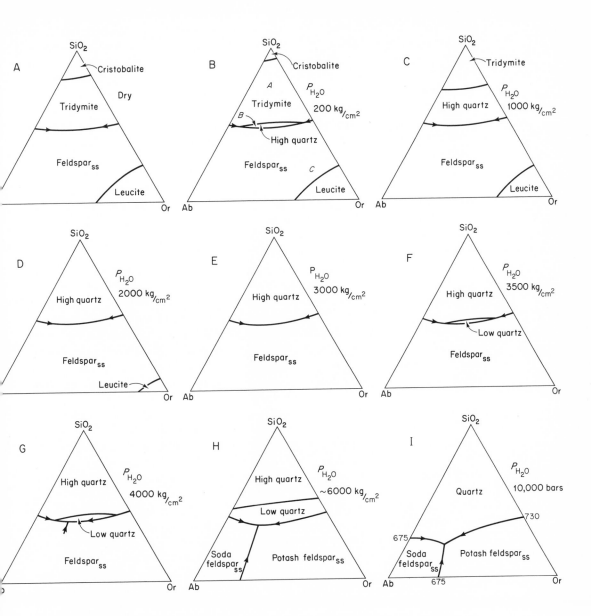

FIGURE 182
Liquidus surfaces in the system NaAlSi$_3$O$_8$—KAlSi$_3$O$_8$—SiO$_2$—H$_2$O at various water pressures. [Sections A–H are from Tuttle and Bowen (1958), and section I is from Luth, Jahns, and Tuttle (1964).]

Raising of the water pressure to 2000 kg/cm² serves to eliminate the primary fields of cristobalite and tridymite completely by lowering liquidus temperatures below their fields of stability. Under these conditions, there is no possibility for the precipitation of cristobalite or tridymite, hence all liquids on the SiO_2 side of the boundary curve precipitate high quartz directly. In addition, the field of leucite is almost entirely eliminated at this pressure. It is easy to see from these diagrams why cristobalite, tridymite, and leucite are encountered almost entirely in rocks formed under near-surface conditions or in extrusive flows. Most intrusive silicic magmas possess sufficient water to depress freezing temperatures below the fields of stability of these high-temperature phases.

At pressures of 3000 kg/cm² the field of leucite is completely eliminated from this system. At the still higher pressure of 3500 kg/cm², the diagram shows that low quartz exists at the liquidus. A pressure of 4000 kg/cm² lowers liquidus temperatures within the system sufficiently to bring the liquidus into contact with the alkali feldspar solvus. This occurs first near the quaternary minimum point and produces a short boundary curve between soda- and potash-rich feldspars. Along the boundary curve two feldspars may be precipitated simultaneously from the melt. For liquids richer in the feldspar components a single feldspar phase will crystallize, as the boundary curve extends only a short distance into the low-temperature trough that continues to the alkali feldspar binary system. At pressures higher than 5000 kg/cm² the feldspar solvus is encountered for all compositions within the silica-poor part of the diagram, and the subtraction curve between the alkali feldspars continues to the $KAlSi_3O_8$—$NaAlSi_3O_8$ binary. The area of complete solid solution in the alkali feldspars has been eliminated, and crystallization of the feldspars allows primary precipitation of two feldspar phases directly from the melt. These two feldspars unmix to some extent during cooling, as can be seen by the slope of the solvus (Fig. 175, A,B), but not to the extent that a large amount of perthite is formed, which is what happens at lower water pressures. This characteristic change in behavior, as well as other features, lead Tuttle and Bowen (1958) to classify granite as *hypersolvus* (feldspar crystallization above the solvus) and *subsolvus* (primary precipitation of two feldspars due to intersection of liquidus surface with the solvus).

At the high pressures of Figure 182, *H*, *I*, a boundary curve, rather than a low temperature trough, separates the two feldspar types. Although this appears to be a simple ternary arrangement with three boundary curves approaching the eutectic, it must be remembered that considerable solid solution still exists between the feldspars. Thus a melt located above either feldspar liquidus surface will *not* precipitate a "pure" feldspar end-member, change in composition away from the corner of the compositional triangle, and arrive at the eutectic in the usual way. In addition, it is important to note that the final liquid does not always reach the quaternary minimum point, despite the disarming simplicity of the liquidus surface.

The detailed cooling sequence can perhaps be best understood from the schematic diagrams in Figures 183 and 184. In the alkali feldspar boundary

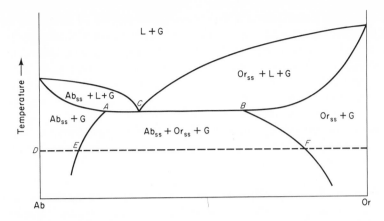

FIGURE 183
Schematic sketch of the system $NaAlSi_3O_8$—$KAlSi_3O_8$—H_2O at high water-pressure, projected onto the Ab-Or side of the temperature-composition prism (Ab_{ss} = albite solid solution, Or_{ss} = orthoclase solid solution, L = liquid, and G = gas).

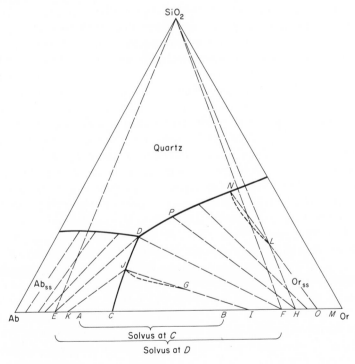

FIGURE 184
Crystallization paths and liquid-crystal tie lines on the schematic sketch of the liquidus of the system $NaAlSi_3O_8$—$KAlSi_3O_8$—SiO_2—H_2O at high water pressures.

system (Fig. 183), points *A* and *B* indicate the limits of feldspar solid solution. Any liquid whose original composition lies between the limits *A* and *B*, will, during crystallization, reach the cotectic minimum *C* and precipitate feldspars *A* and *B* simultaneously. Liquids having compositions more soda-rich than *A* or more potash-rich than *B* will not change sufficiently in composition during equilibrium crystallization to arrive at the cotectic minimum. A single feldspar phase will be produced (which may or may not unmix in the solid state as a function of its cooling rate). In the schematic diagram of the system $NaAlSi_3O_8$ —$KAlSi_3O_8$—SiO_2—H_2O (Fig. 184), the points *A* and *B* are again indicated on the boundary system between $NaAlSi_3O_8$ and $KAlSi_3O_8$. A liquid *C* will simultaneously precipitate feldspars *A* and *B* during cooling. At the quaternary minimum point, *D*, where the temperature is considerably lower than at *C*, the liquid will precipitate feldspars of composition *E* and *F*. These compositions can be determined from Figure 183, where the temperature of *D* can be read from the vertical temperature coordinate and where *E* and *F* are shown on the feldspar solvus.* Intermediate positions on the boundary curve between *C* and *D* (Fig. 184) yield feldspars of compositions between *E* and *A*, and *B* and F.

Any original liquid whose composition falls within the triangle SiO_2—*E*—*F* will change in composition so as to arrive at the eutectic *D* and produce a mixture of phases *E*, *F*, and SiO_2, the proportions depending upon the original composition. A liquid such as *G* cools and initially precipitates a K-rich feldspar such as *H*. With continued cooling, crystallization, and reaction, the feldspar changes in composition to *I* as the liquid composition follows a curved path to the boundary curve at *J*. Here a Na-rich feldspar of composition *K* also precipitates. Additional cooling causes continued crystallization of two feldspars (changing in composition from *K* to *E* and from *I* to *F*) as the liquid changes in composition along the boundary curve from *J* to *D*. At *D*, quartz, and feldspars *E* and *F* precipitate simultaneously until the liquid is used up.

Original liquid compositions that do not lie within the triangle SiO_2—*E*—*F* will not complete crystallization at the quaternary minimum *D*. Upon cooling, a liquid such as *L* will first precipitate a K-rich feldspar such as *M*. Continued cooling causes the liquid composition to follow a curved path to the boundary curve at *N* while the coexistent feldspar changes in composition from *M* to *O*. The liquid then follows the boundary curve, precipitating both quartz and feldspar. The coexistent feldspar changes in composition from *O* to *H* as the liquid migrates to *P*. At *P* the last liquid is used up, and the batch consists of feldspar of composition *H* plus quartz.

Melting paths within this system are of course exactly opposite those of cooling. The first-formed liquids, at low P_{H_2O} (depending upon the Na/K ratio), will melt to form compositions either at the cotectic minimum *or on the cotectic line*. The composition of the first-formed melt will vary also as a

---

*This technique can only be used when the phases at the solvus (the alkali feldspars) exhibit no solid solution with other components in the system. Such solid solution would affect the position of the solvus.

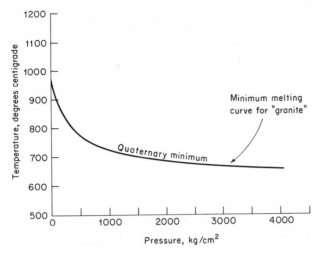

FIGURE 185
The minimum melting curve for "granite," as represented by the quaternary minimum point in the system $NaAlSi_3O_8$—$KAlSi_3O_8$—SiO_2—H_2O. [After Tuttle and Bowen (1958).]

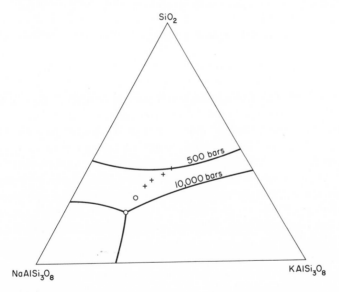

FIGURE 186
The shift in the quaternary minimum point in the system $NaAlSi_3O_8$—$KAlSi_3O_8$—SiO_2—H_2O. Plus signs indicate isobaric minimums, and circles isobaric eutectics. [After Luth, Jahns, and Tuttle (1964).]

function of the water pressure, which shifts the position of the cotectic line and its minimum point. At higher water pressures, initial liquids will form either at the ternary eutectic or along the quartz-feldspar boundary curves.

Tuttle and Bowen (1958) plotted the temperature of the quaternary minimum point as a function of P_{H_2O} (Fig. 185). This curve shows the lowest temperature at which liquid can exist within the system, and has been used as the minimum melting curve for granite in later literature. Similar curves have been determined for many of the common rock types.

In addition, the composition of the minimum melting point has been re-plotted to show its compositional migration as a function of water pressure (Fig. 186). Large numbers of chemically analyzed extrusive and intrusive igneous rocks have been plotted on this type of triangular base; the compositions of extrusive rocks tend to be clustered near the low-pressure minimums, and the compositions of intrusive rocks tend to be clustered near the high-pressure minimums. The analyses nevertheless fall slightly on the K-feldspar side of the minimum points, from which Tuttle and Bowen concluded that the average granitic magma was slightly undersaturated with respect to water. This conclusion, however, has been shown to be incorrect; because the average plagioclase on granitic rocks is oligoclase rather than albite, the melting temperature of the plagioclase component would be higher, causing the minimum points to be shifted slightly toward the K-feldspar-quartz compositions (Smith, 1963), as will be seen in the following section.

FIVE-COMPONENT SYSTEMS

The System $NaAlSi_3O_8$—$KAlSi_3O_8$—$CaAl_2Si_2O_8$—SiO_2—H_2O

Although it has been widely accepted that the previously discussed four-component system is typical of the melts that crystallize to produce rocks of granitic composition, the system does not include the component $CaAl_2Si_2O_8$ (anorthite). The effects of this additional component on melting behavior have been examined in relation to the problem of anatexis by von Platen (1965) and Winkler (1967).

Figure 187, suggested by Winkler (1967), is the generalized diagram of the system at 2000 bars water pressure (an excess of water is assumed). Although many details of the system are not known, sufficient data are available from the work of von Platen (1965), Tuttle and Bowen (1958), Stewart (1957), and Yoder et al. (1957) to permit generalized relations to be shown. The tetrahedron is divided into three primary phase volumes—quartz, solid solutions of K-feldspar-rich compositions, and solid solutions of plagioclase compositions. The base of the tetrahedron corresponds to the "granite" system shown in Figure 182, and the front is the ternary feldspar system, similar to that shown

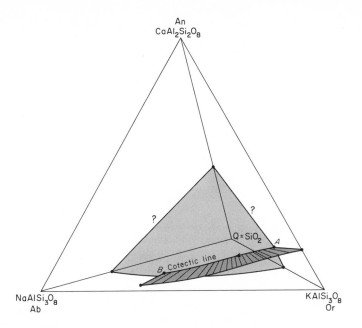

FIGURE 187
Generalized phase relations in the system $NaAlSi_3O_8$—$KAlSi_3O_8$—$CaAl_2Si_2O_8$—SiO_2—H_2O, at a P_{H_2O} of about 2 kb. Abbreviations are Ab = albite, An = anorthite, Or = orthoclase, and Q = quartz. [After Winkler (1967).]

in Figure 177. Compositions within each of the three primary phase volumes begin crystallization with the formation of one of the following: plagioclase, K-rich feldspar, or quartz. These volumes contain three phases—a solid, a liquid, and water vapor. Continued crystallization causes the liquid to change in composition along either a straight or a curved path (depending upon the absence or presence of solid solution). The liquid changes in composition and usually encounters a four-phase surface (two solids, liquid, and gas). Crystallization then continues with the precipitation of two solid phases; the direction of the path of crystallization will be away from the side of the tetrahedron that represents the compositions of the phases crystallizing. This continues until the cotectic line *AB* is encountered. Along this line, quartz, plagioclase, and orthoclase crystallize simultaneously and the liquid shifts in the direction of the arrow toward the albite-orthoclase-quartz triangle. The liquid is consumed at a position on the cotectic that is dependent upon its original composition. Liquids originally richer in the anorthitic component $CaAl_2Si_2O_8$ will freeze at higher temperatures (and closer to *A*) than more albitic ones. Small changes in the $CaAl_2Si_2O_8/NaAlSi_3O_8$ ratio have a large effect on the final melt composition.

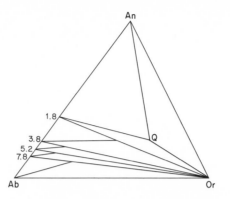

FIGURE 188
Compositional planes within the tetrahedron $NaAlSi_3O_8$—$KAlSi_3O_8$—$CaAl_2Si_2O_8$—SiO_2. The numbers represent albite-anorthite ratios (in weight percent) for each plane. [After von Platen (1965).]

Von Platen (1965) determined minimum melting temperatures and compositions for a group of planes through the system at a fixed pressure of 2000 bars. The planes examined are shown in Figure 188; the reactants consisted of obsidian-anorthite mixtures equivalent to quartz, K-feldspar, and plagioclase of a fixed Ab/An ratio. These compositional planes intersect the three four-phase planes and the five-phase line AB within the tetrahedron. These intersections form three boundary curves and a minimum melting point on each surface. Note that these planes are joins, and are not true 'ternary' slices, because the effects of solid solution place the final liquid composition (along the cotectic AB) somewhat below the chosen compositional slice.

Rather than show the boundary curves on the chosen compositional planes within the tetrahedron, von Platen projected the boundary curves from the anorthite corner of the tetrahedron onto the albite-orthoclase-quartz base (as in Fig. 189) so that an easy comparison can be made between systems containing different amounts of anorthite. All of the data are shown superimposed in Figure 190 for a water pressure of 2000 bars. The temperatures and minimum melting compositions for the various planes are given in weight percent (anorthite disregarded).

Ab/An Ratio	Minimum Melt Temperature	Compositional Ratio		
		Q	Ab	Or
∞	670	34	40	26
7.8	675	40	38	22
5.2	685	41	30	29
3.8	695	43	21	36
1.8	705	45	15	40

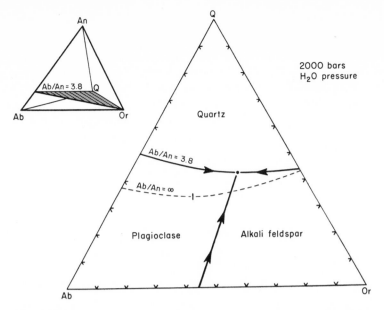

FIGURE 189
The plane Ab/An = 3.8 projected from the anorthite apex onto the Ab-Or-Q base at a P_{H_2O} of 2 kb. The broken line is the An-free cotectic line of Tuttle and Bowen (1958) with the minimum point indicated. The heavier boundary curves apply to the system in which Ab/An = 3.8. [After von Platen (1965) and Winkler (1967).]

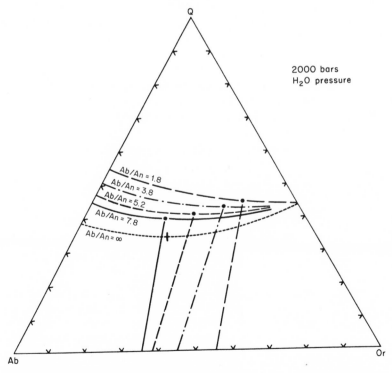

FIGURE 190
Minimum-temperature melts and boundary curves at a P_{H_2O} of 2 kb, from the compositional planes of Figure 188. All of the thermal data have been projected from the An apex to the base for easier comparison. [After von Platen (1965) and Winkler (1967).]

Melts that have high anorthite contents not only have a somewhat higher melting temperature, but show a large shift in composition relative to the Ab-Or-Q base upon which the data are plotted. A pure anorthite (Ab free) cross section, when plotted on the Ab-Or-Q base, must lie along the Q-Or edge (point A in Fig. 187). This system is described for a single water pressure. At a different water pressure, the boundary curves and minimum points occupy different positions.

8

Hydrothermal Melting of Rocks

A wide variety of experiments have been done to study the melting of rocks under water pressure. One of the most significant of these is a study by Yoder and Tilley (1962) on the stability relations of basalts.

Figure 191 shows the melting relationships of an olivine tholeiite with water pressure. The curves labeled "upper limit" for plagioclase, sphene, pyroxene, and olivine indicate the maximum temperatures at which these phases are stable. As these are all anhydrous phases, this limit decreases with increasing water pressure. The amphibole curve, however, increases with water pressure, since amphibole is a hydrous phase.

A rock melt cooling at a water pressure of 5000 bars will consist of magnetite (not indicated on diagram) + liquid + gas until it reaches a temperature of about 1120°C. At this temperature olivine begins to crystallize. At about 1090°C pyroxene also starts to crystallize. Crystallization of these phases continues until the amphibole stability curve is encountered at 965°C. Here amphibole begins to precipitate directly from the melt in addition to forming by reaction of melt with crystals of olivine and pyroxene. The olivine is almost immediately consumed, and at about 25°C lower the pyroxene is eliminated; the stability regions of olivine and pyroxene thus have upper and lower

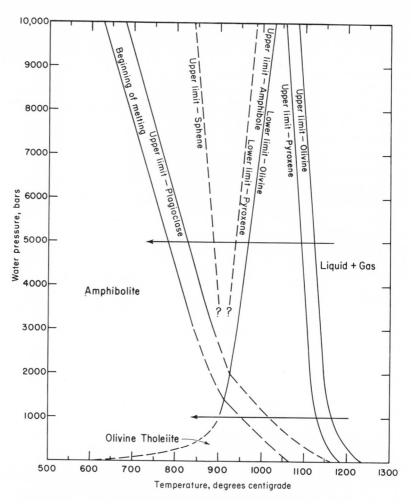

FIGURE 191
Melting relations for an olivine tholeiite as a function of water pressure. Each curve marks the upper or lower limit of the phase indicated. The "beginning of melting" curve shows the minimum temperatures for which liquid can exist in this system. [After Yoder and Tilley (1962).]

temperature limits, and are indicated by the parallel inclined lines in the figure. After elimination of olivine and pyroxene the batch consists of melt, vapor, magnetite, and amphibole. The amphibole continues to precipitate with decreasing temperature. At about 890°C sphene begins to precipitate, and plagioclase at 825°C. The melt is consumed at about 780°C, yielding a mixture of amphibole, plagioclase, sphene, magnetite, and vapor. Heating this assemblage at the same pressure reverses the sequence. The above cooling sequence is uncommon in basaltic melts crystallizing at near-surface conditions, as P_{H_2O} is usually less than P_{conf}.

Crystallization at 1000 bars leads to a different series of reactions. A mixture of liquid, gas, and magnetite begins to precipitate olivine at about 1175°C. It is joined by pyroxene at about 1150°C. Cooling and crystallization continues to about 1000°C, where plagioclase begins to crystallize with olivine and pyroxene. The melt becomes completely solidified at about 950°C. The amphibole stability curve is not encountered until approximately 900°C. As no liquid is present in the system at that temperature, the vapor phase reacts with the pyroxene and olivine to form amphibole. In natural environments the latter process often does not go to completion, as the vapor is commonly lost to the surrounding country rock.

Although it may be obvious, the student should observe that there is a considerable difference between melting and breakdown curves of minerals for whole rock compositions and the univariant curves of single compounds. It is not possible to determine *individual* melting curves for pyroxene, plagioclase, and olivine and then put them all together on the same diagram and assume that they constitute the melting relations of a basalt. As always, the presence of additional components in a system has an effect on the melting points and decomposition temperatures of all other phases in the system. The melting curve of a single phase under various water pressures will always be higher than the melting curve of the same mineral when it is part of a whole rock composition.

9

Systems that Contain Carbon Dioxide

Although most studies of carbonate systems have been limited to subsolidus equilibrium, some investigators have examined liquidus relations as well. Walter, Wyllie, and Tuttle (1962) presented a schematic TX section of the system $MgO—CO_2$ at 1000 bars pressure (Fig. 192). The general arrangement of this system is similar to the hypothetical silicate-water system shown earlier in Figure 158. The technique developed for deriving such a system experimentally consists in taking a particular ratio of the oxide and carbonate phases, sealing them in a nonreactive metal tube, and then subjecting the tube to the proper pressure and temperature; the tube will collapse and remain collapsed until the internal CO_2 pressure and applied external pressure are equal. An increase in the CO_2 pressure of the system raises the temperature stability limits of the solid carbonate-bearing phases. The rise in the dissociation temperature of $MgCO_3$ with increase of P_{CO_2} is shown in Figure 193. The reaction $MgCO_3 \rightleftharpoons MgO + CO_2$ is driven to the left at higher P_{CO_2}, thus raising the dissociation temperature of $MgCO_3$. The curves with negative slopes represent changes in the dissociation temperature when P_{CO_2} is not P_{total}. If, for example, P_{CO_2} has a maximum value of A and additional pressure is exerted by inert gases, the magnesite decomposition curve will be BCD.

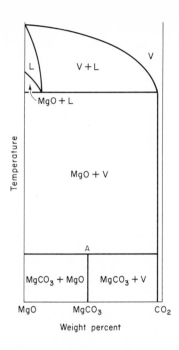

FIGURE 192
Schematic isobaric *TX* section of the system
MgO—CO₂ at a P_{CO_2} of 1 kb (L = liquid, V =
vapor). [After Walter, Wyllie, and Tuttle
(1962).]

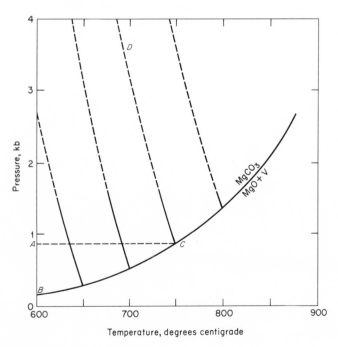

FIGURE 193
Univariant dissociation curve of MgCO₃ as a function of P_{CO_2}. Curves with
negative slopes represent the dissociation curve with P_{CO_2} constant and inert
pressure variable. [After Walter (1963).]

Carbonate minerals may melt incongruently to liquid and vapor, as in the system CaO—CO$_2$ (Fig. 194). Cooling and heating reactions in these systems follow the general binary behavior discussed earlier.

The System CaO—MgO—CO$_2$

The system CaO—MgO—CO$_2$ was examined at subsolidus temperatures by Harker and Tuttle (1955a,b). Compatibility relations as a function of T and P_{CO_2} are shown in Figure 195. At low temperatures the carbonate-containing phase CaMg(CO$_3$)$_2$ is stable with CaCO$_3$ or MgCO$_3$. Increasing the temperature causes the dissociation of MgCO$_3$ to MgO and CO$_2$, as indicated by the first univariant curve. At higher temperatures CaMg(CO$_3$)$_2$ dissociates to CaCO$_3$ + MgO + CO$_2$ (second univariant curve). The final carbonate dissociation is that of CaCO$_3$ to CaO + CO$_2$, which is shown at high temperatures and low P_{CO_2}. Although this diagram shows the generalities of the system, it does not show the considerable solid solution that results at higher CO$_2$ pressures and temperatures.

Solid solution within this system was studied by Graf and Goldsmith (1955) and by Goldsmith and Heard (1961). The subsolidus binary system CaCO$_3$—MgCO$_3$ (Fig. 196) was examined under a CO$_2$ pressure kept high enough to prevent decomposition of any of the carbonate-containing phases. Both one- and two-phase areas are present. Very slight solid solution of Ca is seen in MgCO$_3$; limited solid solution of either Ca or Mg is present for CaMg(CO$_3$)$_2$. Only CaCO$_3$ shows extensive solution of Mg.

The magnesian calcite structure has a random arrangement of Mg atoms in the normal layered Ca sites, whereas the dolomite structure has a regular ordered arrangement of alternate layers of Mg and Ca atoms. Between the compositional areas of these different structures is an immiscibility region (solvus) that narrows with increase in temperature, but persists almost to 1100°C. The adjacent broken line marks the gradual change that takes place at high temperatures from the well-ordered dolomite structure to the disordered calcite. Never has more than a single solid phase been experimentally detected at this transition. If there does exist a narrow two-phase region, as would normally be expected in a standard, first-order transition, then the diagram of the system would resemble the one shown in Figure 197. But if the transition is gradual (as it appears to be) and is the result of constantly increasing thermal disordering of Ca and Mg, then it may be regarded as second order, like the one in Figure 196. This transition is discussed in some detail by Goldsmith and Heard (1961).

The usually accepted decomposition reaction of dolomite at atmospheric pressure is

$$CaMg(CO_3)_2 \xrightarrow[\text{Low } P_{CO_2}]{\Delta T} CaCO_3 + MgO + CO_2$$
$$\text{(dolomite)} \qquad\qquad \text{(calcite)} \quad \text{(periclase)}$$

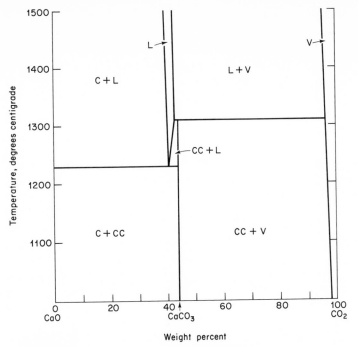

FIGURE 194
Isobaric *TX* section of the system CaO—CO₂ at a P_{CO_2} of 1 kb (C = CaO, CC = CaCO₃, L = liquid, V = vapor). [After Wyllie and Tuttle (1960b).]

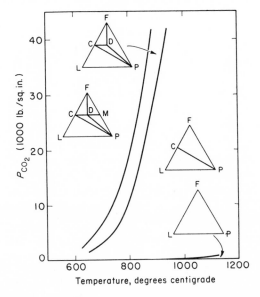

FIGURE 195
The system CaO—MgO—CO₂. Univariant curves for the dissociation of calcite, magnesite, and dolomite are shown with compatibility triangles in divariant areas; solid solution is ignored (F = CO₂, C = calcite, D = dolomite, M = magnesite, L = lime, P = periclase). [After Harker and Tuttle (1955a).]

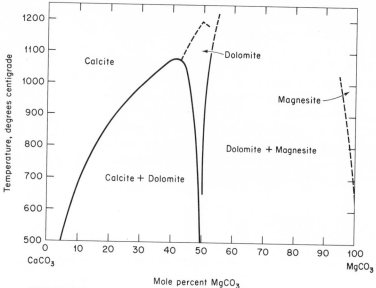

FIGURE 196
The system $CaCO_3$—$MgCO_3$. [After Goldsmith and Heard (1961).]

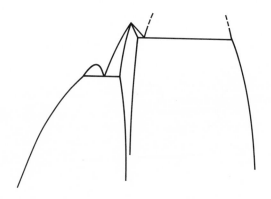

FIGURE 197
Schematic phase diagram illustrating a classical phase change
between calcite and dolomite. [After Goldsmith and Heard (1961).]

This is the reaction shown in Figure 195. If the decomposition occurs at high P_{CO_2}, the dissociation produces a smaller amount of CO_2 and periclase plus a magnesium-containing calcite. This may be written as

$$Ca_1Mg_1(CO_3)_2 \xrightarrow[\substack{\text{High} \\ P_{CO_2}}]{\Delta T} Ca_1Mg_{(1-x)}(CO_3)_{(2-x)} + xMgO + xCO_2$$

The dissociation temperature of dolomite is plotted versus P_{CO_2} in Figure 198, A. Compatibility triangles for the system CaO—MgO—CO₂ appear at several temperatures and pressures (A, B, C, D, E, and F) adjacent to the dolomite dissociation curve. The solid solution of magnesium in calcite increases with both T and P_{CO_2} along the curve, as indicated by the compatibility triangles. The numbers written on the curve state the magnesium content of the magnesian calcite.

The composition of magnesian calcites that are in equilibrium with dolomite (the low-temperature side of the curve) are independent of P_{CO_2}, because neither MgO nor CO₂ is involved in solid solution reactions between magnesian-calcite and dolomite, as is seen in the following equation:

$$Ca_1Mg_x(CO_3)_{(1+x)} \rightleftharpoons Ca_yMg_y(CO_3)_{2y} + Ca_{(1-y)}Mg_{(x-y)}(CO_3)_{(1+x-2y)}$$

(magnesian calcite I)   (dolomite)   (magnesian calcite II)

The reaction is sensitive to temperature, and not P_{CO_2}. This is shown by the identical magnesium content of magnesian-calcites in compatibility triangles A and G.

At temperatures above the dolomite dissociation curve, the stability of magnesian-calcite is controlled by the following reaction:

$$Ca_1Mg_x(CO_3)_{(1+x)} \rightleftharpoons Ca_1Mg_{(x-y)}(CO_3)_{(1+x-y)} + yMgO + yCO_2$$

(magnesian calcite I)   (magnesian calcite II)   (periclase)

Since CO₂ is a product of the reaction, the P_{CO_2} and the temperature control the composition of the magnesian-calcite. The magnesian-calcite compositional dependence on P_{CO_2} is shown by triangles H and I, and the temperature dependence by B and H.

Normally such detailed information cannot be represented in the manner shown in Figure 198, A because of the complex nature of the data and the difficulty of interpolating between the variously described points. A condensed method of showing these data (Fig. 198, B) was devised by Graf and Goldsmith (1955).* The heavy diagonal curve is the thermal decomposition curve of dolomite. Dolomite is stable at lower temperatures (to the right); at higher

---

*This type of plot is commonly used because of its relationship to the Clapeyron equation

$$\frac{dP}{dT} = \frac{\Delta H}{T \Delta V}$$

where ΔH is the heat of transition, T the temperature of transition, and ΔV the volume change. The integrated version of this equation is

$$\frac{d\ln P}{dT} = \frac{\Delta H}{RT^2} \quad \text{or} \quad \left(\frac{dP}{P}\right)\frac{1}{dT} = \frac{\Delta H}{RT^2}$$

For an ideal system, where ΔH is constant, a plot of $\ln P$ versus $1/T$ results in a straight line. Such a straight line allows not only a determination of ΔH, but extrapolation of data past experimentally determined points. The dolomite decomposition curve, however, is nonideal, hence the equation cannot be properly applied. As pointed out by Graf and Goldsmith (1955), several factors could cause deviation from ideality, such as the variability in composition of the magnesian calcites or the nonideal behavior of the CO₂ gas.

204

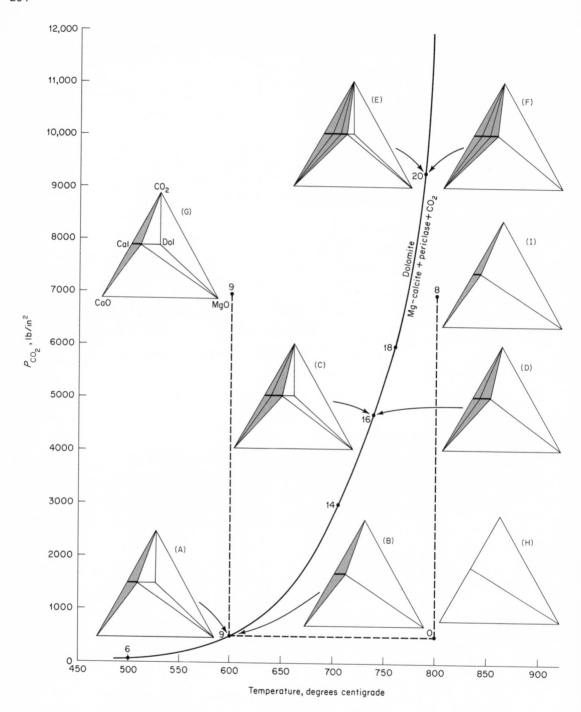

FIGURE 198A
The dissociation curve of dolomite to magnesian calcite, periclase, and CO_2. Compatibility triangles A, B, C, D, E, and F are immediately below and above the dolomite dissociation curve. The lengths of the thick lines within the triangles and the external numbers indicate the limits of magnesium substitution in magnesian calcites. [After Graf and Goldsmith (1955).]

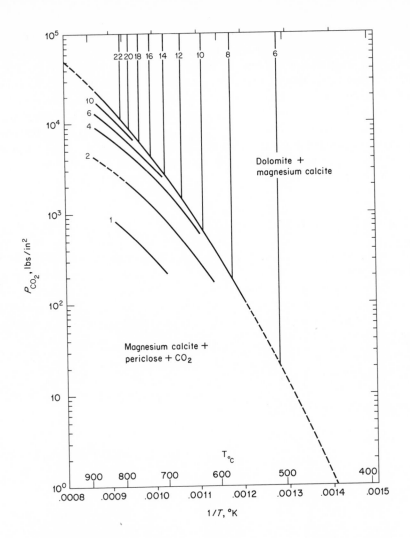

FIGURE 198B
The dolomite dissociation curve. The vertical and inclined numbered lines represent the composition (in mole percent Mg) of calcite that can exist with dolomite, or with periclase and CO_2. [After Graf and Goldsmith (1955).]

temperatures (to the left) it decomposes to magnesian-calcite, periclase, and carbon dioxide. The vertical lines (termed iso-Mg lines) indicate (in mole percent) the magnesium content of magnesian-calcites that may exist with dolomite. Note again that as these lines are vertical, the composition of these magnesian calcites is not changed by P_{CO_2}. The group of curved diagonal lines in the left-hand part of the figure are iso-Mg lines for magnesian-calcite at temperatures above the dissociation of dolomite. These lines show the maximum amount of magnesium substitution at the indicated pressures and temperatures. If the experimental data were complete, these lines would intersect the dolomite decomposition curve at the matching vertical iso-Mg lines.

The System $CaCO_3$—$MgCO_3$—$MnCO_3$

The three-component system $CaCO_3$—$MgCO_3$—$MnCO_3$ was studied by Goldsmith and Graf (1960). The boundary system $CaCO_3$—$MgCO_3$ has already been discussed. The binary $MgCO_3$—$MnCO_3$ shows complete solid solution at the temperatures investigated, whereas the system $CaCO_3$—$MnCO_3$ (Fig. 199) has a two-phase area of limited temperature range between $CaMn(CO_3)_2$ and $MnCO_3$.

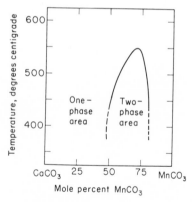

FIGURE 199
The system $CaCO_3$—$MnCO_3$.
[After Goldsmith and Graf (1957).]

The ternary data were presented on four isothermal isobaric compositional triangles, reproduced here in slightly modified form (Fig. 200). Although the effect of solid solution gives them an unusual appearance compared to the typical compatibility triangles of most ternary silicate systems, they are easily interpreted after the first shock.

The four diagrams are subdivided into areas labeled 1, 2, and 3; these numbers refer to the number of phases existing in each region in stable equilibrium. The one-phase fields are regions of complete solid solution. In the two-

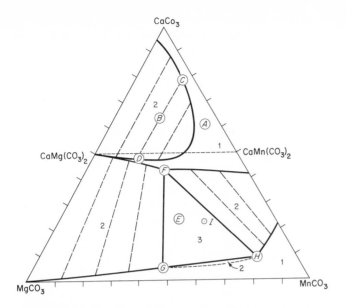

FIGURE 200A
Subsolidus equilibria in the system $CaCO_3$—$MgCO_3$—$MnCO_3$ at 500°C and 10 kb. One-, two-, and three-phase areas are labeled. [After Goldsmith and Graf (1960).]

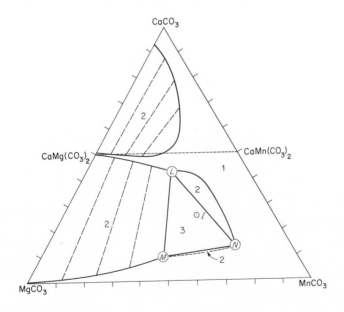

FIGURE 200B
Subsolidus equilibria in the system $CaCO_3$—$MgCO_3$—$MnCO_3$ at 600°C and 10 kb. [After Goldsmith and Graf (1960).]

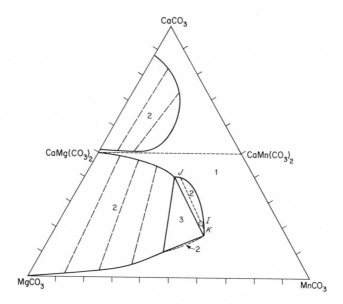

FIGURE 200C
Subsolidus equilibria in the system $CaCO_3$—$MgCO_3$—$MnCO_3$ at 700°C and 10 kb. [After Goldsmith and Graf (1960).]

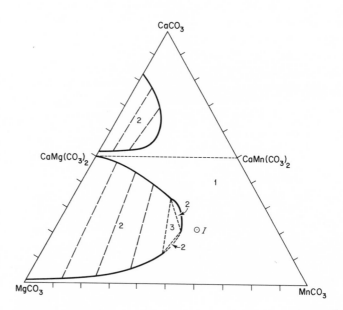

FIGURE 200D
Subsolidus equilibria in the system $CaCO_3$—$MgCO_3$—$MnCO_3$ at 800°C and 10 kb. [After Goldsmith and Graf (1960).]

phase fields, compositions are indicated at the ends of the tentative two-phase tie-lines (broken lines). The single three-phase triangle indicates coexistence of three phases whose compositions are indicated by the apices of the three-phase triangle. Point A (Fig. 200, A) consists of a single phase of composition A. Point B consists of a mixture of phases C and D (the ratio of which is found by the lengths BD and BC). Point E consists of a mixture of the three phases F, G, and H, the ratio of which can be determined by the position of E within the triangle FGH.

A thin two-phase region, although not detected experimentally, is required along the Ca-poor part of the three-phase triangle; this separates the one- and three-phase areas. The join between CaMg(CO$_3$)$_2$ and CaMn(CO$_3$)$_2$ lies partly within the extended two-phase region between dolomite and calcite. The decrease in size of this calcite-dolomite two-phase region with increasing temperature shows this join to be binary at 700° and 800°C and pseudo-binary at 500° and 600°C. The curve outlining the calcite-dolomite two-phase region is continuous from the ordered dolomite structure to the disordered calcite structure; the absence of any distinct break implies that the transition between two structures is second order or higher.

Consideration of a compositional point, such as I (Fig. 200), reveals the sequence of reactions that can occur with changes of temperature in this system. At 800°C, I consists of a single phase of composition I. Cooling to 700°C shifts the various stability fields and puts the point I in a two-phase field; unmixing leads to the formation of phases J and K. At 600°C additional unmixing and diffusion produces a third phase M, and changes J and K to L and N. Additional cooling to 500°C allows the phase compositions L, M, and N to change to F, G, and H. Bringing the system to room temperature would cause additional divergence of the compositions of the three phases if the cooling rate was sufficiently slow to allow diffusion processes to operate.

10

Systems Involving Changes in Oxidation State

An excellent general summary of the effects of changes of oxidation states on phase equilibria is given by Muan (1958). More specific examples of the role of oxygen are described by Eugster (1959), Osborn (1959), Taylor (1964), Wones and Eugster (1965), Garrels and Christ (1965), Hamilton and Anderson (1967), and Ernst (1960).

In many geological systems (of the type discussed in the first five chapters of this book) the vapor phase has been ignored, and equilibrium is discussed in terms of the Condensed Phase Rule ($P + F = C + 1$), where P refers to the number of condensed phases (liquids and solids). This can be done because the vapor pressure of each phase is very low. At equilibrium each constituent exerts a certain partial pressure in the gas phase. Whether or not the equilibrium composition of the gas phase is attained is of little importance, as the amounts of each constituent in the gas phase are quite small, and the composition of the condensed phases is not changed significantly.

In systems involving elements which can exist in more than one oxidation state, the value of the oxygen partial pressure in the surrounding environment can affect significantly the condensed phase composition. For example, an increase in the oxygen partial pressure will increase the amount of dissolved oxygen in the liquid phase, or may cause solid phases of low oxygen content to convert to more oxygenated compounds. Because of the interchange of oxygen between vapor and condensed phases, experimentation conducted on such systems must provide control of both the oxygen partial pressure as well as total pressure.

The mixing of oxygen with inert gases provides a means of controlling the oxygen content of the surrounding atmosphere. A second method of regulating oxygen pressures is to keep the condensed phases in contact with either H_2O or CO_2. These gases will decompose on heating; for example, $2CO_2 \rightarrow 2CO + O_2$. The equilibrium constants of such dissociation reactions are known for various temperatures. Although the O_2 pressure and temperature cannot be varied independently for either of these pure gases, a large measure of variation can be introduced by mixing gases in various ratios or by adding CO to the CO_2 or H_2 to the H_2O. The very nice aspect of this type of atmospheric control is that if some oxygen is taken up or released by the solid or liquid phases, the gas will quickly react to return to its original equilibrium state.

Figure 201 shows the phases of the system $Fe\text{-}Fe_2O_3$ — hematite (Fe_2O_3), wüstite (FeO), metallic iron (Fe), liquid iron oxide, and liquid iron — as a function of temperature and partial pressure of oxygen. The total pressure of the system is one atmosphere. The partial oxygen pressure is plotted as log P_{O_2} to allow a wide range of values to be placed on the same diagram. The light broken lines show the partial pressure of oxygen versus temperature for various P_{H_2O}/P_{H_2} ratios. These lines slope upward and to the right, indicating higher dissociation with increasing temperature. The Fe-containing phases that can exist in equilibrium with the various oxygen partial pressures are indicated. Phase changes that occur at a fixed original P_{H_2O}/P_{H_2} ratio with change in temperature can be found by following the appropriate curved, light-weight, broken line.

This method of portrayal, however, compresses the fields of stability of the intermediate phases in an unnecessary manner. A more readable diagram is derived from the same data by plotting the log P_{H_2O}/P_{H_2} against temperature (Fig. 202); the light broken lines indicate the partial pressure of oxygen, and the solid lines delineate the fields of stability of the various condensed phases. The system is again shown for a total pressure of one atmosphere. Using this diagram, one can trace the phase changes that take place for fixed values of the log P_{H_2O}/P_{H_2} as temperature changes. Thus the point A, at the fixed log P_{H_2O}/P_{H_2} ratio of zero, indicates magnetite at room temperature. When the system is heated to B the partial pressure of oxygen increases to about 10^{-20} atmosphere, and the magnetite converts to wüstite. At a temperature corresponding to C the wüstite melts at a partial oxygen pressure of 10^{-10} atmosphere. With continued heating the liquid oxide remains stable and the partial pressure of oxygen gradually increases.

Although these diagrams effectively present many of the data for the system,

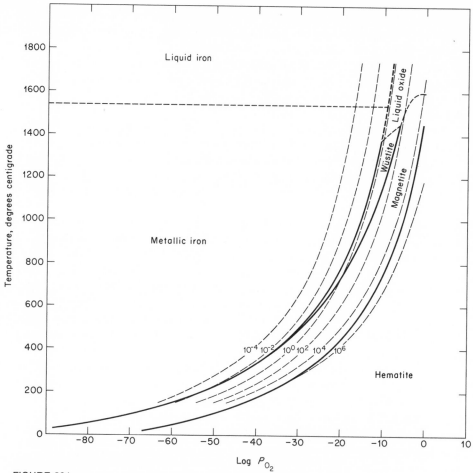

FIGURE 201

The system Fe—Fe$_2$O$_3$. The light, broken lines show the partial pressure of oxygen versus temperature for various ratios of P_{H_2O}/P_{H_2}. Total pressure is 1 atmosphere. [After Muan (1958).]

and relate the experimental approach to equilibrium conditions, they have a limited usefulness, for they do not indicate the phase compositions, as do the conventional temperature-composition diagrams. This problem has been overcome in the diagram of the system Fe-Fe$_2$O$_3$ (Fig. 203). Lines of equal partial pressure of oxygen (the broken lines) indicate the composition of the gas phase in equilibrium with the condensed phases at various temperatures and compositions. The heavy lines are the usual boundary curves, indicating the various one- or two-phase areas. Aside from the broken lines, this diagram is identical to the typical "normal" binary systems discussed earlier. The oxygen isobars are horizontal where they cross a two-phase field because only a single degree of freedom exists when three phases (i.e., vapor and two solids) are in equilibrium in a two-component system. If the bulk composition

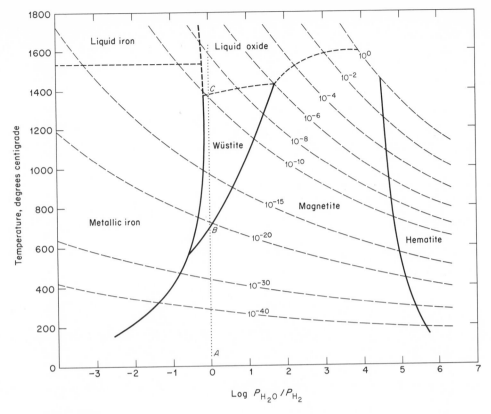

FIGURE 202
The system Fe—Fe$_2$O$_3$ plotted with the log of P_{H_2O}/P_{H_2} against temperature. The light broken lines are partial pressure of oxygen. Total pressure is one atmosphere. [After Muan (1958).]

is variable, the conditions require isobaric invariance. Where the isobaric lines cross a single phase field they are not fixed at a single temperature, but vary with the composition of the condensed phase.

Whether crystallization in such a system takes place in the "normal" anhydrous manner or otherwise depends upon the restrictions imposed on the system. If the partial pressure of oxygen is kept at a constant value, the course of crystallization must proceed along that partial pressure line, and the bulk composition of the condensed phases is changed by gain or loss of oxygen. Consider crystallization at a fixed partial pressure of 10^{-8} atmospheres. The point A (Fig. 203) is the highest temperature at which the 10^{-8} partial pressure line exists. This means that at higher temperatures there is no composition within the system for which this partial pressure can be maintained under equilibrium conditions. Note also that only a particular composition (here pure Fe) can exist at the highest point of the partial pressure line. Therefore, when considering a cooling sequence at a fixed partial pressure, we are restricted to that particular composition (or compositions) which coincides with the temperature

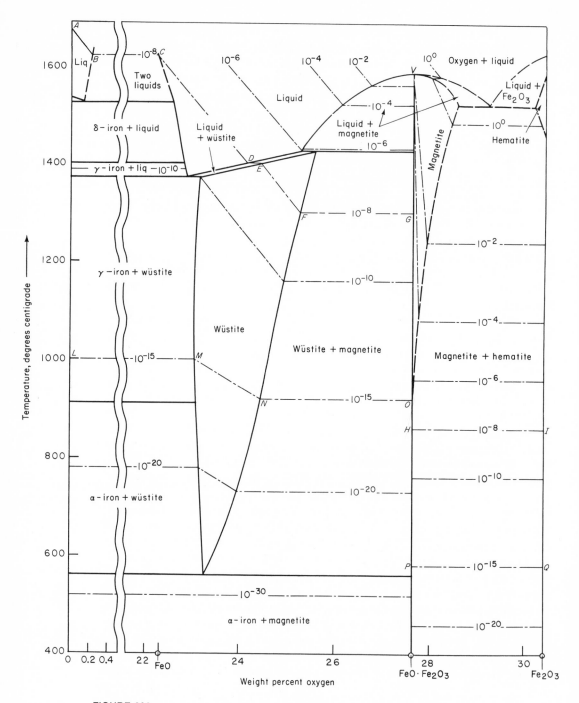

FIGURE 203

The system Fe—O. The heavy lines are boundary curves, and the light broken lines are O_2 isobars. Total pressure is 1 atmosphere. [After Darken and Gurry (1945) and Muan (1958).]

maximum of the desired partial pressure line. With removal of heat, the condensed phases so change in composition as to remain on the indicated partial pressure line. During cooling, the molten iron (A) dissolves a small percentage of oxygen and shifts in composition to point B, somewhat above 1600°C. As heat is removed from the system at this same temperature, the liquid metal dissolves considerable amounts of oxygen, forming a second liquid of composition C. The process continues isothermally until all of the liquid is of composition C, essentially liquid FeO. The relative amounts of the two liquids between B and C depend upon the amount of heat in the system and cannot be determined by examination of the diagram. Continued removal of heat causes the liquid oxide to become oxygen-rich and to migrate to D. Here, at about 1400°C, the liquid crystallizes to form wüstite of composition E. Further cooling causes the wüstite to change to composition F.* Removal of heat at this temperature allows magnetite to form at the expense of wüstite. When the wüstite is eliminated (at G), the magnetite can cool to point H, a temperature near 850°C. At H, hematite forms at the expense of magnetite until the solid consists entirely of hematite (composition I). The hematite cools to room temperature.

A similar cooling sequence further illustrates use of the diagram in Figure 203. Assume a fixed partial pressure of 10^{-15} atmospheres. The highest region of this isobar is along the horizontal line LM. If it is further stipulated that the system contains its maximum heat content, consistent with this partial pressure and temperature, the cooling sequence will begin at point L, pure γ-iron. Removal of heat at this temperature causes conversion of γ-iron to wüstite (point M); with continued cooling the wüstite shifts in composition to N. At this temperature the wüstite converts to magnetite of composition O. The magnetite cools to P, where it converts to hematite of composition Q, which remains stable to room temperature.†

If the system were closed, so that the condensed phases could neither remove nor add oxygen to the surroundings, the cooling sequence would be quite different. To trace a cooling path, one would have to assume a bulk composition of fixed amounts of Fe and O_2—for example, the composition represented by point A in Figure 204. Such a point represents a particular Fe/O ratio, since the Fe—Fe_2O_3 system is part of the larger Fe—O_2 system. With the composition fixed, any cooling path in the system can be traced by using the ordinary procedures described earlier, with the exception that the partial pressure of oxygen decreases with the temperature so as to maintain a fixed condensed phase composition. Initially the sample at A consists of liquid oxide at a particular partial pressure of oxygen (10^{-6} atmosphere in this example). The sam-

---

*The composition of wüstite, usually taken as FeO, has a somewhat variable and lower Fe/O ratio as indicated by the diagram.

†It may be noted that the ordinary presence of metallic iron at room temperature and atmospheric conditions represents a nonequilibrium situation. Given sufficient time, the iron will convert to hematite as indicated by the diagram.

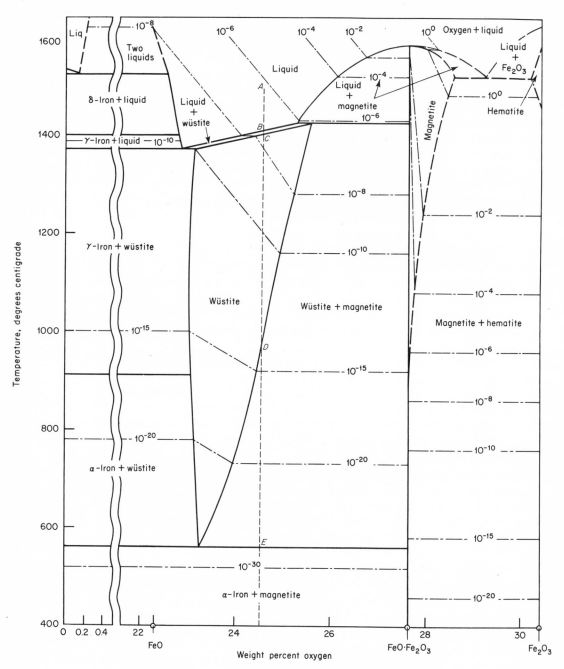

FIGURE 204
The system Fe—O, illustrating a cooling sequence under closed conditions.

ple is cooled to B at about 1400°C, where wüstite crystallizes; as the wüstite is of variable composition, the first-formed crystals are slightly richer in O_2 than the liquid (see the analogous example of plagioclase crystallization on pp. 17–18). The liquid and crystals react as they cool until the wüstite is of the same composition C as the original liquid; here the liquid is consumed, leaving only wüstite and vapor with a P_{O_2} somewhat above 10^{-8} atmosphere. With continued cooling to D, the partial pressure continues to decrease. At D, magnetite begins to form, and magnetite and wüstite coexist until the batch reaches the temperature corresponding to E. This marks the lower stability limit of wüstite; here the wüstite decomposes to form α-iron and additional magnetite. These two phases remain stable down to room temperature. Observe that the total oxygen content of the condensed phases has not changed, as the bulk composition has not shifted.

As in the binary systems, crystallization in ternary systems follows different paths, depending on whether the partial oxygen pressure or the condensed phase composition is held constant. Crystallization within the system FeO—Fe_2O_3—SiO_2 has been discussed in detail by Muan (1955), Osborn (1959), and Lindsley, Speidel, and Nafziger (1968). A brief summary of the approach will be presented here.

Figure 205 shows part of the liquidus surface of the system FeO—Fe_2O_3—SiO_2. In addition to the usual boundary curves and isotherms, lines of equal O_2 pressure are included. Figure 206 shows the compatibility relations of the various solids at liquidus temperatures. Magnetite, in addition to wüstite, shows some solid solution at high temperatures.

Consider first the case of "normal" crystallization under constant total composition of the condensed phases. The same procedures outlined earlier under ternary systems may be used to follow crystallization paths. A liquid of composition A lies above the primary field of wüstite (see enlarged area of Fig. 205 shown in Fig. 207) and lies also above a compatibility line between fayalite and wüstite of composition B. Therefore, if the composition of the condensed phases is kept constant, these two phases must be the final crystalline products. Conjugation lines, which indicate that a liquid such as C exists in equilibrium with these two phases, must be determined experimentally; these are indicated in the figure by the heavy broken lines as CB, EF, C-fayalite, E-fayalite.* During cooling, the liquid A will first precipitate wüstite of composition D. Additional precipitation of different compositions and reaction with earlier precipitated materials causes the liquid to follow a curved path to the boundary curve at E. The wüstite composition in equilibrium with liquid E is F. Along the boundary curve, fayalite precipitates with wüstite of changing composition as the liquid approaches the composition C, where it is completely consumed. As the original composition of liquid A was on the compatibility line B-fayalite, only those two phases can therefore be formed, and the liquid will not reach the ternary invariant point during crystallization.

---

*The various conjugations lines discussed have not been experimentally determined.

218

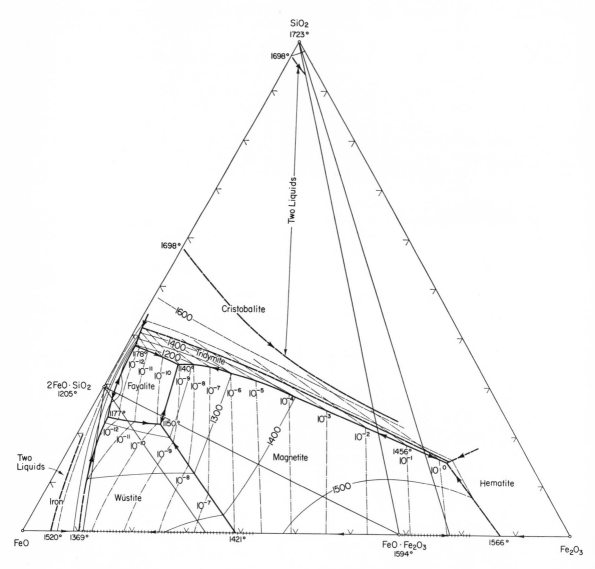

FIGURE 205
The system FeO—Fe₂O₃—SiO₂. The light broken lines indicate equal P_{O_2} (atm) of the gas phase at liquidus temperature. [After Osborn and Muan (1960d).]

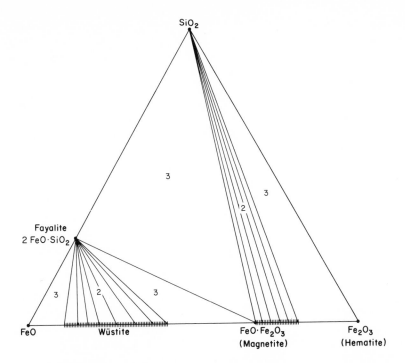

FIGURE 206
Compatibility relations of solid phases at liquidus temperatures in the system FeO—Fe$_2$O$_3$—SiO$_2$. [After Figure 205.]

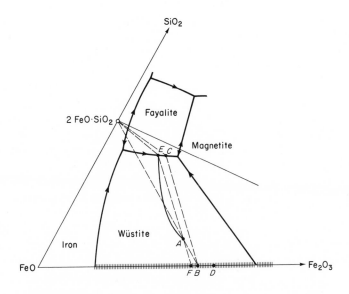

FIGURE 207
The primary field of wüstite in the system FeO—Fe$_2$O$_3$—SiO$_2$, showing a hypothetical crystallization path. [After Muan (1958).]

A liquid of composition A (Fig. 208), upon cooling under conditions of constant total composition of the condensed phases, will first precipitate fayalite, and then migrating directly away from that composition point and arrive at the boundary curve at B, where fayalite and tridymite precipitate together. The liquid then changes composition along the boundary curve until the eutectic C is reached, where magnetite crystallizes with the other two solid phases. The final condensed phase composition is determined by the initial composition, which is within the Alkemade triangle fayalite-silica-magnetite, as shown on Figure 206.

In contrast to the above examples, consider the situation in which these same two starting compositions must crystallize under a fixed partial pressure of oxygen. This requires that the condensed phases react with the surrounding atmosphere, hence the bulk composition of the condensed phases will change by either addition or release of oxygen to the surroundings. The change in the bulk composition of the condensed phases can be examined by recourse to the larger system $Fe—Si—O_2$ (Fig. 209), of which the ternary system $FeO—Fe_2O_3—SiO_2$ is but a small part. Any condensed phase composition within the system can change either toward or away from the oxygen apex. Thus a starting bulk composition A might change to B or C during cooling or heating, D might change to E or F, etc. Along each of these lines the Fe/Si ratio remains constant, with only the oxygen content changing.

The $FeO—Fe_2O_3—SiO_2$ portion of the larger $Fe—Si—O$ system is outlined in Figure 209 by broken lines. This is shown in the usual form of an equilateral triangle in Figure 210. If we wish to find a line on this diagram whose characteristic property is that it contains a fixed Fe/Si ratio and a variable oxygen content, it is only necessary to find compositions on the left and right limbs of the triangle whose Fe/Si ratios are the same and join these points with a straight line. The compositions $2FeO \cdot SiO_2$ and $Fe_2O_3 \cdot SiO_2$ are such a pair, as they both have Fe/Si contents at the same ratio of 2:1. These compositions are plotted in both mole percent and weight percent, and lines are drawn to connect corresponding compositions. Along these connecting lines, the only variable is oxygen content (which increases to the right). Use of weight percentages produces a line that is essentially parallel to the base. Consequently the $FeO—Fe_2O_3—SiO_2$ system is plotted in weight percent, as reactions in which the condensed phases yield or take up oxygen from the surroundings can be shown by migration of the condensed phase bulk composition parallel to the base of the triangle.

Such lines are referred to by Muan (1958) as *oxygen reaction lines* and by Osborn (1959) as *total composition lines*. To avoid semantic problems, such lines will here be referred to as the line X-X'. In Figure 211 a liquid of original composition A lies on the oxygen partial pressure line of 10^{-9} atmosphere. If the partial pressure of the condensed phases is kept constant, the liquid must remain on this particular partial pressure line during the entire crystallization sequence. The liquid is cooled until the liquidus surface is encountered, and wüstite begins to crystallize as the liquid migrates to lower temperatures along the 10^{-9} isobar toward B. During this time the composition of the

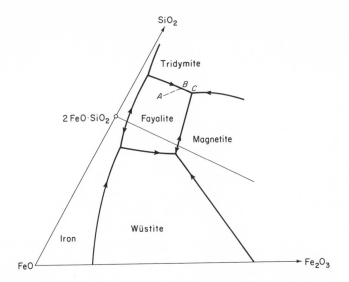

FIGURE 208
The primary field of fayalite in the system $FeO-Fe_2O_3-SiO_2$,
showing hypothetical crystallization path. [After Osborn and Muan (1960d).]

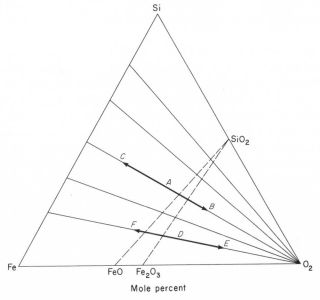

Mole percent

FIGURE 209
The triangle $FeO-Fe_2O_3-SiO_2$ as part of the larger triangle $Fe-Si-O_2$.
Change of oxygen content will cause the composition to migrate either toward or
away from the oxygen apex.

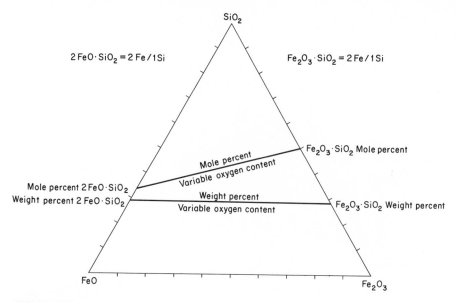

FIGURE 210

The compositions $2FeO \cdot SiO_2$ and $Fe_2O_3 \cdot SiO_2$ have the same Fe/Si ratio and differ only in oxygen content. These compositions are plotted in both mole and weight percent within the compositional triangle FeO—Fe_2O_3—SiO_2. Differences in oxygen content can be represented by a line essentially parallel to the base when compositions are plotted in weight percent.

wüstite in equilibrium with the liquid changes continuously (as could be deduced from a complete set of conjugation lines between liquid and associated wüstite). We can say at least that initially the first wüstite crystals have a composition such as C, and that as the liquid migrates to B, the coexistent wüstite compositions shift to a composition near D. The bulk composition of these two condensed phases shifts along the line X-X', as oxygen is acquired. This composition must necessarily lie between the composition of the coexisting liquid and crystals at any one time. We can therefore say that as the liquid migrates from A to B, and wüstite from C to D, the bulk composition of the condensed phases must also shift from A to E. As BD is a two-phase tie-line, the lengths BE and ED represent the percentages of wüstite D and liquid B, respectively.

As the liquid reaches the boundary curve at B, crystallization must continue along the constant O_2 pressure line. This means that the liquid must cross the boundary curve and follow the isobar (with decreasing temperature), across the primary field of magnetite to F. The reaction that takes place at B is therefore the conversion of all of the previously crystallized wüstite to magnetite.

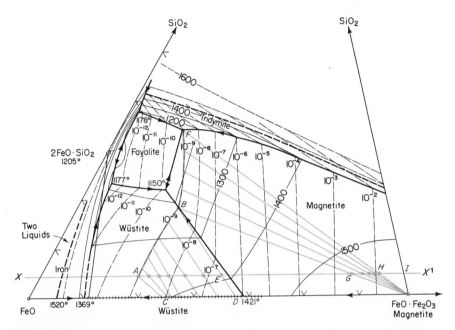

FIGURE 211
An enlarged portion of the system FeO—Fe_2O_3—SiO_2, illustrating crystallization of batch A at a fixed partial-oxygen pressure. The line X-X' has a fixed Fe/Si ratio and variable oxygen content. [After Osborn and Muan (1960d).]

This takes place not only at a fixed P_{O_2}, but also at a fixed temperature, and changes the bulk composition of the condensed phases from E to G. Observe that G not only lies on the line X-X' but is also between the composition of the two phases that are present—the liquid B and magnetite. As the liquid leaves the boundary curve at B, it changes in composition across the primary field of magnetite with continuous precipitation of magnetite.* When the liquid reaches the magnetite-tridymite boundary curve at F (with a condensed phase composition of H), these two phases precipitate until the liquid is consumed. The point F is the lowest temperature on the oxygen isobar, and therefore liquid cannot exist below this point. As these two phases crystallize, the composition of the condensed phases shifts from H to I. The final condensed phase composition must lie at I, because I lies on the line X-X', as well as between the composi-

*Although not indicated on the diagram, the magnetite becomes very slightly richer in oxygen during this migration.

tion points for magnetite and tridymite. As a result of crystallization at fixed P_{O_2}, the condensed phase composition has shifted from A to I.

For another example of crystallization under a fixed partial oxygen pressure, consider a liquid composition that is originally in the field of fayalite (Fig. 212). This composition, again called A for the sake of familiarity, is on a line of variable oxygen content (X-X') to indicate possible changes of condensed phase composition. Point A has a partial pressure of oxygen of 10^{-11} atmosphere and therefore must remain on this isobar during crystallization. As the liquid cools, fayalite precipitates and the liquid moves down the isobar to the boundary curve at B. The bulk composition of the condensed phases has meanwhile shifted along the line X-X' from A to C. Point B is the lowest liquidus temperature of the isobar, so the liquid must be consumed here. Tridymite and fayalite crystallize together until the liquid is eliminated, and the bulk composition of the condensed phases shifts to point D (on the line X-X'). During crystallization under a fixed P_{O_2}, the oxygen content of the crystallizing phases may either decrease (as it does in this example) or increase (as it did in the previous example).

In the quaternary system MgO—FeO—Fe_2O_3—SiO_2, which bears some resemblance to basaltic compositions, the course of crystallization has been demonstrated to be very dependent upon the role of oxygen. If the total composition of the condensed phases is kept constant, an iron-rich differentiate may be produced (with the degree of iron enrichment depending upon the extent of fractional crystallization); some of the iron is incorporated in the silicate phases. If crystallization occurs at constant oxygen pressure, the liquid changes composition in the direction of increasing silica and decreasing iron oxide content. This is due largely to the depletion of the magma in iron by continuous precipitation of magnetite and consequent enrichment in silica. Representation and description of this system and its relationship to natural basaltic melts are beyond the scope of this book, but are discussed in detail by Muan (1958), Osborn (1959), Yoder and Tilley (1962), and Hamilton and Anderson (1967).

Special techniques are required to produce and maintain partial oxygen pressures under hydrothermal conditions. Such techniques are necessary, as many hydrous phases contain elements capable of existing in more than one oxidation state. The most commonly used method for working under specified partial oxygen pressures is described by Darkin and Gurry (1945, 1946), Darkin (1948), and Eugster (1957, 1959). This is accomplished by keeping the condensed phases physically separated, but in chemical contact with a buffering material that will control the gaseous environment.

The buffering material contains an element that is able to exist in two oxidation states under the pressure-temperature conditions of the experiment. If the buffering material is of fixed composition and is allowed to reach equilibrium (by oxidation or reduction), it will fix the surrounding oxygen pressure.

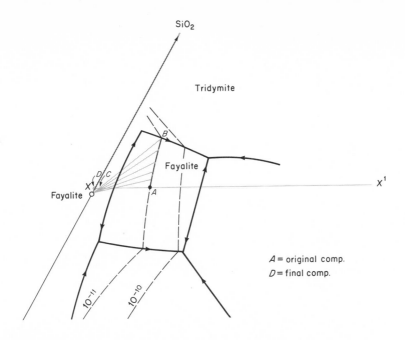

FIGURE 212
Part of the system shown in Figure 211, showing another example of crystallization at a fixed partial oxygen pressure.

The reaction $2Fe_3O_4 + \frac{1}{2}(O_2) \rightleftharpoons 3Fe_2O_3$ has an equilibrium constant for any particular pressure and temperature:

$$K_{P,T} = \frac{(Fe_3O_4)^2(O_2)^{1/2}}{(Fe_2O_3)^3}$$

Equilibrium constants have been determined for a number of such mixtures of materials, and the partial pressure of oxygen as a function of temperature is known (Fig. 213). Moreover, it has been established that the influence of total pressure on the presently used buffers is small. Curves of this type are commonly plotted as the log of the oxygen fugacity against temperature, rather than the log of the partial oxygen pressure against temperature. Fugacity is used to facilitate use of thermodynamic relationships, for which ideal gas behavior is assumed. If the real gas exhibits ideal behavior, its partial pressure and fugacity are identical. Deviations from ideality that are exhibited by real gases are taken account of by the use of fugacity (f) where

$$f = \gamma P$$

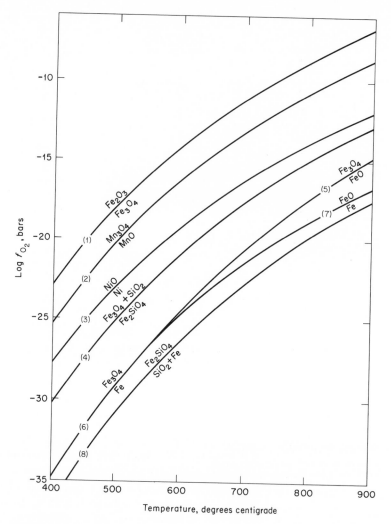

FIGURE 213
Oxygen buffer assemblages plotted as log f_{O_2} against T. [After Ernst (1968).]

Here P is the actual pressure (or partial pressure) of the gas, and γ is the fugacity coefficient. Deviations from ideality are small to negligible for the commonly used buffer mixtures throughout most of the experimental temperature range (with differences increasing at lower temperatures).

The technique used for experimental runs with buffers (Fig. 214) is to place the sample with excess water in a thin-walled, sealed metal capsule that is permeable to hydrogen. This tube is placed within a larger, sealed, thin-walled metal capsule (impermeable to hydrogen) that contains the chosen buffering assemblage and water. The particular buffer to be used will depend upon

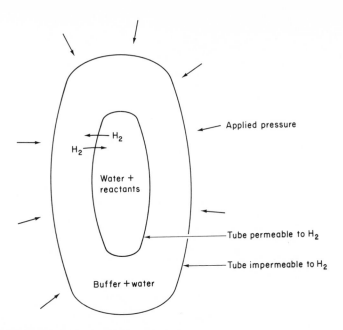

FIGURE 214
Schematic double-tube arrangement for oxygen buffering
in hydrothermal experimentation.

the choice of partial pressure of oxygen. The buffering assemblage in the outer tube is kept separate from the sample in the inner tube to prevent contamination. The double capsule is placed within a pressure vessel, where it is subjected to (water or gas) pressure and temperature. The double tubes will collapse until the internal pressure equals the applied pressure. In a short time buffering materials will come to equilibrium and establish a particular oxygen partial pressure. The water present with the buffer partially dissociates to hydrogen and oxygen; as the buffer controls the oxygen partial pressure, the hydrogen content of the gas phase is likewise controlled. This may be shown as follows. The buffering materials reached equilibrium at a particular partial pressure of oxygen according to relation

$$K_{\text{buffer}} = \frac{(Fe_3O_4)^2(O_2)^{1/2}}{(Fe_2O_3)^3} \tag{1}$$

The H_2O present partially dissociates by

$$H_2O \rightleftharpoons H_2 + \tfrac{1}{2}O_2 \tag{2}$$

$$K_{H_2O} = \frac{(H_2)(O_2)^{1/2}}{(H_2O)} \tag{3}$$

Equation (3) can be rewritten as

$$\frac{(K_{H_2O})(H_2O)}{(H_2)} = (O_2)^{1/2} \tag{4}$$

Substituting equation (4) into equation (1) gives

$$K_{buffer} = \frac{(Fe_3O_4)^2 \dfrac{(K_{H_2O})(H_2O)}{(H_2)}}{(Fe_2O_3)^3} \tag{5}$$

which can be rewritten as

$$\frac{K_{buffer}}{K_{H_2O}} = \frac{(Fe_3O_4)^2(H_2O)}{(Fe_2O_3)^3(H_2)} = K \tag{6}$$

The two equilibrium constants form a third constant K for the reaction. This reaction results in control of the hydrogen content in the outer tube. As the inner tube is permeable to hydrogen, control of hydrogen content in the outer tube leads (by diffusion) to control of hydrogen content in the sample-containing inner tube. As the degree of dissociation of water in both the inner and outer tubes must be the same (because of the ability of hydrogen to cross the interface), the partial pressure of oxygen must be identical in both inner and outer tubes. The reader is referred to more detailed discussions on this general subject by Ernst (1960) and Wones and Eugster (1965).

The use of buffered reactions is illustrated by the stability diagram for the ferruginous biotite annite $KFe_3^{+2}AlSi_3O_{10}(OH)_2$, shown in Figure 215 at a total pressure of 2070 bars. The seven buffer reactions that were used are indicated as curved diagonal broken lines. Use of these buffers limits the experiments to combinations of temperature and oxygen fugacities indicated by the positions of these lines. Thus the stability limits of annite can be determined only at points where the buffer curve and decomposition curve intersect—in this example, at points E, N, D, C, and B. If the stability of annite were independent of oxygen fugacity, the decomposition curve would be a vertical straight line of fixed temperature; in contrast, the diagram not only illustrates that the decomposition temperature of annite varies as a function of oxygen fugacity, but also indicates that annite (which contains Fe^{++} rather than Fe^{+++}) has no field of stability at high oxygen fugacities.

The manner in which such a system varies with total pressure is shown in the $P_{total} - f_{O_2} - T$ diagram in Figure 216. Total pressure increases vertically in the prism, and each of the three horizontal surfaces represents an isobaric surface, the uppermost being equivalent to Figure 215. The nearly vertical lines labeled DDD, EEE, etc. represent the various buffers that were used, and each lettered point at a line-plane intersection is an experimentally determined point (with the exception of line AAA). That these lines are not exactly vertical indicates the small change of oxygen fugacity for each of the buffer reactions with total pressure. The univariant curves of Figure 215 are shown

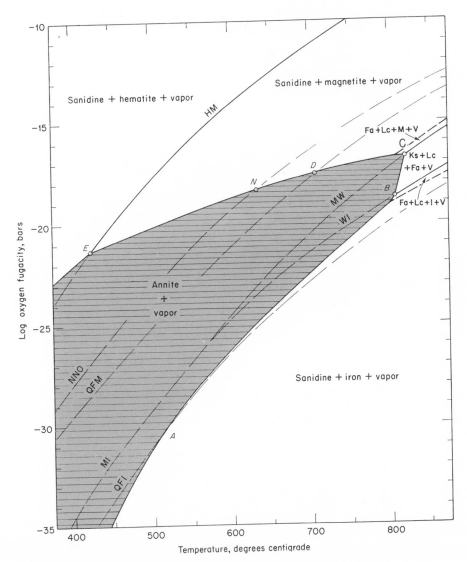

FIGURE 215
The stability relations of annite at a total pressure of 2070 bars. Broken lines
correspond to buffers in Figure 213 (Fa = fayalite, Le = Leucite, M = magnetite,
I = iron, V = vapor, Ks = kalsilite). [After Eugster and Wones (1962).]

230

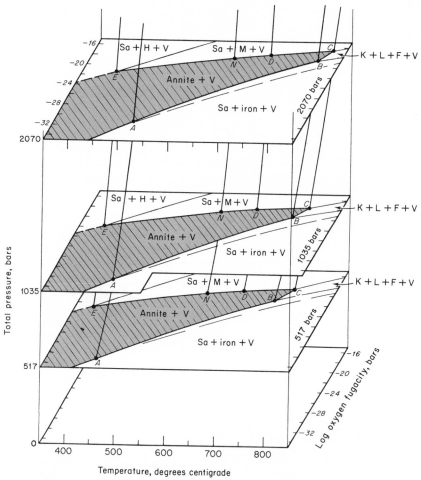

FIGURE 216
The stability relations of annite as a function of temperature, log of the oxygen fugacity, and total pressure (Sa = sanidine, H = hematite, M = magnetite, K = kalsilite, L = leucite, F = fayalite, and V = vapor). [After Eugster and Wones (1962).]

in Figure 216 as divariant surfaces. Invariant points from Figure 215 now become univariant lines. The shaded volume, which is the stability area of annite and vapor, shows increasing values of T and f_{O_2} with increasing total pressure.

Eugster (1959) has suggested that the stability regions of all iron-bearing silicates are dependent upon P_{O_2} as well as T. The fields of stability of ferric silicates would extend to higher values of P_{O_2} rather than ferrous silicates. Phase boundaries representing equilibria without transfer of oxygen must lie parallel to the P_{O_2} axis, whereas others representing oxidation-reduction reactions are inclined to the P_{O_2} axis.

11

Systems that Contain Sulfur

Sulfide equilibrium relations have been a subject of ever-increasing interest for about two decades, particularly to Kullerud and co-workers at the Geophysical Laboratory in Washington, D.C. Much of this work has been summarized by Kullerud (1964, 1967) and Barton and Skinner (1967).

Although a variety of experimental techniques have been utilized, most experiments have been accomplished in evacuated sealed silica glass tubes, and the majority of diagrams show only the condensed phases. Whether the fugacity of sulfur is or is not controlled depends upon the purpose of the investigation.

The Fe—S system (Kullerud, 1967) has been examined in detail by a number of investigators, and is shown in Figure 217. Pyrrhotite ($Fe_{1-x}S$) varies in composition from 50 to 44 mole percent iron. It melts congruently and exists in eutectic melting relationship with iron. Pyrite (FeS_2) melts incongruently at 743°C to form pyrrhotite and an almost pure sulfur liquid. Below 400°C the relationships become more complicated due to the polymorphism of pyrrhotite and the formation of additional compounds. This region has been in dispute for a number of years, primarily because of problems posed by the slow ex-

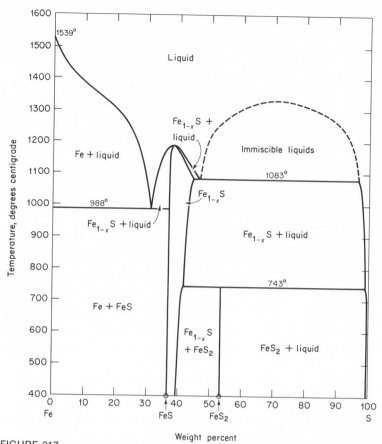

FIGURE 217

The condensed system Fe—S above 400°C (FeS$_2$ = pyrite, Fe$_{1-x}$S = pyrrhotite, FeS = troilite). [After Kullerud (1967).]

perimental reaction rates. The latest diagram of the low-temperature region is shown in Figure 218 (Taylor, 1970).

The system ZnS—FeS, originally described by Kullerud (1953), has been in some dispute in recent years. It is known that Zn does not substitute for Fe in the pyrrhotite structure, but substitution of Fe for Zn in the sphalerite structure is considerable, and increases with temperature. The limits of Fe substitution are shown in Figure 219. At higher temperatures, sphalerite converts to wurtzite; as the nature of intermediate polytypes are not yet well known, this region of the diagram is left vague.

The Fe—Zn—S system at 325°C (Fig. 220) is taken from Barton and Toulmin (1966). Although the central triangle in the diagram is an isothermal section, the attached areas are boundary systems that have been "unfolded" to show phase relations to higher temperatures. They can be readily understood in spite of being rotated to a high angle or turned upside down (as for

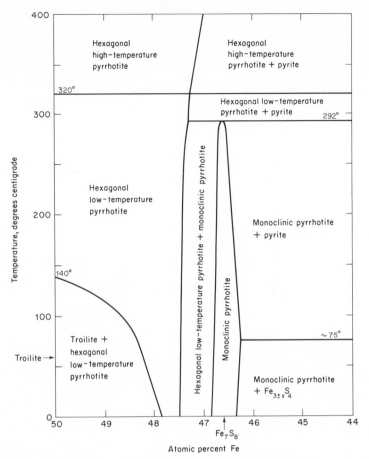

FIGURE 218
Low-temperature phase relations in a part of the system Fe—S. [After Taylor (1970).]

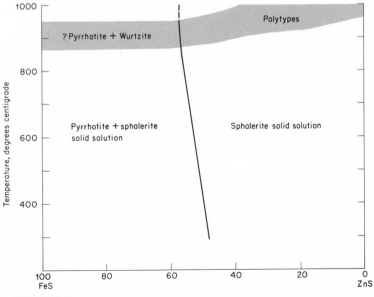

FIGURE 219
Phase relations in the system FeS—ZnS. [After Barton and Toulmin (1966).]

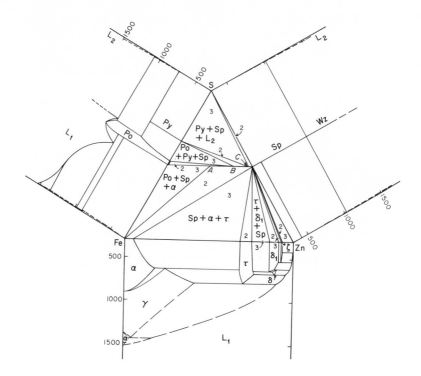

FIGURE 220
The system Fe—Zn—S at 325°C. The *TX* sections of the boundary systems are folded out
from the ternary system. All assemblages have vapor present (Sp = sphalerite, Wz = wurtzite,
Po = pyrrhotite, Py = pyrite, L_1 = metal-rich to sulfide-rich liquid, L_2 = liquid sulfur, and
α, γ, Γ, δ, δ_1, and ζ are zinc-iron phases). Points *A*, *B*, and *C* correspond to similarly labeled
points and lines in Figures 221, 222, and 224. [After Barton and Toulmin (1966).]

Fe—Zn). The true phase diagram buff enjoys studying them almost regardless
of their angle of orientation. Sphalerite and wurtzite are seen to exist stably
with essentially pure S or Zn on the Zn—S binary. Iron, which exists in several
polymorphs, forms intermediate compounds with zinc. Within the ternary
system, the compositional limit of sphalerite extends to point *A*. Various
compositions of sphalerite can exist with all other phases in the system; this
property causes the Fe—Zn—S compositional triangle to be broken into
three-phase triangles and two-phase areas, which have been numbered as
such in Figure 220. A composition within a three-phase triangle consists of
a mixture of the three phases indicated at the apices of the triangle. A composition within a two-phase area consists of two phases whose compositions
lie at the edges of this area; these could be indicated at the ends of experimentally determined tie lines (as in an earlier system shown in Fig. 200, A).

The effect of temperature on this system (Fig. 221) is indicated by a series
of isothermal sections. The limit of solid solution of Fe in sphalerite (indicated

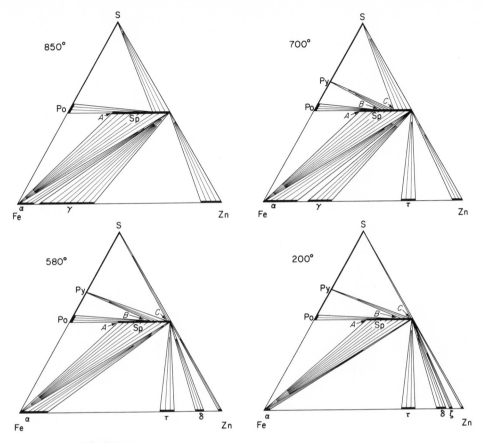

FIGURE 221
Isothermal sections of the condensed phases in the system Fe—Zn—S.
[After Barton and Toulmin (1966).]

by point *A*) increases with temperature, but changes very little. The phases
in equilibrium with various compositions of sphalerite change considerably
with temperature. The variation of point *B*—the composition of sphalerite
(*B*) that exist with both pyrite and pyrrhotite—has been of interest to geo-
thermometry. Barton and Toulmin (1966) show variation of this point as curve
B in the 600°-700°C region (Fig. 222). For the region below this temperature,
two interpretations exist—that of Boorman (1967) and that of Chernshev
and Anfilogov (1968). The data of Chernshev and Anfilogov are incorporated
in Figure 221 at 200°C, as well as in Figure 222. Figure 222 shows phase
relations in the FeS—ZnS—S system projected on the FeS—ZnS boundary.
Also indicated are the various sulfur-rich compounds that exist with different
compositions of sphalerite. Lines *A*, *B*, and *C* correspond to the points *A*, *B*,
and *C* in Figure 221. At higher temperatures, the sphalerite that coexists
with pyrite and pyrrhotite is progressively enriched in zinc.

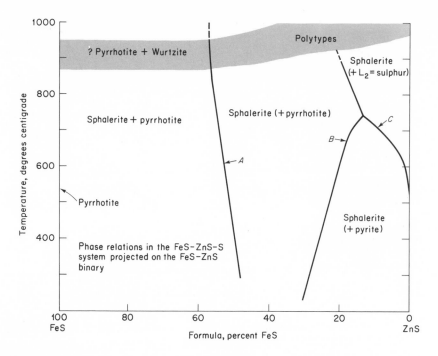

FIGURE 222
Sphalerite compositions in equilibrium with other phases (in parenthesis) in the Fe—Zn—S system. Sphalerite compositions extend from ZnS to line A. Compositions richer in iron (to the left of A) consist of pyrrhotite and sphalerite of composition A. [After Barton and Toulmin (1966) and Chernshev and Anfiligov (1968).]

Examination of Figure 221 shows a swarm of two-phase tie-lines connecting pyrrhotites and sphalerites of various compositions. The most Fe-rich sphalerites, A, exist with the most Fe-rich pyrrhotite, whereas S-rich pyrrhotites are stable with more Zn-rich sphalerites, B. A very small change in the sulfur content of a sample whose composition is between that of pyrrhotite and sphalerite causes a large change in the Fe content of the sphalerite. The compositions of pyrrhotite and sphalerite are therefore dependent upon the partial pressure of sulfur in the system, as is indicated in Figures 223 and 224, in which T is plotted versus $\log f_{s_2}$. For example, in Figure 223, at a temperature of 600°C, a very low sulfur fugacity allows iron to coexist with pyrrhotite of 50 atomic percent iron. Increase of the sulfur fugacity at this temperature eliminates the iron phase and permits only an Fe-rich pyrrhotite. Further increase yields pyrrhotite progressively poorer in Fe. At still higher values pyrrhotite is eliminated in favor of first pyrite, and finally of liquid sulfur. The pinching out of the field of pyrite and sulfur vapor with an increase of temperature above 743°C reflects the elimination of pyrite as a stable phase above this temperature.

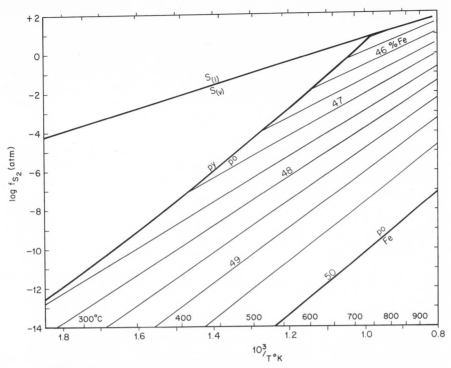

FIGURE 223
Pyrrhotite compositions (in atomic percent Fe) as a function of S_2 fugacity and temperature ($S_{(l)}$ = sulfur liquid, $S_{(v)}$ = sulfur vapor, py = pyrite, po = pyrrhotite). [After Barton and Toulmin (1966).]

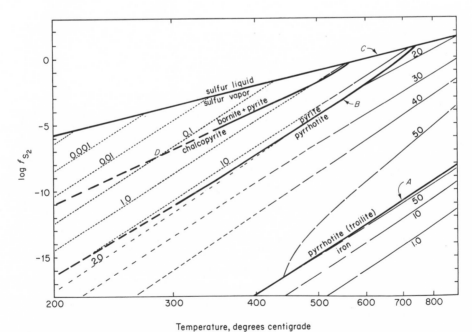

FIGURE 224
Sphalerite compositional variation (in mole percent FeS) in equilibrium with pyrite, pyrrhotite, or iron, as a function of temperature and sulfur fugacity. [After Barton and Toulmin (1966). The chalcopyrite, bornite, pyrite curve is from Barton and Toulmin (1964).]

The significant dependence of sphalerite composition upon temperature and sulfur fugacity is shown in Figure 224. The heavy boundary curves labeled *A*, *B*, and *C* correspond to similarly labeled points in Figures 221 and 222. Intersections of univariant curves representing different mineral assemblages (such as bornite + pyrite = chalcopyrite) may be plotted on Figure 224 and correlated with particular sphalerite compositions to yield a temperature of formation of the total assemblage (assuming no solid solution effects between the sphalerite and the different mineral phases). Thus an assemblage (*D*) of bornite, pyrite, chalcopyrite, and sphalerite containing 0.1 mole percent FeS can exist in equilibrium only at a temperature near 285°C.

12

Systems with Two Volatiles

As experimental techniques have increased in sophistication over the years, systems containing more than a single volatile component have come to be studied. A recent investigation by Naldrett (1969) on parts of the Fe—S—O system is extremely useful in explaining some aspects of the crystallization of sulfide-oxide ore magmas. Naldrett has shown that the composition of a sulfide-oxide melt is dependent on the sulfur and oxygen fugacities of the associated silicate magma before separation, and virtually independent of these fugacities after separation.

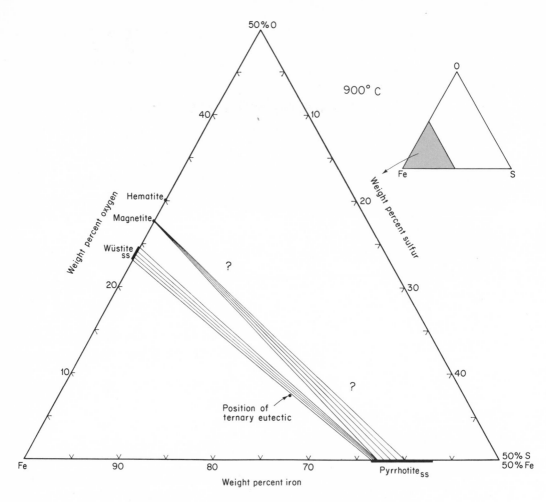

FIGURE 225
Solid phases in equilibrium with vapor in a part of
the system Fe—S—O at 900°C. [After Naldrett (1969).]

The solidus relations of the system Fe—S—O have been discussed by Kullerud (1967). The relations pertinent to the present discussion are reproduced from Naldrett (1969) and shown in Figure 225 for a temperature of 900°C. Various compositions of pyrrhotite exist in equilibrium with either magnetite, or wüstite of variable composition, or with both. Iron-rich wüstite and Fe-rich pyrrhotite form a three-phase compatibility triangle with iron. Liquidus relations for a part of this system are shown in Figure 226. The region of liquid immiscibility between Fe and FeO (seen in the binary Fe—Fe$_2$O$_3$ in Fig. 203) extends well into the ternary system toward FeS. The boundary curves be-

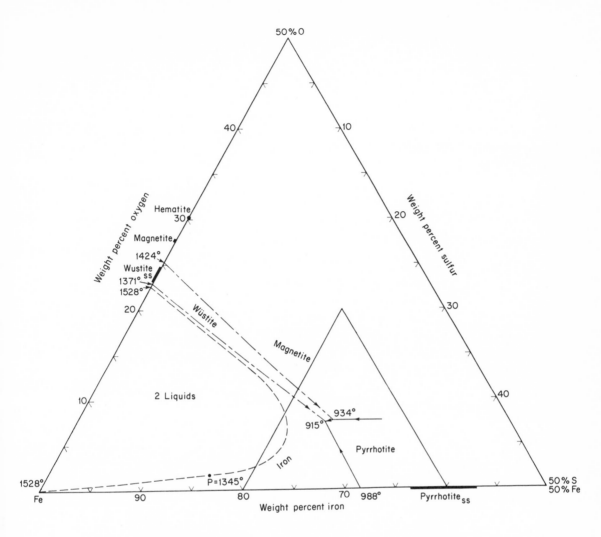

FIGURE 226
Liquidus relations in the Fe–rich portion of the system Fe—S—O.
[After Naldrett (1969).]

tween magnetite, wüstite and pyrrhotite intersect as a tributary reaction point, and those between wüstite, iron, and pyrrhotite at a ternary eutectic. In an enlarged view of this portion of the system (Fig. 227), swarms of dashed tie-lines are indicated between the various compositions of pyrrhotite that crystallize with magnetite, wüstite, or both. The four lines labelled *a, b, c,* and *d* show pyrrhotite compositions that crystallize with magnetite (in the presence of liquid and vapor), whereas the other three, *e, f,* and *g* (which are inclined at a slightly different angle) indicate coexistence of pyrrhotite, wüstite, liquid,

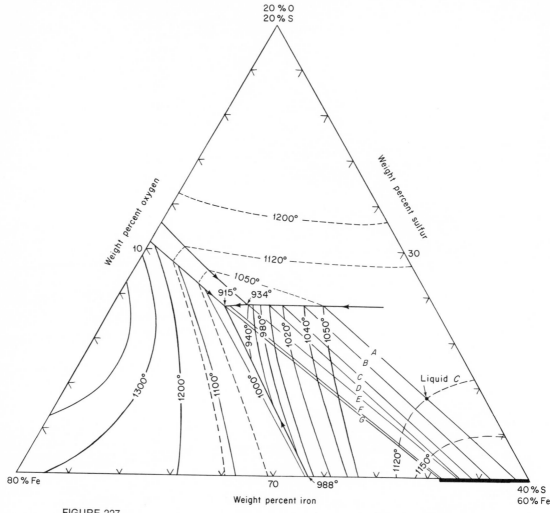

FIGURE 227

Liquidus relations of that part of the Fe—S—O system shown in the small triangle in Figure 226. Diagonal lines *A–G* are Alkemade lines between various compositions of pyrrhotite, wüstite, and magnetite. [After Naldrett (1969).]

and vapor. Isotherms shown as solid lines are taken from Naldrett (1969), and those that are speculative have been added as broken lines.

The vapor that coexists with the condensed phases contains both S_2 and O_2. As the liquids and solids in this system contain both sulfur and oxygen, there must be a relationship between the composition of the condensed phases and the fugacities of both sulfur and oxygen in the coexisting vapor. This is brought out by the 1200°C isothermal section shown in Figure 228. This section covers the same region as Figure 227; the 1200°C isotherm is marked as both a solid line, as well as a line with question marks. A large central

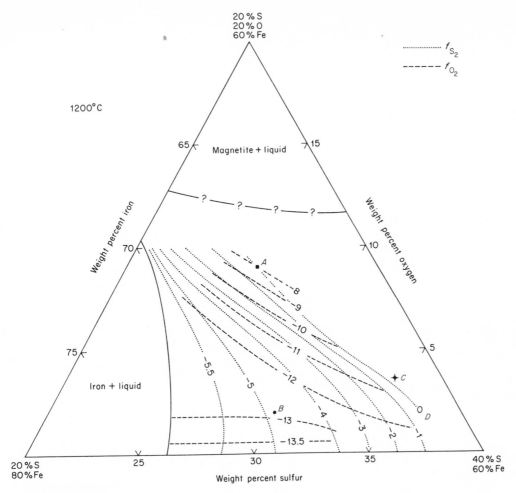

FIGURE 228
Isothermal section (1200°C) of a part of the Fe—S—O system, showing variations in $\log f_{S_2}$ and $\log f_{O_2}$. [After Naldrett (1969).]

liquid region is present at this temperature; any composition within this region consists of liquid and vapor. Two smaller three-phase regions are present within which liquid and vapor coexist with either iron or magnetite. Sulfur and oxygen fugacity isobars are indicated over much of the liquid region. Any specific liquid composition is characterized by a particular combination of sulfur and oxygen fugacities. For point A, for example, $f_{O_2} = 10^{-8}$ bars and $f_{S_2} = 1$ bar; logarithms of these values are plotted in the diagram, where $\log f_{O_2} = -8$ and $\log f_{S_2} = 0$. If the liquid composition is changed, reaction with the vapor phase will cause (under equilibrium conditions) a change in the fugaci-

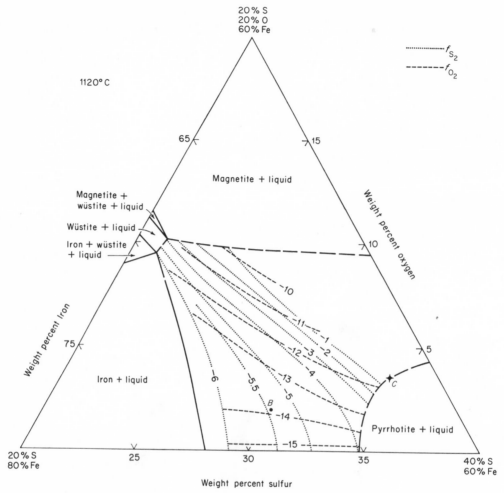

FIGURE 229
Isothermal section (1120°C) of a part of the Fe—S—O system, showing variations in log f_{S_2} and log f_{O_2}. [After Naldrett (1969).]

ties, and vice versa. Isothermal sections at 1120 and 1050°C are shown in Figures 229 and 230; these, along with the isothermal section of Figure 228 are outlined as isotherms in Figure 227. Note carefully that the fugacity isobars in these diagrams are for the vapor phase at a fixed temperature, and cannot be treated in the same way as the isobars in Figure 205 for the system FeO—Fe$_2$O$_3$—SiO$_2$, for those isobars are drawn on the liquidus surfaces and differ in temperature within different parts of the system. In the present system the isobars in each of the three isothermal sections are limited to the particular temperature of that section. The difference can be brought out by considering the same liquid composition in the three sections. The liquid B

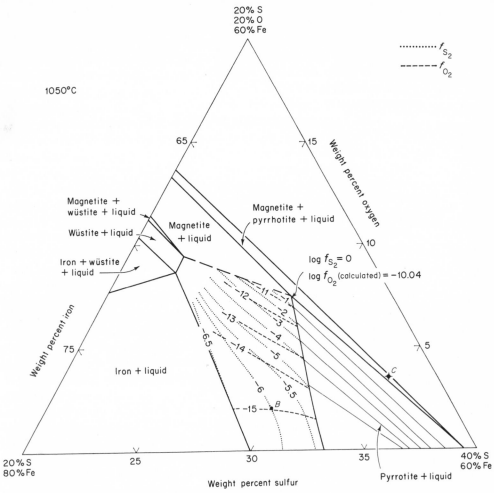

FIGURE 230
Isothermal section (1050°C) of a part of the Fe—S—O system, showing variations in log f_{S_2} and log f_{O_2}. [After Naldrett (1969).]

coexists with vapor of different fugacities at different temperatures, as shown in the table.

Figure	Temp.	Log f_{O_2}	Log f_{S_2}
228	1200	~ −13	~ −5.0
229	1120	~ −14	~ −5.5
230	1050	−15	−6.0

In the previously discussed $FeO—Fe_2O_3—SiO_2$ system (Fig. 205), a particular liquid composition is considered to be on the liquidus surface; such positions fix not only the temperature but the fugacity.

Crystallization of a melt within the system Fe—S—O can take place in a variety of ways, depending upon whether the condensed phases are fixed or variable in composition. Consider first a liquid C (Fig. 227) that crystallizes under conditions of constant composition of the condensed phases, the sort of "normal" crystallization that has been described throughout most of this book. Since the liquid C lies over the two-phase tie-line (a) between magnetite and pyrrhotite, we can conclude that these will be the two solid phases that must form with equilibrium crystallization. At 1200°C (Fig. 228), C consists of liquid coexisting with a vapor phase, where $\log f_{O_2} = 10.5$ and $\log f_{S_2}$ is near 0. With cooling to 1120°C (Figs. 227 and 229), the pyrrhotite liquidus is encountered and pyrrhotite begins crystallization; the fugacities of the vapor phase is given by $\log v_{O_2} = -11.5$ and $\log f_{S_2} = 0$. As the pyrrhotite shows some solid solution, the liquid will follow a curved path to the magnetite-pyrrhotite subtraction curve. The pyrrhotite in equilibrium with the liquid during this stage becomes somewhat richer in Fe as crystallization proceeds. The liquid becomes enriched in oxygen due to depletion of pyrrhotite. When the subtraction curve is reached, both magnetite and pyrrhotite (of changing composition) precipitate as the liquid migrates in the direction of the peritectic. When the magnetite and pyrrhotite compositions fall on the two-phase tie-line that includes the original composition C, the last drop of liquid is consumed. Figure 230 shows the situation just before final solidification.

A quite different cooling sequence occurs if the oxygen and sulfur fugacities of the vapor in contact with the melt are fixed. Assuming a temperature of 1200°C (Fig. 228), $\log f_{S_2} = 0$ and $\log f_{O_2} = -11.5$, and the melt associated with the vapor phase is at D. Cooling to 1120°C with maintenance of the sulfur and oxygen fugacities at the stated values, causes a change in liquid composition to C (Fig. 229), where pyrrhotite begins to crystallize. As this is the only point in the system where liquid and pyrrhotite can exist with a vapor whose fugacities are given by $\log f_{S_2} = 0$ and $\log f_{O_2} = -11.5$, crystallization must be completed at this point and at this temperature. As point C lies over the primary field of pyrrhotite, the liquid crystallizes isothermally to produce only pyrrhotite. As the liquid must maintain its composition at point C during this process, crystallization of pyrrhotite is accompanied by the loss of oxygen from the liquid; this restricts the constantly decreasing amount of liquid to this composition until crystallization is complete. During crystallization the total composition of the condensed phases (liquid and solid) shifts from C to the composition of pyrrhotite being precipitated. Completion of crystallization yields pyrrhotite. The remaining portion of the original bulk composition C has gone into the vapor phase.

A further study on the effect of SiO_2 on the system $FeS—FeO—Fe_3O_4$ at liquidus temperatures has recently been described by MacLean (1969), who discusses the relationships between sulfide liquids and silicate magmas.

Because of the common association of carbonate rocks and sulfide ores, Kullerud (1967) examined the effects of sulfur on various carbonates. Carbonates and sulfur were reacted in evacuated silica tubes in the temperature range from 100° to 300°C. This low temperature was used to ensure lack of thermal decomposition of the carbonate minerals, as well as to simulate ore-forming temperatures. Siderite ($FeCO_3$) and sulfur fall within the quaternary system $Fe—S—C—O$ (Fig. 231). Siderite and sulfur mixtures react either completely or partially to form pyrite (FeS_2) and a gas phase containing CO_2 and SO_2 (in a 2:1 ratio). The various products and reactants are shown on the plane $Fe—S—$"CO_3" (Fig. 232), which has been extracted from the quaternary system (Fig. 231). Because of the compatibility arrangement, which is

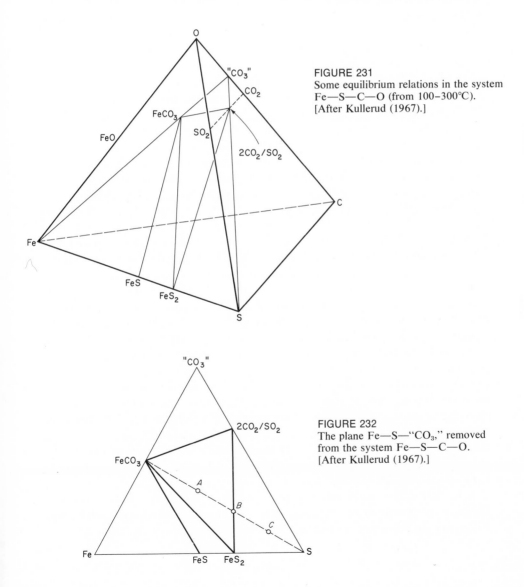

FIGURE 231
Some equilibrium relations in the system $Fe—S—C—O$ (from 100–300°C). [After Kullerud (1967).]

FIGURE 232
The plane $Fe—S—$"CO_3," removed from the system $Fe—S—C—O$. [After Kullerud (1967).]

indicated in both diagrams, different ratios of $FeCO_3$ to S yield the following data

Composition	Products
A	$FeCO_3$, FeS_2, $2CO_2/SO_2$
B	FeS_2, $2CO_2/SO_2$
C	FeS_2, $2CO_2/SO_2$, S

The reaction is

$$2FeCO_3 + 5S \leftrightarrow 2FeS_2 + 2CO_2 + SO_2$$

A similar reaction occurs with cerussite ($PbCO_3$) and sulfur, leading to the formation of galena (PbS), CO_2, and SO_2 according to the reaction

$$2PbCO_3 + 3S \rightleftharpoons 2PbS + 2CO_2 + SO_2$$

The result of the reactions involving both $FeCO_3$ and $PbCO_3$ with S are shown on the tetrahedron Fe—Pb—S—"CO_3" (Fig. 233), which expresses the compatibility relations to at least 400°C. It can be seen that cerussite and pyrrhotite are incompatible, and when mixed in a 1:1 ratio will react to form siderite and galena. Similarly, reaction of cerussite and pyrite can yield siderite, galena, and a gas phase.

Interest in carbonatite magmas and the metamorphism of carbonate-rich sediments has led to the study of a number of systems that contain both CO_2 and H_2O. One of the most significant of these is the system CaO—CO_2—H_2O, investigated by Wyllie and Tuttle (1960b). The boundary systems CaO—H_2O and CaO—CO_2 were discussed earlier and presented in Figures 157 and 193. The remaining boundary system, between H_2O and CO_2, consists of homogeneous vapor. The compounds portlandite ($Ca(OH)_2$) and calcite ($CaCO_3$) exist within the boundary systems; due to the incongruent melting of calcite to form liquid and vapor (at the pressure of 1000 bars used in this study) this forms a pseudo-binary join (Fig. 234). The compositions of the liquid and vapor indicated in Figure 234 are seen in Figure 193 (the binary CaO—CO_2). A field of calcite II is indicated above 945°C; this is a nonquenchable phase, which undergoes only a small structural change, and does not affect the liquidus surface significantly.

The nature of the ternary system can be brought out by considering an isobaric TX prism. Figure 235 shows part of such a prism. The triangular compositional base is labeled as CaO—CO_2—H_2O. The surface containing CO_2—H_2O versus temperature is visualized as being in front, with the other two boundary systems containing CaO extending toward the back. Only the two binary systems, CaO—H_2O and CaO—CO_2, have been plotted and

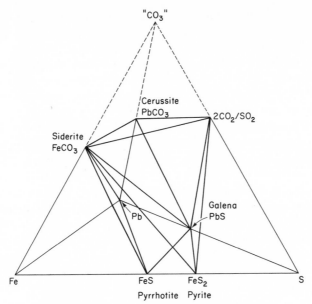

FIGURE 233
Phase relations in the system Fe—Pb—S—"CO₃,"
below about 400°C. [After Kullerud (1967).]

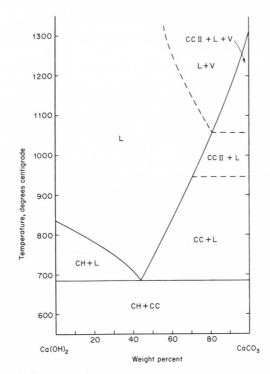

FIGURE 234
The pseudo-binary system CaCO₃—Ca(OH)₂ at 1000 bars (CC =
CaCO₃, CH = Ca(OH)₂, L = liquid, V = vapor). [After Wyllie and
Tuttle (1960b).]

labeled. In the higher-temperature parts of each of these binary systems are areas of liquid + CaO, liquid, liquid + vapor, and vapor. These same regions extend continuously from one CaO-containing binary system, across the ternary, to the opposite CaO-containing binary system. This is emphasized in Figure 236. The vapor region is present as a thin vertical strip along the front of the TX prism, (indicating that very minor amounts of CaO are soluble in vapor of any composition). The lines HF and GF drawn on the back surface of the vapor area indicate the vapor compositions that coexist with various minimum temperature liquids within the system. The region of liquid is present in the shaded region. Between this almost vertical prism of liquid and the thin strip of vapor in the front lies a large area of coexisting liquid and vapor. In Figure 237 a series of two-phase liquid-vapor tie-lines have been drawn to show the composition of vapors existing in equilibrium with minimum-temperature liquids along the boundary curves DE_1 and AE_1, at the base of the liquid field. Corresponding liquid-vapor compositions can be more easily determined by projecting both the boundary curves DE_1 and AE_1, and the coexistent vapor curves GF and HF onto the same two-dimensional coordinate system. This has been done in the perspective TX section (Fig. 238), where both sets of curves have been projected from the CaO temperature axis to the front surface of the TX prism (namely the H_2O—CO_2 temperature surface). The curves HF and GF correspond to vapor compositions, and DE_1 and AE_1 are liquids that exist along the liquid boundary curves. At any chosen temperature, corresponding liquid-vapor compositions can be determined. Thus at 1000°C a liquid containing CO_2 and H_2O in a ratio of 75:25 coexists with a vapor containing CO_2 and H_2O in a ratio of about 20:80.

The system can now be examined in more detail by considering the usual paths of crystallization. Figure 239 shows the isobaric compositional triangle with primary fields and boundary curves outlined. The primary fields labeled CaO, $Ca(OH)_2$, and $CaCO_3$ are "normal" in that they represent thermal surfaces along which these solids may crystallize from a melt. The lower, and largest, field represents the surface of the liquid + vapor region. The swarm of lines through this region are two-phase tie-lines between the coexisting vapor and minimum-temperature liquids; they are the same as the two-phase lines seen in Figure 237. The boundary curves DE_1 and AE_1 are unique, as they indicate liquid compositions that are compatible with a solid and vapor (rather than the usual type of boundary curve such as E_1E_2, CE_2, E_2B which reflect equilibrium between a liquid and two solids).

Figure 240 is an enlarged and distorted portion of the same diagram. Use of Alkemade's rule and the tangent rule allows labeling of the boundary curves as indicated. A liquid composition K, overlying the primary field of $CaCO_3$ cools to precipitate $CaCO_3$. The liquid changes in composition directly away from $CaCO_3$, and encounters the boundary curve between E_1 and E_2 at the saddle point L, where $CaCO_3$ and $Ca(OH)_2$ simultaneously precipitate at a fixed temperature until the liquid is consumed. A different liquid, of bulk composition M, cools to precipitate $CaCO_3$ while changing in composition to the boundary curve at N, where CaO and $CaCO_3$ precipitate together as the

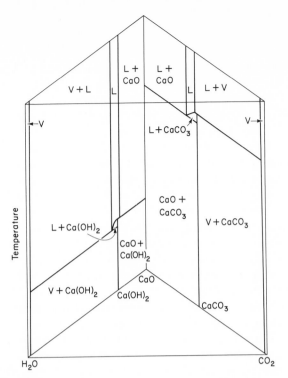

FIGURE 235
Stability relations of the binary systems CaO—H₂O and CaO—CO₂ on the schematic *TX* prism for the system CaO—CO₂—H₂O at 1000 bars. [After Wyllie and Tuttle (1960b).]

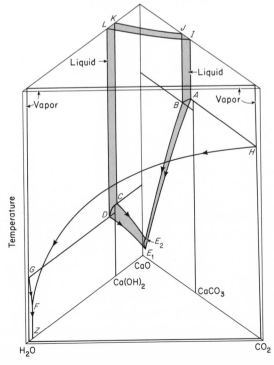

FIGURE 236
The addition of liquid-field volume *LKJIABE₁E₂CD* to the *TX* prism of Figure 235. Minimum-temperature CaO-poor liquids (seen on lines *DE₁* and *E₁A*) can coexist with vapor (seen on lines *GF* and *FH*), respectively.

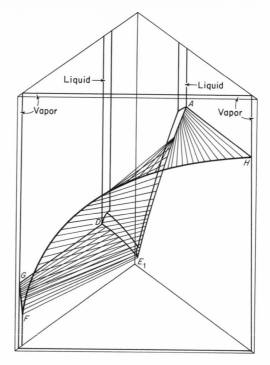

FIGURE 237
The addition of tie lines to Figure 236, showing the
coexistence of liquids DE_1A with vapors GFH.

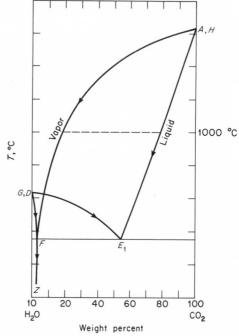

FIGURE 238
Projection of coexistent liquid and vapor compositions from the
CaO edge of the TX prism to the front CO_2-H_2O surface. [After
Wyllie and Tuttle (1960b).]

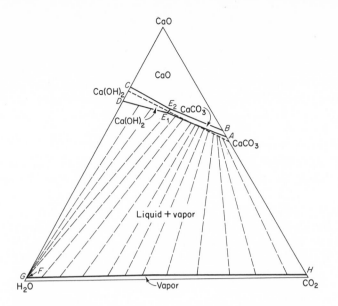

FIGURE 239
Equilibrium diagram for the system CaO—CO₂—H₂O at 1000 bars.
The fields labeled CaO, Ca(OH)₂, and CaCO₃ are liquidus surfaces;
the larger field shows coexisting liquid-vapor compositions. [After
Wyllie and Tuttle (1960b).]

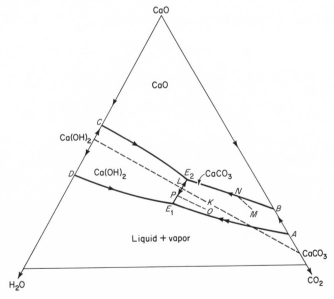

FIGURE 240
The CaO-rich (somewhat distorted) part of Figure 239, showing
labeled boundary curves and paths of crystallization. [After
Wyllie and Tuttle (1960b).]

liquid migrates along the boundary curve to E_2. At the ternary eutectic E_2 the three phases CaO, Ca(OH)$_2$, and CaCO$_3$ precipitate at a fixed temperature until the liquid is consumed. A liquid of bulk composition O cools to crystallize CaCO$_3$ as the liquid changes in composition to P. At P, Ca(OH)$_2$ and CaCO$_3$ simultaneously precipitate as the liquid moves to E_1. At E_1, not only do CaCO$_3$ and Ca(OH)$_2$ continue to precipitate, but they are joined by a vapor phase as well. The simultaneous boiling and crystallization of the melt continues at this temperature until the liquid is eliminated. The final products are CaCO$_3$, Ca(OH)$_2$, and a vapor consisting of essentially pure H$_2$O. The vapor composition, which may be found by reference to the liquid-vapor conjugation line drawn from E_1 in Figure 239, is represented by the point F.

Figure 241, which shows the entire ternary system, illustrates the course of crystallization of a bulk composition Q. First note that Q is within the Alkemade triangle Ca(OH)$_2$—CaCO$_3$—vapor F, indicating that these are the final products. If the composition Q had fallen on the CO$_2$ side of the CaCO$_3$-vapor

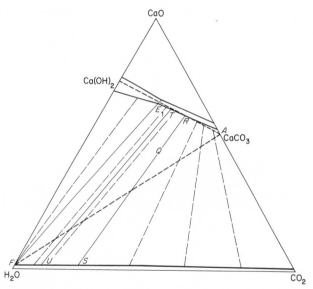

FIGURE 241
Crystallization path of liquid-vapor Q in the CaO-rich portion of the system CaO—CO$_2$—H$_2$O.

F line, then only CaCO$_3$ and a vapor would be the final products. At very high temperatures Q is a vapor (by analogy with the system MgO—CO$_2$ in Fig. 192). Although no information exists on liquid-vapor compositions at temperatures above the boundary curve, it can be assumed that cooling of vapor will produce, in addition, a liquid phase; the liquid and vapor will continuously diverge in composition until the liquid reaches the boundary curve at R. The conjugation lines indicate that the associated vapor is to be found at S. The boundary curve at R (line AE_1) is a reaction curve, along which the liquid precipitates calcite while simultaneously dissolving some of the associated vapor (analogous to the manner in which a congruently melting compound is lost by reaction with liquid to form an incongruently melting compound along a "normal" reaction curve). As the vapor constantly changes in composition, the liquid shifts down the reaction curve with additional precipitation of calcite (i.e., liquid T and vapor U). When the liquid reaches E_1, it is consumed at a fixed temperature while $CaCO_3$ and $Ca(OH)_2$ precipitate simultaneously and a small amount of boiling produces additional vapor of composition F. It is necessary for this final boiling to occur, as the final liquid E_1 does not lie on the compositional line between $CaCO_3$ and $Ca(OH)_2$ and therefore must yield a third phase in addition to the two crystalline phases.

Another method of illustrating phase relationships in this system is by means of isothermal isobaric sections (Fig. 242), as was done by Wyllie and Tuttle (1960b). Cooling or heating reactions can be considered for any composition by plotting a fixed composition in all of the sections and noting the phase assemblages indicated about that point (as was done earlier for the carbonate system shown in Fig. 200).

The ternary system $MgO—CO_2—H_2O$ was examined by Walter, Wyllie, and Tuttle (1962). The boundary system $MgO—CO_2$ has been discussed earlier and is illustrated in Figure 192. The binary $MgO—H_2O$ (Fig. 243) is similar in having a vapor phase, intermediate compound, and restricted field of liquid. Both of these systems (within the pressure range investigated) are different from the $CaO—H_2O$ and $CaO—CO_2$ systems (Fig. 235) in that the intermediate compounds $Mg(OH)_2$ and $MgCO_3$ dissociate to an oxide (MgO) and vapor well below temperatures at which liquids are encountered. This is illustrated in the schematic isobaric TX prism for the ternary system (Fig. 244); a large region of MgO + vapor occupies much of the central portion of the diagram. Liquid is encountered at higher temperatures. The liquid compositions make up a continuous wedge-shaped volume extending between the binary systems $MgO—CO_2$ and $MgO—H_2O$. Minimum liquid temperatures drop from points A and B on the boundary systems to a low point within the ternary at C. Such minimum temperature liquids (ACB) can coexist with MgO and vapor; the vapor compositions that exist with these liquids are given by the curve DEF. As the slope of the vapor curve DEF is different than that of the liquid curve ACB, it follows that liquids with a particular H_2O/CO_2 ratio will exist with vapors of a different H_2O/CO_2 ratio. Put another way, we can say that intermediate vertical TX sections drawn from the MgO temperature axis are not binary slices. This can be demonstrated (Fig. 245) by projecting both the liquid ACB and vapor DEF curves onto the front surface of the TX prism (as was done in Fig. 238 for the $CaO—CO_2—H_2O$ system) or by viewing both curves down the temperature axis on the triangular compositional base (Fig. 246). The lettering in these figures corresponds to that in Fig. 244. Only one of the two-phase liquid-vapor tie-lines (CE), correspond-

256

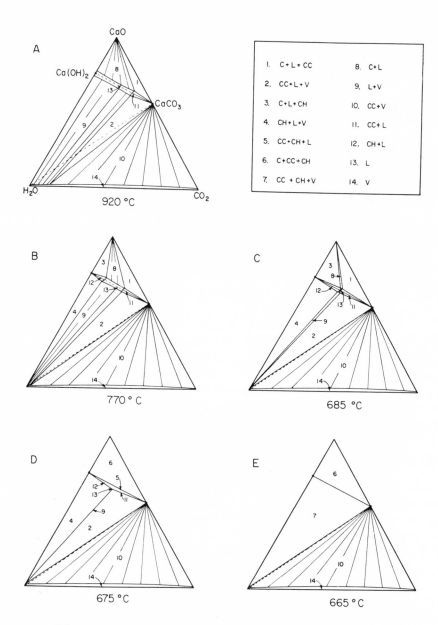

FIGURE 242,A,B,C,D,E
Isobaric isothermal planes for the system CaO—CO$_2$—H$_2$O at 1000
bars. [After Wyllie and Tuttle (1960b).]

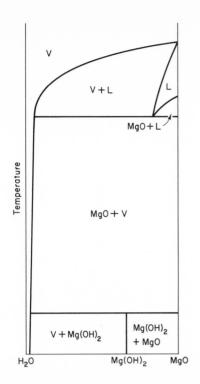

FIGURE 243
Schematic equilibrium relations in the system
MgO—H₂O. [After Walter, Wyllie and Tuttle
(1962).]

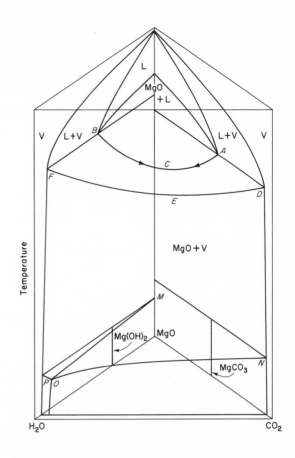

FIGURE 244
The TX prism for the MgO—CO₂—H₂O up to 4 kb. The curve
ACB shows minimum-temperature liquids that coexist with vapor
DEF. [After Walter, Wyllie, and Tuttle (1962).]

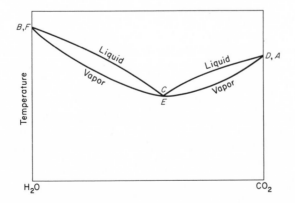

FIGURE 245
Coexistent liquid-vapor compositions projected to the CO₂—H₂O
side of the TX prism for the system MgO—CO₂—H₂O. [After
Walter, Wyllie, and Tuttle (1962).]

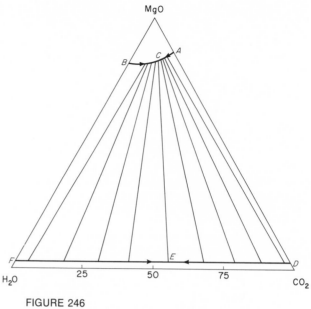

FIGURE 246
Liquid-vapor tie lines in the system MgO—CO₂—H₂O.

ing to the minimum temperature liquid composition C, can be projected (Fig. 246) back through the MgO corner (thus indicating binary conditions for this temperature and composition). All other such liquid-vapor tie-lines within the ternary system have different CO_2/H_2O ratios and therefore cannot be extended through the MgO apex.

As the temperature is lowered through the field of MgO and vapor, first the stability field of $MgCO_3$ is encountered (temperature n in Fig. 244), followed by $Mg(OH)_2$ (at temperature P). The stability relations of these two compounds at 1000 bars are given in Figure 247. Note that the composition of the vapor phase determines both the initial crystallization temperature and the phase compositions during cooling. Ternary relations can best be illustrated by a series of isothermal isobaric sections. Cooling of MgO and CO_2-rich vapor to a temperature somewhat below n (Fig. 244) allows $MgCO_3$ to exist with a variety of vapor compositions, ranging from almost pure CO_2 to a composition such as V_1 (seen in Fig. 248). Vapor compositions existing with MgO extend from almost pure H_2O to V_1. Coexistent vapors and solids are shown by swarms of two-phase tie-lines. Vapor V_1 can exist with both MgO and $MgCO_3$ as indicated by the three-phase triangle. Cooling to lower temperatures allows the range of vapor compositions coexisting with $MgCO_3$ to increase (see V_1, Fig. 249). The variable compositions of V_1 are located along the line n-o in Figure 244. The lower temperature of Figure 250 shows that the stability field of $Mg(OH)_2$ has been encountered. This compound can exist with limited H_2O-rich vapors (to V_2) and results in the presence of a second three-phase triangle within which MgO, $Mg(OH)_2$, and vapor V_2 are compatible.

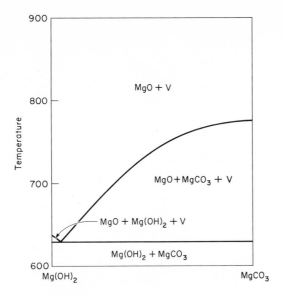

FIGURE 247
The system $MgCO_3$—$Mg(OH)_2$ at 1000 bars.
[After Walter, Wyllie and Tuttle (1962).]

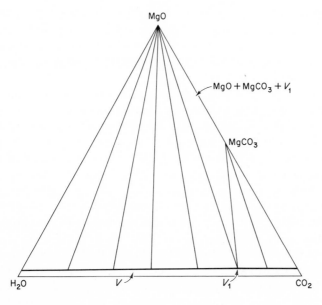

FIGURE 248
Schematic isothermal section at 1 kb, just below temperature n in
Figure 244.

260

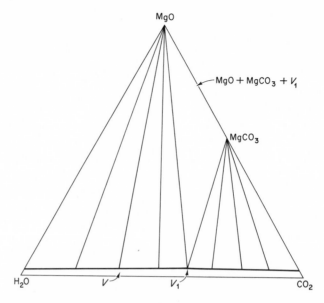

FIGURE 249
Schematic isothermal section at 1 kb, at a temperature lower than depicted in Figure 248, and somewhat above the temperature p of Figure 244.

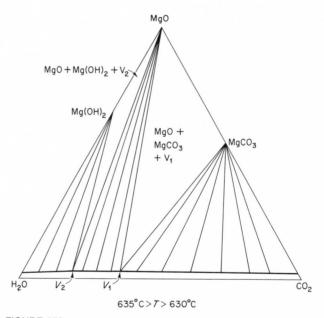

$635°C > T > 630°C$

FIGURE 250
Schematic isothermal section at 1 kb at temperatures between 630 and 635°C. [After Walter, Wyllie, and Tuttle (1962).]

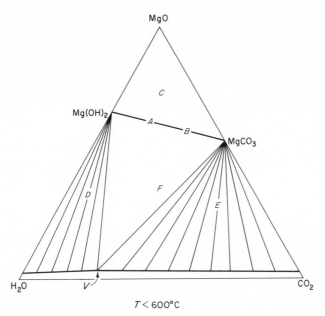

FIGURE 251
Schematic isothermal section at 1 kb at temperatures below 600°C.
[After Walter, Wyllie, and Tuttle (1962).]

Additional cooling (below point O in Fig. 244) permits $MgCO_3$ and $Mg(OH)_2$ to exist together without either excess vapor or MgO (as at points A or B in Fig. 251). The presence of this $Mg(OH)_2$—$MgCO_3$ Alkemade line eliminates the possibility of MgO existing in the presence of vapor. Addition of vapor to MgO will cause the formation of some $Mg(OH)_2$ and $MgCO_3$ (for a bulk composition such as C); addition of larger amounts of vapor to MgO (such as D, E, or F) causes elimination of MgO to form either $Mg(OH)_2$, $MgCO_3$ or both, coexisting with vapor.

Related studies that are of interest but beyond the scope of this book are the system CaO—MgO—CO_2—H_2O (Wyllie, 1965) and CaO—SiO_2—CO_2—H_2O (Wyllie and Haas, 1965, 1966).

Bibliography

Andersen, Olaf (1915), The system anorthite-forsterite-silica, *Am. J. Sci.*, 4th Ser., Vol. 39, pp. 407–454.

Barth, T. F. W. (1951), The feldspar geologic thermometer. *Neues Jahrb. Min. Abh.* Vol. 82, pp. 143–154.

Barth, T. F. W. (1968), Additional data for the two-feldspar geothermometer, *Lithos*, Vol. 1, pp. 305–306.

Barton, Paul B., Jr., and Brian J. Skinner (1967), Sulfide mineral stabilities, in Hubert Lloyd Barnes (ed.), *Geochemistry of hydrothermal ore deposits*, Holt, Rinehart and Winston, Inc., New York, pp. 236–333.

Barton, Paul B., Jr., and Pristley Toulmin, III (1964), Experimental determination of the reaction chalcopyrite + sulfur = pyrite + bornite from 350° to 500°C, *Econ. Geology*, Vol. 59, pp. 747–752.

Barton, Paul B., Jr., and Priestley Toulmin, III (1966), Phase relations involving sphalerite in the Fe—Zn—S system, *Econ. Geology*, Vol. 61, pp. 815–849.

Bell, Peter M., and Eugene H. Roseboom (1969), Melting relationships of jadeite and albite to 45 kilobars with comments on the melting behavior of binary systems at high pressures, in J. J. Papike (ed.), *Pyroxenes and amphiboles: crystal chemistry and phase petrology*, Mineralogical Soc. America Special Paper No. 2, pp. 151–161.

Bergman, A. G., and E. P. Dergunov (1941), The fusion diagram of the system LiF—NaF—MgF$_2$, *Compt. rend. Acad. Sci. URSS*, Vol. 31, pp. 755–756.

Berman, R., and Francis E. Simon (1955), The graphite-diamond equilibrium, *Z. Elektrochem.*, Vol. 59, pp. 333–338.

Boettcher, A. L., and P. J. Wyllie (1967), Hydrothermal melting curves in silicate water systems at pressures greater than 10 kilobars, *Nature*, Vol. 215, pp. 572–573.

Boettcher, A. L., and P. J. Wyllie (1969), Phase relationships in the system $NaAlSiO_4$—SiO_2—H_2O to 35 kilobars pressure, *Am. Jour. Sci.,* Vol. 267, pp. 875–909.

Boorman, R. S. (1967), Subsolidus studies in the ZnS—FeS—FeS_2 system, *Econ. Geology*, Vol. 62, pp. 614–631.

Bowen, N. L. (1913), The melting phenomena of the plagioclase feldspars, *Am. Jour. Sci.*, Vol. 35, 4th Series, pp. 577–599.

Bowen, N. L. (1915), The crystallization of haplobasaltic, haplodioritic, and related magmas, *Am. Jour. Sci.*, Vol. 40, 4th Series, pp. 161–185.

Bowen, N. L. (1928), *The evolution of the igneous rocks*, Princeton Univ. Press, Princeton, New Jersey. [Reprinted by Dover Publications, New York (1956).]

Bowen, N. L., and Olaf Andersen (1914), The binary system MgO—SiO_2, *Am. Jour. Sci.*, 4th Series, Vol. 37, pp. 487–500.

Bowen, N. L., and O. F. Tuttle (1949), The system MgO—SiO_2—H_2O, *Geol. Soc. America Bull.*, Vol. 60, pp. 439–460.

Boyd, F. R., and J. L. England (1963), Some effects of pressure on phase relations in the system MgO—Al_2O_3—SiO_2, Annual Rpt. Dir. Geophys. Lab., *Carnegie Inst. Wash. Year Book 62*, pp. 121–124.

Boyd, F. R., and J. L. England (1965), The rhombic enstatite-clinoenstatite inversion, Annual Rpt. Dir. Geophys. Lab., *Carnegie Inst. Wash. Year Book 64*, pp. 117–120.

Boyd, F. R., and I. D. MacGregor (1964), Ultramafic rocks. Ann. Rpt. Dir. Geophys. Lab., *Carnegie Inst. Wash. Year Book 63*, pp. 152–156.

Bridgman, P. W. (1912), Water in the liquid and five solid forms, under pressure, *Proc. Am. Acad. Arts Sci.*, Vol. 47, pp. 441–558.

Bridgman, P. W. (1958), *The physics of high pressure*, G. Bell and Sons, Ltd., London.

Buckley, H. E. (1951), *Crystal growth*, John Wiley and Sons, New York.

Bundy, F. P. (1962), Melting point of graphite at high pressure: heat of fusion, *Science*, Vol. 137, pp. 1055–1057.

Bundy, F. P., H. P. Bovenkerk, H. M. Strong, and R. H. Wentorf, Jr. (1961), Diamond-graphite equilibrium line from growth and graphitization of diamond, *Jour. Chem. Physics*, Vol. 35, pp. 383–391.

Bundy, F. P., H. T. Hall, H. M. Strong, and R. H. Wentorf, Jr. (1955), Man-made diamonds, *Nature*, Vol. 176, pp. 51–55.

Chernyshev, L. V., and V. N. Anfilogov (1968), Subsolidus phase relations in the ZnS—FeS—FeS_2 system, *Econ. Geology*, Vol. 63, pp. 841–847.

Clark, Sidney P., Jr., J. F. Schairer, and John de Neufville (1962), Phase relations in the system $CaMgSi_2O_6$—$CaAl_2SiO_6$—SiO_2 at low and high pressure, Ann. Rpt. Dir. Geophys. Lab., *Carnegie Inst. Wash. Year Book 61*, pp. 59–68.

Dachille, Frank, and Rustum Roy (1960), High pressure studies of the system Mg_2GeO_4—Mg_2SiO_4 with special reference to the olivine-spinel transition, *Am. Jour. Sci.*, Vol. 258, pp. 225–246.

Darken, L. S. (1948), Melting points of iron oxides on silica; phase equilibria in the system Fe—Si—O as a function of gas composition and temperature, *Jour. Am. Chem. Soc.*, Vol. 70, pp. 2046–2053.

Darken, L. S., and R. W. Gurry (1945), The system iron-oxygen. I. The wüstite field and related equilibria, *Jour. Am. Chem. Soc.*, Vol. 67, pp. 1398–1412.

Darken, L. S., and R. W. Gurry (1946), The system iron-oxygen. II. Equilibria and thermodynamics of liquid oxides and other phases. *Jour. Am. Chem. Soc.*, Vol. 68, pp. 798–816.

Eitel, Wilhelm (1951), *Silicate melt equilibria*, Rutgers Univ. Press, New Brunswick, New Jersey.

Eitel, Wilhelm (1965), *Silicate Science*. Vol. III. *Dry silicate systems*, Academic Press, New York.

Eitel, Wilhelm (1966), *Silicate Science*. Vol. IV. *Hydrothermal silicate systems*, Academic Press, New York.

Ernst, W. G. (1960), Stability relations of magnesioriebeckite, *Geochim. Cosmichim. Acta*, Vol. 19, pp. 10–40.

Ernst, W. G. (1968), *Amphiboles*, Springer-Verlag, New York, p. 125.

Eugster, Hans P. (1957), Heterogeneous reactions involving oxidation and reduction at high pressure, *Jour. Chem. Phys.*, Vol. 26, pp. 1760–1761.

Eugster, Hans P. (1959), Reduction and oxidation in metamorphism, *in* P. H. Abelson (ed.), *Researches in geochemistry*, John Wiley and Sons, New York, pp. 397–426.

Eugster, Hans P., and D. R. Wones (1962), Stability relations of the ferruginous biotite, annite, *Jour. Petrology*, Vol. 3, pp. 82–125.

Fawcett, J. J., and H. S. Yoder, Jr. (1966), Phase Relations of chlorites in the system $MgO-Al_2O_3-SiO_2-H_2O$, *Am. Mineralogist*, Vol. 51, pp. 333–380.

Franco, R. R., and J. F. Schairer (1951), Liquidus temperatures in mixtures of the feldspars of soda, potash, and lime, *Jour. Geology*, Vol. 59, pp. 259–267.

Fyfe, W. H. (1960), Hydrothermal synthesis and determination of equilibrium between minerals in the subliquidus region, *Jour. Geology*, Vol. 68, pp. 553–566.

Garrels, Robert M., and Charles L. Christ (1965), *Solutions, Minerals and Equilibria*, Harper and Row, New York.

Gentile, Anthony L., and Wilfrid R. Foster (1963), Calcium hexaluminate and its stability relations in the system $CaO-Al_2O_3-SiO_2$, *Jour. Am. Cer. Soc.*, Vol. 46, pp. 74–76.

Gibbs, J. Willard (1961), *The scientific papers of J. Willard Gibbs* (2 vols.), Dover Publications, New York.

Goldsmith, Julian R., and Donald L. Graf (1957), The system $CaO-MnO-CO_2$: solid solution and decomposition relations, *Geochim. Cosmochim. Acta*, Vol. 11, pp. 310–334.

Goldsmith, Julian R., and Donald L. Graf (1960), Subsolidus relations in the system $CaCO_3-MgCO_3-MnCO_3$, *Jour. Geology*, Vol. 68, pp. 324–335.

Goldsmith, Julian R., and Hugh C. Heard (1961), Subsolidus phase relations in the system $CaCO_3-MgCO_3$, *Jour. Geology*, Vol. 69, pp 45–74.

Graf, Donald L., and Julian R. Goldsmith (1955), Dolomite-magnesian calcite relations at elevated temperatures and CO_2 pressures, *Geochim. Cosmochim. Acta*, Vol. 7, pp. 109–128.

Green, D. H., and A. E. Ringwood (1967), An experimental investigation of the gabbro to eclogite transformation and its petrological applications, *Geochim. Cosmochim. Acta*, Vol. 31, pp. 767–833.

Greenwood, H. J. (1961), The system $NaAlSi_2O_6-H_2O-Argon$: total pressure and water pressure in metamorphism, *Jour. Geophys. Res.*, Vol. 66, pp. 3923–3946.

Greenwood, H. J. (1963), The synthesis and stability of anthophyllite, *Jour. Petrology*, Vol. 4, pp. 317–351.

Greig, J. W. (1927), Immiscibility in silicate melts, *Am. Jour. Sci.*, 5th Ser., Vol. 13, pp. 1–44.

Greig, J. W., and T. F. W. Barth (1938), The system $Na_2O \cdot Al_2O_3 \cdot 2SiO_2$ (nepheline, carnegieite) – $Na_2O \cdot Al_2O_3 \cdot 6SiO_2$ (albite), *Am. Jour. Sci.*, v. 35A, pp. 93–112.

Hamilton, D. L., and G. M. Anderson (1967), Effects of water and oxygen pressures on the crystallization of basaltic magmas, *in* H. H. Hess and Arie Poldervaart (eds), *Basalts*, Interscience, New York, pp. 445–482.

Harker, R. I., and O. F. Tuttle (1955a), Studies in the system $CaO—MgO—CO_2$. I. The thermal dissociation of calcite, dolomite, and magnesite, *Am. Jour. Sci.*, Vol. 253, pp. 209–224.

Harker, R. I., and O. F. Tuttle (1955b), Studies in the system $CaO—MgO—CO_2$. II. Limits of solid solution along the binary join, $CaCO_3—MgCO_3$, *Am. Jour. Sci.*, Vol. 253, pp. 274–282.

Herold, Paul G., and W. J. Smothers (1954), Solid state equilibrium relations in the system $MgO—Al_2O_3—SiO_2—ZrO_2$, *Jour. Am. Cer. Soc.*, Vol. 37, pp. 351–353.

Hill, Robin, E. T., and A. L. Boettcher (1970), Water in the earth's mantle: melting curves of basalt-water and basalt-water-carbon dioxide, *Science*, Vol. 167, pp. 980–981.

Kennedy, George C. (1961), Large Pressure Units, *in* F. P. Bundy, W. R. Hibbard, Jr., and H. M. Strong (eds.), *Progress in very high pressure research*, John Wiley and Sons, New York, p. 314.

Kennedy, George C., and William T. Holser (1966), Pressure-volume-temperature and phase relations of water and carbon dioxide, *in* Sidney P. Clark, Jr. (ed.), *Handbook of physical constants*, G.S.A. Memoir 97, pp. 371–383.

Kennedy, George C., G. J. Wasserburg, H. C. Heard, and R. C. Newton (1962), The upper three-phase region in the system $SiO_2—H_2O$, *Am. Jour. Sci.*, Vol. 260, pp. 501–521.

Kitahara, S., S. Takenouchi, and G. C. Kennedy (1966), Phase relations in the system $MgO—SiO_2—H_2O$ at high temperatures and pressures, *Am. Jour. Sci.*, Vol. 264, pp. 223–233.

Korzhinskii, D. S. (1959), *Physiochemical basis of the analysis of the paragenesis of minerals*, Consultants Bureau, New York.

Kullerud, G. (1953), The FeS—ZnS system, a geological thermometer, *Norsk. Geol. Tidsskr.*, Vol. 32, pp. 61–147.

Kullerud, G. (1964), Review and evaluation of recent research on geologically significant sulfide-type systems, *Fort. der Min.*, Vol. 41, pp. 221–270.

Kullerud, G. (1967), Sulfide Studies, *in* P. H. Abelson (ed.), *Researches in geochemistry*, John Wiley and Sons, New York, pp. 286–321.

Kushiro, Ikuo (1968), Composition of magmas formed by partial zone melting of the earth's upper mantle, *Jour. Geophys. Res.*, Vol. 73, pp. 619–634.

Levin, Ernest M., Carl R. Robbins, and Howard F. McMurdie (1964), *Phase diagrams for ceramists*, The American Ceramic Society, Columbus, Ohio.

Levin, Ernest M., Carl R. Robbins, and Howard F. McMurdie (1969), *Phase diagrams for ceramists, 1969 supplement*, The American Ceramic Society, Columbus Ohio.

Lindsley, Donald H. (1968), Melting relations of plagioclase at high pressure, New York State Museum and Science Service Memoir 18, pp. 39–46.

Lindsley, Donald H., and R. F. Emslie (1968), Effect of pressure on the diopside curve in the system diopside-albite-anorthite, Ann. Rpt. Dir. Geophys. Lab., *Carnegie Inst. Wash. Year Book 66*, pp. 479–480.

Lindsley, Donald H., D. H. Speidel, and R. H. Nafziger (1968), $P—T—f_{O_2}$ relations for the system $Fe—O—SiO_2$, *Am. Jour. Sci.*, Vol. 266, pp. 342–360.

Luth, William C., Richard H. Jahns, and O. Frank Tuttle (1964), The granite system at pressures of 4 to 10 kilobars, *Jour. Geophys. Res.,* Vol. 69, pp. 759–773.

McLachlan, Dan, Jr., and Ernest G. Ehlers (1971), Effect of pressure on the melting temperature of metals, *Jour. Geophys. Res.* Vol. 76, pp. 2780–2789.

MacLean, Wallace M. (1969), Liquidus phase relations in the FeS—FeO—Fe_3O_4—SiO_2 system and their application in geology, *Econ. Geology,* Vol. 64, pp. 865–884.

Mahan, Bruce H. (1964), *Elementary chemical thermodynamics,* W. A. Benjamin, New York.

Muan, Arnulf (1955), Phase equilibria in the system FeO—Fe_2O_3—SiO_2, *Jour. Metals, AIME Transactions,* Vol. 203, pp. 965–967.

Muan, Arnulf (1958), Phase equilibria at high temperatures in oxide systems involving changes in oxidation states, *Am. Jour. Sci.,* Vol. 256, pp. 171–207.

Morse, S. A. (1970), Alkali feldspars with water at 5 kb. pressure, *Jour. Petrology,* Vol. 11, pp. 221–253.

Naldrett, A. J. (1969), A portion of the system Fe—S—O between 900 and 1080°C and its application to sulfide ore magmas, *Jour. Petrology,* Vol. 10, pp. 171–201.

Newton, M. S., and G. C. Kennedy (1968), Jadeite, analcite, nepheline, and albite at high temperatures and pressures, *Am. Jour. Sci.,* Vol. 266, pp. 728–734.

Newton, Robert C. (1966), Kyanite-sillimanite equilibrium at 750°C, *Science,* Vol. 151, pp. 1222–1225.

Newton, Robert C., and J. V. Smith (1967), Investigation concerning the breakdown of albite at depth in the earth, *Jour. Geology,* Vol. 75, pp. 268–286.

Niggli, Paul (1954), Rocks and mineral deposits, W. H. Freeman and Company, San Francisco.

Nishioka, Uasburo (1935), The equilibrium diagram of the ternary system $CaO \cdot TiO_2 \cdot SiO_2$—$CaO \cdot SiO_2$—$CaO \cdot Al_2O_3 \cdot 2SiO_2$, *Kinzoku-no-Kenkyu* (Jour. Research Inst. for Metals), Vol. 12, pp. 449–58.

O'Hara, M. J. (1963), The join diopside-pyrope at 30 kilobars, Ann. Rpts. Dir. Geophys. Lab., *Carnegie Inst. Wash. Year Book 62,* pp. 116–118.

O'Hara, M. J. (1968), The bearing of phase equilibria studies in synthetic and natural systems on the origin and evolution of basic and ultrabasic rocks, *Earth Science Reviews,* Vol. 4, pp. 69–133.

O'Hara, M. J., and H. S. Yoder, Jr. (1967), Formation and fractionation of basic magmas at high pressures, *Scottish Jour. Geology,* Vol. 3, Part I, pp. 67–117.

Osborn, E. F. (1959), Role of oxygen pressure in the crystallization and differentiation of basaltic magma, *Am. Jour. Sci.,* Vol. 257, pp. 609–647.

Osborn, E. F., and Arnulf Muan (1960a), Phase Equilibrium Diagrams of Oxide Systems. The system "FeO"—Al_2O_3—SiO_2. Plate 9, The American Ceramic Society and Edward Orton Jr. Ceramic Foundation, Columbus, Ohio.

Osborn, E. F., and Arnulf Muan (1960b), Phase Equilibrium Diagrams of Oxide Systems. The system MgO—Al_2O_3—SiO_2. Plate 3. American Ceramic Society and the Edward Orton Jr. Ceramic Foundation, Columbus, Ohio.

Osborn, E. F., and Arnulf Muan (1960c), *Phase equilibrium diagrams of oxide systems. The system CaO—Al_2O_3—SiO_2.* Plate 1. The American Ceramic Society and the Edward Orton Jr. Ceramic Foundation, Columbus, Ohio.

Osborn, E. F., and Arnulf Muan (1960d), *Phase equilibrium diagrams of oxide systems. The system FeO—Fe_2O_3—SiO_2.* Plate 6. The American Ceramic Society and the Edward Orton Jr. Ceramic Foundation, Columbus, Ohio.

Osborn, E. F., and J. F. Schairer (1941), The ternary system pseudo-wollastonite-akermanite-gehlenite, *Am. Jour. Sci.,* Vol. 239, pp. 715–763.

Osborn, E. F., and D. B. Tait (1952), The system diopside-forsterite-anorthite, *Am. Jour. Sci.,* Bowen Volume, pp. 413–433.

Ostrovsky, I. A. (1966), PT-diagram of the system SiO_2—H_2O, *Geol. Jour.,* Vol. 5, Pt. 1, pp. 127–134.

Ostrovsky, I. A. (1967), On some sources of errors in phase-equilibria investigations at ultra-high pressure; phase diagram of silica, *Geol. Jour.,* Vol. 5, Pt. 2, pp. 321–328.

Perel'man, Fanya Moiseerna (1966), *Phase diagrams of multicomponent systems, Geometric methods,* Consultants Bureau, New York.

Phillips, Bert, and Arnulf Muan (1959), Phase equilibria in the system CaO-iron oxide-SiO_2 in air, *Jour. Am. Cer. Soc.,* Vol. 42, pp. 413–423.

Prince, A. T. (1943), The system albite-anorthite-sphene, *Jour. Geology,* Vol. 51, 1–16.

Rankin, G. A., and Fred E. Wright (1915), The ternary system CaO—Al_2O_3—SiO_2; with optical study, *Am. Jour. Sci.,* 4th Ser., Vol. 39, pp. 1–79.

Ricci, John E. (1951, 1966), *The phase rule and heterogeneous equilibrium,* D. Van Nostrand, Inc., New York, Dover Publications, New York.

Ringwood, A. E. (1969), Phase transformations in the mantle, *Earth and Planetary Science Letters* (5), pp. 401–412.

Ringwood, A. E. (1970), Phase transformations and the constitution of the mantle, *Phys. Earth Planetary Interiors,* 3, pp. 109–155.

Ringwood, A. E., and D. H. Green (1966), An experimental investigation of the gabbro-eclogite transformation and some geophysical implications, *Tectonophysics,* Vol. 3, pp. 383–427.

Ringwood, A. E., I. D. MacGregor, and F. R. Boyd (1964), Petrologic constitution of the upper mantle, Ann. Rpt. Dir. Geophys. Lab., *Carnegie Inst. Wash. Year Book 63,* p. 150.

Robertson, E. C., F. Birch, and G. J. F. MacDonald (1957), Experimental determination of jadeite stability relations to 25,000 bars, *Am. Jour. Sci.,* Vol. 255, pp. 115–137.

Roy, Della M., and Rustum Roy (1955), Synthesis and stability of minerals in the system MgO—Al_2O_3—SiO_2—H_2O, *Am. Mineralogist,* Vol. 40, pp. 147–178.

Schairer, J. F. (1950), The alkali-feldspar join in the system $NaAlSiO_4$—$KAlSiO_4$—SiO_2, *Jour. Geology* (Univ. Chicago Press), Vol. 58, pp. 512–517.

Schairer, J. F., and N. L. Bowen (1947), Melting relations in the systems Na_2O—Al_2O_3—SiO_2 and K_2O—Al_2O_3—SiO_2, *Am. Jour. Sci.,* Vol. 245, pp. 193–204.

Schairer, J. F., and N. L. Bowen (1948), The system anorthite-leucite-silica, *Comm. Geol. Finlande Bull.,* No. 140, pp. 67–87.

Schairer, J. F., and H. S. Yoder, Jr. (1961), Crystallization in the system nepheline-forsterite-silica at one atmosphere pressure, Ann. Rpt. Dir. Geophys. Lab., *Carnegie Inst. Wash. Year Book 60,* pp. 141–144.

Schairer, J. F., and H. S. Yoder, Jr. (1967), The system albite-anorthite-forsterite at 1 atmosphere, Ann. Rpt. Dir. Geophys. Lab., *Carnegie Inst. Wash. Year Book 65,* pp. 204–209, 273.

Schreyer, W., and F. Seifert (1970), Pressure dependence of crystal structures in the system MgO—Al_2O_3—SiO_2—H_2O at pressures up to 30 kilobars, *Phys. Earth Planetary Interiors,* Vol. 3, pp. 422–430.

Schreyer, W., and H. S. Yoder, Jr. (1964), The system Mg cordierite-H_2O and related rocks, *Neues Jahrb. Mineral.*, Abh. 101, pp. 271–342.

Smith, F. Gordon (1963), *Physical geochemistry*, Addison-Wesley, Reading, Mass.

Smith, J. V. (1969), Crystal structure and stability of the $MgSiO_3$ polymorphs; physical properties and phase relations of the Mg, Fe pyroxenes, Pyroxenes and amphiboles: crystal chemistry and phase petrology, *Mineralogical Soc. America*, Spec. Paper 2, pp. 3–29.

Stewart, D. B. (1957), The system $CaAl_2Si_2O_8$—SiO_2—H_2O, Ann. Rpt. Dir. Geophys Lab., *Carnegie Inst. Wash. Year Book 56,* pp. 214–216.

Stewart, John W. (1967), *The world of high pressure*, Van Nostrand Momentum Book No. 17, D. Van Nostrand, Princeton, New Jersey.

Taubeneck, William H., and Arie Poldervaart (1960), Geology of the Elkhorn Mountains, northeastern Oregon. 2. Willow Lake intrusion, *Geol. Soc. America Bull.*, Vol. 71, pp. 1295–1322.

Taylor, L. A. (1970), Low temperature phase relationships in the Fe—S system, Ann. Rpt. Dir. Geophys. Lab., *Carnegie Inst. Wash. Year Book 68,* pp. 1968–1969.

Taylor, R. W. (1964), Phase equilibria in the system FeO—Fe_2O_3—TiO_2 at 1300°C., *Am. Mineralogist*, Vol. 49, pp. 1016–1030.

Turner, Francis J. (1968), *Metamorphic petrology*, McGraw-Hill Book Company, New York.

Turner, Francis J., and John Verhoogen (1960), *Igneous and metamorphic petrology* (2nd ed.), McGraw-Hill Book Company, New York.

Tuttle, O. F., and N. L. Bowen (1958), Origin of granite in the light of experimental studies in the system $NaAlSi_3O_8$—$KAlSi_3O_8$—SiO_2—H_2O, *Geol. Soc. America*, Memoir 74.

Tuttle, O. F., and J. V. Smith (1958), The nepheline-kalsilite system II. Phase relations, *Am. Jour. Sci.*, Vol. 256, pp. 571–589.

Von Platen, Hilmar (1965), Experimental anatexis and genesis of migmatites, *in* W. S. Pitcher and Glenys W. Flinn (eds.), *Controls of metamorphism*, Oliver and Boyd, Edinburgh, Scotland, pp. 202–218.

Walter, L. S. (1963), Data on the fugacity of CO_2 in mixtures of CO_2 and H_2O, *Am. Jour. Sci.*, Vol. 261, pp. 151–156.

Walter, L. S., P. J. Wyllie, and O. F. Tuttle (1962), The system MgO—CO_2—H_2O at high pressures and temperatures, *Jour. Petrology*, Vol. 3, pp. 49–64.

Winkler, H. G. F. (1967), *Petrogenesis of metamorphic rocks* (2nd ed.), Springer-Verlag, New York.

Wones, D. R., and H. P. Eugster (1965), Stability of biotite: experiment, theory, and application, *Am. Mineralogist*, Vol. 59, pp. 1228–1272.

Wyllie, Peter J. (1965), Melting relationships in the system CaO—MgO—CO_2—H_2O with petrological applications, *Jour. Petrology,* Vol. 6, pp. 101–123.

Wyllie, Peter J., and J. L. Haas, Jr. (1965), The system CaO—SiO_2—CO_2—H_2O. I. Melting relationships with excess vapor at 1 kilobar pressure, *Geochim. Cosmochim Acta*, Vol. 29, pp. 871–893.

Wyllie, Peter J., and J. L. Haas, Jr. (1966), The system CaO—SiO_2—CO_2—H_2O. II. The petrogenic model, *Geochim. Cosmochim. Acta*, Vol. 30, pp. 525–543.

Wyllie, Peter J., and O. F. Tuttle (1960a), Experimental investigation of silicate systems containing two volatile components. I. Geometrical considerations, *Am. Jour. Sci.*, Vol. 258, pp. 498–517.

Wyllie, P. J., and O. F. Tuttle (1960b), The system CaO—CO$_2$—H$_2$O and the origin of carbonates, *Jour. Petrology,* Vol. pp. 1–46.

Yoder, H. S., Jr. (1952), The MgO—Al$_2$O$_3$—SiO$_2$—H$_2$O system and the related metamorphic facies, *Am. Jour. Sci.*, Bowen Vol., pp. 569–627.

Yoder, H. S., Jr. (1955), Role of water in metamorphism, in *The Crust of the Earth*, Geol. Soc. America Special Paper 62, pp. 505–524.

Yoder, H. S., Jr. (1965), Diopside-anorthite-water at five and ten kilobars and its bearing on explosive vulcanism, Ann. Rpt. Dir. Geophys. Lab., *Carnegie Inst. Wash. Year Book 64*, pp. 82–89.

Yoder, H. S., Jr., D. B. Stewart, and J. V. Smith (1957), Ternary feldspars, Ann. Rpt. Dir. Geophys. Lab., *Carnegie Inst. Wash. Year Book 56*, pp. 206–214.

Yoder, H. S., Jr., and C. E. Tilley (1962), Origin of basalt magmas: an experimental study of natural and synthetic rock systems, *Jour. Petrology*, Vol. 3, pp. 342–532.

Yoshiki, B., S. Koide, and M. Waki (1953), Modification of alkaline earth feldspars and their stability relations, Reports of the Research Laboratory, *Asahi Glass Company, Ltd.*, Vol. 3, pp. 137–147.

Indexes

Author Index

Systems Index

CHEMICAL AND MINERALOGICAL LISTING

The systems included here are listed according to two methods of naming. In one, the mineral names of the end-member components are used, when possible (with H_2O listed as "water"), and the mineral names within each system are arranged alphabetically.

In the other, the chemical names are used, the components of a system being arranged according to simplicity and valence. The most simple components (those with the least number of oxides) are listed before the more complex (i.e., H_2O—$NaAlSi_3O_8$), and the components are arranged in order of increasing valency of the cations (i.e., MgO—Al_2O_3—SiO_2)—in much the same way that we write the oxides of complex chemical compounds. Within each valence group the components are arranged in alphabetical order (i.e., $KAlSi_3O_8$—$NaAlSi_3O_8$).

General Index